Handbook of Research on Smart Power System Operation and Control

Hassan Haes Alhelou
Tishreen University, Syria

Ghassan Hayek
Tishreen University, Syria

A volume in the Advances in Computer and Electrical Engineering (ACEE) Book Series

IGI Global
DISSEMINATOR OF KNOWLEDGE

Published in the United States of America by
IGI Global
Engineering Science Reference (an imprint of IGI Global)
701 E. Chocolate Avenue
Hershey PA, USA 17033
Tel: 717-533-8845
Fax: 717-533-8661
E-mail: cust@igi-global.com
Web site: http://www.igi-global.com

Library of Congress Cataloging-in-Publication Data

Names: Alhelou, Hassan Haes, 1988- editor. | Hayek, Ghassan, 1958- editor.
Title: Handbook of research on smart power system operation and control / Hassan Haes Alhelou and Ghassan Hayek, editors.
Description: Hershey PA : Engineering Science Reference, [2019] | Includes bibliographical references.
Identifiers: LCCN 2018042436| ISBN 9781522580300 (hardcover) | ISBN 9781522580317 (ebook)
Subjects: LCSH: Smart power grids. | Microgrids (Smart power grids)
Classification: LCC TK3105 .H36 2019 | DDC 621.31--dc23 LC record available at https://lccn.loc.gov/2018042436

This book is published in the IGI Global book series Advances in Computer and Electrical Engineering (ACEE) (ISSN: 2327-039X; eISSN: 2327-0403)

British Cataloguing in Publication Data
A Cataloguing in Publication record for this book is available from the British Library.

For electronic access to this publication, please contact: eresources@igi-global.com.

Advances in Computer and Electrical Engineering (ACEE) Book Series

Srikanta Patnaik
SOA University, India

ISSN:2327-039X
EISSN:2327-0403

Mission

The fields of computer engineering and electrical engineering encompass a broad range of interdisciplinary topics allowing for expansive research developments across multiple fields. Research in these areas continues to develop and become increasingly important as computer and electrical systems have become an integral part of everyday life.

The **Advances in Computer and Electrical Engineering (ACEE) Book Series** aims to publish research on diverse topics pertaining to computer engineering and electrical engineering. **ACEE** encourages scholarly discourse on the latest applications, tools, and methodologies being implemented in the field for the design and development of computer and electrical systems.

Coverage

- Computer Architecture
- Power Electronics
- Electrical Power Conversion
- Sensor Technologies
- Circuit Analysis
- Chip Design
- Optical Electronics
- Analog Electronics
- VLSI Design
- Programming

IGI Global is currently accepting manuscripts for publication within this series. To submit a proposal for a volume in this series, please contact our Acquisition Editors at Acquisitions@igi-global.com or visit: http://www.igi-global.com/publish/.

The Advances in Computer and Electrical Engineering (ACEE) Book Series (ISSN 2327-039X) is published by IGI Global, 701 E. Chocolate Avenue, Hershey, PA 17033-1240, USA, www.igi-global.com. This series is composed of titles available for purchase individually; each title is edited to be contextually exclusive from any other title within the series. For pricing and ordering information please visit http://www.igi-global.com/book-series/advances-computer-electrical-engineering/73675. Postmaster: Send all address changes to above address.

Titles in this Series

For a list of additional titles in this series, please visit: www.igi-global.com/book-series

Global Virtual Enterprises in Cloud Computing Environments
N. Raghavendra Rao (FINAIT Consultancy Services, India)
Engineering Science Reference • copyright 2019 • 281pp • H/C (ISBN: 9781522531821) • US $215.00 (our price)

Advancing Consumer-Centric Fog Computing Architectures
Kashif Munir (University of Hafr Al-Batin, Saudi Arabia)
Engineering Science Reference • copyright 2019 • 217pp • H/C (ISBN: 9781522571490) • US $210.00 (our price)

New Perspectives on Information Systems Modeling and Design
António Miguel Rosado da Cruz (Polytechnic Institute of Viana do Castelo, Portugal) and Maria Estrela Ferreira da Cruz (Polytechnic Institute of Viana do Castelo, Portugal)
Engineering Science Reference • copyright 2019 • 332pp • H/C (ISBN: 9781522572718) • US $235.00 (our price)

Advanced Methodologies and Technologies in Network Architecture, Mobile Computing, and Data Analytics
Mehdi Khosrow-Pour, D.B.A. (Information Resources Management Association, USA)
Engineering Science Reference • copyright 2019 • 1857pp • H/C (ISBN: 9781522575986) • US $595.00 (our price)

Emerging Innovations in Microwave and Antenna Engineering
Jamal Zbitou (University of Hassan 1st, Morocco) and Ahmed Errkik (University of Hassan 1st, Morocco)
Engineering Science Reference • copyright 2019 • 437pp • H/C (ISBN: 9781522575399) • US $245.00 (our price)

Advanced Methodologies and Technologies in Artificial Intelligence, Computer Simulation, and Human-Computer Interaction
Mehdi Khosrow-Pour, D.B.A. (Information Resources Management Association, USA)
Engineering Science Reference • copyright 2019 • 1221pp • H/C (ISBN: 9781522573685) • US $545.00 (our price)

Optimal Power Flow Using Evolutionary Algorithms
Provas Kumar Roy (Kalyani Government Engineering College, India) and Susanta Dutta (Dr. B. C. Roy Engineering College, India)
Engineering Science Reference • copyright 2019 • 323pp • H/C (ISBN: 9781522569718) • US $195.00 (our price)

Advanced Condition Monitoring and Fault Diagnosis of Electric Machines
Muhammad Irfan (Najran University, Saudi Arabia)
Engineering Science Reference • copyright 2019 • 307pp • H/C (ISBN: 9781522569893) • US $225.00 (our price)

IGI Global
DISSEMINATOR OF KNOWLEDGE

701 East Chocolate Avenue, Hershey, PA 17033, USA
Tel: 717-533-8845 x100 • Fax: 717-533-8661
E-Mail: cust@igi-global.com • www.igi-global.com

Editorial Advisory Board

List of Contributors

Table of Contents

Detailed Table of Contents

Demand-side management (DSM) is a strategy enabling the power supplying companies to effectively manage the increasing demand for electricity and the quality of the supplied power. The main objectives of DSM programs are to improve the financial performance and customer relations. The idea is to encourage the consumer to use less energy during peak hours, or to move the time of energy use to off-peak times. The DSM controls the match between the demand and supply of electricity. Another objective of DSM is to maintain the power quality in order to level the load curves. In this chapter, a genetic algorithm is used in conjunction with demand-side management techniques to find the optimal scheduling of energy consumption inside N buildings in a neighborhood. The issue is formulated as multi-objective optimization problem aiming at reducing the peak load as well as minimizing the energy cost. The simulations reveal that the adopted strategy is able to plan the daily energy consumptions of a great number of electrical devices with good performance in terms of computational cost.

Using microgrid generation technologies is proposed in order to organize reliable power supply to rural areas. The concept of microgrid based on RES is considered as one of the realization forms of the distributed energy paradigm. In this chapter, there are the principles of generating complex formation in any given microgrid considering the specifics of the region, consumption patterns, and the potential of renewable energy sources in a given area. The algorithm for meeting the challenges of forming the structure of the microgrid generating structure is shown. The criteria for selection of power generation sources when solving the issue of their inclusion in the microgrid is proposed. The chapter also suggests the design of the micro gas turbine that is able to operate on biogas.

It is critical for today's power system to remain in a state of equilibrium under normal conditions and severe disturbances. Power imbalance between the load and the generation can severely affect system stability. Therefore, it is necessary that these imbalance conditions be addressed in the minimum time possible. It is well known that power system frequency is directly proportional to the speed of rotation of synchronous machines and is also a function of the active power demand. As a consequence, when active power demand is greater than the generation, synchronous generators tends to slow down and the frequency decreases to even below threshold if not quickly addressed. One of the most common methods of restoring frequency is the use of under frequency load shedding (UFLS) techniques. In this chapter, load shedding techniques are presented in general but with special focus on UFLS.

Smart grids have become an urgent need to overcome the challenges of the 21st century. To transit the traditional grid to smart one, there must be a well thought out plan, called road map, which is also being carefully developed by organizations according to standards for deploying smart networks. Most studies focused on modernizing distribution networks because it was passive and technologically poor. Two approaches to developing distribution networks were presented. The smart grid modernization was also presented from social and psychological perspectives.

Electric vehicles were proposed as a good solution to solving energy crisis and environmental problems caused by the traditional internal combustion engine vehicles. In the last years due to the rapid development of the electric vehicles, the problem of power grid integration was addressed. In order to not put additional pressure onto the power grid several new technologies were developed. This chapter presents the smart grid technology, vehicle-to-grid concept, and electric vehicles grid integration. These technologies made possible the integration of electric vehicles without any major changes in the power grid. Moreover, electric vehicles integration brought new benefits to the power grid like better integration of renewable energy.

Microgrid (MG) is the vital technology that can be utilized to supply electricity to rural areas by fulfilling various aspects of electricity such as sustainability and reliability. Further, MG technology can also be used as localized generation sources and back up supply source. As MG can be worked in interconnected mode, various issues related to interconnection with utility grid are raised. Several issues such as technical, regulatory, and operational are associated with grid integration. Therefore, this chapter deals with the issues that are associated with the grid integration of microgrid.

Power quality problems can cause processes and equipment to malfunction or shut down. And the consequences can range from excessive energy costs to complete work stoppage. Obviously, power quality is critical. There are many ways in which a power feed can be poor quality, and so no single figure can completely quantify the quality of a power feed. In this chapter, the authors present all definitions, classifications, and problems related to power quality. Finally, they do a comparison between the practical measurements and standards related to power quality.

In this chapter, wide area measurement systems (WAMS), which are one of the cornerstones in modern power systems, are overviewed. The WAMS has great applications in power system monitoring, operation, control, and protection systems. In the modern power systems, WAMS is adopted as a base for the modern monitoring and control techniques. Therefore, an introduction of WAMS is firstly provided. Then, phasor measurement unit (PMU), which is the base of WAMS, is described. Afterward, the most recent developments in power system estimation, stability, and security techniques, which are based on WAMS, are introduced. Later, general system setup for WAMS-based under-frequency load shedding (UFLS) is provided. Finally, the required communications infrastructures are comprehensively discussed.

This chapter introduces EMC installation for variable speed drive systems. As an introduction, EMC standards have been mentioned in order to define the requirements characteristics, besides the fundamentals of static, electric, magnetic, and electromagnetic fields. Both inductive and capacitive coupling have been discussed to deal with shielding. Finally, VSDS emission and electromagnetic interferences were studied with installation requirement in VSDS (supply cables, cable between converter and motor, control cables, earthing requirements, and grounding).

Presently, economic operation of micro-grid is a major concern in smart grid environment. This is a very complex problem, which can be solved with the help of various meta-heuristic techniques. Therefore, this chapter provides a comparative analysis of four different renowned meta-heuristic techniques with reference to the problem of optimal operation of micro-grid (MG). Genetic algorithm (GA), particle swarm optimization (PSO), differential evaluation (DE), and firefly (FF) algorithm are utilized for this purpose.

This chapter describes the upcoming technology for electrical power systems that gives the appropriate solution for the integration of the distributed energy resources. In this chapter, different categories of smart grids have been classified, and the advantages, weakness, and opportunities of each one, are given in addition to determining its own operating conditions. Micro-grids are the most common kind of smart grid. It has been classified under different criteria, such as architecture with different topology (connected mode, island mode, etc.) and demand criteria (simple micro grids, multi-DG, utility) and by capacity into simple micro-grid, corporate micro-grid, and independent micro-grid, and by AC/DC type to DC micro-grids, AC micro-grids, Hybrid micro-grids. Finally, most familiar Micro-grid components have been discussed such as an energy management system along with several types of control and communication systems in addition to the economic study of a micro-grids.

Buildings maintenance has received increasing international attention in various fields of scientific research. As a result, there has been a change in the maintenance of buildings from the preventive to the predictive approach. This is done through an evaluation model to support and assist the management of the facility in selecting alternatives and making appropriate decisions in maintenance according to building status and maintenance budget. This chapter investigated the reasons for the electrical maintenance of the university buildings and the degree of importance of each element of electrical maintenance through the design of a questionnaire in which the electrical components were divided into elements and then each element was linked to all maintenance items that related to it. At the end of the research, mathematical models were developed; these models help to forecasting the electrical maintenance items and distribution of the maintenance budget, and to verify the validity of these models, they have been applied to study the case of dorm buildings in Tishreen University.

This chapter introduces an analysis of power quality of variable-speed drive systems (VSDS). In this chapter, several issues have been discussed as VSDS in terms of cost and effectiveness, VSDS control loops (open loop, closed loop), and power quality in VSDS. Harmonics standards and harmonics in VSDS were discussed in this chapter in addition to highlighting the effects of harmonics on power factor, crest factor, and other power quality specifications. The solutions used to mitigate the harmonics in VSDS were discussed in detail. Finally, simulation of the conventional VSDS model and VSDS with one harmonic mitigation solution in order to clarify the usefulness of using this solution on power quality specifications were discussed.

The demand for electricity is increasing day by day due to technological advancements. According to the demand, the size of the grid is also increasing rapidly in the past decade. However, the traditional centralized power grid has many drawbacks such as high operating cost, customer satisfaction, less reliability, and security. Distribution generation has less pollution, high energy efficiency, and flexible installation than traditional generation. It also improves the performance of the grid in peak load and reliability of supply. The concept of micro-grid has been raised due to the advent of new technologies and development of the power electronics and modern control theory. Micro-grid is the significant part of the distribution network in the future of smart grid, which has advanced and flexible operation and control pattern, and integrates distributed clean energy.

Microgrids are the most innovative area in the electric power industry today. A microgrid can operate in grid-connected or islanded mode. In islanded mode, microgrids can provide electricity to the rural areas with lower cost and minimum power losses. Several methods have been proposed in the literature for the successful operation of a microgrid. This chapter presents an overview of the major challenges and their possible solutions for planning, operation, and control of islanded operation of a microgrid. Microgrids are the most innovative area in the electric power industry today. Moreover, microgrids provide local voltage and frequency regulation support and improve reliability and power capacity of the grid. The most popular among the control strategies based on droop characteristics, in addition a central controller is described within a hierarchical control scheme to optimize the operation of the microgrid during interconnected operation. Microgrid control methods, including PQ control, droop control, voltage/frequency control, and current control methods are formulated.

Foreword

I am delighted to write the foreword to this book, as its scope and content provide commercial and technical enterprises with the essential ingredients for implementing and managing power and energy systems techniques and management schemes. The recent advances in the power and energy technology, especially in the power system operation, control, optimization, renewable, energy management through analysis and simulation allowed a potentially enormous market for energy management and conversion. With advances in the latest field of electrical engineering including: operation and control, energy system security, energy efficiency, optimization and renewable energy resources have contributed to the solution of recent energy problems for any nation and the environmental issues. However, now the major challenge is to maintain the high energy efficiency during operation and to explore sustainable energy resources. To deal with the technical challenges, two major areas in power and energy technologies are being identified:

1. Modern Power System Operation
2. Modern and Smart Power System Control

This handbook provides answers to many challenging questions dealing with power and energy system optimization. It addresses a variety of issues related to the energy management and the recently developed optimization techniques. This handbook comprised of 19 chapters divided into two parts as per the information providing according to the above areas. I recommend this handbook to researchers and practitioners in the field, and for scientists and engineers involved in power system operation and control. I really appreciate the efforts of all the Editors to compile this book. The managing editor Dr. Hassan Haes Alhelou and his team have meticulously collected the chapters, reviewed and place them in appropriate way for better in depth understanding. I believe the readers of power and energy system will be benefited from the work presented in this book.

D. Che
University of Rome "Tor Vergata", Italy

Preface

In recent years, the world has witnessed a rapid growth in the demand for electrical energy and consequently the serious energy crises having direct impact on economics, society and development of any country. With the development of new technologies, the modern energy systems are changing rapidly and reveal increasingly multifaceted features. Nowadays, energy conversion and supply is not the only mission of power utilities rather energy stability, reliability, security and efficiency has become the major concern. This is to manage, control and operate the energy resources for a long-term future and to minimize their impacts on our ecology and environmental system. Energy efficiency in planning and operation of power systems allows maximizing the benefits of utilities and customers concurrently. Though, saving of single unit of energy is termed as energy resource so it contributes to the mitigation of carbon emissions for the globe's climate. Therefore, energy optimization also expected to secure their resources by loss reduction in operation. In order to find the solution of the above-mentioned objectives, new technologies have continuously emerged with improved efficiency. With the implementation of modern technologies, it further helps to explore the renewable energy resources and their optimum usages. Although we achieved much, still there is a huge gap between present energy generation and the demand. To meet the load demand, we are dependent on the conventional resources since they are available continuously whereas, renewable energy sources such as solar and wind are available at the different time of the day. Therefore, for reliable operation dependence on conventional fossil fuels is still sturdy in most parts of the world. On the other hand, predominant reliance on fossil fuel-based electrical energy use is a direct cause for a serious issue of the global warming. Therefore, promotion of the use of renewable energy sources, for meeting electrical energy demand, is an important strategy in order to enhance the energy security of any country. In this connection, wind and solar based electrical power generation have gained the attention of the researchers all over the world. This is enabling in high penetration of the renewable sources with the main electrical grid. Power generation from these renewable resources is intermittent because of its dependence on the environmental conditions. As a result, the power generation from solar and wind systems keep on fluctuating and has a direct impact on the voltage magnitude, supply frequency, and waveform, and hence on the quality and quantity of supplying power to the interconnected grid system. The world energy requirement would be 70% higher in 2030 than today's demand, even at the constant rate of growth. On the other hand, conventional resources are depleting day by day which raises the serious concern for energy utilization and optimization to improve operating efficiency. As a fact, this is the energy consumption per capita which describes the living standard in the country. Therefore, we need to explore more and more options for the energy resources, which are naturally available, and optimization techniques to improve energy efficiency. The existing problems of modern energy systems indicate clearly that the available means of planning and operating the energy systems are far from perfect solution, and there is a large potential for the improvements. This book demonstrates the potential of energy systems engineering-based approach to systematically quantify different options at different levels of complexity (i.e. planning, operation, control and utilization) through state-of-the-art modeling, simulation, control, and optimization-based frameworks. The successful implementation of these approaches in a number of real-life case studies highlights further the significance of this integrated system-wide approach. This book presents the importance of fundamental and applied research in power and energy systems applications by developing mechanisms for the transfer of the new methodology, which is applicable to the real-time problem. In recent years, power and energy is the major area for research. For optimal utilization of electrical power, we need to explore the several optimization techniques. The recent developments in the computational intelligence found to be a most effective tool for obtaining the optimal operation of power systems. In power system, loads are not of any specific type and they vary with the state of the economy of system operation. Therefore, it becomes very difficult to identify the optimal configuration of distribution system under the change in loading conditions. In this scenario, energy efficiency and auditing plays an important role in the energy management system. The automatic control of the power component improves the reliability and the bi-directional communication involves the smart control of system operation, which is termed as the smart grid.

DESCRIPTION OF THE BOOK

This handbook aims to be an essential reference source, building on the available literature in the field of power system operation and control, providing further research opportunities in this field. This specific text is expected to provide the primary and major resources necessary for researchers, academicians, students, faculties, and scientists, across the globe, to adopt and implement new inventions in thrust area of power generation from conventional and non-conventional resources and their utilization and energy management. Therefore, the handbook of research on power system operation and control is to provide a platform to share up-to-date scientific achievements in the core as well as related fields. There are three main objectives underlying this book:

1. Identifying and exploring the scope of different operation and control methods.
2. Identifying the scope of operation and control in modern power systems with high share of renewable energy resources.
3. Identifying the various operation and control schemes/algorithms/approaches/techniques for implementation in future power generation, transmission, and utilization.

This handbook aims to be an essential reference source, building on the available literature in the field of modern and future power system operation and control, providing further research opportunities in this field. This specific text is expected to provide the primary and major resources necessary for researchers, academicians, students, faculties, and scientists, across the globe, to adopt and implement new inventions in thrust area of power generation from conventional and non-conventional resources and their utilization and energy management. According to the application domain and nature, the chapters in this book are categorized into two sections:

Section 1: Electric Power Systems Operation
Section 2: Energy Power Systems Control

From recent trends, it can be observed that research is mainly focused on electric energy systems operation and control, renewable energy resource, and the computational intelligence. In the world of smart grid, application of different methods such as artificial intelligence has become a trend. Three domains covered in this book will provide the vast knowledge to the researcher and it will also give common source of information to the power engineers. Energy efficiency is the burning issue nowadays. Researchers are highly involved in this area and this is due to the fact that energy demand is exponentially growing whereas, energy resources are depleting day by day. This requires optimal use of electrical power.

In order to find the optimal operation of the electrical system, we have to explore several optimization techniques. This book will be a source of motivation to the researcher and can exchange the knowledge of energy efficiency, optimization techniques and integration of DG for better utilization of the resources. In order to meet the load demand and to improve the energy efficiency researchers are working in the area of integration of distributed generation, micro-grid, automation in power distribution, hybrid electric vehicle, and synchronized operation of solar power, battery and the grid supply for the development of smart power system for the home. They have developed several optimization techniques involving extensive search, analytical approach, and the computational or artificial intelligence.

In recent years, the computational intelligence based on nature inspired meta-heuristic approaches are finding wider acceptance for optimization of discrete problems particularly for large systems with no fixed solution under different operating conditions. In the first part, book chapters are describing the application of the different algorithm for optimal power systems operation. Then, the most recent and highly recommended power system control methods are introduced. A brief description of the chapters in this book is given as follows.

Chapter 1

Protection of transmission system is crucial for the secured and stable operation of power system. In the past few years, the growth in the power demand is high because of the modernization of society which indirectly influences the power market. In some situations, demand and supply are almost equal. So, in that case the system will be going too overstressed. Under stress conditions, the operating parameters of power system violate their limits and this is a major cause behind unwanted system failure. Past anatomy reports of blackouts reveal that under system stressed conditions and fault events are difficult for the system protection schemes to distinguish. Different solutions are proposed by several researchers to mitigate this issue. As back-up protection is the main influenced scheme behind these blackouts which provides delayed protection, help of wide area monitoring devices can be taken to better perform this task. Wide-area backup protection is an indisputable subdivision of power system protection. As third-zone of distance relay is highly affected by system stressed events such as voltage instability and load encroachment. This seizes the intended operation of distance relay and impuissant to discriminate these events from symmetrical fault. It happens only because of their symmetric nature, which is a leading cause behind power system blackout. Thus, relay finds challenges in discriminating the events for its accurate and reliable operation. An adaptive and exact distinction technology is essential for discrimination. This chapter introduces a synchrophasor technology to monitor the change in voltage angle between two interconnected areas to enhance the back-up protection function so that percentage of future blackouts can be reduced. The deviation is constant for stable system operation and system stressed conditions but is highly variable during symmetrical fault condition. So that it can provide better solution towards power system blackout.

Chapter 2

In this chapter, an amalgamation of Artificial Bee Colony (ABC) algorithm and Artificial Neural Network (ANN) approach is recommended for optimizing the location and capacity of distribution generations (DGs) in distribution network. The best doable place in the network have been approximated using ABC algorithm by means of the voltage deviation, power loss and real power deviation of load buses and the DG capacity is approximated by using ANN. In this, single DG and two DGs have been considered for calculation of doable place in the network and capacity of the DGs to progress the voltage stability and reduce the power loss of the system. The power flow of the system is analyzed using iterative method (The Newton-Raphson load flow study). From which; the bus voltages, active power, reactive power, power loss and voltage deviations of the system have been achieved. The proposed method is tested in MATLAB and the results are compared with Particle Swarm Optimization (PSO) algorithm, ANN and hybrid PSO & ANN methods for effectiveness of the proposed system.

Chapter 3

In this chapter, the dynamic performance of a wind-diesel-generator (WDG) system has been studied against wind and load perturbations. The wind perturbation is modeled by simulating base, ramp, gust, and random wind. An optimized cascade tilt-integral-derivative (CC-TID) controller is provided to the test system for producing desired control signal to regulate the blade pitch angle of wind turbine. To confirm the efficacy of CC-TID controller, the output results are compared to that of PI and PID-controller. The optimum gains of the proposed controllers are obtained through Levy-embedded grey wolf optimization, whale optimization algorithm, drone squadron optimization, and search group algorithm. To show the effectiveness, the output results are compared to the results of genetic algorithm and particle swarm optimization tuned controllers. A thyristor control series compensator (TCSC) is provided to WDG model for increasing the damping of system oscillations. Analysis of the presented results confirms the supremacy of CC-TID-TCSC controller over other controllers considered in this chapter.

Chapter 4

The major purpose of uninterruptible power supply (UPS) systems is to supply regulated sinusoidal voltage at constant frequency and amplitude. UPS systems are gaining much popularity as a means of providing clean and continuous electricity to critical loads during any disturbances in main grid. Modern equipments are sensitive to power fluctuation and requires back up power supply for optimal performance. This chapter introduces a set of possible solutions so that uninterrupted power supply can be provided to emergency feeders and critical loads such as hospitals and communication systems. Different network configurations can be applied to micro-grid system for obtaining an uninterrupted power supply. Various hybrid energy and modern UPS systems for micro-grid along with their control techniques have been elucidated. A comparative assessment of all UPS technologies on the basis of cost, performance and efficiency of the system has been presented.

Chapter 5

Demand Side Management (DSM) is a strategy enabling the power supplying companies to effectively manage the increasing demand for electricity and the quality of the supplied power. The main objectives of DSM programs are to improve the financial performance and customer relations. The idea is to encourage the consumer to use less energy during peak hours, or to move the time of energy use to off-peak times. The DSM controls the match between the demand and supply of electricity. Another objective of DSM is to maintain the power quality in order to level the load curves. In this chapter, a genetic algorithm is used in conjunction with demand side management techniques to find the optimal scheduling of energy consumption inside N buildings in a neighborhood. The issue is formulated as multi-objective optimization problem aiming at reducing the peak load as well as minimizing the energy cost. The simulations reveal that the adopted strategy is able to plan the daily energy consumptions of a great number of electrical devices with good performance in terms of computational cost.

Chapter 6

Using microgrid generation technologies is proposed in order to organize reliable power supply to rural areas. The concept of microgrid based on RES is considered as one of the realization forms of the distributed energy paradigm. In this chapter there is the principles of generating complex formation in any given microgrid considering the specifics of the region, consumption patterns and the potential of renewable energy sources in a given area. The algorithm for meeting the challenges of forming the structure of the microgrid generating structure is shown. The criteria for selection of power generation sources when solving the issue of their inclusion in the microgrid is proposed. The paper also suggests the design of the micro gas turbine which is able to operate on biogas.

Chapter 7

It is critical for today's power system to remain in a state of equilibrium under normal conditions and severe disturbances. Power imbalance between the load and the generation can severely affect system stability. Therefore, it is necessary that these imbalance conditions be addressed in the minimum time possible. It is well known that power system frequency is directly proportional to the speed of rotation of synchronous machines and is also a function of the active power demand. As a consequence, when active power demand is greater than the generation, synchronous generators tends to slow down and the frequency decreases to even below threshold if not quickly addressed. One of the most common methods of restoring frequency is the use of under frequency load shedding (UFLS) techniques. In this report load shedding techniques are presented in general but with special focus on UFLS.

Chapter 8

Smart grids have become an urgent need to overcome the challenges of the 21st century. To transit the traditional grid to smart one, there must be a well thought out plan, called road map, which is also being carefully developed by organizations according to standards for deploying smart networks. Most studies focused on modernizing distribution networks because it was passive and technologically poor. Two approaches to developing distribution networks were presented. The Smart Grid modernization was also presented from social and psychological perspectives.

Chapter 9

Electric vehicles were proposed as a good solution to solving energy crisis and environmental problems caused by the traditional internal combustion engine vehicles. In the last years due to the rapid development of the electric vehicles, the problem of power grid integration was addressed. In order to not put additional pressure onto the power grid several new technologies were developed. This chapter presents the smart grid technology, vehicle to grid concept and electric vehicles grid integration. This technology made possible the integration of electric vehicles without any major changes in the power grid. Moreover electric vehicles integration brought new benefits to the power grid like better integration of renewable energy.

Chapter 10

Microgrid (MG) is the vital technology, which can be utilized to supply electricity to rural areas with fulfilling various aspects of electricity such as sustainability and reliability. Further, MG technology can also be used as localized generation sources and back up supply source. As MG can be worked in interconnected mode, various issues related to interconnection with utility grid raised. Several issues such as technical, regulatory, and operational are associated with grid integration. Therefore, this chapter deals with the issues which are associated with the grid integration of micro grid.

Chapter 11

Power quality problems can cause processes and equipment to malfunction or shut down. And the consequences can range from excessive energy costs to complete work stoppage. Obviously, power quality is critical. There are many ways in which a power feed can be poor quality, and so no single figure can completely quantify the quality of a power feed. In this chapter will be presented all definitions, classifications and problems related to power quality. Finally, a comparison will be done between the practical measurements and standards related to power quality.

Chapter 12

In this chapter wide area measurement systems (WAMS) which are one of the most important cornerstone in modern power systems are overviewed. The WAMS has a great applications in power system monitoring, operation, control and protection systems. In the modern power systems, WAMS is adopted as a base for the modern monitoring and control techniques. Therefore, an introduction of WAMS is firstly provided. Then, phasor measurement unit (PMU) which is the base of WAMS is described. Afterward, the most recent developments in power system estimation, stability and security techniques which are based on WAMS are introduced. Later, general system setup for WAMS based under frequency load shedding (UFLS) is provided. Finally, the required communications infrastructures are comprehensively discussed.

Chapter 13

This chapter introduces EMC installation for variable speed drive systems. As an introduction, EMC standards have been mentioned in order to define the requirements characteristics, besides the fundamentals of static, electric, magnetic, and electromagnetic fields.

Both inductive and capacitive coupling have been discussed, in addition to deal with shielding. Finally, VSD_S emission, and electromagnetic interferences were studied with installation requirement in VSDS (supply cables, cable between converter and motor, control cables, earthing requirements, and grounding).

Chapter 14

Presently, economic operation of Micro grid is a major concern in smart grid environment. It is a very complex problem, which can be solved with the help of various Meta heuristic techniques. Therefore this chapter provides a comparative analysis of four different renowned Meta heuristic techniques with reference to the problem of optimal operation of micro grid. Genetic algorithm (GA), Particle Swarm Optimization (PSO), Differential Evaluation (DE), and firefly (FF) algorithm are utilized for this purpose.

Chapter 15

This chapter describes the new upcoming technology for electrical power systems that gives appropriate solution for the integration of the distributed energy resources. During this chapter, A different categories of smart grids have been classified, and giving the advantages, weakness, and opportunities of each one, besides determining its own operation conditions. Micro grids are most common kind of smart grid. It has been classified with different criteria. Such as architecture with different topology (connected mode, island mode, etc.), and demand criteria (simple micro grids, multi- DG, utility) and by capacity into simple micro grid, corporate micro grid, and independent micro grid, and by AC/DC type to DC micro grids, AC micro grids, Hybrid micro grids. Finally, most familiar Microgrid's components have been discussed such as energy management system, and several types of control and communication systems in addition to economic study of micro grid.

Chapter 16

Buildings maintenance has received increasing international attention in various fields of scientific research. As a result, there has been a change in the maintenance of buildings from the preventive to the predictive approach. This is done through an evaluation model to support and assist the management of the facility in selecting alternatives and making appropriate decisions in maintenance according to building status and maintenance budget.

This study investigated the reasons for the electrical maintenance of the university buildings and the degree of importance of each element of electrical maintenance through the design of a questionnaire in which the electrical components were divided into elements and then each element was linked to all maintenance items that related to it. At the end of the research mathematical models were developed, these models help to forecasting the electrical maintenance items and distribution of the maintenance budget, and to verify the validity of these models, they have been applied to study the case of dorm buildings in Tishreen University.

Chapter 17

This chapter introduce an analysis of power quality of variable speed drive systems VSD_S. During this chapter several issues have been discussed as VSDS in terms of cost and effectiveness, VSDS control loops (open-loop, close loop) and power quality in VSDS. Harmonics standards and Harmonics in VSD_S were discussed in this chapter.in addition to highlighting the Effects of Harmonics on Power Factor, Crest Factor, and Other Power Quality Specifications.

The solutions use to mitigate the harmonics in VSD_S were discussed in details. Finally, Simulation of the Conventional VSDS Model and VSDS with one harmonic mitigation solutions in order clarify the useful of using this solution on power quality specifications.

Chapter 18

The demand for electricity is increasing day by day due to technological advancements. According to the demand, the size of the grid is also increasing rapidly in the past decade. However, the traditional centralized power grid has many drawbacks such as high operating cost, customer satisfaction, less reliability and security. Distribution generation has less pollution, high energy efficiency and flexible installation than traditional generation. It also improves the performance of the grid in peak load and reliability of supply. The concept of Micro-Grid has been raised due to the advent of new technologies and development of the power electronics and modern control theory. Micro-grid is the significant part of the distribution network in the future of smart grid, which has advanced and flexible operation and control pattern, and integrates distributed clean energy.

Chapter 19

In response to the ever increasing energy demand, integrating distributed energy resource-based microgrid will be the most promising power system improvement in the near future. Microgrid system implementation provides significant advantages for both electric utility provider and end customer user. This chapter gives a review on the current key issues on control strategies of microgrid islanded mode operation, about microgrid, types of microgrid. Brief descriptions are provided for typical microgrid control methods, PQ control, droop control, voltage/frequency control, and current control, which are associated with microgrid mode of operation. Finally, research conclusions of the important microgrid control requirements for future development are also described.

The *Handbook of Research on Smart Power System Operation and Control* contains 19 chapters of high-quality contributions from international leading researchers in the field of power and energy systems.

The book chapters are divided into two sections depending upon the methodology and the approach described by the authors for the energy optimization and or management. First parts mainly consist of optimization techniques for voltage and frequency control, stability and reliability improvement in the electrical system's operation, control and planning. Recently, several heuristic and meta-heuristic optimization techniques are presented by the researchers in which genetic algorithm is the popular approach whereas other newly developed approaches such as binary whale optimization, grey wolf optimization, symbiotic organism search, JAYA (based on victory means Jaya), teaching learning based optimization and artificial neural network found to give significantly improved results in the power and energy optimization. However, simulation results have also appeared in some chapters showing the time response for dynamic operation. The second part consists of management and conversion techniques for improvement in energy efficiency and security. Since energy utilization is mainly by motors load and lighting. It is observed that the energy management and efficiency is the major concern in the present scenario when conventional resources are depleting day by day.

Acknowledgment

This handbook of research on power system operation and control is an outcome of the inspiration and encouragement given by many individuals for whom these words of thanks are only a token of our gratitude and appreciation for them.

Our sincere gratitude goes to the peoples who contributed their time and expertise to this book. We highly appreciate their efforts in achieving this project. The editors would like to acknowledge the help of all the people involved in this project and, more specifically, the editors would like to thank each one of the authors for their contributions and the editorial board/reviewers regarding the improvement of quality, coherence and the content presentation of this book.

Second, Editors would like to express their sincere thanks to Jordan Tepper, Jan Travers, Maria Rohde, Mariah Gilbert, Lindsay Wertman, Courtney Tychinski, and other individuals of IGI Global for their continuous support and giving us an opportunity to edit this book.

The editors are thankful to our family members for their prayers, encouragement, and care shown towards us during the completion of this handbook of research on power system operation and control. Thank you all!!!

We also express our gratitude to the GOD for all the blessings!!!

Hassan Haes Alhelou
Tishreen University, Syria

Ghassan Hayek
Tishreen University, Syria

Chapter 1

Synchrophasor-Based Wide Area Protection Scheme for System Stressed Conditions to Avoid Power System Blackout:

Wide Area Protection Scheme for System Stressed Conditions

Kasimala Venkatanagaraju
National Institute of Technology Raipur, India

Monalisa Biswal
National Institute of Technology Raipur, India

ABSTRACT

Protection of transmission system is crucial for the secured and stable operation of power system. Under stress conditions, the operating parameters of power system violate their limits. From the past anatomy reports of several blackouts, it is clear that equipment, control, and protective relay failures are the major causes behind large power system failure. From study, it is also revealed that failure of back-up protection is more prone during system stressed conditions. In transmission system, third-zone of distance relay is highly affected by system stressed events such as voltage instability and load encroachment. As third-zone protection is a delayed protection scheme, with the help of wide area measurement system better protection function can be provided to reduce future percentage of blackout. In this chapter, a detailed discussion about the existing solutions is provided to mitigate the issue of system stressed conditions and a synchrophasor technology-based approach is provided. Results for different cases are provided to show the efficacy of the proposed method.

DOI: 10.4018/978-1-5225-8030-0.ch001

INTRODUCTION

Generally, the electric power system comprises electrical components that are equipped to supply, transmit and distribute the electric power for the utilization. Generators supply the power by the utilization of natural resources. The transmission system transmits power from generating stations to distribution sub-stations, there the electric power is distributed among the connected utilities. Hence, the power flow is unidirectional. As compared with the last two decades the demand for the utilization of electricity increased due to rapid growth in population. Even though the expansion of the existing interconnected power system network is restricted due to economic constraint. Adoption of distributed generating units such as solar power (Zamani,2018; Makdisie, 2018) and wind power makes the direction of power flow become bidirectional that requires advanced power flow controlling devices. Implementation of smart grid concepts in the electric power network provides secure operation with the combination of electric vehicles and high penetration level of renewable energy resources (Alhelou, 2015; Alhelou, 2016; Alhelou, 2018; Alhelou, 2018; Alshahrestani,2018). However, it increases the use of effective communication system for the system visualization. Interconnections between the areas provide reliable operations of the system during emerging conditions but the number of parallel lines is increased which cause congestion as shown in Figure 1. Planned or forced outage of generating units causes unbalancing between generation and load demand. This leads to a significant change in system frequency which is tuned by load frequency controllers (Fini, 2016; Alhelou, 2018; Nadweh,2018; Njenda 2018). As the consequences of all these makes the operation of power system very closer to its security limits.

Whenever a fault or abnormal conditions occur and that persists for a long time on such system, then it will damage certain portion of the system. Thus, causes an imbalance between the electric power supply and load demand. Consequently, it may lead to initiation of cascaded events which make failure or collapse of the power system i.e. blackout. As per the North American Electric Reliability Corporation (NERC) reports, the possible failures that are continually supporting the occurrence of blackouts in the real time power system listed in a flow chart as shown in Figure 2. Suppose if anyone among these failures present in the modern power system, then the system gets stressed and gives rise to cascading events. Thus, the propagation of cascaded events may lead to blackout scenario. To prevent future blackout, a

Figure 1. An interconnected power system network

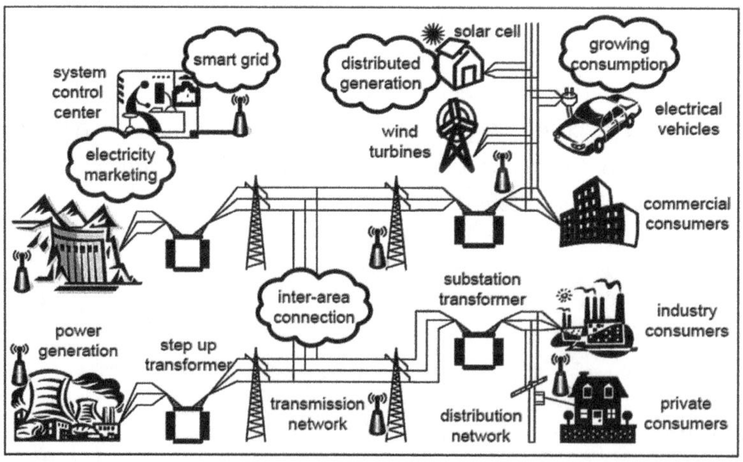

Figure 2. Possible power system failures

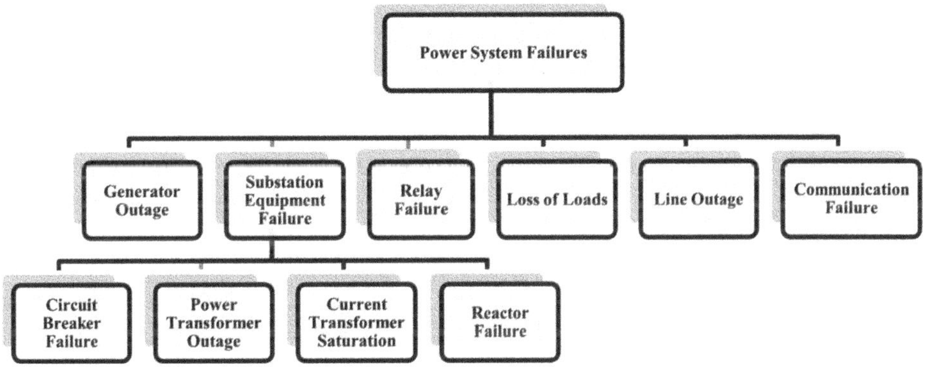

detailed steady state and transient study of power system is essential. Therefore, it is required to study and extract the valuable information to understand the initializing events, causes and other aspects of blackout.

A detailed list about the worldwide blackout, including the year, location and relative causes is mentioned in Table 1. In this table, a complete description about the power system blackout from 1965 to 2017 has been provided. These locations include the United States, Thailand, Japan, Brazil, India, Italy, Turkey, Indonesia, Australia, Bangladesh, Pakistan, Sri Lanka and Kenya. According to this investigation, it has been observed that blackouts or power system failures are more probable to happen in abnormal conditions. This kind of detailed study on past blackouts gives a chance to find out the actual causes behind them and they are listed in Figure 3. Every cause which may be relevant to supply system or transmission system or distribution system has its own significance to contribute a large disturbance.

So, there is a need to protect the power system from possible failures otherwise the risk associated with them is volatile. Protection is an indisputable part of the power system and it plays a vital role over the subsections of the power system such as generation, transmission and distribution. Every component present in the subsections of power system is collectively protected. The main aim of the protection system is to disconnect the affected part from the healthy part, so that system will feel the least effect of abnormality. In some situations, the individual protection system designed for the power system components may fail due to unpredictable events.

Here, the authors mainly focus on protection of the transmission system which is the subsection of the entire power system. The transmission system transfers electric power through the transmission lines. Depending upon length of the transmission lines they are segregated into short, medium and long transmission lines. In order to protect them for stable operation of the power system, a suitable protection scheme need to employ. Distance protection scheme is the best choice for the transmission lines protection. Distance relay is one of the most prominent protective elements equipped with the transmission system. Its operation is always dependent on the impedance measured from fault point to fault location. Based on the type of transmission line protection, distance relays are classified as reactance relay, impedance relay and mho relay. Reactance, impedance and mho relays are used for protecting short, medium and long transmission lines.

Mho type distance relay deals with the bulk amount of electric power transfer through the high voltage and extra high voltage long transmission lines. It has three independent protective zones such as Zone-1,

Table 1. List of power system blackouts

S. No	Location	Year	Initializing Events
1	United States	1965	Overloading on the transmission line results unintentional line trip.
2	United States	1977	Severe weather conditions such as thunder storms and lightning strikes tripped transmission lines.
3	Thailand	1978	Generator failure
4	Japan	1987	Remarkable peak load demand due to extreme hot weather condition.
5	United States	1996	Tripping of transmission line due to a flashover caused by tree.
6	Brazil	1999	Tripping of transmission line due to lightening stroke at substation
7	India	2001	Tripping of transmission line due to a flashover
8	Italy	2003	Tree flashover hit the tie line on heavily loaded line.
9	Turkey	2003	Loss of nuclear generation.
10	United States	2003	Insufficient reactive power demand.
11	Indonesia	2005	Tripping of transmission line.
12	Turkey	2006	Tripping of several high voltage lines due to overload.
13	Australia	2007	Tripping of transmission line due to bush fire in extreme weather conditions.
14	Brazil	2009	Tripping of transmission line due to adverse weather conditions.
15	India	2010	Tripping of transmission line due to flashover.
16	India	2012	Tripping of transmission line due to overload.
17	Bangladesh	2014	Transmission line tripping due to inability of substation to withstand more than 400MW.
18	Turkey	2015	Transmission line tripping due to overload results loss of synchronism.
19	Pakistan	2016	Short circuited the national electricity grid due to militants attack on a transmission line.
20	Sri Lanka	2016	Substation tripping.
21	Kenya	2016	Failure of transformer at generating station.

Figure 3. Causes of power system failures

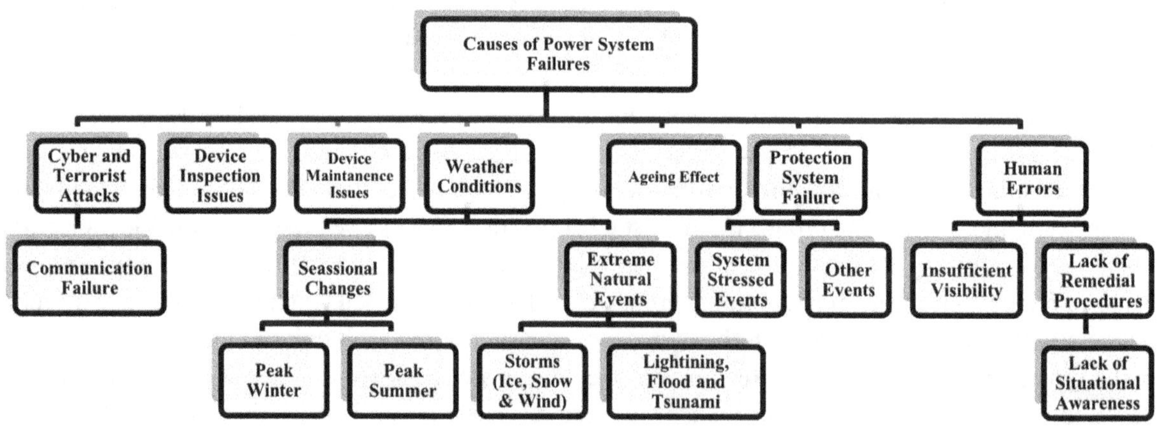

Figure 4. Stepped characteristics of distance relays

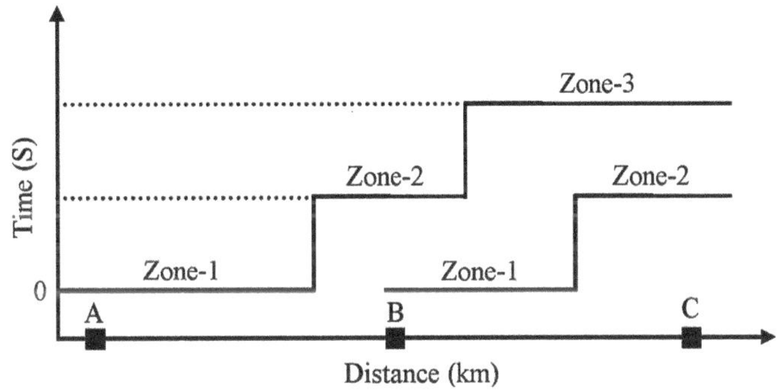

Figure 5. Operational characteristics of Mho type distance relay

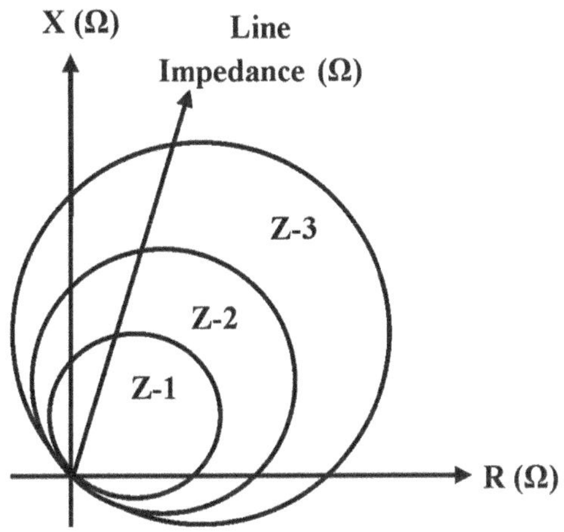

Zone-2 and Zone-3 with directional characteristics as shown in Figure 4. The operating characteristic of Mho relay is a circle passing through the origin as illustrated in Figure 5.

Protective zone settings of the Mho relay are considered as Zone -1 covers 80% of the primary protected line and no time delay to activate the relay. The main reasons behind 80% -90% zone-1 setting of distance relay are current transformer (CT) saturation, line capacitance and infeed/outfeed condition. Zone-2 covers 100% of primary protected line and 20% of adjacent shortest line with time delay is 0.35 sec. If it further increases beyond the set value, it faces a problem of overlapping with zones. Zone-3 provides protection to 100% of primary protected line and 100% of longest adjacent line with a 1 sec time delay. Since, Zone-3 covers large reach area and long-time delay inherently. This relay provides primary as well as secondary or backup protection to them but the rest of the protection schemes only meant for primary protection.

Backup Protection

Power system is highly integrated in nature and thus the demand spreads over all the connected area. Sometimes all the different entities of the power system operate very close to their limits. Under such a condition, a clear discrimination between normal and abnormal event is difficult. Backup protection

Figure 6. Backup protection

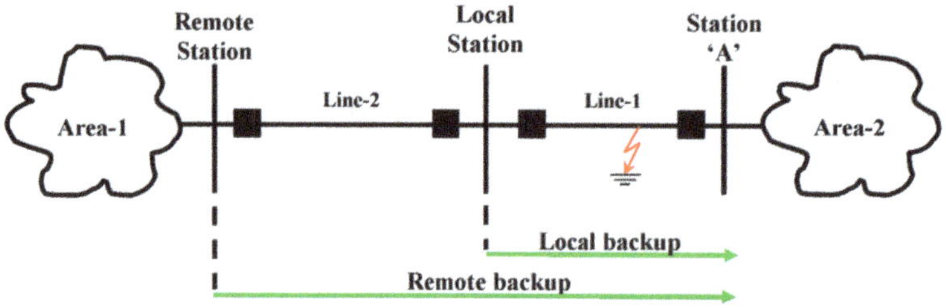

in power system plays a vital role in providing adequate performance to limit the spread of a disturbance that will never give a chance for the occurrence of cascaded events may lead to blackout. When a disturbance is accompanied by any one of the primary protecting element failures then the backup protection is initiated to clear that disturbance. It may initiate not only in single contingency but also multiple contingencies scenarios. Basically, backup protection is set with certain intentional time delay (NERC-SPCS, 2011) to full fill the specified task and it provide local backup and remote backup which are illustrated in Figure 6. A fault occurred nearby area-2 on the line-1 then the primary relay detects and clears the fault immediately. If the primary relay fails to operate then the local back relay clears the fault at the same time. Even though, if the entire protection scheme fails, then remote backup relays clear the fault after certain time delay.

1. **Local Backup:** The philosophy of local backup is to detect all failures in the primary protection system and take local measures to correct the complexity. It provides backup protection by the addition of protection systems locally at a substation. If any of the power system components such as CT, potential transformer (PT), breaker trip coils, batteries and dc circuitry will fail, then the failed component is backed up by another component at the substation. In order to ensure the best local backup, the local backup system should have an ability to sense all the faults associated with the components of the primary protection system.
   Local backup is categorized as relay backup and breaker backup. In relay backup, whenever the primary relay fails to operate, then an additional identical relay is used for detecting the fault and trip the circuit breaker instantaneously. Whereas in breaker backup, when a protective relay initiates trip signal corresponding to a fault the circuit breaker fails to trip. After some time delay the backup relay initiates trip signal to all the adjacent circuit breakers, which are connected to a bus bar. This kind of local backup protection is implemented where the remote backup application is restricted.
2. **Remote Backup:** The primary protection scheme has a relay, CT, PT, circuit breaker and battery. In remote backup, backup relays are located at remote station and are set with intentional time delay to operate. They provide backup when the primary protection scheme fails to operate. This kind of backup protection is simple, economic and never affected with the factors.

System Stressed Events

1. **Voltage Instability:** Voltage instability is a balanced phenomenon and it refers to the inability of an electric power system to maintain acceptable voltage at all the busses in the system under nor-

mal operating conditions and after being subjected to a disturbance. The main factors contributing voltage instability are

a. Heavy loads
b. Sudden loss of load in an area
c. Reactive power deficiency
d. Transformation of system state
e. Periodic change in bus voltage profile
f. Bulk power long distance transmission through highly inductive transmission lines

Any one of these causes a significant part of the system enters into a state of voltage instability results an uncontrollable decline in voltage (Shehu, 2009).

2. **Load Encroachment:** The load fed by the transmission line also appears to the distance relay as impedance. During normal operating condition of the power system, the impedance decreases as the load increases. In general, the load impedance remains large enough that it does not encroach the circular characteristics of the distance relay. The potential causes of load encroachment are

a. Shifting of load flows under steady state
b. Change in transmission network configuration
c. Sudden increase of large load

Whenever this kind of phenomenon exists in the system then the load impedance become small enough that encroaches distance protective zones. This phenomenon is referred as load encroachment or static encroachment.

Other Events

1. **Hidden Failure:** Hidden failure in relaying systems is also one of the most critical factors causing major power system disturbances. It is defined as a permanent defect that will cause a relaying system to incorrectly and inappropriately remove circuit elements as a direct consequence of another switching event (Surachet, 1994). Hidden failures are related directly or indirectly to hardware or software malfuncion. A relaying system consists of many components and accessories which may fail in their own way depending on their nature and characteristics, rendering relay misoperations. Furthermore, a particular failure of a component occurs in two different relaying schemes may have different effects on the power system. Subsequently, each component of a complete relaying system needs to be analyzed to determine its significant impact on the power system. So, in a general form hidden failure in the protection system caused by the following factors.

a. Hardware or software failure
b. Out dated settings
c. Human errors

It is not a frequent problem and it may go unidentified for a long time. But the risk associated with its effects in power systems may be catastrophic considering that the product of probability times consequence is a measure of risk.

2. **CT Saturation:** Current transformer is one of the instrument transformers and is used for measuring the current flowing in an element of the power system. In real time, the CT reduces the higher value of measured current to lower value i.e. 5A or 1A which is essential to operate the distance relay. The CT used for distance protection is quite different as compared with the CT for instrumentation. Generally, the distance relay needs to operate reliably at steady state current and fault current. So that the CT has the ability to provide a correct ratio up to multiple times the primary current.

Even though, the best CT that has the accuracy up to 10-15 times the rated current is preferred for reliable operation of the relay (Badri ram, 2011), the CT core gets saturated due to burden on its secondary winding and the fault current that exceeds the CT primary rating. Thus, giving erroneous values that will influence the distance protective relay operation.

3. **Noisy Input Signal:** The actual input signal given to the distance relay is contaminated with noise that contains high frequency components. In order to determine the impedance seen by the relay require fundamental frequency components. But, in certain cases the input signal having both fundamental and high frequency components. Thereby, the relay can be misinterpreted as a fault and may lead to maloperation of the relay.

Impact of System Stressed Events on Distance Protection

Especially, voltage instability and load encroachment immensely affect the reliable operation of the distance protective relay. This means that the third protective zone is heavily affected due to its large reach and time delay. Voltage instability occurred due to reactive power demand and change in system conditions may lead the system to uncontrollable state i.e. voltage collapse. On the other hand, load encroachment is also occurred due to an excessive load demand and loss of transmission line. Further, it can cause over loading on the transmission system. In both the scenarios, the impedance measured by the distance relay is always less than its setting and the corresponding impedance trajectories enter into the third protective zone as shown in Figure 7. At this instant, the distance relay takes wrong decision and trips the power system component unintentionally. Further, it causes cascading events which may lead to blackout. Again discrimination of symmetrical fault from voltage instability and load encroachment is difficult for distance relay at its third operating zone. It happens only because of their balanced phenomena. Thus, causes maloperation of the distance relay, which will initiate the cascading events leading to blackouts.

The NERC revealed that almost 70% of the disturbances in the power system are contributed by relaying system and third zone maloperation is one of the most notable causes of power system blackouts (NERC, 2003). From the post-mortem reports of several worldwide power system blackouts(WSCC, 1996; NERC, 2003; RAE, 2003; US-CPSOTF, 2004; UCTE-RAE, 2004; UCTE, 2004; NERC, 2005; IEEE PSRC WG D6, 2005; AER, 2007; UCTE, 2007; Australian, 2007; Žaneta,2008; Atputharajah,

Figure 7. Impedance trajectories during fault and system stressed condition

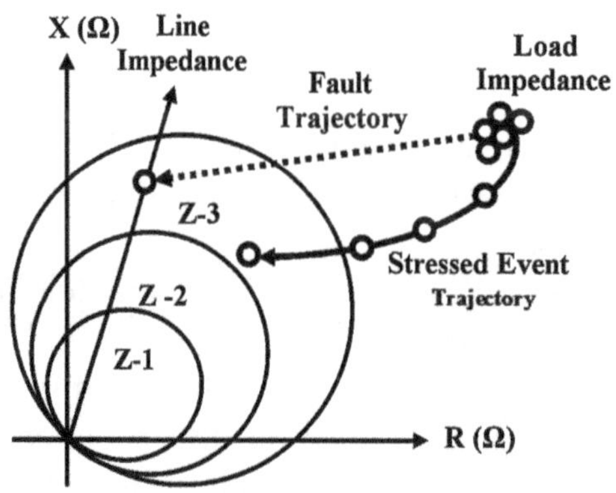

2009; NRPC, 2010; Michael, 2011; Power grid, 2012; Zeng, 2015; Veloza. 2016), it has been pointed out that the blackouts have great impact on population, electrical supply systems, economy, life supporting systems in hospitals and traffic systems.

Issues Related to Conventional Wide-Area Protection

A lot of sensing, communicating, monitoring and controlling elements are spread throughout the power system usually are equipped to handle such undesired operations and bring back the system to its steady state. For better operation of the power system, a supervisory control and data acquisition (SCADA) technology was implemented in earlier days. The generalized real time power system operational paradigm is shown in Figure 8. From Figure 8, it is understood that the system information collected from different sensors are fed to the EMS system and again to the SCADA system. The monitoring, control and protection task assigned by supervisory unit is again translated to the required system through different RTUs. Every control center is equipped with energy management system (EMS) and all the collected data processed by different algorithms i.e. running as engines in the EMS.

Figure 8. Power system operational paradigm

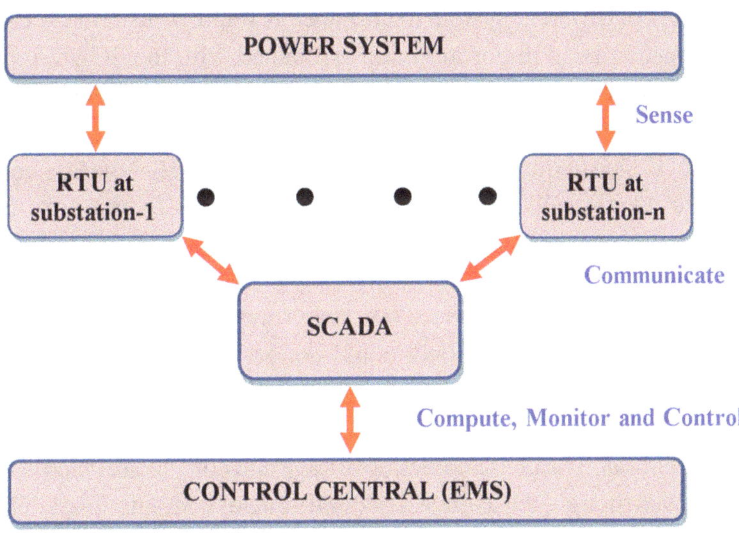

Figure 9. The traditional measurement and monitoring infrastructure using SCADA technology

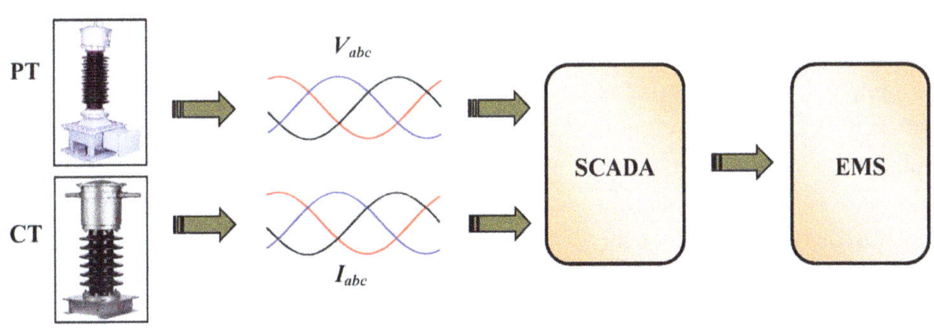

Figure 9, shows the traditional way of measuring and monitoring the power system using SCADA technology. The analog voltage (V_{abc}) and current (I_{abc}) signals measured by the PT and CT are given to the SCADA system. Thus, the SCADA system gives the acquired data to EMS after every 4 to 6 sec. All the algorithms are performed as a part of EMS and the computed results were displayed for monitoring the system state. Based on system state, the power system operator can decide what control actions need to be taken for the system. However, the traditional SCADA technology suffering with distinguished issues. They are

Issue 1: All the algorithms, optimization techniques and power system security analysis tools are not viable during an emergency condition because they require certain time delay for computing corresponding programs. In addition, propagation of a large blackout is difficult to incorporate with them and may require heuristic methodologies (IEEE PES-PSRC-SPSWG C-6, 2002).

Issue 2: When certain system stressed events such as voltage instability and load encroachment takes place, the corresponding data received once in every 4 to 6 sec from the SCADA system is not enough to take right control actions.

Issue 3: The SCADA system is unable to capture the oscillations that are happening in the input signal of a particular system. So that it might take wrong control actions. Therefore, it is essential to update the data transfer rate so that exact information about power system dynamics can be captured. Hence, better analysis of different system disturbance is required to achieve faster control.

Issue 4: Time synchronization is an important issue associated with the SCADA system. Due to geographical distance, the substation which is closer to the control center will get the data much quicker than the substation located far away. Moreover, these substations do not have internal time synchronization clock. Due to this the data acquired by the SCADA system is not precisely captured at the same time from various locations. Therefore, it is very difficult to recognize what time the data belong to and may lose the data.

Issue 5: The power system is so dynamic, especially during contingencies and disturbance scenarios. At this situation, if the control center does not have time synchronized data from the wide-area perspective of the system then the operator could not get the right data about the power system. Hence, time synchronized data is required to visualize the entire power system.

Issue 6: The SCADA technology provides the magnitude of voltage and current as direct measurements for analyzing AC networks. The phase angles of voltage and current are calculated by the computer-based algorithms. Even though, those are not accurate due to computational delays associated with the algorithms.

In order to overcome the aforementioned issues related to the traditional SCADA technology, a synchronize measurement technology with high communication and huge data handling capability is introduced to monitor and control the wide area power system. The unique feature of this technology is, it has ability to sample the data of analog voltage and current signals and is synchronized with global positioning system (GPS). Finally, it computes the consistent frequency components from different locations.

Synchrophasor based wide area protection scheme is proposed in this book chapter. The main objective of the proposed scheme is to achieve reliable operation of the distance relay during system stressed conditions and also avoid the consequences of the power system blackout. It gives an ideal measurement system with which to monitor and control the power system during system stressed conditions. In this scheme, the rate of change of differential impedance (RCDI) of a two-area interconnected system is considered for monitoring the power system. The impedance magnitude deviation between two areas is almost very small during system stressed conditions whereas, in symmetrical fault condition it becomes large. Here, the RCDI is a key indicator for discriminating symmetrical fault from system stressed conditions.

LITERATURE REVIEW

Based on the philosophy of backup protection and significant impact of system stressed events on distance protection, several researchers, power engineers and academicians are proposed numerous techniques or solutions to prevent occurrence of cascaded events which may lead to blackout. All those solutions are presented here with reference to the local end and remote end information.

Local End Information Based Solutions

1. **Rate of Change of Voltage (ROCOV):** An adaptive relaying algorithm is proposed (Jonsson, 2003) based on rate of change of voltage. Initially, it utilizes the apparent impedance for determining the operating zone and secondly, the rate of change of voltage is used for discriminating three-phase fault from voltage instability.

$$\frac{\Delta V}{\Delta t} \leq \left(\frac{\Delta V}{\Delta t}\right)_{fault\ maximum} \tag{1}$$

$$\frac{\Delta V}{\Delta t} > \left(\frac{\Delta V}{\Delta t}\right)_{fault\ maximum} \tag{2}$$

where, ΔV and Δt are the change in voltage and time. The above Eq. (1) describes the initiation of three-phase fault and similarly, Eq. (2) also describes the existence of voltage instability phenomena. Accordingly, this algorithm averts the unintentional trips of a distance relay during voltage instability. But, it cannot be applicable in the system where the protected line generates reactive power. Moreover, it may fail due to dynamics of the power system. Sometimes anyone of the protected line gets overloaded due to unexpected load demand or loss of its neighboring line.

2. **Voltage Stability Index (VSI):** According to load flows, the over loaded lines are indexed with VSI (Arya, 2008) and it becomes 0.5 at voltage collapse. The impedance seen by the relay makes necessary relationship with its operational characteristics that emphasizes security as follows.

$$\frac{V_m}{I_m} \geq \left|\frac{B}{A}\right| \text{As voltage stability} \tag{3}$$

$$\frac{V_m}{I_m} < C_1 \left|\frac{B}{A}\right| \text{Alarm activated} \tag{4}$$

$$\frac{V_m}{I_m} < C_2 \left|\frac{B}{A}\right| \text{Load shedding activated} \tag{5}$$

where V_m and I_m are the voltage and currentsignals at bus 'm'. C_1 and C_2 are the constants corresponds to alarm and trip actions. $\left|\frac{B}{A}\right|$ is the radius of the most inner circle in R-X plane. However, this approach faces the difficulty with the relay setting due to the algebraic manipulations. Further, the relay margins are pleased with the formulation of a combinatorial optimization and control strategies (Song, 2009) against voltage collapse.

All these techniques meant only for the detection of voltage instability and symmetrical fault but not the impedance fault. It means that they may fail in detection of impedance fault results third zone maloperation of distance relay.

3. **Voltage and Current Criteria:** In order to achieve such difficulty during voltage degraded conditions, a local end measurements-based algorithm is propounded (Sharifzadeh, 2014) with the use of rate of change of voltage $\left(\frac{\Delta V}{\Delta t}\right)$ and change in current with respect to change in voltage magnitude $\left(\frac{\Delta I}{|\Delta V|}\right)$.

$$\frac{\Delta V}{\Delta t} > \frac{\Delta V}{\Delta t}\max : \mathit{No\,fault\,has\,occured} \tag{6}$$

$$\frac{\Delta V}{\Delta t} \leq \frac{\Delta V}{\Delta t}\max : \mathit{Low\,impedance\,fault\,has\,occured} \tag{7}$$

From the Eq. (6) and (7), the voltage criterion is unable to detect and discriminate high impedance fault from voltage degraded event due to its low current control capability and identical characteristics of the events. Thereafter, it is achieved by considering the current criteria as follows;

$$\frac{\Delta I}{|\Delta V|} > \frac{\Delta I}{|\Delta V|}\min : \mathit{High\,impedance\,fault\,has\,occured} \tag{8}$$

This algorithm provides faster detection and discrimination of impedance faults from voltage degraded event.

4. **Wavelet Packet Transform (WPT):** Similarly, the voltage and current signals are sampled and are employed to wavelet packet transform (WPT) for extracting their high frequency coefficients (Mahari, 2015). The WPT voltage and current coefficients are used for detecting high impedance

fault and faulted zone. If the absolute value of consecutive WPT voltage coefficients difference violets the threshold then the counter starts counting the samples. If the counted samples greater than the number of samples per cycle, the impedance fault has detected within 0.005 sec. Meanwhile, the faulted zone is also identified, when the averages of the sum values of WPT current coefficients exceeds the threshold. This algorithm is also used for detecting the faulty phase. However, the wavelet domain decomposed signal mismatches with the actual signal. Thus, erroneous information causes relay maloperation.

5. **Anti-Encroachment Zone (AEZ):** Steady state security analysis and adaptive anti-encroachment zone (AEZ) are used to avoid inadequate tripping caused by the load encroachment (Jin, 2008). The security of the system is analyzed by considering the duel contingency criteria. This analysis gives the information about the most vulnerable relays during load encroachment event which are indexed with load encroachment index (LEI) as follows.

$$LEI = \sum_{i \in A} \frac{\left|Z_{pi}\right|}{\left|Z_i\right|} \tag{9}$$

where A is the group of distance relays, Z_{pi} and Z_i are the pick-up and apparent impedance of distance relay 'i' respectively. On the basis of security analysis, the corresponding single or dual parameter AEZ model will be designed through online to prevent relay maloperation due to load encroachment event.

$$Z_{set} < \left|Z\right| \text{and } \varphi_{set} > \left|\varphi_Z\right| \text{dual parameter model} \tag{10}$$

$$\varphi_{set} > \left|\varphi_Z\right| \text{single parameter model} \tag{11}$$

Implementation of single parameter model is very easier than dual parameter model and is restricted due to large blocking region. However, it requires high speed communication systems for getting reliable operation and may fail to detect fault during over loaded conditions.

6. **Positive Sequence Impedance Angle (PSIA) and Total DC Component (TDC):** The positive sequence impedance angle (PSIA) and the total DC component (TDC) based algorithm is suggested (Jithin, 2017).Thus, the two criterions are expressed as

$$\text{Positive sequence impedance (PSI) } Z\angle\theta_P = \frac{V\angle\delta}{I\angle\varphi} \tag{12}$$

$$\text{PSIA}\left(\theta_P\right) = \delta - \varphi \tag{13}$$

$$\text{DC component of k-phase current } \left(I_{DC,K}\right) = \frac{\sqrt{2}}{N}\sum_{n=0}^{N-1} I_k\left(n\right) \tag{14}$$

Total DC component of three phases (TDC) $= I_{DC,R} + I_{DC,Y} + I_{DC,B}$ (15)

Where $V\angle\delta$ and $I\angle\varphi$ are the voltage and current phasors. $I_{DC,R}$, $I_{DC,Y}$ and $I_{DC,B}$ are the three phase DC component of currents. During load encroachment condition, the estimated PSI falls within zone-3 and the proposed criterions PSIA and TDC less than their threshold. Consequently, the relay should be blocked. Whenever a three-phase fault occurs then the PSI encroaches zone-3 and the corresponding PSIA and TDC exceeds their threshold. Accordingly, an unblocking signal is generated after 800 msec for relay tripping.

7. **Steady State and Transient Components (TC):** Steady state and transient components (TC) based fault detector is postulated to visualize the actual behavior of the power system through state diagram (Kim, 2005).The sum value of TC is denoted as

$$H_{sum} = \sum_{k=2}^{N/2} \left| X\left(k\right) \right| \quad (16)$$

where H_{sum} is the sum value of TCs and $X\left(k\right)$ is the Discrete Fourier Transform (DFT) of the input signal. During voltage instability and load encroachment, the input signal has only steady state components and the impedance seen (Z_{seen}) by the relay lies within the third zone ($Z_{3,set}$). Consequently, the signal 'S' is generated for blocking the relay. However, a three-phase fault occurs during voltage instability or load encroachment, transient component become significant and the value of H_{sum} exceeds the threshold. Thus, the signal 'T' is generated for initiating trip signal.

$$Z_{seen} < Z_{3,set} \text{ and } H_{sum} < Threshold \text{ voltage instability(or) load encroachment} \quad (17)$$

$$Z_{seen} < Z_{3,set} \text{ and } H_{sum} > Threshold \text{ initiation of fault} \quad (18)$$

Even though this technique improves the performance of the relay but also it may fail in the extraction of high frequency transient components precisely during parallel feeders switching.

8. **Transient Monitoring Function (TMF):** All these detection techniques may fail due to CT saturation and noisy condition. In order to achieve that transient monitoring function (TMF) and the positive sequence impedance angle $\left(\varnothing_z\right)$ based fault detection algorithm is introduced (Nayak, 2015). With the existence of dc components, the TMF is defined as a sum of absolute values of difference between the re-constructed and actual current signal (d_k) over a cycle. The maximum value *TMF* associated with all the three phases is achieved and is used as an index (g) for the detection of three phase fault.

$$TMF = \sum_{k=1}^{N} \left| d_k \right| \quad (19)$$

$$g = \max\left(TMF_a, TMF_b, TMF_c\right) \tag{20}$$

The value of '$\varnothing_z$' is obtained from the line voltage and current corresponding to the busses which are heavily loaded with reactive loads. The detection and discrimination of three phase fault from voltage instability and load encroachment events is very simple as follows;

$$g < g^{th},\ \varnothing_z < \varnothing_{th} \text{ and } Z_1 < Z_3 \text{ voltage instability (or) load encroachment} \tag{21}$$

$$g > g^{th},\ \varnothing_z > \varnothing_{th} \text{ and } Z_1 < Z_3 \text{ three-phase fault} \tag{22}$$

However, the presence of noise in the actual signal influences the proposed thresholds. A lot of steps are involved to determine detection parameters and switching of over loaded lines may immensely affected the angle '$\varnothing_z$'.

9. **Discrete Wavelet Transform (DWT):** In the same way an adaptive indices-based detection technique is proposed to provide a secure decision for discriminating critical events (Jose, 2018). Half cycle DFT is employed for extracting the magnitude of DC offset of three phase currents. The maximum value of decaying dc component magnitude is considered as the first index (ψ). Similarly, discrete wavelet transform (DWT) is used for calculating the energy of the first level detail coefficients (ξ). Thus, the maximum value of the energy among the three phases is chosen as the second index (ξ_m).

$$\psi = \max\left(Idc_a, Idc_b, Idc_c\right) \tag{23}$$

$$\xi_m = \max\left(\xi_a, \xi_b, \xi_c\right) \tag{24}$$

Based on the proposed indices the three-phase fault is detected and discriminated from voltage instability and load encroachment events as followed.

$$\psi > \psi_{th}\ and\ \xi_m > \xi_{th} \text{ three-phase fault} \tag{25}$$

$$\psi < \psi_{th}\ and\ \xi_m < \xi_{th} \text{ voltage instability (or) load encroachment} \tag{26}$$

This technique enhances the detection functionality of distance relay during critical events.

REMOTE END INFORMATION BASED SOLUTIONS

1. **Synchrophasor Based Voltage Instability Monitoring Index (SVIMI):** A synchrophasor based voltage instability monitoring index (SVIMI) is assigned to determine the weakest load bus in the

wide area power system (Sodhi, 2012). The SVIMI is estimated based on the weighted sum of voltage deviation from its reference (DFR) and consecutive voltage deviation (CVD) of a load bus.

$$VIMI_k = W_1(k)\frac{DFR_k}{DFR_{\max}} + W_2(k)\frac{CVD_k}{CVD_{\max}} \tag{27}$$

$$SVIMI_k = \max\left(VIMI_k^l\right);\ l \in \text{set of load buses} \tag{28}$$

where $VIMI_k$ is the voltage instability monitoring index at the instant$-k$. $W_1(k)$ and $W_2(k)$ are the weighing factors at the instant$-k$. When the voltage deviates at the acceptable limits then the SVIMI approaches to 1 and gives the alarm for the initiation of immediate control actions. However, it is difficult to set the maximum value of VIMI in real time application and detection of fault is also not possible.

2. **Support Vector Machine (SVM):** A supervisory control scheme (Seethalekshmi, 2012) is proposed to enhance the conventional distance protection. In this, support vector machine (SVM) based fault and disturbance classifiers are used for detecting the fault and discriminating the system stressed conditions. Simultaneously, the most vulnerable relays during system stressed conditions are also identified based on their relay ranking index (RRI). SVM based fault classifier generates blocking signal for voltage instability and trip signal for faults. On the other hand, SVM based disturbance classifier distinguishes the system stressed events. As compared with the other conventional methods, this scheme provides better accuracy in disturbance classification and quicker in decision making. However, it is not viable due to the prerequisite of huge training data set and is incapable to distinguish fault from hidden failures.
3. **Synchrophasor Measurement Technique:** An index based synchrophasor measurement technique (Kundu, 2014) is introduced, which utilizes both the current deviation and the estimated apparent impedance for detecting stressed event and fault as follows.

$$\text{Disturbance detector index } I_{dis_i}(k) = I_{sum_i}(k) - I_{sum_i}(k-1) \tag{29}$$

$$I_{dis_i}(k)\begin{cases} = 0 & for\ normal\ operation \\ > 0 & for\ voltage\ instability\ or\ load\ encroachment \end{cases}$$

$$\text{Fault detector index } f_d = \sum Z_{d_{jln}} \times W_j \tag{30}$$

$$f_d = \begin{cases} + & for\ no\ fault\ condition \\ - & for\ fault\ condition \end{cases}$$

where $I_{sum_i}(k)$ is the phasor sum of ith- group currents at k^{th}-instant, W_j is the weight associated with j[th] measurement and $Z_{d_{jln}}$ is the impedance of the line - l connected to n^{th} bus in a group. Even though, this technique able to discriminate the fault from voltage instability but, it is difficult to assess the optimal placement of phasor measurement units (PMUs) and latency issue with the communication.

4. **Phasor Measurement Units (PMUs) Based Technique:** A synchrophasor assisted protection algorithm is evoked (Dipak, 2017).In this algorithm, the data associated with the over loaded buses is collected from the respective PMUs. Thus, the data is used for calculating the positive sequence impedance seen by each relay. The load encroachment event is identified, when the positive sequence impedance seen by the back-up relays ingression into zone-3 and the adjacent primary relays could not ingression into zone-2 or 1. Accordingly the corresponding back-up relays are blocked. Moreover, it can be able to detect symmetrical and asymmetrical faults and rescind the block signal, when the positive sequence impedance seen by the primary relays encroaches zone-1.
5. **Vulnerable Relay Index (VRI) and Monitoring Index (MI):** Similarly, a novel index based protective scheme is proposed by using synchrophasor measurement device (Bolandi, 2017). It delineates the vulnerable relay index (VRI) and monitoring index (MI) as follows.

$$VRI_{TH-i} = 0.3373 \times \left|\bar{Z}_{ij(pu)}\right|^2 \tag{31}$$

$$VRI_i = \frac{\left|RM_i\right|}{\Delta M_i\left(V_{i,j},\theta_{i,j}\right)} \tag{32}$$

$$VRI_i < VRI_{TH-i} \text{ vulnarable relay need to be monitored} \tag{33}$$

$$MI_{mn} = \arg\left(\Delta\bar{S}_m\right) + \arg\left(\Delta\bar{S}_n\right) = \angle\Delta\bar{S}_m + \Delta\bar{S}_n \tag{34}$$

where RM and ΔM are the relay margin and sensitivity, which are extracted from positive sequence voltage phasors. $\Delta\bar{S}_m$ and $\Delta\bar{S}_n$ are the phasor quantities of fault component complex power (FCCP). During load encroachment, the relays which are the most susceptible to zone-3 maloperation are identified and ranked by comparing their *VRI* with VRI_{TH} .Among these the most critically operated relays are again indexed with MI for continuous monitoring of their performance. The index MI has an ability to discriminate the symmetrical fault from load encroachment event as follows.

$$-180^0 - \theta < \left(MI_{mn} : \arg\left(\Delta\bar{S}_m\right) + \arg\left(\Delta\bar{S}_n\right)\right) < -180^0 + \theta \text{ symmetrical fault} \tag{35}$$

$$0^0 - \theta < \left(MI_{mn} : \arg\left(\Delta\bar{S}_m\right) + \arg\left(\Delta\bar{S}_n\right)\right) < 0^0 + \theta \text{ load encroachment} \tag{36}$$

All these methods are computed with the help of either local end information or remote end information. Using local end information although reliable and faster protection system can be developed but such an information-based technique is also sensitive to primary zone faults. As back-up protection operate with a time delay of minimum 800 msec, within that time period remote end information can be accessed to provide better back-up protection against system stressed events. Monitoring of power system with the help of synchrophasor technology can provide better solution towards power system blackout. Any relative change between two areas of power system can be monitored and based on magnitudes of impedance deviation; accurate decision can be taken to differentiate system stressed conditions from symmetrical fault.

SYNCHROPHASOR TECHNOLOGY

Need of Synchrophasor Technology

In a two-area interconnected power system network as shown in Figure 10, the power flow between 'Area-1' and 'Area-2'through the transmission system. The magnitude of power is normally dependent on the magnitude of their respective bus voltages V_1 and V_2 but, it is also dependent on the sign of the angular difference between the voltage phasors i.e.$(\phi_1 - \phi_2)$.

$$P_{12} = \frac{V_1 V_2}{X_L} \sin\left(\varnothing_1 - \varnothing_2\right) \tag{37}$$

As stated in eq. (37), the angle information plays a vital role in power transmission and needs to be monitored at the important transmission corridors in real time to maintain the healthy operation of power system. Any wide deviation in the angular difference value may lead to unstable power system operation and it is one of the root causes for the occurrence of the blackout at United Sates and Canada on August 14, 2003(UCTE, 2004). The angular separation between the buses just before the occurrence of blackout is 30^0. But this has further led to separation of buses and the angular separation reached at 160^0. So, with angular deviation monitoring, the percentage of such a blackout in future can be reduced.

Figure 10. Simple two-area interconnected power system network

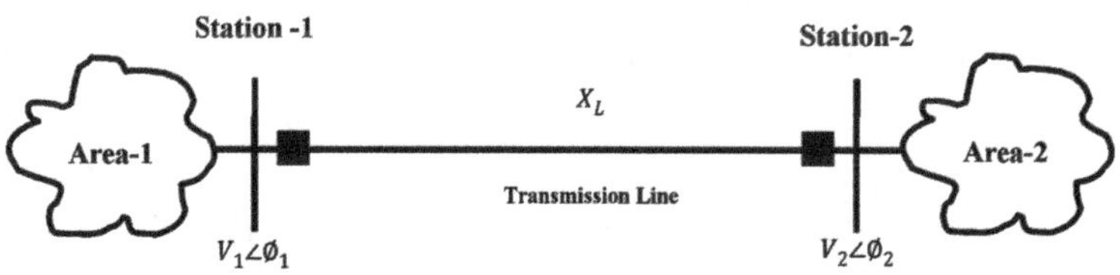

Figure 11. Synchronous measurement technology

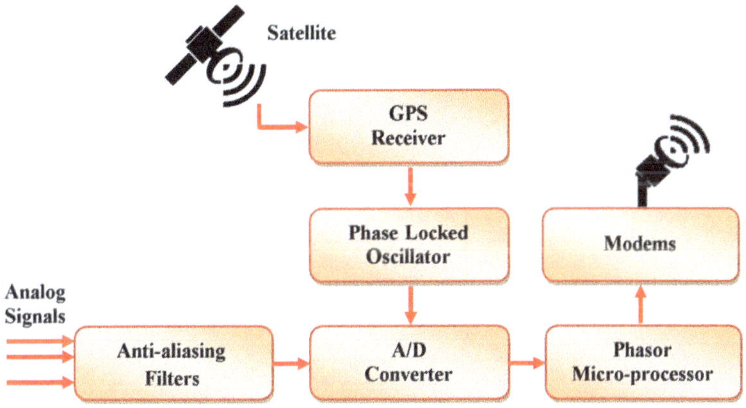

Synchrophasor Measurement Unit (SMU)

Synchrophasor based technology is well established and it offers an ideal measurement system. Such a technology can offer advantage during system stressed conditions to monitor and control the power system. This technology provides positive sequence measurements with time synchronization which are useful to compare the measurements accurately at different parts of the wide area network. The time synchronization is attained from the global positioning system (GPS). Figure 11 shows that block diagram of a synchrophasor measurement unit (SMU). The major functional elements of the SMU are GPS receiver, anti-aliasing filters, A/D converter, phase locked oscillator, phasor micro-processor and modems.

The GPS receiver receives the broadcast time from the GPS system, then it is delivered to phase locked oscillator for generating clock pulses. Simultaneously, the input analog signals are processed through an antialiasing filter to prevent aliasing phenomena. Thus, the analog signals are converted into digital signal by analog to digital converter. There the digital signals are time stamped through the phase locked oscillator. Further the time stamped signals delivered to phasor micro-processor for determining phasors of the three phase signals using recursive Discrete Fourier Transform (DFT). The phasors information corresponding to certain part of the wide area system transfer through modems for system monitoring and controlling at the control center (Mallikarjuna, 2017).

Application of SMUs in Wide-Area Power System

Due to the recent development in digital relaying, information and communication technology with the help of synchrophasor application, information of a particular area of power system can be transmitted to another area for the better monitoring and protection. A wide-area measurement system (WAMS)is a system that provides a time-synchronized view of electrical situations over a widespread geographical area, thereby improving the situational awareness of the energy management system (EMS) of a power grid. With this enhanced situational awareness, utilities would be able to react quickly to contingencies, and avoid large-power system blackouts.

Figure 12. Architecture of the WAMS in real-time

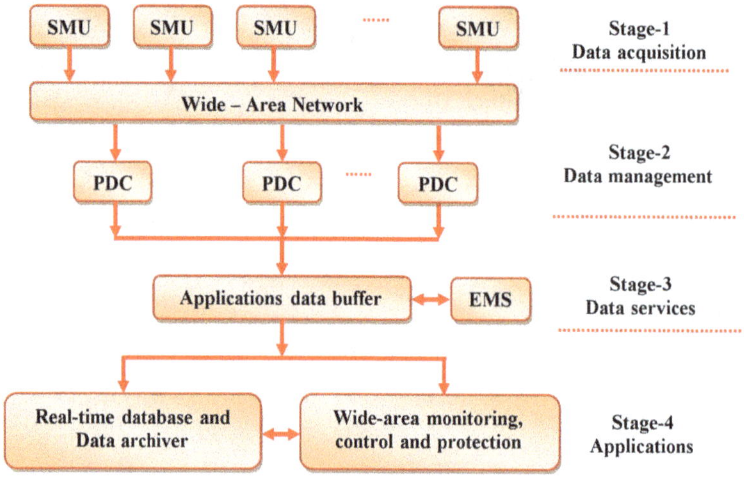

Figure 12 shows the general architecture of the WAMS (Yee, 2013). At stage-1, all the SMUs acquired the voltage and current phasors information from their respective areas and such information is always time stamped with high-precision internal clocks and the GPS. At stage-2, the information acquired by the SMUs is forwarding to the PDC through wide-area network. The PDC enables the correlation of phasor measurements across the wide area. At stage-3, application data buffer gathers the information from PDCs and is processes through the EMS to know the system state. It monitors the information for error, losses and synchronization. Further the information supplying to the next stage in the required format. At stage-4, the real-time database and data archiver takes the responsibility for documenting the information for past incident analysis and assessment. Simultaneously, the computed information is useful for various applications like system monitoring, control, and protection activities.

The main issue with synchrophasor measurement technology is the delay in data transfer rate. So, primary protection is challenging using synchrophasor technology, but with such secondary protection or backup protection can be provided. As third zone of distance relay is highly affected by system stressed conditions, thus with the help of synchrophasor technology the performance of such a unit can be improved.

SYNCHROPHASOR BASED WIDE AREA PROTECTION SCHEME

During normal operation of the power system, the voltages at each bus and current flowing between the buses through the transmission lines are lies within their prescribed limits. Consequently, the impedance measured by the distance relay is also concentrating at outside the protective zones. But, whenever the power system exposes to any stressed events or short circuit faults the system become stressed and the corresponding bus voltages and currents are drastically distorted. Therefore, the impedance seen by the relay located at each bus deviates from its set value and the differential impedance between the buses is significantly fluctuating. Especially, a three-phase fault during voltage instability or load encroachment, the system becomes unbalanced and the apparent impedance measured by the distance relay enters into third protective zone. Thus, the relay misinterprets the stressed events as a fault and maloperates due to their identical nature. Sometimes these kinds of relay maloperations lead to propagate as a major power system blackout.

In view of security of wide-area power system, the current and conventional backup protection algorithms are not the best choice because the functions of each distance relay hardly coordinated with each other. Therefore, a synchrophasor based wide-area protection scheme is proposed based on rate of change of differential impedance (RCDI) between two-areas to avoid relay maloperations and the occurrence of the power system blackout. The corresponding flow chart of the proposed algorithm is as shown in Figure 13. There are few computational steps needed to understand how the proposed algorithm can able to detect and discriminate stressed events from the three-phase fault.

Step 1: Acquire the synchrophasor measurement information from strategically located SMUs in the wide-area power system.
Step 2: Compute positive sequence impedance at each relay location with the help of positive sequence voltage and currents.
Step 3: Check whether the computed positive sequence impedance lies within the third zone or not. If it lies within the third zone, then go to the next step otherwise go to step-2.

Figure 13. Flowchart for the proposed algorithm

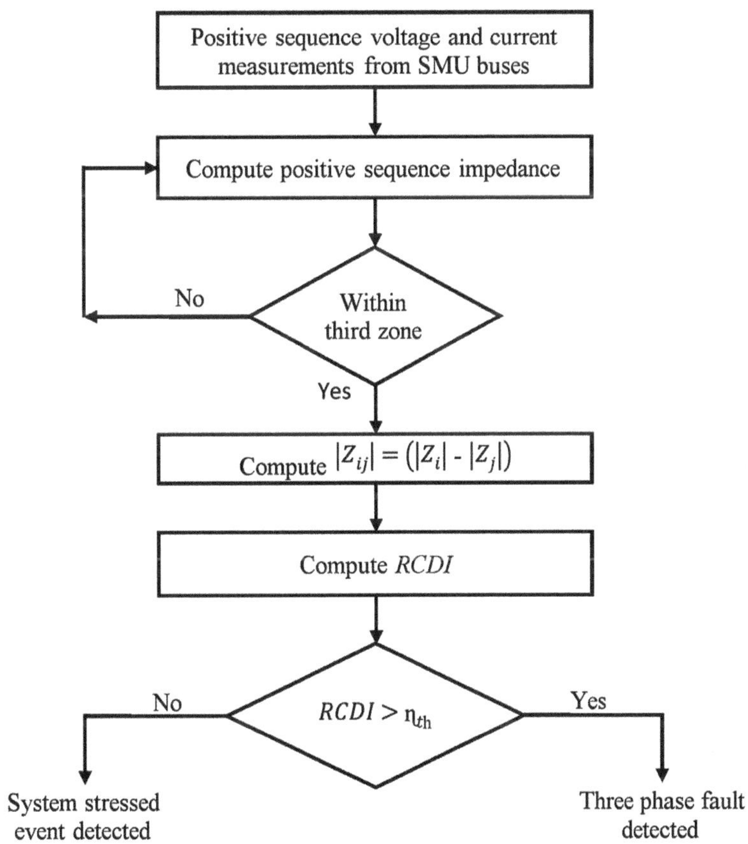

Step 4: Compute magnitude of differential positive sequence impedance by subtracting the magnitude of positive sequence impedance corresponding to relay '*j*' from the magnitude of positive sequence impedance corresponding to relay '*i*'.

$$\left|Z_{ij}\right| = \left(\left|Z_{i}\right| - \left|Z_{j}\right|\right) \tag{38}$$

Step 5: Compute the rate of change of magnitude deviation in positive impedance between two areas as

$$RCDI = \frac{d\left|Z_{ij}\right|}{dt} \tag{39}$$

Step 6: Check whether the computed *'RCDI'* is exceeds the threshold 'η_{th}' or not. If it exceeds threshold then the three-phase fault detected otherwise it identified as a stressed event.

$$RCDI > \eta_{th} \tag{40}$$

For the studies of third zone maloperation of distance relay, it is required to use a simple test system on which different system stressed events and fault during such events can be created. A simple two-area, four machine system (Kundur, 1994) is being used as the test system for the proposed algorithm is given in Figure 14.

Figure 14. A simple two-area, four machine system.

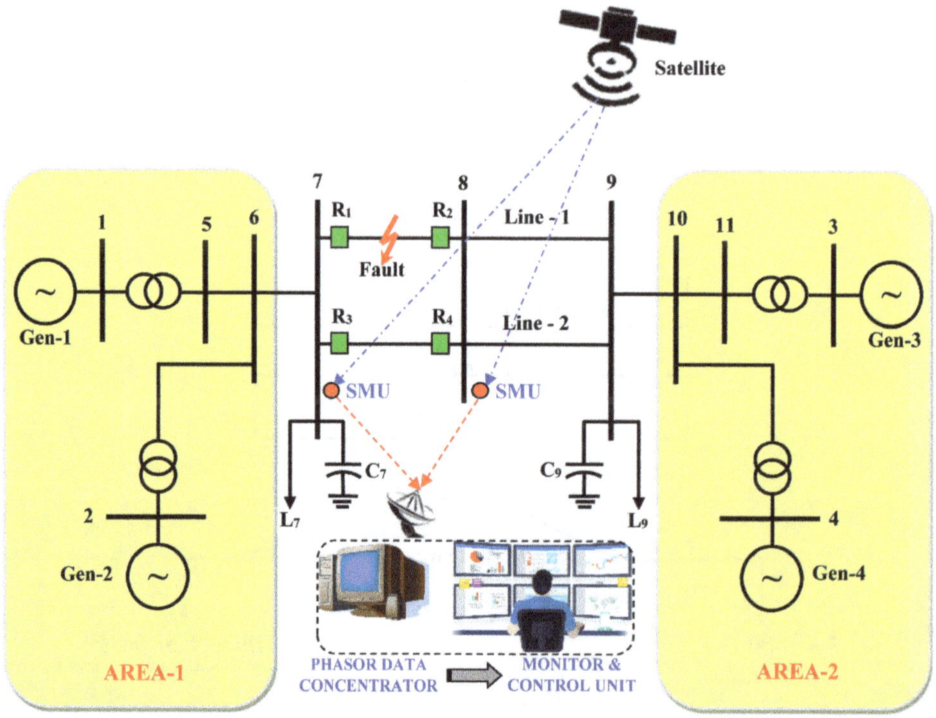

The parameters of a simple two-area, four machine system are provided in the Appendix. Modeling of the test system and creation of fault, voltage instability and load encroachment events have been done with 1 kHz sampling frequency in the PSCAD/EMTDC software. Meanwhile, the generated sample data is incorporated in the MATLAB software to prove the efficacy of the proposed algorithm.

To test the proposed algorithm, different test conditions are created on the test system which includes faults and different system stressed condition such as voltage instability and load encroachment. During these conditions the positive sequence impedance data have been taken from SMUs through the relays R_1 and R_2. Further the collected data processed through PDC and forward to monitor and control unit. There the proposed algorithm is computing and identify the system state. Thus, the numerical relays differentiate the symmetrical fault from system stressed events according to the algorithm and give appropriate signals to the circuit breakers to prevent the system from voltage instability and load encroachment.

Performance During Voltage Instability

This condition is created on the simple test system as increasing the reactive power of the loads L_7 and L_9 which are connected at either of the areas with a step change of 350 MVAR for every 1 sec from 2.5 sec onwards. Hence, it creates a condition of voltage instability for both therelaysR_1 and

Figure 15. Measurements at bus-7 (a) Positive sequence voltage. (b) Positive sequence current. (c) Positive sequence impedance.

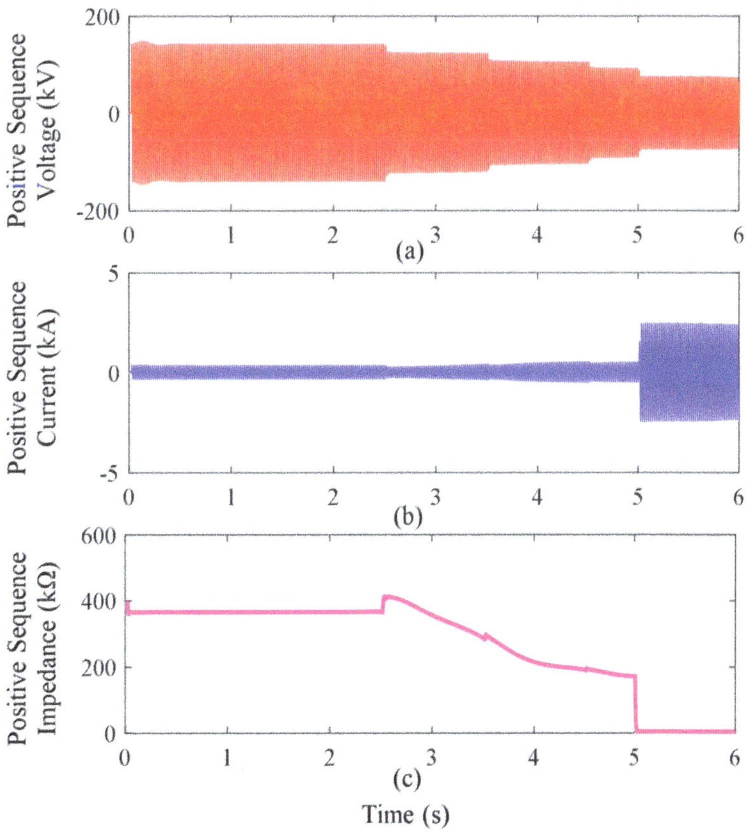

Figure 16. Measurements at bus-8 (a) Positive sequence voltage. (b) Positive sequence current. (c) Positive sequence impedance.

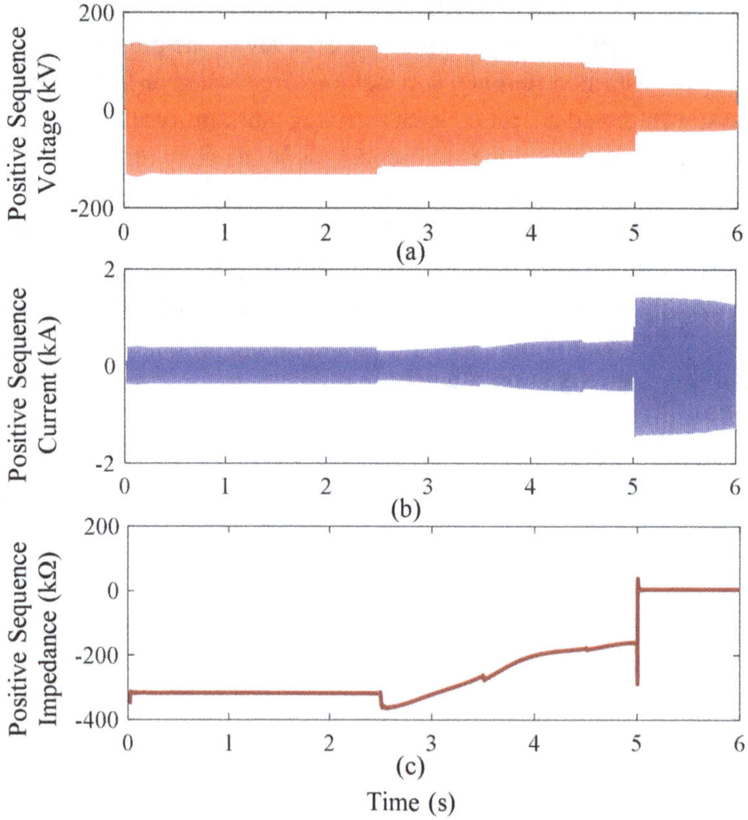

R_2. Since the system is under stress, the impedance seen by the relay either R_1 or R_2 will enter into the third protective zone. At this instant it gets the wrong idea about the occurred event and will unnecessarily send trip signal to its corresponding circuit breaker.

To validate performance of the proposed RCDI, a three-phase fault is created at 5 sec during voltage instability, on line-1 which is connected between the bus-7 and bus-8. During voltage instability, the voltages are decreased and currents also increased progressively from 2.5 see to 4.5 see as shown in Figure 15 and 16. But, for fault during voltage instability there is drastic changes occurred in both voltage and current and reaches to uncontrollable state. Based on these parameters the positive sequence impedances are computed at each bus and their magnitudes become almost zero after the fault inceptions illustrated in Figure 15 (c) and 16 (c).

The magnitude of positive sequence impedances at bus-7 and bus-8 are subtracted from each other to get differential impedance existed between two areas. Therefore, the magnitude of RCDI is determined by the use of differential impedance as seen in the Figure 17(c) and 17(d). From the Figure 17(c), it is observed that the magnitude of differential impedance continuously decreased from 2.5 see to 4.5 see during voltage instability. At 5 see its magnitude completely reaches to zero that indicates multiples of rated current flowing through the line-1 and bulk amount of power is exchanged between the two areas which may impact on the system security criterion. To detect a

Figure 17. Measurements at bus-8 (a) Positive sequence impedance at bus-7. (b) Positive sequence impedance at bus-8. (c) Differential Impedance between bus-7 and bus-8. (d) Rate of change of differential impedance.

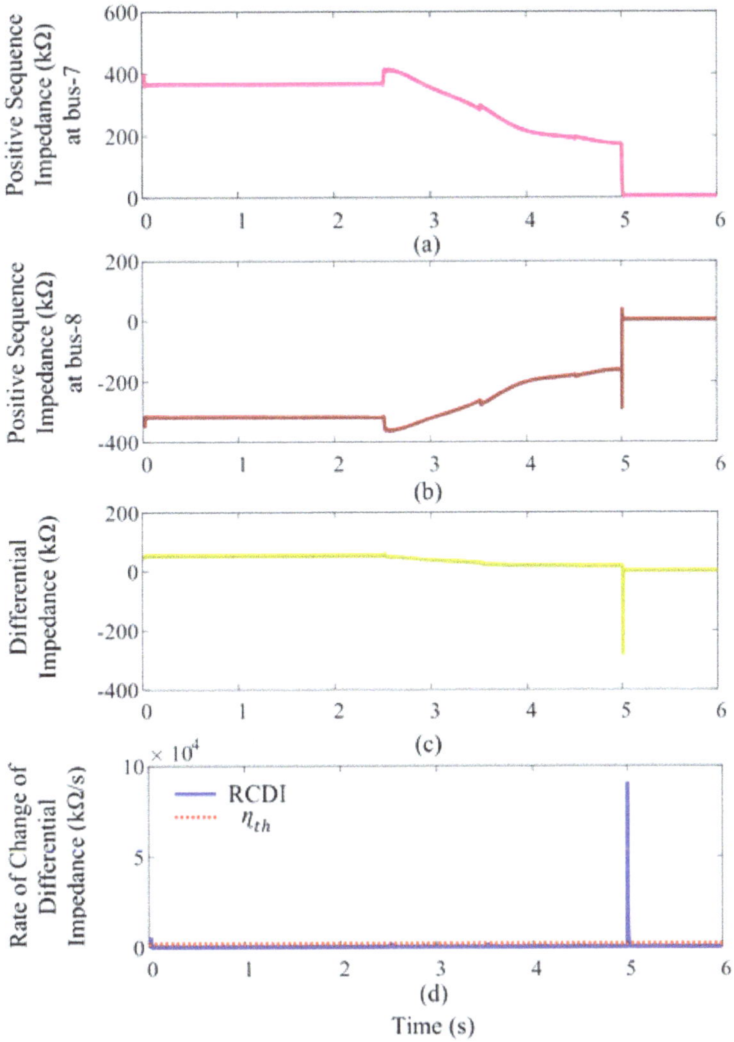

fault existed in the system and distinguish it from voltage instability proper threshold is required. In this algorithm suitable threshold is defined after validation of different events that are created on the same system.

The rate of change of differential impedance magnitude is compared with the threshold. During voltage instability, its magnitude is very less and falls below the threshold whereas in three-phase fault scenario the magnitude of RCDI is several times the normal condition. At this point of time the RCDI exceeds the threshold and three-phase fault is detected within a half cycle and discriminated from voltage instability.

Performance During Load Encroachment

This condition is created on the two-area, four machine test system by increasing the active power required by the loads L_7 and L_9 with a step change of 200MW for every 0.3 sec from 2.5 sec to 5.2 sec. Thus, it creates load encroachment condition for the relays which are connected at nearer to the load buses i.e. R_1 and R_2. During this period the magnitude of voltages degraded and corresponding currents also increased from their normal values periodically. The positive sequence impedances at bus-7 and bus-8 are computed with the help of measured voltages and currents. During voltage instability, it deviates from its steady state value results the active power flow between the buses gradually increased as shown in Figure 18 and 19. Here, a three-phase fault is created during load encroachment condition at 5.6 sec on line-1 which is connected between the buses 7 and 8. At this situation also both the voltage and currents gets affected and reached to out of control. Consequently, the magnitude of positive sequence impedance becomes zero which will allow bulk amount of active power flow through the lines. Thus, the test system gets stressed more and more and will operate at its maximum limits. Sometimes, it may affect the system security criterion.

Figure 18. Measurements at bus-7 (a) Positive sequence voltage. (b) Positive sequence current. (c) Positive sequence impedance.

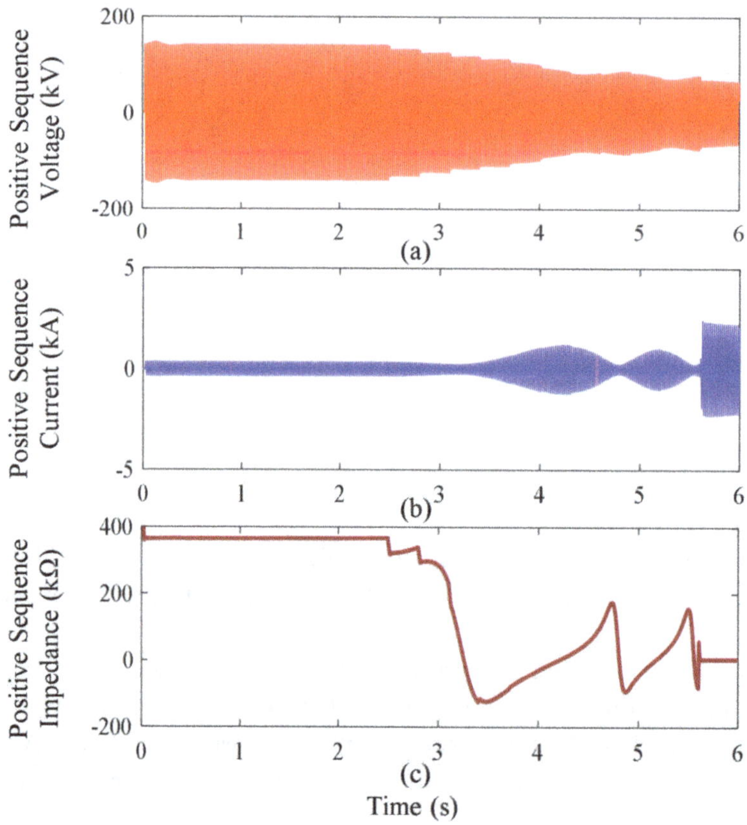

Figure 19. Measurements at bus-7 (a) Positive sequence voltage. (b) Positive sequence current. (c) Positive sequence impedance.

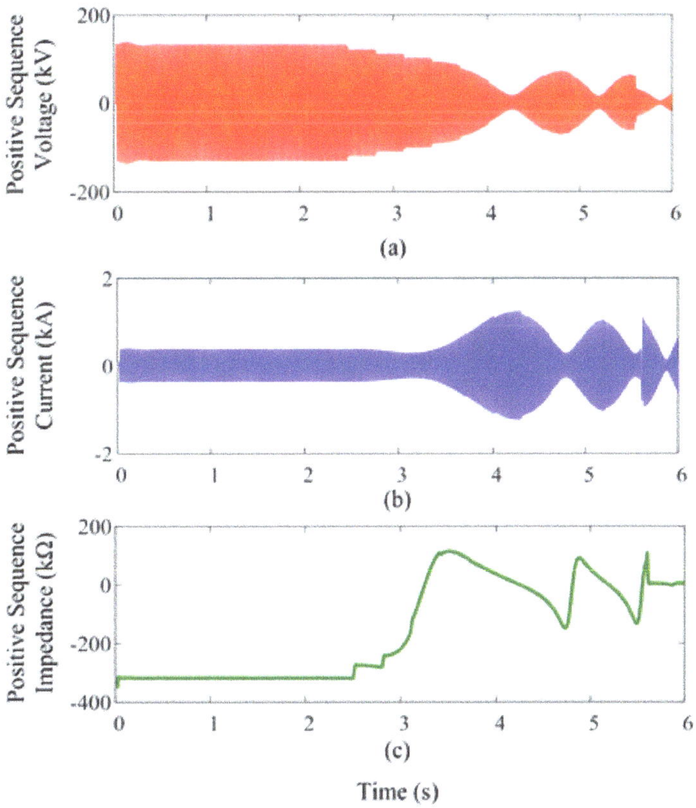

The positive sequence impedances computed at both the buses 7 and 8 are considered to determine differential impedance existing between them as shown in Figure 20 (c). Further, it was differentiated with respect to time to determine the status of the system. Figure 20 (d) describes the magnitude of RCDI compared with the threshold continuously at both the scenarios. During load encroachment scenario the magnitude of RCDI lies well below that of the threshold, whereas in fault scenario its magnitude becomes large and it crosses the threshold at a particular instant. Here, a three-phase fault is detected within a half cycle and discriminated from the load encroachment condition.

FUTURE RESEARCH DIRECTIONS

- Synchrophasor based RCDI scheme can also be used to detect third zone fault during another power system stressed condition i.e. power swing.
- New angular based techniques, computed through wide area measurement system can also be implemented for the better monitoring and protection of the large integrated power network during system stressed conditions.

Figure 20. Measurements at bus-8 (a) Positive sequence impedance at bus-7. (b) Positive sequence impedance at bus-8. (c) Differential Impedance between bus-7 and bus-8. (d) Rate of change of differential impedance.

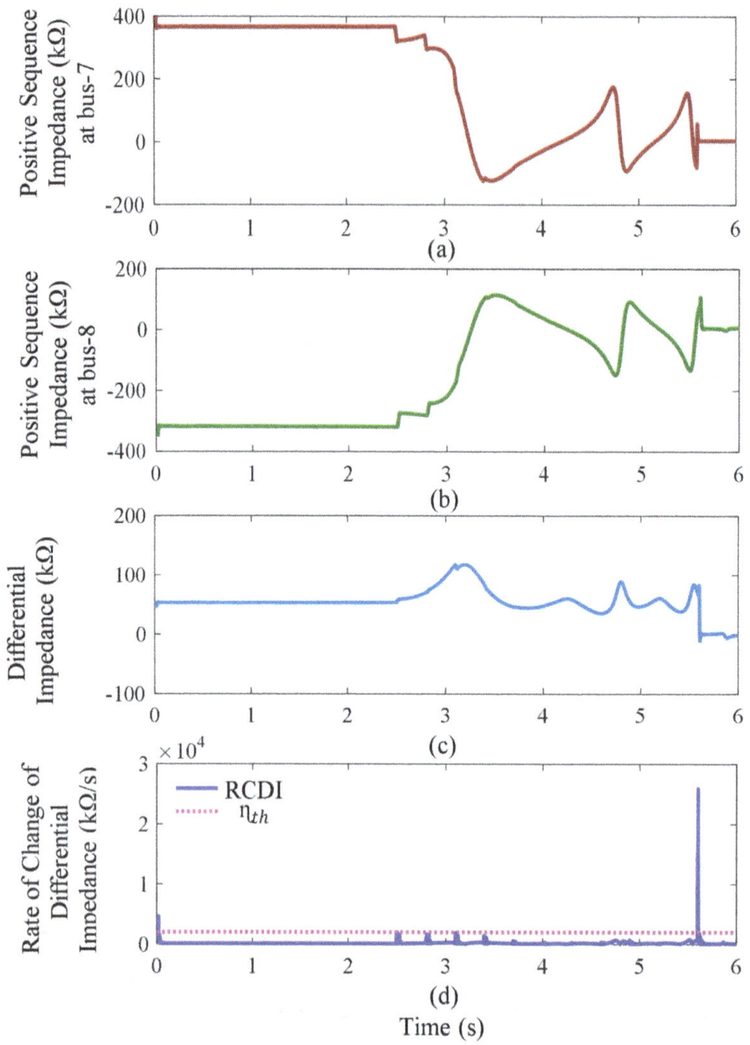

CONCLUSION

The third zone protection of distance relay suffers highly due to power system stressed conditions. This causes unintentional tripping of the system components which may further lead to blackout. To avert the occurrence of such events in the power system, a synchrophasor based wide-area protection scheme is proposed. With the help of synchrophasor measurements, the rate of change of differential impedance (RCDI) is calculated for obtaining the enhanced back-up protection. The algorithm provides an ideal operation of distance relay during voltage instability and load encroachment conditions. This scheme monitors the magnitude deviation in positive sequence impedance at both ends of protected line. For any three-phase fault during system stressed events; the magnitude of RCDI is significant and thus a clear discrimination of system stressed events from three-phase fault is possible. The effectiveness of the proposed scheme is validated through the simulated test cases. From the simulated results it is evident that the three-phase fault detected within a half cycle and it is reliable for wide-area backup protection.

ACKNOWLEDGMENT

The authors are thankful to SERB, New Delhi for sponsoring the project, "Advanced protection to prevent zone-3 maloperation during stress system conditions" through which the work is carried out.

REFERENCES

AER. (2007). *Final the events of 16 January 2007 investigation report.* Retrieved from https://www.aer.gov.au

Alhelou, H., Hamedani-Golshan, M. E., Zamani, R., Heydarian-Forushani, E., & Siano, P. (2018). Challenges and opportunities of load frequency control in conventional, modern and future smart power systems: A comprehensive review. *Energies*, *11*(10), 1–35. doi:10.3390/en11102497

Alhelou, H. H. (2018). Fault detection and isolation in power systems using unknown input observer. In *Advanced Condition Monitoring and Fault Diagnosis of Electric Machines.* Hershey, PA: IGI Global publisher.

Alhelou, H. H., Golshan, M., & Fini, M. (2018). Wind Driven Optimization Algorithm Application to Load Frequency Control in Interconnected Power Systems Considering GRC and GDB Nonlinearities. *Electric Power conponents and Syst.*

Alhelou, H. H., & Golshan, M. E. H. (2016). Hierarchical plug-in EV control based on primary frequency response in interconnected smart grid. *24th Iranian Conference on Electrical Engineering (ICEE)*, 561-566. 10.1109/IranianCEE.2016.7585585

Alhelou, H. H., Golshan, M. H., & Askari-Marnani, J. (2018). Robust sensor fault detection and isolation scheme for interconnected smart power systems in presence of RER and EVs using unknown input observer. *International Journal of Electrical Power & Energy Systems*, *99*, 682–694. doi:10.1016/j.ijepes.2018.02.013

Alhelou, H. H., Hamedani-Golshan, M. E., Heydarian-Forushani, E., Al-Sumaiti, A. S., & Siano, P. (2018, September). Decentralized Fractional Order Control Scheme for LFC of Deregulated Nonlinear Power Systems in Presence of EVs and RER. In *2018 International Conference on Smart Energy Systems and Technologies (SEST)* (pp. 1-6). IEEE. 10.1109/SEST.2018.8495858

Alhelou, H. S. H., Golshan, M. E. H., & Fini, M. H. (2015). Multi agent electric vehicle control based primary frequency support for future smart micro-grid. *Smart Grid Conference (SGC)*, 22-27.

Alshahrestani, A., Golshan, M. E. H., & Alhelou, H. H. (2018, November). WAMS Based Online Estimation of Total Inertia Constant and Damping Coefficient for Future Smart Grid Systems. In *Smart Grid Conference (SGC)* (pp. 1-5). Academic Press.

Arya, L. D., Choube, S. C., & Shrivastava, M. (2008). Technique for voltage stability assessment using newly developed line voltage stability index. *Energy Conversion and Management*, *49*(2), 267–275. doi:10.1016/j.enconman.2007.06.018

Atputharajah, A., & Tapan Kumar, S. (2009). Power system blackouts - Literature review. *Fourth International Conference on Industrial and Information Systems (ICIIS).*

Australian Energy Regulator. (2007). *Final the events of 16 January 2007 investigation report*. Retrieved from https://www.aer.gov.au

Badri Ram, & Vishwakarma, B. N. (2011). *Power system protection and switchgear.* McGraw Higher Ed.

Bolandi, T. G., Haghifam, M. R., & Khederzadeh, M. (2017). Real-time monitoring of zone 3 vulnerable distance relays to prevent maloperation under load encroachment condition. *IET Generation, Transmission & Distribution*, *11*(8), 1878–1888. doi:10.1049/iet-gtd.2016.0486

Dipak, P., Mallikarjuna, B., Jayakumar Reddy, R., Jaya Bharata Reddy, M., & Mohanta, D. K. (2017). Synchrophasor assisted adaptive relaying methodology to prevent zone-3 mal-operation during load encroachment. *IEEE Sensors Journal*, *17*(23), 7713–7722.

Fini, M. H., Yousefi, G. R., & Alhelou, H. H. (2016). Comparative study on the performance of many-objective and single-objective optimisation algorithms in tuning load frequency controllers of multi-area power systems. *IET Generation, Transmission & Distribution*, *10*(12), 2915–2923. doi:10.1049/iet-gtd.2015.1334

IEEE PES-PSRC-SPSWG C-6. (2002). *Wide Area Protection and Emergency Control.* Retrieved from http://www.pes-psrc.org

IEEE PSRC WG D6. (2005). *Power swing and out-of-step considerations on transmission lines.* New York: IEEE.

Jin, M., & Sidhu, T. S. (2008). Adaptive load encroachment prevention scheme for distance protection. *Electric Power Systems Research*, *78*(10), 1693–1700. doi:10.1016/j.epsr.2008.02.016

Jithin, K. K., Hareesh, S. V., Raja, P., & Selvan, M. P. (2017). Design and implementation of an algorithm for diagnosis of load encroachment in EHV Lines. *Energy Procedia*, *117*, 519–526. doi:10.1016/j.egypro.2017.05.178

Jonsson, M., & Daalder, J. E. (2003). An adaptive scheme to prevent undesirable distance protection operation during voltage instability. *IEEE Transactions on Power Delivery*, *18*(4), 1174–1180. doi:10.1109/TPWRD.2003.817501

Jose, T., Biswal, M., Venkatanagaraju, K., & Malik, O. P. (2018). Integrated approach based third zone protection during stressed system conditions. *Electric Power Systems Research*, *161*, 199–211. doi:10.1016/j.epsr.2018.04.011

Kim, C. H., Heo, J. Y., & Aggarwal, R. K. (2005). An enhanced zone 3 algorithm of a distance relay using transient components and state diagram. *IEEE Transactions on Power Delivery*, *20*(1), 39–46. doi:10.1109/TPWRD.2004.837827

Kundu, P., & Pradhan, A. K. (2014). Synchrophasor-assisted zone-3 operation. *IEEE Transactions on Power Delivery, 29*(2), 660–667. doi:10.1109/TPWRD.2013.2276071

Kundur, P. (1994). *Power System Stability and Control.* New York: Mc-Graw-Hill.

Mahari, A., & Seyedi, H. (2015). High impedance fault protection in transmission lines using a WPT-based algorithm. *Electric Power and Energy Systems, 67*, 537–545. doi:10.1016/j.ijepes.2014.12.022

Makdisie, C., Haidar, B., & Alhelou, H. H. (2018). An optimal photovoltaic conversion system for future smart grids. In *Handbook of Research on Power and Energy System Optimization* (pp. 601–657). IGI Global. doi:10.4018/978-1-5225-3935-3.ch018

Mallikarjuna, B., Vardhan Varma, P. V., Samir, S. P., Jaya bharatareddy, J., & Mohanta, D. K. (2017). An adaptive supervised wide-area backup protection scheme for transmission lines protection. *Protection and Control of Modern Power Systems, 2*(22), 1-16.

Michael, B., Münch, V., Aichinger, M., Kuhn, M., Weymann, M., & Schmid, G. (2011). Power blackout risks. *CRO Forum.*

Nadweh, S., Hayek, G., Atieh, B., & Haes Alhelou, H. (2018). Using Four – Quadrant Chopper with Variable Speed Drive System Dc-Link to Improve the Quality of Supplied Power for Industrial Facilities. *Majlesi Journal of Electrical Engineering.*

Nayak, P. K., Pradhan, A. K., & Bajpai, P. (2015). Secured zone 3 protection during stressed condition. *IEEE Transactions on Power Delivery, 30*(1), 89–96. doi:10.1109/TPWRD.2014.2348992

NERC. (2003). *NERC recommendations to blackout-prevent and mitigate the impacts of future cascading blackouts.* Retrieved from https://www.nerc.com

NERC. (2005a). *Protection system review program - Beyond zone 3.* Retrieved from https://www. nerc.com

NERC. (2005b). *Rationale for the use of local and remote (zone 3) protective relaying backup systems - A report on the implications and uses of zone-3 relays.* Retrieved from https://www.nerc.com

NERC-SPCS. (2011). *Reliability guideline: Transmission system phase backup protection.* Retrieved from https://www. nerc.com

Njenda, T. C., Golshan, M. E. H., & Alhelou, H. H. (2018, November). WAMS Based Intelligent Under Frequency Load Shedding Considering Online Disturbance Estimation. In *Smart Grid Conference (SGC)* (pp. 1-5). Academic Press.

Njenda, T. C., Golshan, M. E. H., & Alhelou, H. H. (2018, November). WAMS based Under Frequency Load Shedding Considering Minimum Frequency Prediction and Extrapolated Disturbance Magnitude. In *Smart Grid Conference (SGC)* (pp. 1-5). Academic Press.

NRPC. (2010). *Report of the inquiry committee on grid disturbance in Northern region on 2nd January 2010.* Retrieved from http://www.nrpc.gov.in

Powergrid. (2012). *Report of the enquiry committee on grid disturbance in northern region on 30th July 2012 and in Northern, Eastern and North-Eastern Region on 31st July 2012.* Tech Rep. Grid_ENQ_REP_16_8_12, 2012, New Delhi, India.

Seethalekshmi, K., Singh, S. N., & Srivastava, S. C. (2012). A classification approach using support vector machines to prevent distance relay maloperation under power swing and voltage instability. *IEEE Transactions on Power Delivery*, *27*(3), 1124–1133. doi:10.1109/TPWRD.2011.2174808

Sharifzadeh, M., Lesaniand, H., & Sanaye-Pasand, M. (2014). A new algorithm to stabilize distance relay operation during voltage-degraded conditions. *IEEE Transactions on Power Delivery*, *29*(4), 1639–1647. doi:10.1109/TPWRD.2013.2285502

Shehu, A. A. (2009). *Voltage stability and distance protection zone (Master's thesis)*. Chalmers University of Technology. Retrieved from http://webfiles.portal.chalmers.se/et/MSc/ShehuAbbaAliyuMSc.pdf

Sodhi, R., Shrivastav, S. C., & Singh, S. N. (2012). A simple scheme for wide area detection of impending voltage instability. *IEEE Transactions on Smart Grid*, *3*(2), 818–827. doi:10.1109/TSG.2011.2180936

Song, H., Lee, B., & Ajarappu, V. (2009). Control strategies against voltage collapse considering undesired relay operations. *IET Generation, Transmission & Distribution*, *3*(2), 164–172. doi:10.1049/iet-gtd:20080055

Surachet, T. (1994). *Analysis of power system disturbances due to relay hidden failures* (PhD thesis). Virginia Polytechnic Institute.

UCTE. (2004). *Final report on the August 14, 2003 blackout in the United States and Canada*. Retrieved from https://www.energy.gov

UCTE. (2007). *Final report system disturbance on 4 November 2006*. Retrieved from https://www.ucte.org

UCTE-RAE. (2004). *Final report of the investigation committee on the 28 September 2003 blackout in Italy*. Retrieved from http://www.rae.gr

US-CPSOTF. (2004). *Final report on the August 14, 2003 Blackout in the United States and Canada*. Author.

Veloza, O. P., & Santamaria, F. (2016). Analysis of major blackouts from 2003 to 2015: Classification of incidents and review of main causes. *The Electricity Journal*, *29*(7), 42–49. doi:10.1016/j.tej.2016.08.006

WSCC. (1996). *Disturbance report for the power system outage that occurred on the western interconnection August 10, 1996. 1548 PAST*. Salt Lake City, UT: The Council.

Yee Wei, L., & Marimuthu, P. (2013). WAKE: Key management scheme for wide-area measurement systems in smart grid. *IEEE Communications Magazine*, *51*(1), 34–41. doi:10.1109/MCOM.2013.6400436

Zamani, R., Hamedani-Golshan, M. E., Haes Alhelou, H., Siano, P., & Pota, R, H. (. (2018). Islanding Detection of Synchronous Distributed Generator Based on the Active and Reactive Power Control Loops. *Energies*, *11*(10), 2819. doi:10.3390/en11102819

Žaneta, E., & Anton, B. (2008). Blackout in the power system. *AT&P Journal PLUS2*.

Zeng, B., Shaojie, O., Jianhua, Z., Hui, S., Geng, W., & Ming, Z. (2015). An analysis of previous blackouts in the world: Lessons for china's Power industry. *Renewable & Sustainable Energy Reviews*, *42*, 1151–1163. doi:10.1016/j.rser.2014.10.069

KEY TERMS AND DEFINITION

Dependability: Dependability is the degree of certainty that the relay will operate correctly.

Distance Protection: A fault in a transmission line will result in the decrease of line impedance which is compared with a pre-defined threshold value. The trip signal will be sent to the breaker if the measured impedance is smaller than the threshold.

Protective Relay: It is an intelligent electronic devices (IEDs) which receive measured signals from the secondary side of CTs and VTs and detect whether the protected unit is in a stressed condition (based on their type and configuration) or not. A trip signal is sent by protective relays to the circuit breakers to disconnect the faulty components from power system if necessary.

R-X Plot: A graphical method of showing the characteristics of a relay element in terms of the ratio of voltage to current and the angle between them.

Reach: The maximum distance from the relay location to a fault for which a particular relay will operate. The reach may be stated in terms of miles, primary.

Reliability: A measure of the degree of certainty that the relay, or relay system, will perform correctly.

Selectivity: The property by which only the faulty element of the system is isolated, and the remaining healthy sections are left intact.

Zone of Protection: It is the segment of a power system in which the occurrence of assigned abnormal conditions should cause the protective relay system to operate.

APPENDIX

Table 2.

230-KV, 50 Hz Two-Area Four Machine System Data				
Generator				
Parameters (pu)				
$X_d = 1.8$	$X_q = 1.7$	$X_l = 0.2$	$X_d^{'} = 0.3$	$X_q^{'} = 0.55$
$X_d^{''} = 0.25$	$X_q^{''} = 0.25$	$R_a = 0.0025$	$T_{d0}^{'} = 8.0s$	$T_{q0}^{'} = 0.4s$
$T_{d0}^{''} = 0.03s$	$T_{q0}^{''} = 0.05s$	$A_{Sat} = 0.015$	$B_{Sat} = 9.6$	$\Psi_{T1} = 0.015$
$H_{G1} = 6.5$	$H_{G2} = 6.5$	$H_{G3} = 6.175$	$H_{G4} = 6.175$	$K_D = 0$

Power Ratings of the Generating Units							
Gen - 1		**Gen - 2**		**Gen - 3**		**Gen - 4**	
P(MW)	**Q(MVAr)**	**P(MW)**	**Q (MVAr)**	**P(MW)**	**Q(MVAr)**	**P(MW)**	**Q(MVAr)**
700	185	700	235	719	176	700	202

Per Unit Voltage Ratings of the Generating Units			
Gen - 1	**Gen - 2**	**Gen - 3**	**Gen - 4**
$1.03\angle 20.2^0$	$1.01\angle 10.5^0$	$1.03\angle -6.8^0$	$1.01\angle -17.0^0$

Step-Up Transformer	
Parameters	
Impedance	*0+j0.15 pu*
MVA rating	900 MVA
Primary voltage rating	20 kV
Secondary voltage rating	230 kV
Off-nominal ratio	1.0

Transmission Lines		
Parameters		
$r = 0.0001 pu / km$	$x_L = 0.001 pu / km$	$b_C = 0.00175 pu / km$

Line Lengths							
BUS 5-6	**BUS 6-7**	**BUS 7-8(1)**	**BUS 7-8(2)**	**BUS 8-9(1)**	**BUS 8-9(2)**	**BUS 9-10**	**BUS 10-11**
25 km	10 km	110 km	110 km	110 km	110 km	10 km	25 km

Loads and Reactive Power Supply			
Ratings			
BUS No.	**P_L (MW)**	**Q_L (MVAr)**	**Q_C (MVAr)**
7	967	100	200
9	1767	100	350

Chapter 2
Enrichment of Distribution System Stability Through Artificial Bee Colony Algorithm and Artificial Neural Network

Gummadi Srinivasa Rao
V. R. Siddhartha Engineering College (Autonomous), India

Y. P. Obulesh
VIT University, India

B. Venkateswara Rao
V. R. Siddhartha Engineering College (Autonomous), India

ABSTRACT

In this chapter, an amalgamation of artificial bee colony (ABC) algorithm and artificial neural network (ANN) approach is recommended for optimizing the location and capacity of distribution generations (DGs) in distribution network. The best doable place in the network has been approximated using ABC algorithm by means of the voltage deviation, power loss, and real power deviation of load buses and the DG capacity is approximated by using ANN. In this, single DG and two DGs have been considered for calculation of doable place in the network and capacity of the DGs to progress the voltage stability and reduce the power loss of the system. The power flow of the system is analyzed using iterative method (The Newton-Raphson load flow study) from which the bus voltages, active power, reactive power, power loss, and voltage deviations of the system have been achieved. The proposed method is tested in MATLAB, and the results are compared with particle swarm optimization (PSO) algorithm, ANN, and hybrid PSO and ANN methods for effectiveness of the proposed system.

DOI: 10.4018/978-1-5225-8030-0.ch002

INTRODUCTION

Losses are a very key role when constructing and arrangement of the power system. Losses are predictable in every set of links; however, the quantity can fluctuate considerably depending on the planning of the power system. The power flows in the system decide the loss. One of the largest consumer markets in the world is the electric power industry. The cost of electricity is estimated at around 50% for fuel, 20% for generation, 5% for transmission and 25% for distribution. Distribution systems must deliver electricity to each customer's service entrance at an appropriate voltage rating. The X/R ratio for distribution levels is low as compared to transmission levels, causing high power losses and a drop in voltage magnitude along radial distribution lines. Studies have indicated that just about 13% of the total power generated is consumed as real power losses at the distribution level. Such non-negligible losses have a direct impact on the financial issues and overall efficiency of distribution utilities. The installation of Distributed Generation (DG) units is becoming more famous in distribution systems due to their overall positive impacts on power networks such as energy competence, deregulation, diversification of energy sources, ease of finding sites for smaller generators, shorter erection times and lesser investment costs of smaller plants, and the nearness of the generation plant to heavy loads, which decreases transmit costs. (K. Varesi, 2011) Hence the allotment of DG units gives a possibility to decrease power loss (S. A. Hosseini, M. Karami and S. S. KarimiMadahi, 2011 &NareshAcharya, PukarMahat and N. Mithulananthan, 2006& Nadweh et al,2018).

The addition of Distributed Generation (DG) units changes the load features of the distribution system, which slowly becomes an active load network and involves changes in the power flows. The performance of the network by addition of each DG can be determined by performing the load flow solution. For that reason, it is required to build up mathematical optimization that can be implemented in the network to decrease the power loss and to maintain the voltage magnitudes at each bus within the acceptable limits. Hence the author is interested in the area of optimization methods in the domain of Smart Micro-Grid and power system operation and control. The different optimization methods for improvement of performance of the network are already developed such as Genetic Algorithm (GA), Particle swarm optimization (PSO), Artificial Neural Network (ANN) and Artificial Bee Colony (ABC) etc. are supportive for optimizing the DG size and location in decreasing the power loss and for enhancement of voltage profile (F. S. Abu-Mouti, El-Hawary, 2011 &H. Nasiraghdam and S. Jadid, 2012& Madisie et al, 2018).A hybrid technique which is the amalgamation of Particle Swarm Optimization (PSO) and Artificial Neural Network (ANN) has been implemented to find out the optimal location and rating of DG to diminish the power loss in the network and voltage profile enhancement at all buses (F. S. Abu-Mouti and M. E. El-Hawary, 2009 & Gummadi SrinivasaRao and Y.P.Obulesh, 2013).In 2016 (Hassan Haes Alhelouand M. E. H. Golshan, 2016) A high penetration level of RERs causes some problems to the grid operator, e.g., lack in primary reserve. This paper proposes a new scheme to provide necessary primary reserve from electric vehicles by using hierarchical control of each individual vehicle. The proposed aggregation scheme determines the primary reserve and contracts it with system operator based on electricity market negotiation.

A comprehensive literature reviews and state of arts in nature inspired optimization algorithm could be found in (Mehdi Khosrow-Pour, 2018). In this Incorporating Nature-Inspired Paradigms in Computational Applications is a critical scholarly resource that examines the application of nature-inspired paradigms on system identification.In this year (H. HaesAlhelou,M.E. HamedaniGolshan and J. Askari-Marnani, 2018), propose the use of unknown input observer for detection of faults in interconnected smart power systems in presence of renewable energy resources and electric vehicles. In this fault detection and isolation scheme has been proposed for smart grids. Due to proper isolation of faults the stability of the system has been improved.In 2018 (Carlo Makdisie, BadiaHaidar and Hassan HaesAlhelou, 2018) authors propose the optimal photovoltaic system based on neuro fuzzy control for future smart grids. The proposed PV conditioner with the neuro-fuzzy control compensates the nonlinear and unbalanced loads of the electrical power systems then improve the system performance.

In the last decade, there are several major directions for optimization technology development (ÖmürTosun, 2014, Alhelou et al, 2018), use Artificial Bee Colony Algorithm for solving single and multi-objective ptimization problems. (MasoudHajiakbariFini, Gholam Reza Yousefi, and Hassan Hae-sAlhelou, 2016) present the various optimization techniques for tuning and load frequency control for multi area power system by considering single and multi objective functions.(Hassan S. Haes Alhelou ; M.E.H. Golshan ; Masoud HajiakbariFini, 2015) proposes a new scheme to provide necessary primary reserve from electric vehicles by using multi-agent control of each individual vehicle. The proposed scheme determines the primary reserve based on vehicle's information such as initial state of charge (SOC), the required SOC for the next trip, and the vehicle's departure time. Which is useful for improve the performance the system consists of more number of distribution generation sources.

In this book chapter it is extended to optimize rating and position of two DGs using a novel hybrid technique which is the combination of Artificial Bee Colony (ABC) and Artificial Neural Network (ANN) to improve the voltage profile and to decrease the system loss. The performance of this method has been compared with other optimization methodologies such as PSO, ANN and hybrid PSO & ANN to reveal the effectiveness of the proposed method.

PROBLEM FORMULATION

The DG unit is positioned in an optimal approach, the power loss and instability troubles have been reduced in the distribution system. For this reason, a combined approach is anticipated for optimizing the placement and size of DGs for improvement of voltage profile and lessening in the power loss of the system. Thus the considered problem is nonlinear optimization problem.

The problem is loss minimization is taken as an objective function and stated as follows,

$$O = \sum_{q=1}^{N} P_L \tag{1}$$

where, $'O'$ represents the objective function of the system, P_L = Active Power loss of the system

The objective function is subjected to equality and inequality constrictions such as the real & reactive power balance and the bus voltage limits.By the placement of DGs, the active power loss in the network is premeditated using the subsequent equation.

$$P_L = \sum_{m=1}^{N}\sum_{n=1}^{N}[\alpha_{mn}(P_m P_n + Q_m Q_n) + \beta_{mn}(Q_m P_n + P_m Q_n)] \tag{2}$$

where,

$$\alpha_{mn} = \frac{r_{mn}}{v_m v_n} \cos(\delta_1 - \delta_m) \tag{3}$$

$$\beta_{mn} = \frac{r_{mn}}{v_m v_n} \sin(\delta_m - \delta_n) \tag{4}$$

$$Z_{mn} = r_{mn} + jx_{mn} \tag{5}$$

P_m, Q_m = m[th] bus real and reactive power injection
N = number of buses.

Voltage Boundary Limitation

The magnitudes of the bus voltages should be within functional limits can be represented as,

$$V_{\min} \leq V_m \leq V_{\max} \tag{6}$$

Here, $V_{\min}$ = Lower bound bus voltage,
$V_{\max}$ = upper bound bus voltage,
V_m = root mean square value of the mth bus voltage.

The difference between the reference voltage and the voltage of the particular bus is called voltage deviation V_m on can be calculated as follows,

$$V_{dev} = 1 - V_m \tag{7}$$

Here, V_{dev} = voltage deviation, V_m = m[th] node voltage and $m = 1,2,3....N$

Active and Reactive Power Constraint

The active and reactive power for insertion at buses are calculated using,

Real power injection = Real power generation$_m$ – Real power demand$_m$ (8)

$$P_m = P_{DGm} - P_{Dm} \tag{9}$$

$$Q_m = Q_{DGm} - Q_{Dm} \tag{10}$$

Here, P_{DGm}, Q_{DGm}, P_{Dm} and Q_{Dm} are indicated as the active power injection, reactive power injection, real & reactive power demands at m[th] bus subject to the lower & upper limits of power generation constraints of DGs at bus m,

$$P_{Gm,\min} \leq P_{Gm} \leq P_{Gm,\max} \tag{10}$$

$$Q_{Gm,\min} \leq Q_{Gm} \leq Q_{Gm,\max} \tag{11}$$

HYBRID ABC-ANN APPROACH

In proposed hybrid approach, ABC computes an optimal position of the DG systems and ANN is sized the DGs rating. The best possible position is estimated by means of the voltage deviation, power loss and real power deviation of load buses. Then, by using ANN, the exact ratings of DGs are calculated to progress the voltage stability and reduce the power loss of the system.

Artificial Bee Colony (ABC)

The ABC algorithm is used to optimize the location of the DGs. The bus voltages, line data and voltage limits have been considered the input. It consists of a set of possible solutions (V_i) that are represented by the location of the food sources. This algorithm consists of four stages, such as initialization stage, employed bee stage, onlooker bee stage and scout bee stage (Gummadi Srinivasa Rao and Y. P. Obulesh, 2015). In a multi dimensional search, the employed bees choose food sources depending on the experience of themselves. The onlooker bees choose food resources based on their nest mates experience and adjust their positions. Scout bees fly and choose the food sources randomly without using experience (Cheng-Jian Lin and Shih-ChiehSu, 2012 & Alshahrestani et al, 2018 & Njenda et al, 2018 & Alhelou et al, 2018). The nectar amount of the food source stands for the fitness of the solution.

Description of ABC Algorithm

Step 1: Initialization

Generate the input values such as, bus voltage, line data and voltage limits in the population.

$$V_m = \{V_0^j, V_1^j, ...V_P^j\}, 0 \leq j \leq d-1 \tag{12}$$

Here, V_m= m[th] node bus voltage of the population in p[th] position,
d = 1, 2, 3...., n., d is the dimensional space and the inputs are specified by the minimum and maximum values.

Step 2: Fitness Function

Evaluate the fitness value of each bus and then calculate the best voltage values.

$$\text{Fitness function F(m)} = \begin{cases} if\ f_m \le 0, select\ the\ particular\ bus; \\ otherwise, update\ the\ bus\ positions \end{cases} \quad (13)$$

Here, $f_m = 1 - V_m$, V_m = m[th] node bus voltage.
The bus voltage is calculated Based on fitness function.

Step 3: Employed Bee Phase

The bus positions are revised using the equation,

$$Y_{m,m} = V_{m,n} + \phi_{m,n}(V_{m,n} - V_{k,n}) \quad (14)$$

Here, $Y_{m,n}$ =new value of the nth position and
φ = randomly produced number in the range [-1, 1].
Then evaluate the fitness values and apply the greedy selection between them $Y_{m,n}$ and $V_{m,n}$.

The probability values for the solutions $V_{m,n}$ can be determined by means of their fitness values using the equation,

$$\Pr_m = \frac{F_m}{\sum_{m=1}^{d} F_m} \quad (15)$$

$\Pr_m$ = probability of the mth bus value.

Step 4: Onlooker Bee Phase

Generate the new positions Y_m for the onlookers from the solutions V_m, selected depending upon the probability value P_m and calculate them. Then, the fitness function (maximum voltage deviation) is determined for the new position. In order to select the best bus, apply greedy selection for the onlooker bee between V_m andY_m

Step 5: Scout Bee Phase

The abandoned solution, if exist, and replace it with a new randomly produced solution V_m. Memorize the best food source position achieved. The particular bus has been selected for an optimal location finding process. This practice is continual until the maximum iteration is reached otherwise the practice is terminated. When the procedure is done the optimum locations are determined. The flowchart for the suggested ABC algorithm is illustrated in Figure 1.

Artificial Neural Network (ANN)

ANN has only one input, and three outputs, contains two stages. (Training stage and testing stage) (Partha, Kayal and Chandan Kumar Chanda, 2013).The training of the neural network is done with back propagation algorithm. The testing is done by giving bus number as input to the neural network and the ratings of DGs are obtained as outputs. The structure of ANN is given in Figure 2

Training Algorithm (Back Propagation)

Initialize the weights of all the neurons of the network. The neuron weights of the hidden layer and the output layer are initiated in the particular interval $\left[w_{\min}, w_{\max}\right]$. The outputs of the network are stated as shown,

$$V_b = \sum_{d=1}^{h} [W_{2d1} * V_b(d)] \tag{16}$$

here

$$V_b(d) = \frac{1}{1 + \exp(-w_{11d} * c)} \tag{17}$$

$$P_b = \sum_{d=1}^{h} [W_{2d1} * P_b(d)] \tag{18}$$

here

$$P_b(d) = \frac{1}{1 + \exp(-w_{11d} * c)} \tag{19}$$

Figure 1. Flow chart for finding an optimal location using ABC algorithm

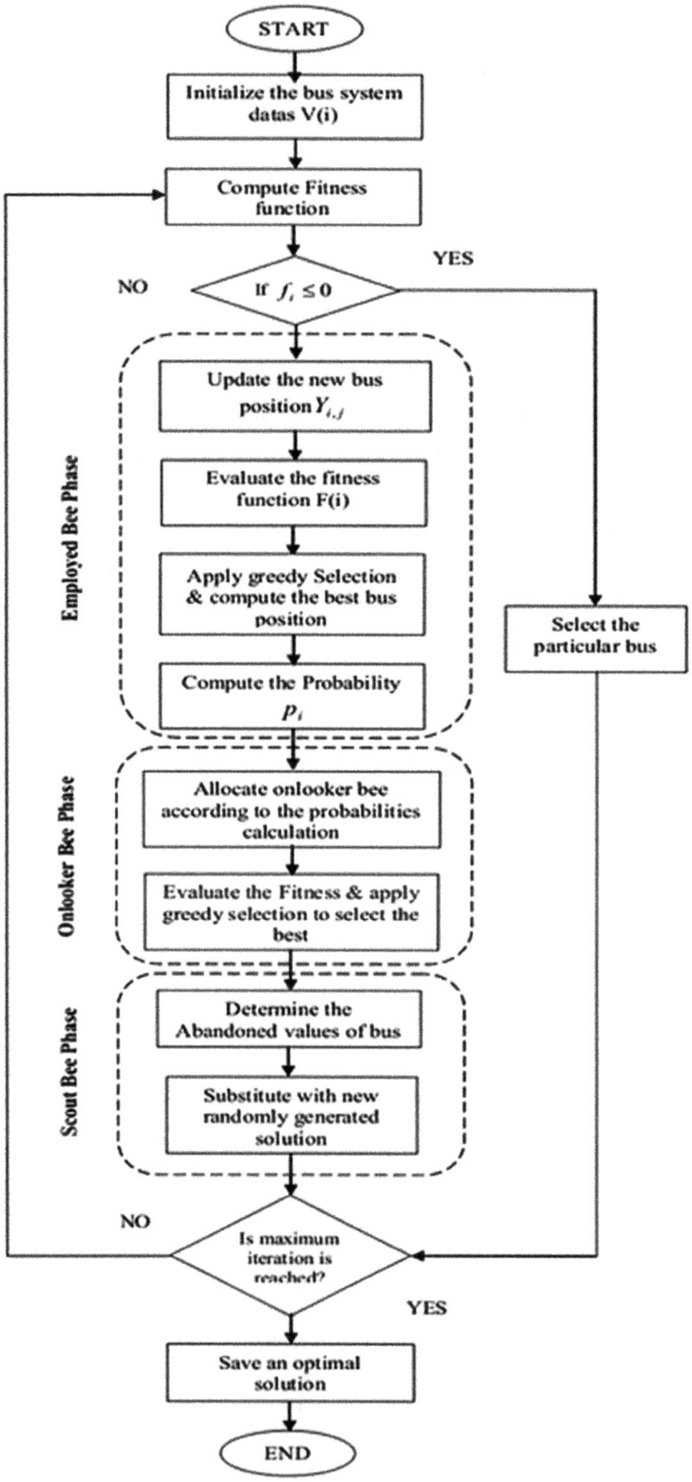

Figure 2. Structure of ANN for training

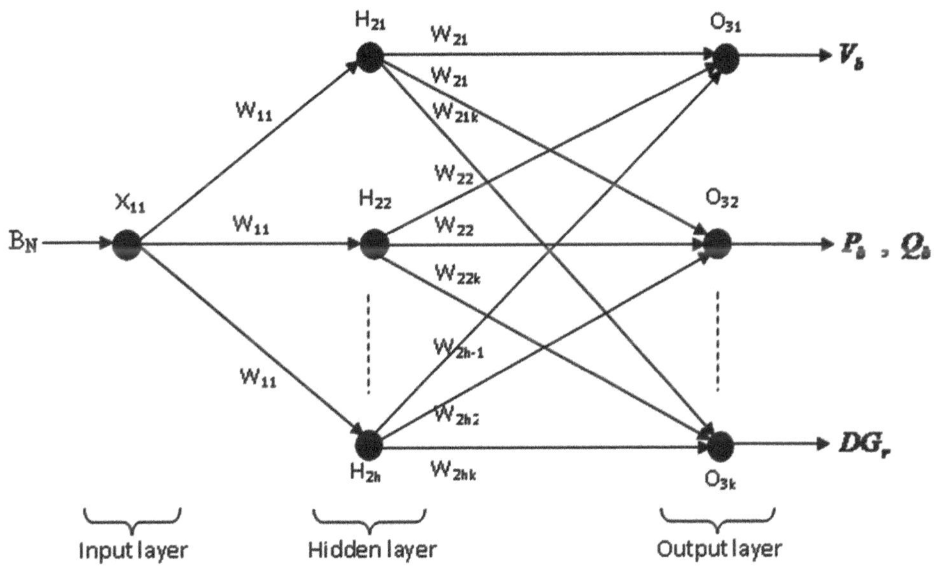

$$Q_b = \sum_{d=1}^{h} [W_{2d2} * Q_b(d)] \tag{20}$$

here

$$Q_b(d) = \frac{1}{1 + \exp(-w_{11d} * c)} \tag{21}$$

$$DG_r = \sum_{d=1}^{h} [W_{2d3} * DG_r(d)] \tag{22}$$

here

$$DG_r(d) = \frac{1}{1 + \exp(-w_{11d} * c)} \tag{23}$$

From the above equations,
'c' is corresponded to an input variable,
$(w_{211}, w_{212}, \ldots w_{2hk})$= The hidden layer to the output layer weights.
Here, B_N = Input data (bus number) is applied to the ANN and h = number of hidden layers.

The weights are adjusted to all the neurons by giving the input and outputs to the network in the training process.

RESULTS AND DISCUSSION

The effectiveness of the anticipated amalgam optimization methodology is tested on IEEE 30 bus system and realized in MATLAB software package. Later fitting only DG and earlier fitting only DG, the voltage profile at various buses are given in Table1. The single week bus is recognized from this with the support of voltage deviation which is bus number 19 using hybrid ABC - ANN method. The voltage of the weak bus earlier fitting DG is 1.019p.u and it is improved to 1.0018p.u later fitting DG at bus number 19.

The bus voltage shape earlier and later fitting DG is revealed in the Figure 3. It is observed that the voltages at various optimal buses are approximately nearer to 1p.u. after placing the DG.

The optimal rating of DG and power loss in the system earlier fitting to the only DG later fitting to the only DG for all the expected optimization methodologies are given in Table 2 and Table 3 respectively. From Table 2, it is observed that the optimal rating of DG is 5.986MW using the anticipated hybrid ABC - ANN method and from Table 3, it is observed that the power loss of the test system is

Figure 3. Comparison of bus voltages when earlier and later fitting of DG using ABC-ANN methodology

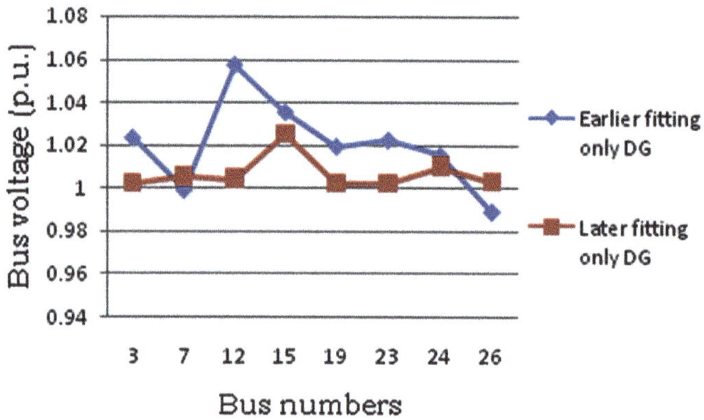

Table 1. Bus voltage profile of optimally positioned buses earlier to and later fitting of only DG for a variety of methodologies

Most Favorable Bus Number	Voltage Profile					Voltage Deviation			
	Earlier to Fitting DG	Later to Fitting DG							
		ANN	PSO	Hybrid PSO-ANN Scheme	Hybrid ABC - ANN Scheme	ANN	PSO	Hybrid PSO-ANN Scheme	Hybrid ABC - ANN Scheme
3	1.023	1.026	1.024	1.027	1.002	0.040	0.066	0.062	0.040
7	0.999	1.005	1.005	1.005	1.005	0.022	0.062	0.121	.0001
12	1.057	1.060	1.060	1.041	1.004	0.058	0.201	0.047	0.025
15	1.035	1.041	1.041	1.038	1.025	0.040	0.094	0.141	0.017
19	1.019	1.027	1.025	1.024	1.002	0.024	0.079	0.097	0.091
23	1.022	1.031	1.032	1.012	1.002	0.005	0.068	0.141	0.022
24	1.015	1.025	1.026	1.010	1.010	0.008	0.071	0.024	0.015
26	0.989	1.002	1.004	1.005	1.003	0.012	0.023	0.058	0.011

reduced to 7.904MW from the normal power loss of the system is 10.714MW by fitting of DG with a capacity of 5.986MW at bus 19.Similarly, after fitting only DG at bus number 19 with an optimal rating of 7.754MW, at bus number 26 with an optimal rating of 25.0213MW, and at bus 19 with an optimal rating of 7.086MW, the loss are reduced to 10.214 MW, 9.501MW, 8.057MW by means of PSO, ANN and hybrid PSO - ANN respectively.

Reduction of power loss for methodologies is revealed in the Figure 4 and it is perceived that the hybrid ABC - ANN methodology decreases the power loss from 10.714MW to 7.904MW. Table 4, Table 5 and Table 6 gives the bus voltages before and after installing DGs, DG capacity and power loss of the system before and after installing two DGs by using ANN method, PSO method and hybrid PSO - ANN methods respectively. From Table 4, it is noticed that the feeble buses are 21 and 30 and after fitting the two DGs at those buses, the loss is condensed to 9.144MW, from Table 5, it is noticed that the feeble buses are 19 and 12 and fitting the two DGs at those buses, the loss is condensed to 9.914MW and from Table 6, it is noticed that the feeble buses are 4 and 23 and after fitting the two DGs at those buses, the loss is condensed to 4.435MW.Bus voltages of the test system earlier to and later fitting of DGs at two bus locations for joint ABC -ANN approach is demonstrated in Figure 5. It can be observed that the voltage profile at the buses is enriched after fixing two DGs and the power loss is reduced to 2.168MW.

The bus voltages earlier and later fitting of two DGs, power loss and an most favourable capacity of DG units are revealed in Table 7. It is noticed that the feeble buses by using united ABC-ANN methodology are (16,4), (4,23), (21,30), (19,12), (12,10), (21,20), and (14,15). From which combinations, it is observed that the optimal bus combinations for placement of two DGs is bus number 14 and bus number 15. The power loss of the test system after placing two DG sat these locations is reduced to 2.168MW from the normal power loss of 10.714 MW before installing the DGs. The voltage profile at these weak buses has been improved to (0.989, 0.100), (1.010, 0.990), (1.019, 0.979), (1.015, 0.990), (0.989, 0.999), (1.004, 0.999), and (1.009, 1.002) respectively after installing the two DGs. As shown in Table 7.

The drop of power loss for all recommended methodologies is shown in Figure 6. From Figure 6, it is observed that the suggested ABC-ANN method is very useful in dipping the power loss as compared to the previous ANN, PSO and hybrid PSO-ANN methodologies.

Figure 4. Assessment of loss reduction in only DG for various methodologies

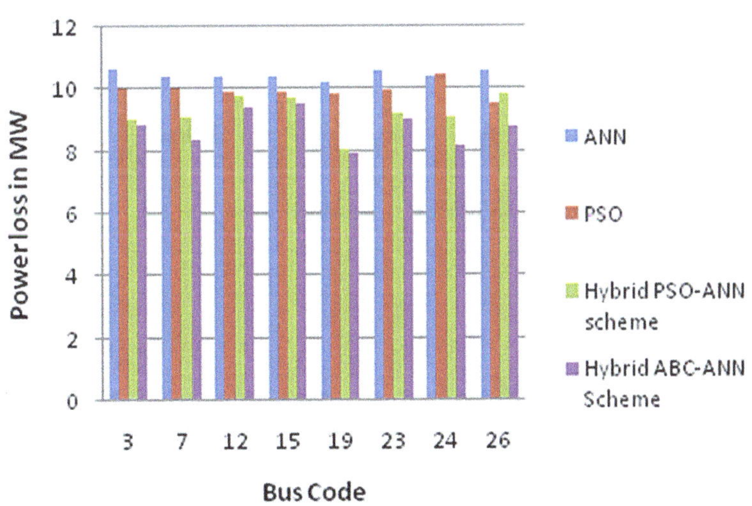

Table 2. Best possible sizes of DG units by different methodologies

Most Favorable Bus Number	Load Power (MW)	DG Capacity (MW)			
		ANN	PSO	Hybrid PSO & ANN Scheme	Hybrid ABC & ANN Scheme
3	2.4	3.002	5.986	5.127	2.037
7	22.8	13.273	3.027	3.020	2.064
12	11.2	7.924	8.011	7.935	5.999
15	8.2	5.999	5.127	20.069	4.036
19	9.5	7.754	8.029	7.086	5.986
23	3.2	8.926	14.031	7.764	3.036
24	8.7	8.030	15.037	7.024	6.010
26	3.5	4.001	25.021	20.069	3.036

Table 3. Power loss earlier and later fitting of only DG for a variety of methodologies

Most Favorable Bus Number	Power Loss (MW)				
	Earlier to Placing of DG	Later Fitting Single DG			
		ANN	PSO	Hybrid PSO-ANN Scheme	Hybrid ABC - ANN Scheme
3	10.714	10.630	10.014	9.015	8.836
7		10.392	10.030	9.077	8.381
12		10.418	9.883	9.784	9.381
15		10.404	9.866	9.738	9.526
19		10.214	9.865	8.057	7.904
23		10.577	9.941	9.241	9.028
24		10.414	10.430	9.063	8.156
26		10.581	9.501	9.828	8.783

Figure 5. Bus voltages earlier to and later fitting of DGs at two bus locations for joint ABC - ANN approach

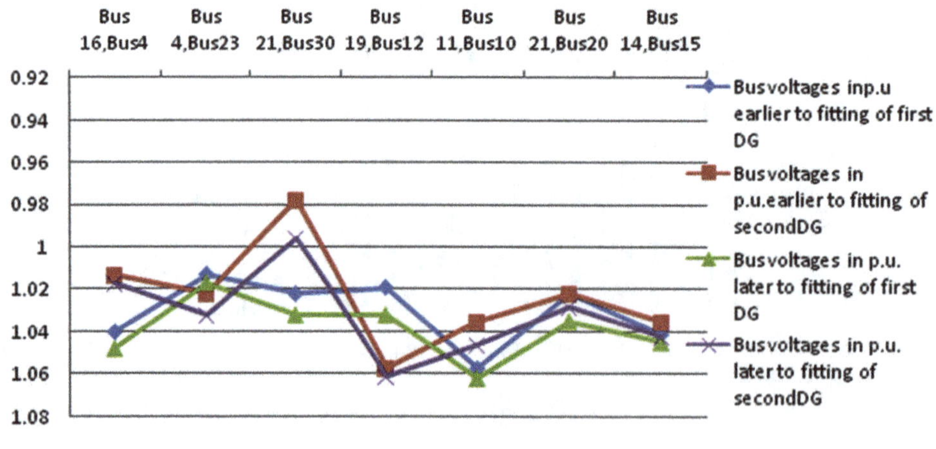

Figure 6. Total power losses for all recommended methodologies

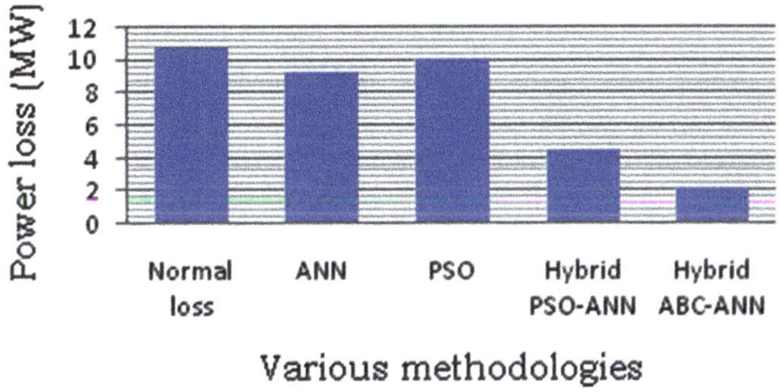

Table 4. Power loss and bus voltages earlier to and later fitting of two DG units by ANN methodology

Most Favorable Bus Numbers		Bus Voltages Earlier to Fitting of DGs		Bus Voltages Later Fitting of DGs		DG Rating		Power Loss Earlier to Fitting of DGs	Power Loss Later Fitting of DGs
B1	B2	p.u.	p.u.	p.u.	p.u.	MW	MW	MW	MW
16	4	1.040	1.014	1.047	1.017	2.528	4.164	10.714	10.305
4	23	1.013	1.022	1.017	1.031	3.873	1.952		10.295
21	30	1.022	0.978	1.031	0.996	9.271	4.995		9.144
19	12	1.019	1.057	1.028	1.060	6.017	7.814		10.019
12	10	1.057	1.035	1.060	1.046	9.864	4.326		10.446
21	20	1.022	1.022	1.031	1.031	7.019	1.026		9.531
14	15	1.041	1.035	1.039	1.041	13.687	5.018		10.154

Table 5. Power loss and bus voltages earlier to and later fitting of two DG units by PSO methodology

Most Favorable Bus Numbers		Bus Voltages Earlier to Fitting of DGs		Bus Voltages Later to Fitting of DGs		Max PSO Voltage Deviation	DG Rating		Power Loss Earlier to Fitting of DGs	Power Loss Later to Fitting of DGs
B1	B2	p.u.	p.u.	p.u.	p.u.	p.u.	MW	MW	MW	MW
16	4	1.040	1.013	1.048	1.017	0.080	2.539	6.148	10.714	10.534
4	23	1.013	1.022	1.017	1.032	0.153	13	2.367		10.616
21	30	1.022	0.978	1.032	0.996	0.038	30.012	7.683		10.059
19	12	1.019	1.057	1.032	1.061	0.043	5.150	7.469		9.914
12	10	1.057	1.035	1.062	1.046	0.147	5.150	6.821		10.126
21	20	1.022	1.022	1.035	1.028	0.068	1.582	14.104		10.289
14	15	1.041	1.035	1.045	1.042	0.186	7.268	6.183		10.573

Table 6. Power loss and bus voltages earlier to and later fitting of two DG units by hybrid PSO - ANN methodology

Most Favorable Bus Numbers		Bus Voltages Earlier to Fitting of DGs		Bus Voltages Later Fitting of DGs		Max PSO Voltage Deviation	DG Capacity		Power Loss Earlier to Fitting of DGs	Power Loss Later Fitting of DGs
B1	B2	p.u.	p.u.	p.u.	p.u.	p.u.	MW	MW	MW	MW
16	4	1.049	1.019	1.040	1.013	0.101	12.807	2.000	10.714	8.702
4	23	1.019	1.033	1.013	1.022	0.052	1.820	12.807		4.435
21	30	1.032	0.996	1.022	0.978	0.092	7.764	6.024		9.389
19	12	1.031	1.062	1.019	1.057	0.083	6.019	5.310		8.841
12	10	1.061	1.051	1.057	1.035	0.061	32.106	21.950		7.353
21	20	1.032	1.0318	1.022	1.022	0.152	2.184	1.024		8.860
14	15	1.029	1.043	1.019	1.035	0.120	4.974	5.310		4.801

Table 7. Power Loss and bus voltages earlier to and later fitting of two DG units by hybrid ABC - ANN methodology

Most Favorable Bus Numbers		Bus Voltages Earlier Fitting of DGs		Bus Voltages Later Fitting of DGs		DG Ratings		Load Power at the Buses		Power Loss Earlier Fitting of DGs	Power Loss Later Fitting of DGs
B1	B2	p.u.	p.u.	pu	pu	MW (B1)	MW (B2)	MW	MW	MW	MW
16	4	1.040	1.013	0.989	1.000	2.060	1.901	3.500	7.600	10.714	8.112
4	23	1.013	1.022	1.010	0.990	1.104	2.177	7.600	3.200		4.274
21	30	1.022	0.978	1.019	0.978	7.000	5.141	17.500	10.600		8.478
19	12	1.019	1.057	1.016	0.989	5.002	0.173	9.500	11.200		8.375
12	10	1.057	1.035	0.989	0.999	3.020	6.019	11.200	5.800		6.242
21	20	1.022	1.022	1.004	0.999	1.088	0.999	17.500	2.200		2.410
14	15	1.041	1.035	1.009	1.002	2.000	3.000	6.200	8.200		2.168

FUTURE RESEARCH DIRECTIONS

Placing of DGs in restructured power system creates challenges to the Independent System Operator (ISO) in finding the optimal transmission pricing and congestion management in each transaction of DG. Hence, this work may expand to new methodologies for conduction lines pricing and overcrowding organization.

CONCLUSION

In this chapter, the performance of IEEE 30 bus test system for a hybrid ABC-ANN optimization methodology has been verified for demonstrating the effectiveness of proposed methodology as compared to other methodologies like ANN, PSO and hybrid PSO-ANN. Optimal locations of DGs have been determined with the support of voltage deviation by applying the ABC algorithm and optimal ratings of the DGs have been found by using ANN approach.

The power loss in the system is reduced to 7.904MW from 10.214MW when only DG is placed at bus number 19 with an optimal capacity of 5.986MW using ABC-ANN approach. It is noticed that, this power loss is less as compared to other methodologies implementation such as ANN, PSO and PSO-ANN and also voltage profile has been improved and also proves that the proposed methodology increases the system voltage profile and power loss of the system is decreased from 10.714MW to 2.1681MW when two DGs are placed at optimal locations such as at bus numbers 14 & 15 with an optimal capacities of 2MW and 3MW respectively

Comparison has been made among PSO, ANN, hybrid PSO-ANN with the hybrid ABC-ANN optimization methodologies. It can be concluded that the hybrid ABC-ANN is more professional in dropping the loss as compared to the other methodologies.

REFERENCES

Abu-Mouti, F. S., & El-Hawary, M. E. (2011). Optimal Distributed Generation Allocation and Sizing in Distribution Systems via Artificial Bee Colony Algorithm. *IEEE Transactions on Power Delivery*, *26*(4), 2090–2101. doi:10.1109/TPWRD.2011.2158246

Abu-Mouti, F. S., & El-Hawary, M. E. (2009). Modified artificial bee colony algorithm for optimal distributed generation sizing and allocation in distribution systems. Proceedings of IEEE Conference on Electrical Power & Energy, 1-9. doi:10.1109/EPEC.2009.5420915

Acharya, N., Mahat, P., & Mithulananthan, N. (2006). An analytical approach for DG allocation in primary distribution network. *Electrical Power and Energy Systems*, *28*(10), 669–678. doi:10.1016/j.ijepes.2006.02.013

Alhelou, H., Hamedani-Golshan, M. E., Zamani, R., Heydarian-Forushani, E., & Siano, P. (2018). Challenges and Opportunities of Load Frequency Control in Conventional, Modern and Future Smart Power Systems: A Comprehensive Review. *Energies*, *11*(10), 2497. doi:10.3390/en11102497

Alhelou, H. H. (2018). *Fault Detection and Isolation in Power Systems Using Unknown Input Observer. In Advanced Condition Monitoring and Fault Diagnosis of Electric Machines* (p. 38). Hershey, PA: IGI Global.

Alhelou, H. H., Golshan, M., & Fini, M. (2018). Wind Driven Optimization Algorithm Application to Load Frequency Control in Interconnected Power Systems Considering GRC and GDB Nonlinearities. *Electric Power Components and Syst.*

Alhelou, H. H., & Golshan, M. E. H. (2016, May). Hierarchical plug-in EV control based on primary frequency response in interconnected smart grid. In *Electrical Engineering (ICEE), 2016 24th Iranian Conference on* (pp. 561-566). IEEE. 10.1109/IranianCEE.2016.7585585

Alhelou, H. H., Golshan, M. H., & Askari-Marnani, J. (2018). Robust sensor fault detection and isolation scheme for interconnected smart power systems in presence of RER and EVs using unknown input observer. *International Journal of Electrical Power & Energy Systems*, *99*, 682–694. doi:10.1016/j.ijepes.2018.02.013

Alhelou, H. H., Hamedani-Golshan, M. E., Heydarian-Forushani, E., Al-Sumaiti, A. S., & Siano, P. (2018, September). Decentralized Fractional Order Control Scheme for LFC of Deregulated Nonlinear Power Systems in Presence of EVs and RER. In *2018 International Conference on Smart Energy Systems and Technologies (SEST)* (pp. 1-6). IEEE. 10.1109/SEST.2018.8495858

Alhelou, H. S. H., Golshan, M. E. H., & Fini, M. H. (2015, December). Multi agent electric vehicle control based primary frequency support for future smart micro-grid. In *Smart Grid Conference (SGC)* (pp. 22-27). Academic Press.

Alshahrestani, A., Golshan, M. E. H., & Alhelou, H. H. (2018, November). WAMS Based Online Estimation of Total Inertia Constant and Damping Coefficient for Future Smart Grid Systems. In *Smart Grid Conference (SGC)* (pp. 1-5). Academic Press.

Fini, M. H., Yousefi, G. R., & Alhelou, H. H. (2016). Comparative study on the performance of many-objective and single-objective optimisation algorithms in tuning load frequency controllers of multi-area power systems. *IET Generation, Transmission & Distribution*, *10*(12), 2915–2923. doi:10.1049/iet-gtd.2015.1334

Hosseini, & Karami, & KarimiMadahi. (2011). Optimal capacity, location and number of distributed generation at 20 kv substations. *Australian Journal of Basic and Applied Sciences*, *5* (10), 1051-1061.

Khosrow-Pour. (2018). *Incorporating Nature-Inspired Paradigms in Computational Applications*. IGI Global.

Lin & Su. (2012). Using an Efficient Artificial Bee Colony Algorithm for Protein Structure Prediction on Lattice Models. *International Journal of Innovative Computing, Information and Control, 8*(3B), 2049-2064.

Makdisie, C., Haidar, B., & Alhelou, H. H. (2018). An Optimal Photovoltaic Conversion System for Future Smart Grids. In *Handbook of Research on Power and Energy System Optimization* (pp. 601–657). IGI Global. doi:10.4018/978-1-5225-3935-3.ch018

Nadweh, S., Hayek, G., Atieh, B., & Haes Alhelou, H. (2018). Using Four – Quadrant Chopper with Variable Speed Drive System Dc-Link to Improve the Quality of Supplied Power for Industrial Facilities. Majlesi Journal of Electrical Engineering.

Nasiraghdam & Jadid. (2012). Optimal hybrid PV/WT/FC sizing and distribution system reconfiguration using multi-objective artificial bee colony (MOABC) algorithm. *International Journal of Solar Energy, 86*(10), 3057–3071.

Njenda, T. C., Golshan, M. E. H., & Alhelou, H. H. (2018, November). WAMS Based Intelligent Under Frequency Load Shedding Considering Online Disturbance Estimation. In *Smart Grid Conference (SGC)* (pp. 1-5). Academic Press.

Njenda, T. C., Golshan, M. E. H., & Alhelou, H. H. (2018, November). WAMS based Under Frequency Load Shedding Considering Minimum Frequency Prediction and Extrapolated Disturbance Magnitude. In *Smart Grid Conference (SGC)* (pp. 1-5). Academic Press.

Partha, K., & Chanda, C. K. (2013). A simple and fast approach for allocation and size evaluation of distributed generation. *International Journal of Energy and Environmental Engineering*, *4*(1), 1–9.

Rao, G. S., & Obulesh, Y. P. (2013). Optimal Location of DG for Maintaining Distribution System Stability: A Hybrid Technique. *International Journal of Power and Energy Conversion*, *4*(4), 387–403. doi:10.1504/IJPEC.2013.057036

Rao & Obulesh. (2015). ABC and ANN based minimization of power loss for distribution system stability. *Proceedings of 10th IEEE conference on Industrial Electronics and Applications (ICIEA)*, 772-777.

Tosun. (2014). *Artificial Bee Colony Algorithm. In Encyclopedia of Business Analytics and Optimization* (pp. 1–14). IGI Global.

Varesi, K. (2011). Optimal allocation of dg units for power loss reduction and voltage profile improvement of distribution networks using PSO algorithm. *World Academy of Science, Engineering and Technology*, *60*, 1938–1942.

KEY TERMS AND DEFINITIONS

Artificial Bee Colony (ABC) Algorithm: Artificial bee colony (ABC) algorithm is an optimization technique that simulates the foraging manners of honey bees, and has been effectively applied to a variety of practical problems. ABC belongs to the assembly of swarm intelligence algorithms.

Artificial Neural Network (ANN): Artificial neural networks (ANN) are the pieces of a computing system designed to simulate the way the human brain analyzes and processes information. ANN has self-learning capabilities that enable them to produce better results.

Distribution Generations (DG): It is an approach that makes use of small-scale technologies to generate electricity nearer to the end users. In many cases, distributed generators can provide lower-cost electricity and higher power consistency.

Optimization: It is the action of making the finest or most successful use of a situation or resource.

Particle Swarm Optimization (PSO) Algorithm: It is a computational method that optimizes a problem by iteratively trying to improve a candidate solution with regard to a given measure of quality. PSO is a metaheuristic as it makes few or no assumptions about the problem being optimized and can search very large spaces of candidate solutions.

Voltage Stability: It refer to the ability of power system to maintain steady state voltages at all buses in the power system after subjected to a faults from a given initial operating point.

APPENDIX A: IEEE 30 BUS TEST SYSTEM SINGLE LINE DIAGRAM

Figure 7. Single line diagram of IEEEE 30 bus system

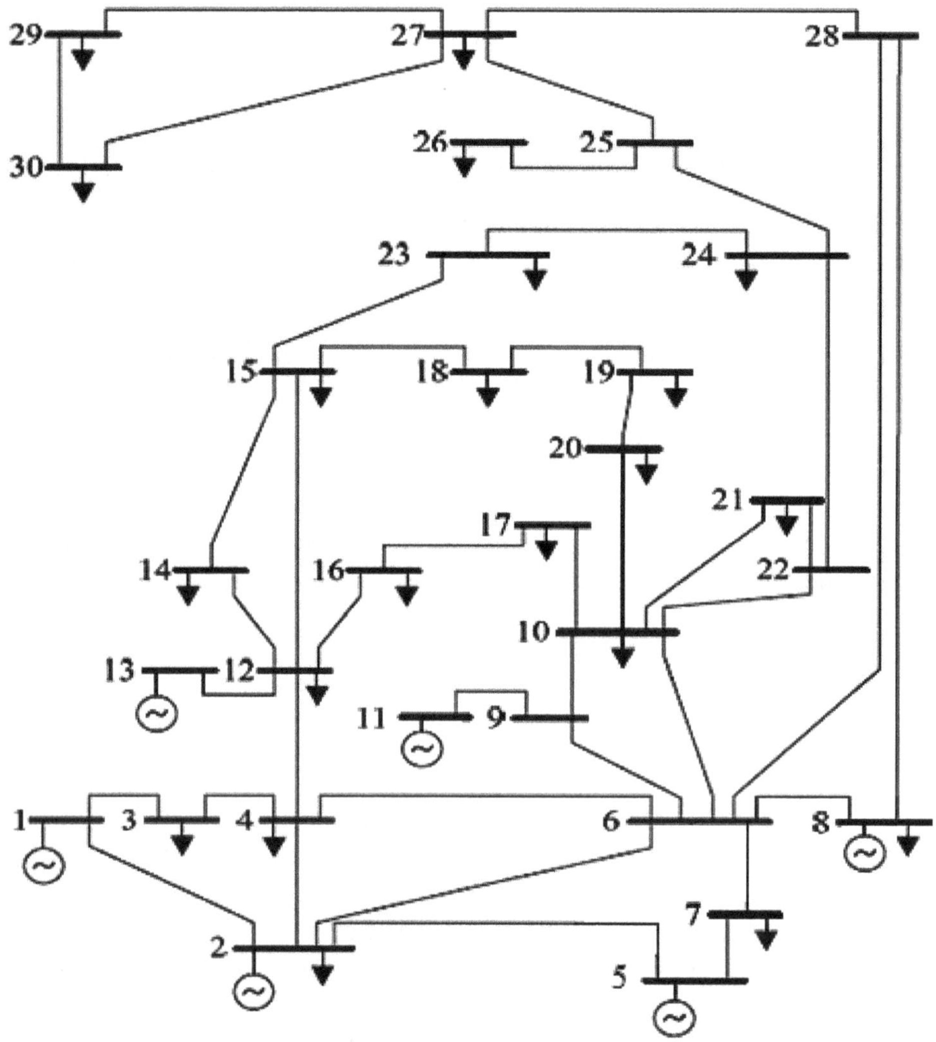

APPENDIX B: BUS DATA OF IEEE 30 BUS SYSTEM

Table 8 provides the bus data of 30 bus system.

Table 8. Bus data

Bus No	Type	Vsp	Theta	Pgi	Qgi	Pli	Qli	QMin	Qmax
1	1	1.060	0	0	0	0	0	0	0
2	2	1.043	0	40	50	22	13	-40	50
3	3	1.000	0	0	0	2	1	0	0
4	3	1.060	0	0	0	8	2	0	0
5	2	1.010	0	20	37	94	19	-40	40
6	3	1.000	0	0	0	0	0	0	0
7	3	1.000	0	0	0	23	11	0	0
8	2	1.010	0	12	37	30	30	-10	40
9	3	1.000	0	0	0	0	0	0	0
10	3	1.000	0	0	19	6	2	0	0
11	2	1.082	0	12	12	0	0	-6	24
12	3	1.000	0	0	0	11	8	0	0
13	2	1.071	0	15	11	0	0	-6	24
14	3	1.000	0	0	0	6	2	0	0
15	3	1.000	0	0	0	8	3	0	0
16	3	1.000	0	0	0	4	2	0	0
17	3	1.000	0	0	0	9	6	0	0
18	3	1.000	0	0	0	3	1	0	0
19	3	1.000	0	0	0	10	3	0	0
20	3	1.000	0	0	0	2	1	0	0
21	3	1.000	0	0	0	18	11	0	0
22	3	1.000	0	0	0	0	0	0	0
23	3	1.000	0	0	0	3	2	0	0
24	3	1.000	0	0	4	9	7	0	0
25	3	1.000	0	0	0	0	0	0	0
26	3	1.000	0	0	0	4	2	0	0
27	3	1.000	0	0	0	0	0	0	0
28	3	1.000	0	0	0	0	0	0	0
29	3	1.000	0	0	0	2	1	0	0
30	3	1.000	0	0	0	11	2	0	0

APPENDIX C: LINE DATA OF IEEE 30 BUS SYSTEM

Table 9 provides the line data of 30 bus system.

Table 9. Line data

Line No.	From Bus	To Bus	Series Impedance		Half Line Charging Susceptance (p.u)	Tap Setting	MVA Rating
			R	X			
1	1	2	0.01920	0.05750	0.02640	--	130
2	1	3	0.04520	0.18520	0.02040	--	130
3	2	4	0.05700	0.17370	0.01840	--	65
4	3	4	0.01320	0.03790	0.00420	--	130
5	2	5	0.04720	0.19830	0.02090	--	130
6	2	6	0.05810	0.17630	0.01870	--	65
7	4	6	0.01190	0.04140	0.00450	--	90
8	5	7	0.04600	0.11600	0.01020	--	70
9	6	7	0.02670	0.08200	0.00850	--	130
10	6	8	0.01200	0.04200	0.00450	--	32
11	6	9	0.00000	0.20800	0.00000	1.01550	65
12	6	10	0.00000	0.55600	0.00000	0.96290	32
13	9	11	0.00000	0.20800	0.00000	--	65
14	9	10	0.00000	0.11000	0.00000	--	65
15	4	12	0.00000	0.25600	0.00000	1.01290	65
16	12	13	0.00000	0.14000	0.00000	--	65
17	12	14	0.12310	0.25590	0.00000	--	32
18	12	15	0.06620	0.13040	0.00000	--	32
19	12	16	0.09450	0.19870	0.00000	--	32
0	14	15	0.22100	0.19970	0.00000	--	16
21	15	17	0.08240	0.19320	0.00000	--	16
22	15	18	0.10700	0.21850	0.00000	--	16
23	18	19	0.06390	0.12920	0.00000	--	16
24	19	20	0.03400	0.06800	0.00000	--	32
25	10	20	0.09360	0.20900	0.00000	--	32
26	10	17	0.03240	0.08450	0.00000	--	32
27	10	21	0.03480	0.67490	0.00000	--	32
28	10	22	0.07270	0.14990	0.00000	--	32
29	21	22	0.01160	0.02360	0.00000	--	32
30	15	23	0.10000	0.20200	0.00000	--	16
31	22	24	0.11500	0.17900	0.00000	--	16
32	23	24	0.13200	0.27000	0.00000	--	16
33	24	25	0.18850	0.32920	0.00000	--	16
34	25	26	0.25440	0.38000	0.00000	--	16
35	25	27	0.10930	0.20870	0.00000	--	16
36	28	27	0.00000	0.36900	0.00000	0.95810	65

continued on following page

Table 9. Continued

Line No.	From Bus	To Bus	Series Impedance		Half Line Charging Susceptance (p.u)	Tap Setting	MVA Rating
			R	X			
37	27	29	0.21980	0.41530	0.00000	--	16
38	27	30	0.32020	0.60270	0.00000	--	16
39	29	30	0.23990	0.45330	0.00000	--	16
40	8	28	0.06360	0.20000	0.02140	--	32
41	6	28	0.01690	0.05990	0.00650	--	32

Chapter 3
Dynamic and Stability Analysis of Wind-Diesel-Generator System With Intelligent Computation Algorithm:
Computation Algorithms Applied to WDG System

Dipayan Guha
https://orcid.org/0000-0002-2603-6955
Motilal Neheru National Institute of Technology Allahabad, India

Provas Kumar Roy
https://orcid.org/0000-0002-3433-5808
Kalyani Government Engineering College, India

Subrata Banerjee
National Institute of Technology Durgapur, India

ABSTRACT

In this chapter, the dynamic performance of a wind-diesel-generator system has been studied against wind and load perturbations. The wind perturbation is modeled by simulating base, ramp, gust, and random wind. An optimized cascade tilt-integral-derivative (CC-TID) controller is provided to the test system for producing desired control signal to regulate the blade pitch angle of wind turbine. To confirm the efficacy of CC-TID controller, the output results are compared to that of PI- and PID-controller. The optimum gains of the proposed controllers are explored employing Levy-embedded grey wolf optimization, whale optimization algorithm, drone squadron optimization, and search group algorithm. To show the effectiveness, the output results are compared to the results of genetic algorithm and particle swarm optimization tuned controllers. A thyristor control series compensator (TCSC) is provided to WDG model for increasing the damping of system oscillations. Analysis of the presented results confirm the supremacy of CC-TID-TCSC controller over other controllers provided in this chapter.

DOI: 10.4018/978-1-5225-8030-0.ch003

INTRODUCTION

Electrical energy is environmentally the most favorable form of energy, the generation routed through the burning of fossil fuels or through nuclear reaction or use of hydro resources. However, with the swift fall of natural resources and degradation of environmental conditions, generation of power through renewable energy resources (RER) has gained ample attention from the researchers over the worldwide (Das et al., 1999). RER comprises solar power, wind power, geothermal power, wave power, tidal power, hydro-power, biomass power etc. Wind and photovoltaic (PV) power generation are two of the most prevalent renewable sources used in the hybrid power system (HPS) due to their clean, abundant, inexpensive, and eco-friendly features. However, PV power generation is costly, has poor conversion efficiency, and lower power density as compared to wind power generation (WPG). Since the power generated from the WTG changes abruptly, hence to balance the intermittent characteristic, a diesel generator (DG) is coordinated with WTG to fulfill the required of load profile. The proposed wind-diesel-generator (WDG) system is trustworthy since the diesel set behaves as a cushion to overlook the variation of load demand and wind speed and meet the deficit demand (Das et al., 1999; Gampa and Das, 2015). Due to an intermittent characteristic of WTG, the unbalance power generation and load demand causes frequency fluctuation that may lead to the problem of instability. Thus the control of frequency and power is emergent for the successful operation of the coordinated WDG system.

In the state-of-art, numerous control methods have been reported for betterment of the dynamic performance of an isolated and/or interconnected WDG system following wind and load fluctuations. A comprehensive review of the challenges and opportunities in frequency control of the power system are presented in (Alhelou, 2018a). A load frequency control (LFC) of HPS with non-scheduled wind plant has been discussed in (Aziz et al., 2018). Uhlen *et al.* presented a robust control algorithm for WDG system using multivariable frequency domain techniques (Uhlen et al., 1994). A coordinated control of fuel cell (FC) and aqua-electrolyzer (AE) to solve the frequency and power fluctuation problem in micro-grid (MG) is discussed in (Ngamroo, 2012; Nadweh, 2018; Njenda, 2018; Zamani, 2018). The usefulness of electric vehicle control on primary frequency response of smart grid is discussed in (Alhelou *et al.*, 2015; Alhelou *et al.*, 2016). A model based fault detection scheme for measurement of variables in frequency control of power system with unknown input observer is discussed in (Alhelou *et al.*, 2018b; Alhelou 2018c).

A proportional-integral-derivative (PID) control action or its variants are mostly utilized for frequency and power deviation minimization due of its numerous advantages (Kamwa, 1990; Bhatti et al., 1997; Senjyu et al., 2005; Tah and Das, 2016; Shankar and Mukherjee, 2016; Guha et al., 2018). But, the performance of PID-controller is deteriorated with system uncertainties and random perturbations. Moreover, the PID-controller starts functioning only after the control variable deviates from the set-point level. Cascade control is an alternative mechanism for enhancing the dynamic performance of a closed-loop control system by introducing secondary measurement and secondary feedback action. This structure has high disturbance rejection ability and good set-point tracking facility. A cascade PID-proportional derivative (PID-PD) controller with two-degree-of-freedom (2DOF) has been utilized for LFC of a hydrothermal power system in (Raju et al., 2018). In (Guha *et al.*, 2018), the dynamic performances of a HPS have been closely investigated by employing 3DOF-PID controller. The superiority of the 3DOF-PID controller was established by performing a comparative analysis with the results yielded by other methods available in the literature. Performances of cascade controller were compared with PID and 2DOF-PID controllers. The utility of fractional order (FO) calculus has recently received imputes due to its suppleness and design methodologies. A combined control scheme for voltage and frequency profiles with FO-controller has been studied in (Rajbongshi and Saikia, 2017). The utilization of FO-controller has been identified by (Pan and Das, 2015; Delassi et al., 2015). Lurie has proposed another variant of PID-controller with FO-calculus called 'tilt-integral-derivative (TID)' controller (Lurie, 1994). Unlike PID-controller, the proportional gain is replaced by a factor of '$s^{-\frac{1}{n}}$' in TID-controller. The effectiveness of integral-tilt-derivative (I-TD) controller in frequency stabilization is presented in (Kumari and Shankar, 2018). The basic idea is to use the merits of PID-controller and FO-calculus simultaneously in TID-controller. Surprisingly, no study has been performed for the WDG system employing a cascade and/or tilt controller.

The usefulness of fuzzy logic controller (FLC) and artificial neural network (ANN) has been identified in the literature for accelerating frequency response profile of power system. An optimized FLC for the hybrid WDG system has been presented in (Mahto and Mukherjee, 2017; Ganguly et al., 2017). Chedid *et al.* developed an adaptive neuro-fuzzy controller for WDG system and compared the results with the conventional controller (Chedid et al., 2000). A control algorithm with conventional and intelligent (FLC, ANN) controllers has been demonstrated in (Chedid et al., 1999). The effectiveness, robustness, and tracking performances of the designed controller have been closely inspected for the simulated model. Makdisie *et al.* presents PV coordinated nuro-fuzzy control system for optimal power conversion in smart grid (Makdisie *et al.*, 2018). Though FLC controller shows better results as compared to the conventional controller, still there is no well-defined mathematical rule for selecting membership functions, rule base, defuzzification methodologies, and inference mechanism (Ganguly et al., 2017).

Owing to the nonlinear time-varying characteristics, the HPS experiences low-frequency electromechanical oscillations. However, to supply quality of power with reliability, these oscillations have to be damp out speedily. To provide active compensation, flexible AC transmission systems (FACTS) are incorporated in the power system with conventional frequency control scheme. FACTS devices are not only ensuring good power transmission through tie-line, it enhances controllability and stability margin of the system. Application of battery energy storage system (BESS), flywheel energy storage system (FESS), and ultracapacitor (UC) has been identified in (Lee and Wang, 2008; Howlader et al., 2012) for active compensations of frequency and power oscillations. The use of ultra-capacitor as an auxiliary power source in a HPS is presented in (Sami *et al.*, 2017). The effectiveness of superconducting magnetic energy storage (SMES) for frequency and power compensation has been discussed in (Singh et al., 2013). However, the structure of SMES is complicated, require high maintenance, and costly. The effect of a thyristor control phase shifter (TCPS) has been presented in (Abraham et al., 2007). In (Abraham et al., 2007), the performance of TCPS has been studied on a two-area reheat thermal power plant. In (Morsali et al., 2017), a coordinated FOPID and thyristor control series compensator (TCSC) has been developed for LFC of a mixed power plant.

To derive better dynamic stability of WDG system, the appropriate selections of controller gains is highly demanded. In the recent past, computational optimization techniques are extensively utilized for parameter optimization because of its robustness, flexibility, and versatility of exploring probable global solutions. Genetic algorithm (GA) (Gampa and Das, 2015), particle swarm optimization (PSO) (Gampa and Das, 2015), grey wolf optimization (GWO) (Guha et al., 2018), teaching learning-based optimization (TLBO) (Guha et al., 2018), quasi-oppositional harmony search algorithm (QOHSA) (Shankar and Mukjerjee, 2016; Kumar and Shankar, 2017), cultural algorithm (Raju et al., 2018), JAYA (Guha et al., 2018), symbiotic organism search (SOS) (Guha et al., 2018), ant colony optimization (ACO) (Kaliannan *et al.*, 2017), multi-verse optimization (Guha et al., 2017) etc. have been successfully implemented for attaining the desired performance of the system. Bahgaat *et al.* have studied frequency instability problem with two new variants of PSO algorithm, i.e., adaptive weight PSO and adaptive accelerated coefficient based PSO. In (Rahman et al., 2015), biogeography-based optimization (BBO) is employed for optimizing controller parameters. A comparative study on single and multi-objective optimization techniques has been studied in (Fini *et al.*, 2016). A hybrid PSO-LEVY flight algorithm based fuzzy PID controller for automatic generation control of multi-area power systems has been presented in (Barisal *et al.*, 2017). The optimal design of state feedback controller for LFC of power system using hybrid evolutionary algorithm is discussed in (Singh, 2017). Despite the expansion in the area of computational algorithms, there are some additional avenues to accelerate the searching mechanism, stabilities and convergence rate in the optimization problem.

The purpose of the chapter describe herein is to derive an effective, robust, and intelligent computational algorithms for appraising the dynamic performance of a WDG system following load and wind power perturbations. The proposed test system has 150KW WTG worked in parallel with DG for matching the load demand of 350KW. The motivations of the proposed work take it shapes from the following aspects.

- A proposal of utilizing RESs is an alternative source to cope up with rapidly decreasing natural fuels, global warming, environmental complication, and rural electrification issues.
- The concern of supplying stable and reliable power supply keeping in mind the shortcomings of the production of wind power. The wind generator is performed at its rated loading and diesel generator is employed as a compensation unit.
- The appropriate sharing of power in the neighboring control areas without allowing much fluctuation in power may be addressed.
- For assuring stable operation of WDG, the applicability of CC-TID controller as a secondary controller is devised.
- The continual progression and noted advancement of computational intelligence have necessitated the use of same in a constrained optimization problem. Hence application of intelligent algorithm in a hybrid power system for parameters optimization is considered as an integral part of the research.
- The shortcomings of conventional controllers may recognize.

The salient contributions of this chapter are summarized below.

- To realize a hybrid WDG system for performing small-signal stability analysis under the action of load and wind power variations. Diesel engine speed governor and wind pitch controller are considered while deriving WDG model.
- To demonstrate dynamic performances, the model of wind power perturbation is derived by considering base wind speed (step), rapid (ramp) wind flow, gust wind, and associated wind noises.
- To design and implement a cascade tilt-integral-derivative (CC-TID) controller for dynamic stability analysis. The proposed CC-TID controller includes both the advantages of fractional-order (FO) calculus and cascade-control (CC) algorithm. The performance of CC-TID controller is compared with PI- and PID-controller in order to confirm the efficacy of CC-TID controller.
- To explore optimum gains of the proposed controller for frequency and power stabilization, four evolutionary computation algorithms such as Levy-embedded grey wolf optimization (LGWO),

whale optimization algorithm (WOA), drone squadron optimization (DSO), and search group algorithm (SGA) are applied in this chapter.

- To evaluate optimum controller gains, two different objective functions are calculated in this chapter.
- To demonstrate supremacy, the outputs of WDG system are compared with GA and PSO optimized controllers.
- To enrich the damping of system oscillation, thyristor control series capacitor is developed and integrated in the WDG model as a frequency stabilizer.

The remaining of the chapter is documented as follows. The mathematical model of the investigated WDG system is described in the ensuing section followed by the mathematical model of CC-TID controller. A brief overview of the proposed optimization techniques is given in Section 3. Section 4 shows the model of TCSC as a frequency stabilizer. Simulation results and comparative study are presented in Section 5. Finally, Section 6 concludes the current work with future scopes.

MODELING AND METHODOLOGIES

The WDG system considered in this chapter is comprised of the following subsystems (Das et al., 1999; Gampa and Das, 2015).

1. Wind turbine generator (WTG) with blade pitch control
2. Diesel dynamic system
3. Generator model

The schematic diagram of a WDG coordinated system is presented in Fig. 1(a) (Das et al., 1999; Gampa and Das, 2015). During start-up and synchronization, a minimum wind speed is needed, while diesel governor is used for controlling the dynamics of diesel generator (DG). The output power of wind turbine generator (WTG) can be regulated by controlling the blade pitch angle of the WTG using hydraulic pitch actuator.

Modeling of Wind Disturbance Model

To revise the dynamics of a coordinated WDG system, initially the model of wind perturbation is developed. To simulate the behavior of wind perturbation, following four types of wind flow is simulated. Mathematically, the model of wind speed $\left(V_w\right)$ is expressed as

$$V_w = V_{base} + V_{gust} + V_{ramp} + V_{noise} \tag{1}$$

where V_{base} is base wind represented by a step function of magnitude k_B (a constant value); V_{gust} is guest-speed represented by (2); V_{ramp} indicates ramp type wind calculated by using (3); V_{noise} is a random

Figure 1. (a) Block diagram of a coordinated WDG system, (b) Wind model with wind power perturbation, (c) Transfer function model of isolated WDG system

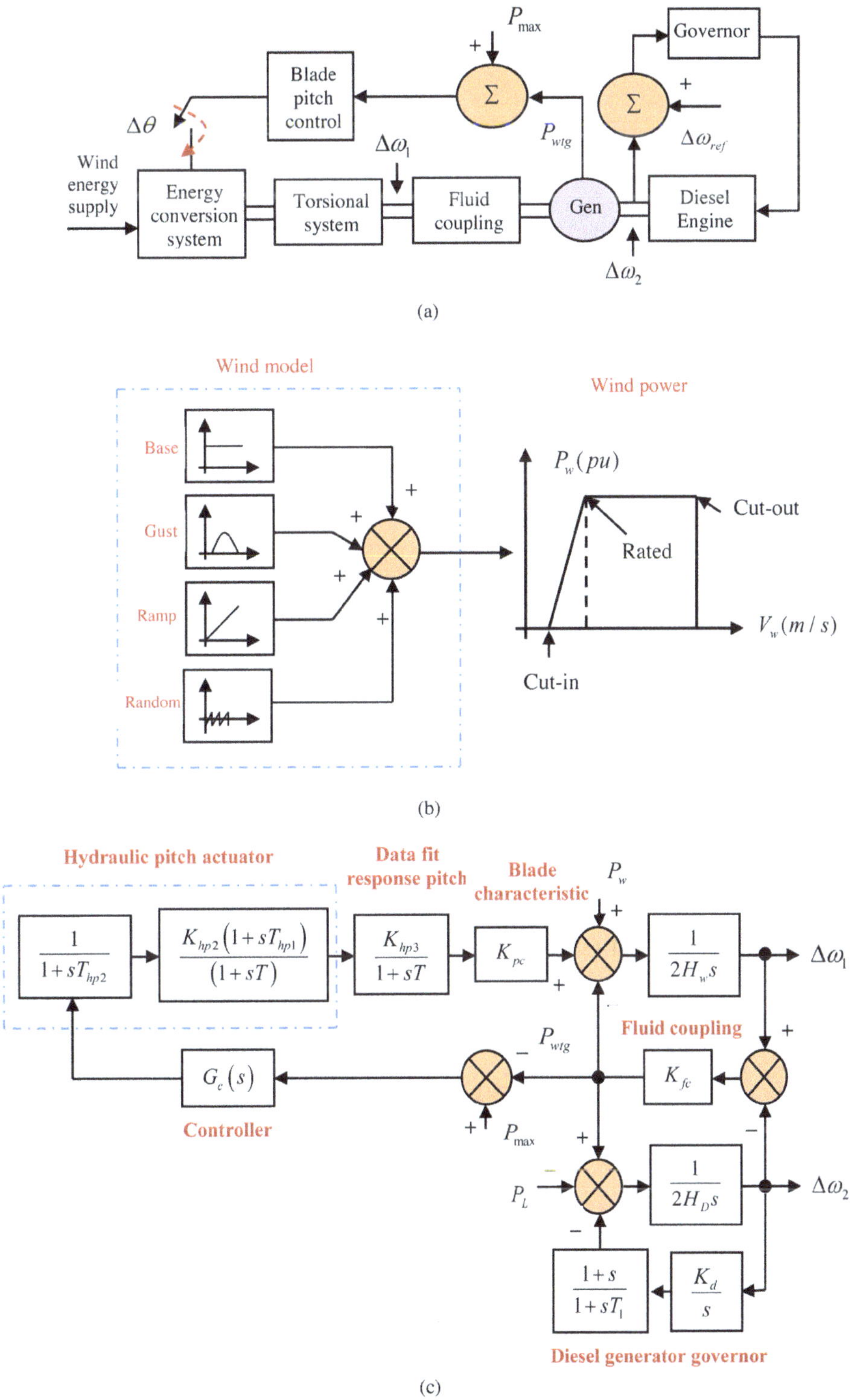

noise of wind computed by using (4). Fig. 1(b) illustrates the typical model of wind disturbance projected for assessing the dynamics of WDG (Das et al., 1999; Gampa and Das, 2015).

$$V_{gust} = \begin{cases} 0 & t < T_{g1} \\ \left(\frac{MGWS}{2}\right)\left(1 - \cos\left(2\pi\left[\left(\frac{t}{T_g}\right) - \left(\frac{T_{g1}}{T_g}\right)\right]\right)\right) & T_{g1} < t < \left(T_g + T_{g1}\right) \\ 0 & t > \left(T_g + T_{g1}\right) \end{cases} \tag{2}$$

$$V_{ramp} = \begin{cases} 0 & t < T_{r1} \\ MRWS\left(1 - \frac{\left(t - T_{r2}\right)}{\left(T_{r1} - T_{r2}\right)}\right) & T_{r1} < t < T_{r2} \\ 0 & t > T_{r2} \end{cases} \tag{3}$$

$$V_{noise} = 2\sum_{i=1}^{N}\left(\frac{2K_N F^2 \left|\Omega_i\right|}{\pi^2\left(1 + \left(F\Omega_i/\mu\pi\right)^2\right)^{\frac{4}{3}}}\right)\cos\left(\Omega_i t + \varphi_i\right) \tag{4}$$

where MGWS is maximum gust wind speed (= 12 m/s); MRWS is maximum ramp wind speed (= 10 m/s); t is time in sec; φ_i indicates a uniformly distributed random variable within $\left(0, 2\pi\right)$; K_N is surface drag coefficient (= 0.004); F is wind turbine scale (= 2000 m); μ is mean speed of noise wind (= 7.5 m/s).

Model of Wind Turbine Generator (WTG)

The power coefficient $\left(C_p\right)$ and wind velocity are considered while modeling WTG. The power coefficient is further determined by the tip speed ratio $\left(\gamma\right)$ and blade pitch angle $\left(\beta\right)$. The dynamics of wind blade may define by using (5) (Das et al., 1999; Gampa and Das, 2015).

$$C_p = \frac{1}{2}\left(\frac{V_w}{\omega_B} - 0.0228\beta^2 - 5.6\right)^{-0.17\left(\frac{V_w}{\omega_B}\right)} \tag{5}$$

The wind power is calculated by using (6).

$$P_w = \frac{1}{2}\rho A_B C_p V_w^{\,3} \tag{6}$$

where ρ is air density (= 1.25 Kg/m^3); A_B is swept blade area (= 1735 m^2); ω_B is wind turbine blade velocity. To control the blade pitch angle of WTG, the hydraulic pitch actuator produces desired control signal. The linear approximated model of hydraulic pitch actuator is shown in (7) (Das et al., 1999; Gampa and Das, 2015).

$$\frac{\Delta H_1(s)}{\Delta U(s)} = \frac{K_{hp2}\left(1 + sT_{hp1}\right)}{\left(1 + sT_{hp2}\right)\left(1 + s\right)} \tag{7}$$

The block 'data-fit-pitch' maintain gain-phase characteristic of WTG system. The diesel generator is coordinated with WTG to provide extra load demand. The operation of a diesel generator is under the control of diesel governor. The T.F form of diesel governor is expressed in (8) (Das et al., 1999; Gampa and Das, 2015).

$$\frac{\Delta P_f(s)}{\Delta \omega_2(s)} = -\frac{K_d\left(1 + s\right)}{s\left(1 + sT_1\right)} \tag{8}$$

The error in power generation and the power generated from WTG are defined in (9).

$$\begin{aligned} \Delta P &= P_{\max} - P_{wtg} \\ P_{wtg} &= k_{fc}\left(\Delta\omega_1 - \Delta\omega_2\right) \end{aligned} \tag{9}$$

The linearized model of WDG system with turbine blade pitch and diesel governor control schemes is displayed in Fig. 1(c). To study the dynamics of the WDG model, the state-variable model is developed by using (10) and shown in (11). Nominal values of system parameters of Fig. 1(c) are taken from (Das et al., 1999; Gampa and Das, 2015) and presented in Table 1.

$$\dot{x} = Ax + \mathrm{B}u + Gw \tag{10}$$

Table 1. Nominal value of coordinated WDG parameters

Parameter	Value	Parameter	Value	Parameter	Value	Parameter	Value
H_w	3.52 sec	K_d	16.5 puKW/Hz	K_{ph3}	1.40	P_{max}	10 sec
H_D	8.7 sec	K_{hp2}	1.25	T_{hp1}	0.60 sec	K_{pc}	0.08
V_{WB}	7 m/sec	k_{fc}	16.2 puKW/Hz	T_{hp2}	0.041 sec	T_1	0.025 sec

$$A=\begin{bmatrix} \frac{-1}{T_{p_2}} & 0 & 0 & 0 & 0 & 0 & 0 \\ (K_{p_2}-\frac{K_{p_2}T_{p_1}}{T_{p_2}}) & -1 & 0 & 0 & 0 & 0 & 0 \\ 0 & K_{p_3} & -1 & 0 & 0 & 0 & 0 \\ 0 & 0 & \frac{K_{pc}}{2H_\omega} & \frac{-K_{fc}}{2H_\omega} & \frac{K_{fc}}{2H_\omega} & 0 & 0 \\ 0 & 0 & 0 & \frac{K_{fc}}{2H_d} & \frac{-K_{fc}}{2H_d} & 0 & \frac{1}{2H_d} \\ 0 & 0 & 0 & 0 & -K_d & 0 & 0 \\ 0 & 0 & 0 & 0 & \frac{-K_d}{T_1} & \frac{1}{T_1} & \frac{-1}{T_1} \end{bmatrix};B=\begin{bmatrix} \frac{1}{T_{p_2}} \\ \frac{K_{p_2}T_{p_1}}{T_{p_2}} \\ 0 \\ 0 \\ 0 \\ 0 \\ 0 \end{bmatrix};G=\begin{bmatrix} 0 & 0 \\ 0 & 0 \\ 0 & 0 \\ \frac{1}{2H_\omega} & 0 \\ 0 & \frac{-1}{2H_d} \\ 0 & 0 \\ 0 & 0 \end{bmatrix} \quad (11)$$

where $A, B, \& G$ are system matrix, input matrix, and disturbance matrix, in order; $x, u, \& w$ are state vector, input vector, and disturbance vector, in order. A cascade tilt-integral-derivative (CC-TID) controller is designed to identify and quantify the undamped system oscillations following wind perturbation and load fluctuation. The proposed model of CC-TID controller is briefly discussed in the ensuing section.

Controller Structure

Cascade controller is commonly utilized in a multi-loop control system for good set-point tracking and better disturbance rejection. Unlike conventional controller, cascade controller has two loops called 'primary or inner or slave' loop and 'outer or secondary or master' loop. Inner loop responds much faster than the outer loop such that disturbance appears in the loop can be easily diminished before it affects the other parts of the process. The merits of the cascade controller over the single-loop controller are documented in (Dash et al., 2015). Fig. 2(a) illustrates a block diagram of the control system with a cascade controller. The controlled output of the closed-loop system is calculated using (12).

Figure 2. Schematic diagram of controller, (a) Cascade controller, (b) TID-controller with derivative filter

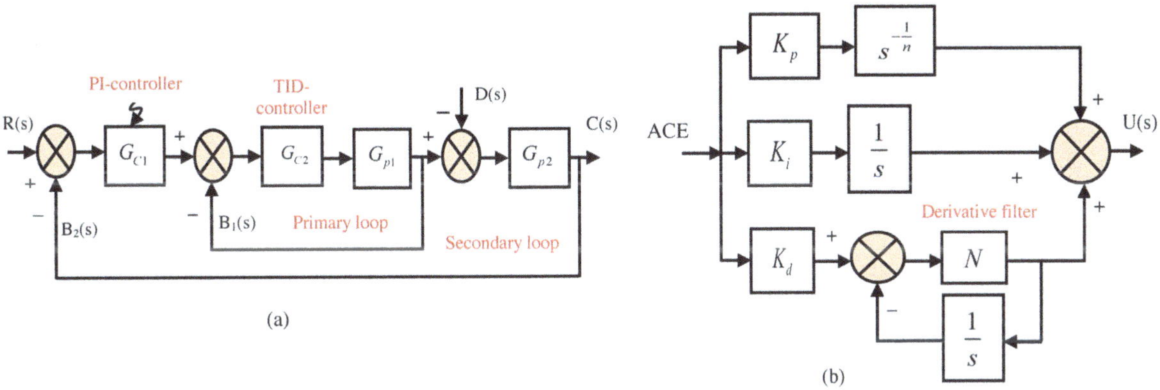

$$C(s) = -\frac{G_{p2}(s)}{1+G_{p1}(s)G_{c2}(s)+G_{p1}(s)G_{p2}(s)G_{c1}(s)G_{c2}(s)}D(s) + \frac{G_{p1}(s)G_{p2}(s)G_{c1}(s)G_{c2}(s)}{1+G_{p1}(s)G_{c2}(s)+G_{p1}(s)G_{p2}(s)G_{c1}(s)G_{c2}(s)}R(s) \quad (12)$$

In (12), $R(s)$ & $D(s)$ are reference and disturbance inputs to the plant, respectively; $G_{c1}(s)$ & $G_{c2}(s)$ are transfer functions of master and slave controllers, respectively.

A tilt integral derivative controller (TID) has an analogous structure of PID-controller except the proportional gain (k_p) is replaced by a block of a transfer function (T.F) $k_p s^{-\frac{1}{n}}$, where $n(\neq 0)$ is called 'tilt parameter' and chosen in between (2,3). This structure is referred to as 'tilt controller' (Lurie, 1994). It alludes in the literature that the tilt controller is more efficient to show good performance as compared to PID-controller (Lurie, 1994). The purpose of applying tilt compensation in LFC loop is to provide an improve feedback loop compensation so that better and optimal responses can derive. It further helps to keep the stability of the system under external and/or internal perturbation. The general layout of TID-controller is displayed in Fig. 2(b). The T.F of TID-controller is calculated as in (13).

$$G_{c2}(s) = k_p s^{-\frac{1}{n}} + \frac{k_i}{s} + k_d \frac{sN}{s+N} \quad (13)$$

where k_i, k_d are integral, and derivative gains, in order; *N* is low-pass filter cut-off frequency. In this work, authors have the aim to develop and apply a novel cascade TID-controller for LFC analysis. To accomplish, TID- controller is used in the inner loop, while PI-controller acts as a master controller in the proposed cascade structure. The optimum settings of the controller are calculated by employing SSA, which is briefly discussed in the ensuing section.

Objective Function

The appropriate selection of the objective function for optimizing the controller settings is very important to derive best results from the evolutionary algorithm. Two objective functions have been considered in this chapter for optimizing the proposed controller parameters. To guarantee the fairness of the comparison made, same objective functions as defined in (Gampa and Das, 2015) is considered. The proposed objective functions are defined as

$$\begin{cases} J_1 = \left|\xi\right| & again, \xi = \max\left(real\left(\lambda_i\right)\right) \\ J_2 = \int_0^T \left(P_{\max} - P_{WTG}\right)^2 dt & \end{cases} \tag{14}$$

where ξ = damping factor; λ is Eigen values of WDG model; Δf = frequency deviation; $P_{\max}$ = maximum output power of WTG; P_{WTG} = output power of WTG; T = maximum simulation time. Since the value of ξ determines the degree of stability, therefore, the value of J_1 has to be maximize. Conversely, the second objective function $\left(J_2\right)$, which is defined with integral square error (ISE) criterion, has to be minimized.

OVERVIEW OF OPTIMIZATION

In this section, four powerful optimization algorithms like LGWO, WOA, SGA, and DSO are studied and modeled for fine tuning of controller parameters to assess the performances of modeled WDG system. In the following sub-sections, a brief introduction and mathematical models of computation algorithms are elaborated.

Levy-Embedded Grey Wolf Optimization (LGWO)

Inspiration

The GWO algorithm is derived by simulating the social hierarchy and hunting activity of grey wolves (Mirjalili et al., 2014; Heidari and Pahlavani, 2017). GWO algorithm has been recognized as a robust optimization tool in the optimization area. Four categories of wolves like alpha (α), beta (β), delta (δ), and omega (Ω) are modeled for imitating the leadership hierarchy and exploring the most favorable region in the search space. Moreover, hunting mechanism in GWO involves searching for prey, encircling prey, and invasion prey. Though the operation of GWO algorithm for searching global optimum solutions is identified to be best in comparison to other evolutionary techniques (Guha et al., 2016a), further a modification has been done in (Heidari and Pahlavani, 2017) for avoiding local solutions and improving the diversity of wolves.

Modeling of GWO

In conventional GWO algorithm, the position of wolves in the search space is determined as

$$\vec{\delta} = \left|\vec{C}.\overrightarrow{X_p}(l) - \vec{X}(l)\right| \tag{15}$$

$$\vec{X}(l+1) = \overrightarrow{X_p}(l) - \vec{A}.\vec{\delta} \tag{16}$$

where l is the present generation and $\overrightarrow{X_p}(l)$ indicates the present location of prey. The coefficients end and r_1, r_2 are computed with the use of (17) and (18), respectively (Mirjalili et al., 2014; Heidari and Pahlavani, 2017).

$$\vec{A} = 2\vec{a} * rand - \vec{a} \text{ where } a = 2\left(1 - \frac{l}{T}\right) \tag{17}$$

$$\vec{c} = 2 * rand \tag{18}$$

The term $\vec{A}$ is the search direction matrix, and it is linearly decreasing from 2 to 0; $l \,\&\, T$ are running iteration and highest iteration count, in order. In the hunting stage, which is mainly headed by the α-wolf, the locations of other wolves are modified w.r.t the position of α wolf. Although alphas are the major agents in this cycle, rarely betas and deltas take part in the hunting phase. So far we have computed the candidate solution in terms of α, β, and δ, but the correct location of prey has not been calculated. Hence to compute the optimum location of prey, three fittest outcomes are saved as alpha, beta, and delta and other wolf positions including omega are modified by using (19) (Mirjalili et al., 2014; Heidari and Pahlavani, 2017).

$$\begin{cases} \overrightarrow{\delta_\alpha} = \left|\overrightarrow{C_\alpha}.\overrightarrow{X_\alpha}(l) - \vec{X}(l)\right| \\ \overrightarrow{\delta_\beta} = \left|\overrightarrow{C_\beta}.\overrightarrow{X_\beta}(l) - \vec{X}(l)\right| \\ \overrightarrow{\delta_\delta} = \left|\overrightarrow{C_\delta}.\overrightarrow{X_\delta}(l) - \vec{X}(l)\right| \end{cases} \tag{19}$$

where $for\ i = 1 : n_p$, and $eco(i,:) = rand(1,n).*(ub - lb) + lb$ are the locations of α, β, and δ wolves, respectively; $\overrightarrow{C_\alpha}, \overrightarrow{C_\beta}$, and $\overrightarrow{C_\delta}$ are three randomly created matrices; *if* is the present population of wolves. The distance within the current wolf and prey is computed by using (19). Finally, the location of wolves within the defined search area has been updated by using (20) and (21) (Mirjalili et al., 2014; Heidari and Pahlavani, 2017).

$$fitness\_new1 < fitness(i) \tag{20}$$

$$\begin{aligned} & futness(i) = fitness\_new1 \\ & eco(i,:) \xleftarrow{update} econew1(X_{i,new}) \end{aligned} \tag{21}$$

where $\overrightarrow{A_\alpha}$, $\overrightarrow{A_\beta}$, and $\overrightarrow{A_\delta}$ are three random vectors. It is seen from (19) that the step size of omega wolves is changed towards the location of α, β, and δ wolves. (20) and (21) are utilized to change the final position of ω-wolves w.r.t prey position.

Modification of GWO With Levy-Flight (LF)

To avoid immature convergence, levy-flight distribution has been incorporated in conventional GWO algorithm. This help GWO to redistributes the wolves nearby places of prey to confine the generation from the loss of diversity and emphasizes on the global searching process (Heidari and Pahlavani, 2017). In (Heidari and Pahlavani, 2017), three modifications have been described for improving the searching process of GWO-algorithm. These are (Heidari and Pahlavani, 2017), (i) the function of δ-wolves in the social hierarchy is taken care by the other wolves, (ii) LF-strategy is employed in hunting phase, and (iii) greedy selection strategy is also engaged in modified LGWO algorithm. LF-distribution has been used to imitate the random position of wolves in the search area. This is modeled by using (22) (Heidari and Pahlavani, 2017).

$$L(s,\gamma,\mu) = \begin{cases} \sqrt{\dfrac{\gamma}{2\pi}} \exp\left[-\dfrac{\gamma}{2(s-\mu)}\right]\left(\dfrac{1}{s-\mu}\right)^{3/2} & 0 < \mu < s < \infty \\ 0 & s \le 0 \end{cases} \tag{22}$$

where s is variable; μ is shift parameter; $\gamma(>0)$ is scalar factor. Further, the LF-distribution can be described by using Fourier series as given in (23) (Heidari and Pahlavani, 2017).

$$F(k) = \exp\left[-\alpha |k|^\beta\right], 0 < \beta \le 2; \alpha \in (-1,1) \tag{23}$$

where α is skewness and scaling factor; β is Levy-index for controlling the computation stability. With LF distribution, the position of wolves is updated by using (24) (Heidari and Pahlavani, 2017).

$$\vec{x}_{new} = \begin{cases} 0.5 \times \left(\vec{x}_\alpha - \vec{A}_1 \vec{D}_\alpha + \vec{x}_\beta - \vec{A}_2 \vec{D}_\beta\right) + rand(size(\dim)) \oplus Levi(\beta) & |A| > 0.5 \\ 0.5 \times \left(\vec{x}_\alpha - \vec{A}_1 \vec{D}_\alpha + \vec{x}_\beta - \vec{A}_2 \vec{D}_\beta\right) & |A| > 0.5 \end{cases} \tag{24}$$

where ' dim ' is the dimension of control variable. The better position of wolves in individual iteration is further improved for the use of next iteration and the worse ones are discarded by performing greedy search (GS) strategy. This operation is performed by using (25) (Heidari and Pahlavani, 2017).

$$\vec{x}(l+1)=\begin{cases}\vec{x}(l) & f\left(\vec{x}_{new}(l)\right)>f\left(\vec{x}(l)\right)\ \&\ r_{new}<p\\ \vec{x}_{new}(l) & otherwise\end{cases} \tag{25}$$

where $f\left(\vec{x}(l)\right)$ is fitness value corresponds to last position of wolves; $f\left(\vec{x}_{new}(l)\right)$ is fitness value corresponds to the position of wolves calculated by using (24); $r_{new}\ \&\ p$ are two random numbers calculated in between $(0,1)$. By integrating *LF* and GS strategies in conventional GWO algorithm, the searching capability of GWO algorithm is enhanced. This helps pioneer wolf to survive and communicate with other wolves during hunting. Flowchart of GWO with LF and GS is displayed in Fig. 3. For more details on of LGWO algorithm, readers are referred to (Heidari and Pahlavani, 2017).

Figure 3. General flowchart of GWO algorithm with LF and GS strategies

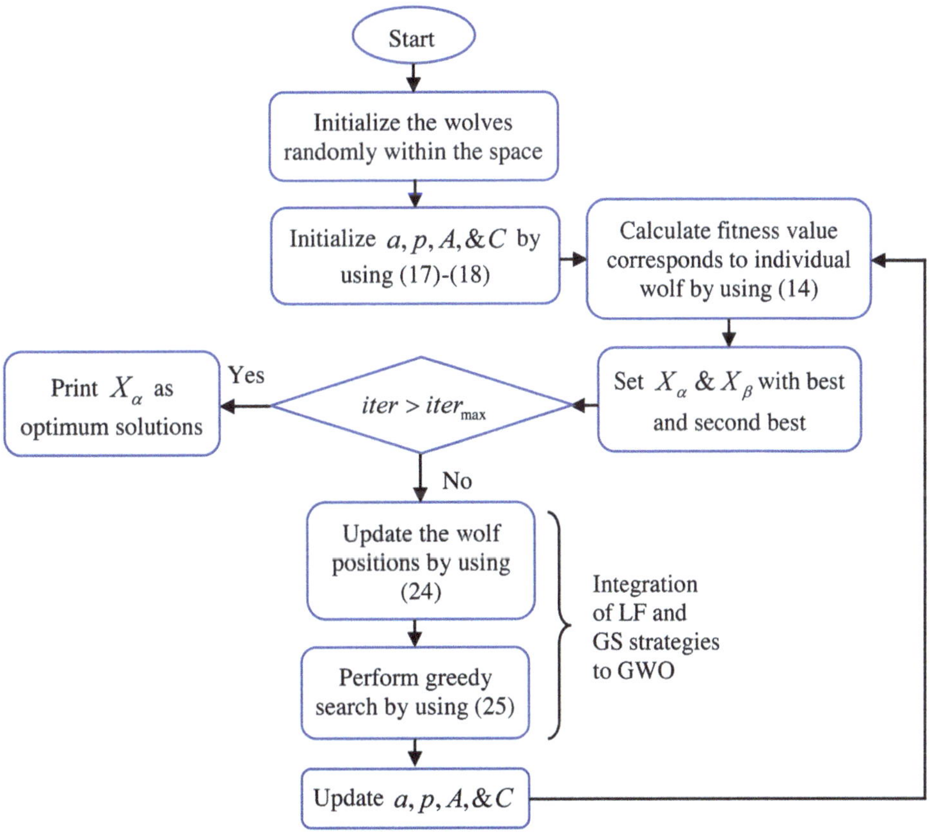

Whale Optimization Algorithm (WOA)

Motivation

Whales are defined as the largest mammals in the universe. There are seven major species of this mammal namely killer, minke, sei, humpback, right, finback and blue (Mirjalili and Lewis, 2016). It has been found through the research that whales have few identical cells in the specified section of their brain similar to human beings called spindle cells. This cells guide whales for judgment, emotions, and doing other social activities (Mirjalili and Lewis, 2016).

The key characteristic of whale is their social behavior. In WOA, independent foraging activity of humpback whales called bubble-net feeding is simulated. It is viewed that humpback whale trace an 'increasingly shrinking circle or a 9-shaped' route, as depicted from Fig. 4(a), during hunting (Mirjalili and Lewis, 2016). The 'bubble-net feeding' technique of humpback whale is mathematically designed in the ensuing sub-section.

Modeling of WOA

The hunting stage in WOA comprises three major activities, such as (i) encircling prey, (ii) spiral bubble-net feeding maneuver, and (iii) searching prey.

Encircling Prey

Initially, humpback whale recognizes the position of prey and encircling them. Since initial generation of control variables is entirely random in search area, hence, WOA consider present best solution as a location of target prey. To update the position remaining search agents w.r.t target position (26) and (27) are simulated in (Mirjalili and Lewis, 2016).

Figure 4. (a) Bubble-net feeding of humpback whale, (b) Flowchart of WOA

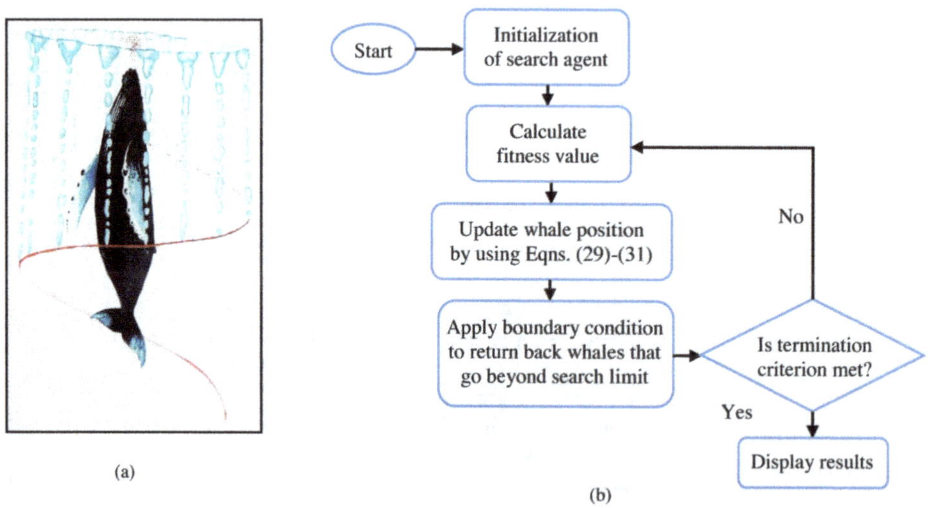

$$\vec{\delta} = \left|\vec{C}.\vec{x}^{best}(l) - \vec{x}(l)\right| \tag{26}$$

$$\vec{x}(l+1) = \vec{x}^{best}(l) - \vec{A}.\vec{\delta} \tag{27}$$

where $\vec{x}$ is present location, $\vec{x}^{best}$ is the fittest outcome computed so far, l shows the present iteration. The values of $\vec{A}$ and $\vec{C}$ are calculated employing (28) (Mirjalili and Lewis, 2016).

$$\begin{cases} \vec{A} = 2 * \vec{a} * \overrightarrow{rand} - \vec{a} \\ \vec{C} = 2 * \overrightarrow{rand} \end{cases} \tag{28}$$

where $\vec{a}$ is search direction matrix linearly decreasing from 2 to 0 with iteration. Thus, the position of the search agents w.r.t target prey for hunting is updated by using (29). In (26), $\overrightarrow{\delta_{i,new}}\left(= \left|\vec{x}^{best}(l) - \vec{x}(l)\right|\right)$ is the distance within the ith whale to prey; b is a constant.

$$\vec{x}(l+1) = \overrightarrow{\delta_{i,new}}.e^{b.rand}.\cos\left(2\pi * rand\right) + \overrightarrow{x^{best}}(l) \tag{29}$$

The whales usually swim around prey within shrinking circle and follow a spiral path while hunting, which can be simulated by using (30).

$$\vec{x}(l+1) = \begin{cases} \overrightarrow{x^{best}}(l) - \vec{A}.\vec{\delta} & if\ rand < 0.5 \\ \overrightarrow{\delta_{i,new}}.e^{b.rand}.\cos\left(2\pi * rand\right) + \overrightarrow{x^{best}}(l) & if\ rand \geq 0.5 \end{cases} \tag{30}$$

Bubble-Net Feeding

To illustrate the bubble-net feeding strategy, following two activities were presented in (Mirjalili and Lewis, 2016).

Shrinking Encircling Mechanism

This is achieved by decreasing the value of $\vec{a}$ from 2 to 0 with iteration. The fluctuations of $\vec{A}$ is also minimized by reducing $\vec{a}$.

Spiral Updating Mechanism

The variation of $\vec{A}$ within $(-1,1)$ is considered for exploitation. Humpback whale randomly searches prey in the line of position of other whales. The position of whales is randomly modified w.r.t chosen whale instead of best whale derived so far. Mathematically, it is defined as in (31).

$$\begin{cases} \vec{\delta} = \left| \vec{C} \,.\, \vec{x}_{rand}(l) - \vec{x}(l) \right| \\ \vec{x}(l+1) = \vec{x}_{rand}(l) - \vec{A} \,.\, \vec{\delta} \end{cases} \tag{31}$$

where $\vec{x}_{rand}$ is the random location of whales. The flowchart of WOA is shown in Fig. 4(b).

Search Group Algorithm (SGA)

Search group algorithm (SGA) is a relatively new population-based evolutionary algorithm derived by Howlader *et al.* in 2015 (Howlader et al., 2015). The basic idea of SGA is to hold a natural balance among the exploration and exploitation strategies while discovering the global optimum solution. At first iteration, SGA attempts to explore the most promising search area within the defined boundary, and then SGA refines the generated solutions as iterations processed. The operation of SGA can be explained by dividing its operations into five steps, such as initialization, selection of search group, mutation, creation of family in each search group, and selection of new search group (Howlader et al., 2015).

Initialization

Like other evolutionary algorithms (EAs), SGA initializes the population randomly within the defined search area. To generate initial population, (32) is employed (Howlader et al., 2015).

$$\begin{aligned} & for\, k = 1 : n_p \\ & \quad for\, m = 1 : \mathrm{d} \\ & \qquad P_{km} = x_m^{up} + rand * \left(x_m^{up} - x_m^{lb} \right); \\ & \quad end \\ & end \end{aligned} \tag{32}$$

where n_p is population size; 'd' indicates the dimension of control parameters; x_m^{up} & x_m^{lb} are upper and lower bounds of mth control variable; P_{km} is population of mth control variable in k^{th} iteration.

Selection of Search Group

After initialization, the fitness value of individual generation has been calculated and best solutions among them are marked as elite solutions. Based on the solutions, a search group 'S' is constituted by following a standard tournament process (Howlader et al., 2015).

Mutation

To improve the global search ability, mutation phase is performed by using (33) (Howlader et al., 2015).

$$x_m^{mut} = E\left[S_{:,m}\right] + t\varepsilon\sigma\left[S_{:,m}\right] \tag{33}$$

where x_m^{mut} indicates m^{th} design variable of a given mutated individual; $E\ \&\ \sigma$ are mean and standard deviation, respectively; ε indicates the mutation variable; t controls how far a mutated individual is generated from the mean value of the population; $S_{:,m}$ is m^{th} column of search group matrix.

Generation of Families in Each Search Group

The families are set including members of search group and the individual generated. Each members of the generated search family is perturbed by using (34) (Howlader et al., 2015).

$$x_m^{new} = S_{km} + \alpha\varepsilon \tag{34}$$

where α is control size of perturbation. It helps SGA to control the exploration phase. The value of α is updated by using (35) (Howlader et al., 2015).

$$\alpha^{l+1} = b\alpha^{l} \tag{35}$$

where b is parameter of SGA; α^{l} controls the distance that a new individual is produced from its search group member.

Choice of New Search Group

SGA is comprised of both global and local stages. In the first iteration of $iter_{\max}$, SGA explore most of the promising search space so as to form a search group including members of the individual family. When the current iteration is greater than $iter_{\max}$, SGA forms a search group with best solutions taking form all families. The latter phase is known as 'local phase'. In this mode SGA exploits the area of the current optimization space. SGA makes it different from other algorithms as follows.

- The proposed optimization technique is derived by computing the mean and standard deviation of the location of current search group members in a given iteration.

- The algorithm involves local and global phases to update the members in a search group.

The proposed SGA requires prior initialization of some algorithm specific control parameters, these are: α^l (controls both exploration and exploitation phases) $\in (0.75, 2.25)$; b indicates the process that α^l reduces with iteration; n_{mut} (number of mutations at individual generation): 3% of the population; n_g (number of search group members): 20% of the population. The flowchart of SGA is depicted in Fig. 5.

Figure 5. Flowchart of SGA

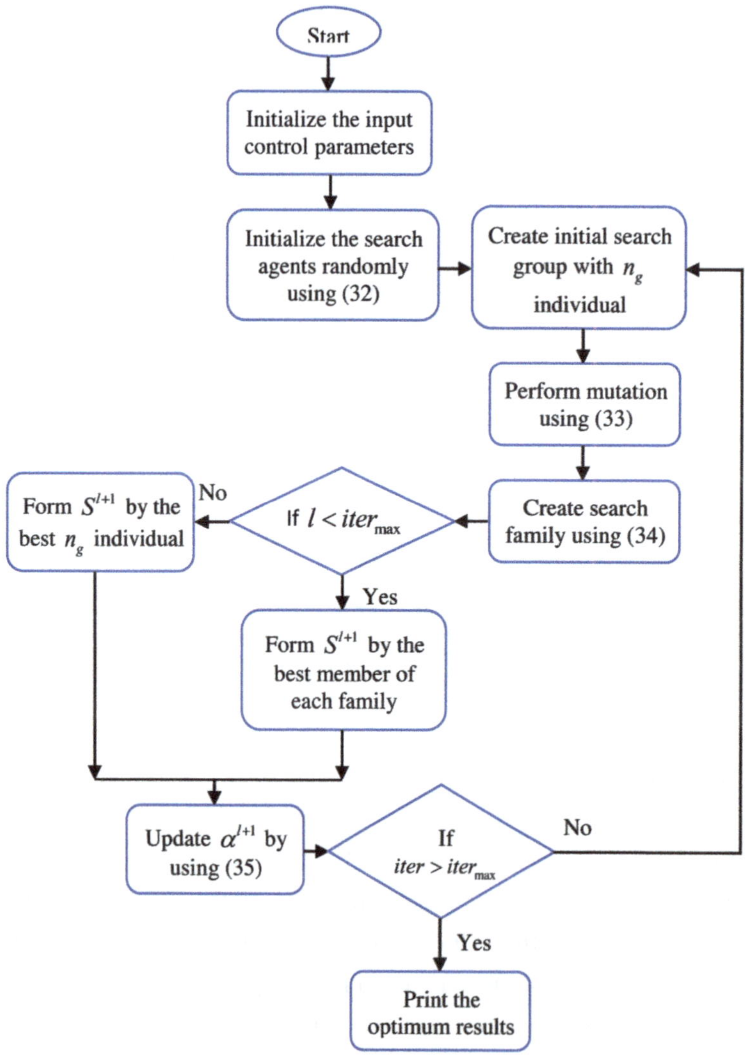

Drone Squadron Optimization (DSO)

Inspiration

The DSO algorithm has been derived by simulating artifact self-adaptive behavior of drone while exploring the landscape in a given region. Unlike other aforesaid computational algorithms, DSO is not inspired by natural phenomenon. The algorithmic operation of DSO can be simulated by following two core aspects (Melo and Banzhaf, 2017).

1. The semiautonomous drones, which moves over the landscape to explore (formation of the team)
2. The command center, which processed the retrieved data and modifies the drones' firmware (calculation lowest fitness value).

The initial generation in DSO-algorithm is saved into an array with a dimension of $\left(n_p \times \dim\right)$ called coordinates. Here, n_p & $\dim$ are population size and dimension of the control variable, respectively. The DSO-algorithm is mainly comprised of (i) drone squadron with the diverse team and (ii) a command center to control the drone position while exploring and computing firmware for controlling drone locations. In DSO, the firmware is distinct rules or configurations set by the users to evolve the population.

Mathematical Model of DSO

Command Center

This is the most important and intelligent phase in DSO algorithm. Command center generates and provides the order to the drones for scanning the landscape. In DSO, scanning landscape is equivalent to calculation of fitness function. The command center modifies the firmware (distinct mechanism employed for evolving the search agents) to update the drone positions (Melo and Banzhaf, 2017).

The Firmware

To modify the position of search agents, firmware is updated in DSO with the help of (36) (Melo and Banzhaf, 2017).

$$\begin{cases} P = departure + offset(); \\ TC = calculate\left(P\right); \end{cases} \tag{36}$$

where P is the perturbation model calculated by using (37); $diparture$ is the fittest location in the search area; $offset$ indicates a function that returns the actual perturbation movement; TC is trial coordinates. The '$departure$ coordinate' and '$offset$ function' are calculated by using (37) (Melo and Banzhaf, 2017).

$$\begin{cases} P_1 = \overrightarrow{GBC} + \left(c_1 \times \left(\overrightarrow{GBC} - \overrightarrow{CBC_{drone}}\right)\right) \\ P_2 = \overrightarrow{CBC_{drone}} + \left(G(0,1) \times \left(\sqrt{U(0,1)_{\dim}} + \overrightarrow{CBC_{drone}}\right)\right) \end{cases} \tag{37}$$

where GBC & CBC are global best coordinate and current best coordinates, respectively; c_1 is user defined function; $G(0,1) \& U(0,1)$ are Gaussian and uniform distribution functions, respectively. In (37), the 1st part in either equation is representing '$departure$' and second part is '$offset$'.

Drone Movement

The drone uses an independent system to compute the target location, shift towards them, and gather messages which is to be delivered to command center. In this step, DSO generates the target position of individual drone in each team called 'team coordinator' $\left(TmC_{team,drone}\right)$. After individual drone go through the perturbation stage, it produces trial coordinates $\left(TC_{drone}\right)$. After this, recombination takes place with the better coordinates derived so far for generating TmC or no recombination. Finally the drone is allows to move within a particular perimeter. The violation of boundary condition can determine by using (38) (Melo and Banzhaf, 2017). In (38), 'ub & lb' are the upper and lower limits of control variable, in order.

$$violation_{team} = \sum_{i=1}^{n_p} \sum_{j=1}^{\dim} \begin{cases} \left|TmC_{team,drone,j} - ub_j\right| \\ \left|lb_j - TmC_{team,drone,j}\right| \end{cases} \tag{38}$$

Firmware Update

The command center measure the quality of team by taking (i) the fitness function value, and (ii) the degree of out-of-bound coordinates. To update the firmware, (39) is employed with *violation* (Melo and Banzhaf, 2017).

$$\begin{aligned} & for\, i = 1 : t\,(no.of\ team) \\ & \quad TeamQuality = Rank_i + violation_i; \\ & end \end{aligned} \tag{39}$$

Local Stagnation Detection

The drone may generate the same landscape while exploring a specific region. In (40), convergence avoidance mechanism has been performed to avoid local stagnation. For more exploration in DSO to detect local stagnation, (40) is employed (Melo and Banzhaf, 2017).

$$\begin{aligned}&if\\&\quad f\left(TmOFV_{drone,best}\right) < f\left(CBOFV_{drone}\right) or\, U\left(0,1\right) < P_{aw}\\&\quad CBC_{drone} \xleftarrow{update} TmC_{drone,best}\\&end\end{aligned} \tag{40}$$

where $U\left(0,1\right)$ is randomly generated number within $(0,1)$ through uniform distribution; P_{aw} is probability of acceptance of worst output. The flowchart of DSO-algorithm is depicted in Fig. 6. DSO-algorithm requires few user defined control variables for its operation. To demonstrate the performance of DSO algorithm, the pseudo codes correspond to *team* are given in Algorithm-1.

Algorithm-1: Performance by the team

$$\begin{aligned}&for\, t = 1 : n_p\\&\quad for\, drone = 1 : \dim\\&\qquad \begin{cases} P = departure + offset(); \\ TC = calculate\left(P\right); \end{cases}\\&\qquad \mathrm{Re}\, combine\, TC\, \&CBC\, to\, form\, TmC_{team,drone}\\&\qquad violation_{team} = \sum_{i=1}^{n_p}\sum_{j=1}^{\dim}\left\{\begin{vmatrix} TmC_{team,drone,j} - \max_j \\ min_j - TmC_{team,drone,j} \end{vmatrix}\right.\\&\qquad TmOFV_{team,drone} \xleftarrow{update} fitness\, value\\&\quad end\\&end\end{aligned}$$

THYRISTOR CONTROL SERIES COMPENSATOR (TCSC)

TCSC is a capacitive reactance compensator that constitutes of series capacitor bank shunted with a thyristor controlled reactor (TCR) to offer smoother changeable capacitive reactance. The schematic diagram of TCSC is shown in Fig. 7(a) (Mondal et al., 2014). The series reactance is regulated by changing the firing angle of TCR. TCSC is worked as a promising series FACTS controller and provides sufficient damping to the power system oscillations to enrich the power transfer capacity through tie-line. TCSC is employed in transmission line to serve the following functions.

1. It is a cheaper FACTS device to deliver MW power capacity vie tie-line.
2. When TCSC is connected with transmission line and engaged with a power oscillation damper (POD), the inter-area oscillation is easily die out.
3. Allow fast and continual variation in the transmission line impedances.
4. Employed in lengthy transmission line to enhance the power transfer capacity by lowering the bus admittances.

Figure 6. Flowchart of DSO algorithm

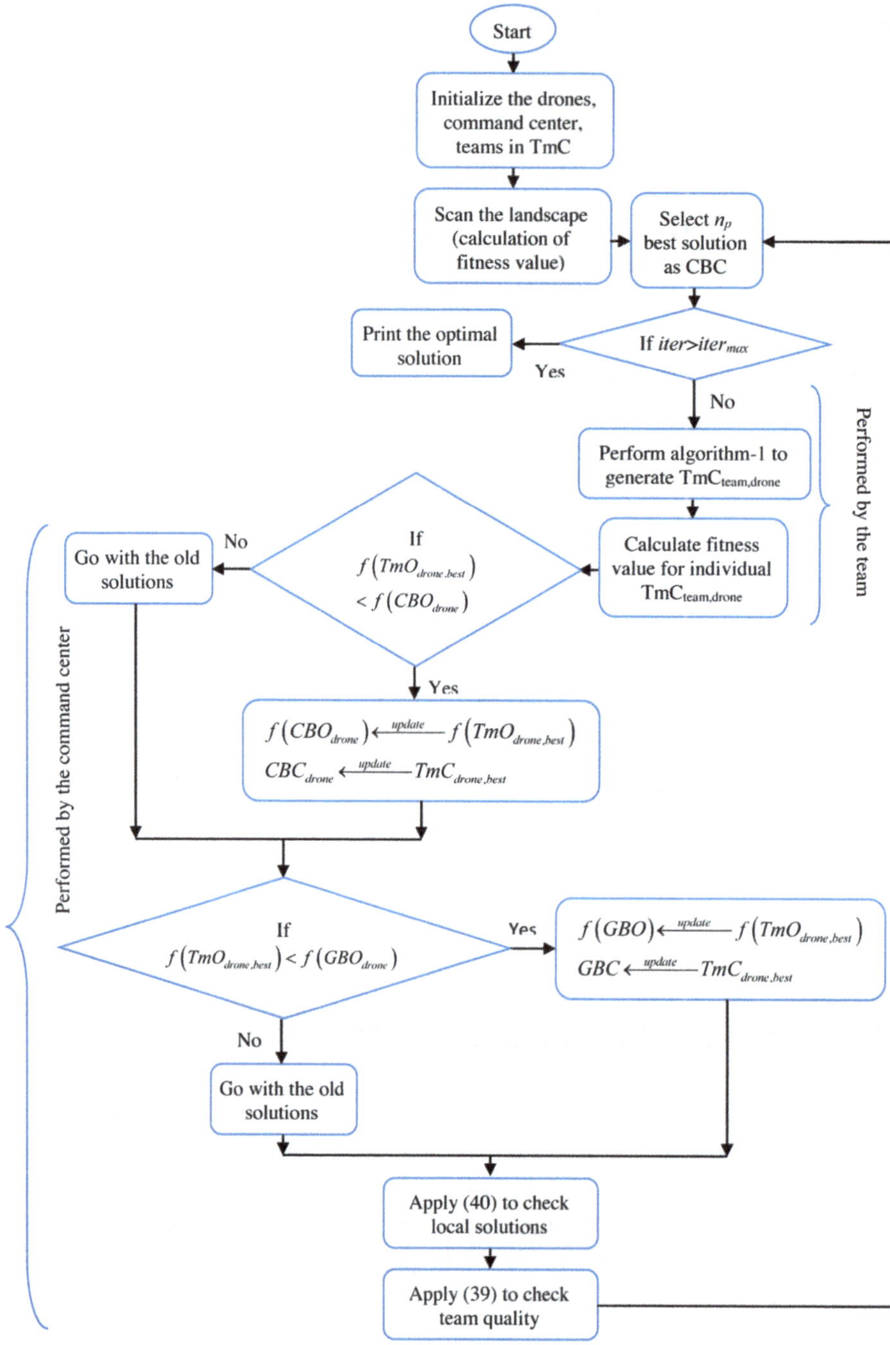

Taylor series expansion is employed to model TCSC controller for the present study (Zare et al., 2015; Guha et al., 2016b). The active power flow via the tie-line with TCSC can define by (41).

$$P_{12} = \frac{|V_s||V_r|}{X_{12} - X_{TCSC}} \sin\delta_{12} = \frac{|V_s||V_r|}{X_{12}\left(1 - \frac{X_{TCSC}}{X_{12}}\right)} \sin\delta_{12} \tag{41}$$

where X_{12} and X_{TCSC} are the reactance's of tie-line and TCSC, respectively, $|V_s|$ and $|V_r|$ are the voltage magnitude of area-1 and area-2, in order, δ_1 and δ_2 are the voltage angle of area-1 and area-2, in order.

Let's assume that $K_c = \frac{X_{TCSC}}{X_{12}}$, known as compensation ratio, then (41) can be rewritten as

$$P_{12} = \frac{|V_1||V_2|}{X_{12}} \sin\delta_{12} + \frac{K_c}{1 - K_c} \frac{|V_1||V_2|}{X_{12}} \sin\delta_{12} \tag{42}$$

In (42), the tie-line power is partitioned into two parts to calculate the performance of TCSC in tie-line power flow individually (Zare et al., 2015; Guha et al., 2016b). For a small variation of δ_1, δ_2, K_c from their nominal results δ_1^0, δ_2^0, K_c^0, the incremental change in tie-line power flow is found by (43).

$$\Delta P_{12} = \frac{|V_1||V_2|}{X_{12}} \cos\delta_{12}^0 \sin\left(\Delta\delta_{12}\right) + \frac{1}{\left(1 - K_c^0\right)^2} \frac{|V_1||V_2|}{X_{12}} \sin\delta_{12}^0 \Delta K_c \tag{43}$$

For a small load perturbation, $\sin\left(\Delta\delta_{12}\right) \approx \Delta\delta_{12}$, and therefore,

$$\Delta P_{12} = \frac{|V_1||V_2|}{X_{12}} \cos\delta_{12}^0 \left(\Delta\delta_{12}\right) + \frac{1}{\left(1 - K_c^0\right)^2} \frac{|V_1||V_2|}{X_{12}} \sin\delta_{12}^0 \Delta K_c \tag{44}$$

Let's assume that

$$T_{12} = \frac{|V_1||V_2|}{X_{12}} \cos\delta_{12}^0 \tag{45}$$

$$c = \frac{1}{\left(1 - K_c^0\right)^2} \frac{|V_1||V_2|}{X_{12}} \sin\delta_{12}^0 \tag{46}$$

Consequently, (44) reduced to

$$\Delta P_{12} = T_{12}\Delta\delta_{12} + c\Delta K_c \tag{47}$$

Since

$$\begin{cases} \Delta\delta_1 = 2\pi\int \Delta f_1 dt \\ \Delta\delta_2 = 2\pi\int \Delta f_2 dt \end{cases} \tag{48}$$

Taking Laplace transform of (47) leads to

$$\begin{aligned} &\Delta P_{12}(s) = \frac{2\pi T_{12}}{s}\left(\Delta f_1(s) - \Delta f_2(s)\right) + c\Delta K_c(s) \\ &\Rightarrow \Delta P_{12}(s) = \Delta P_{12}^0(s) + \Delta P_{TCSC}(s) \end{aligned} \tag{49}$$

where $\Delta P_{12}^{\ 0}(s) = \frac{2\pi T_{12}}{s}\left(\Delta f_1(s) - \Delta f_2(s)\right)$ and $\Delta P_{TCSC}(s) = c\Delta K_c(s)$. It is noteworthy from (49) that the tie-line power flow is controlled by regulating $\Delta K_c(s)$ term. The term $\Delta P_{TCSC}(s)$ shows the consequences of TCSC in tie-line power sharing. By Taylor series expansion, the term $\Delta P_{TCSC}(s)$ is modeled as in (50) (Zare et al., 2015; Guha et al., 2016b).

$$\Delta P_{TCSC}(s) = \Delta K_c + \Delta K_c^{\ 2} + \Delta K_c^{\ 3} + \Delta K_c^{\ 4} + \Delta K_c^{\ 5} + - - \tag{50}$$

where

$$\Delta K_c = \frac{K_{TCSC}}{1 + sT_{TCSC}} \frac{1 + sT_{1,TCSC}}{1 + sT_{2,TCSC}} \frac{1 + sT_{3,TCSC}}{1 + sT_{4,TCSC}} \Delta Error \tag{51}$$

The T.F structure of TCSC controller is illustrated in (51). In (51), $\Delta Error$, input to TCSC, is the frequency error of the respective area and control output is the thyristor conduction angle $\left(\Delta X_{TCSC}\right)$. The linear approximated model of TCSC is portrayed in Fig. 7(b). In Fig. 7(b), the term σ_0 shows the initial conduction angle of TCR. The washout block basically represents a high pass filter with a time constant T_w. The washout block prevents the steady changes in speed with the steady change of frequency. It is reported in (Mondal et al., 2014) that the value of T_w is not so critical from the viewpoint of small-signal analysis and therefore, it is usually chosen between 1 to 20 sec. In the present study, T_w is considered as 10 sec. The two-stage lag-lead compensation block provides adequate phase lead between input and output to compensate the phase lag. The terms K_{TCSC} and T_{TCSC} in Fig. 7(b) are defined as the gain and time delay of TCSC, respectively.

Figure 7. (a) Schematic diagram of TCSC, (b) Transfer function model of TCSC as a frequency stabilizer

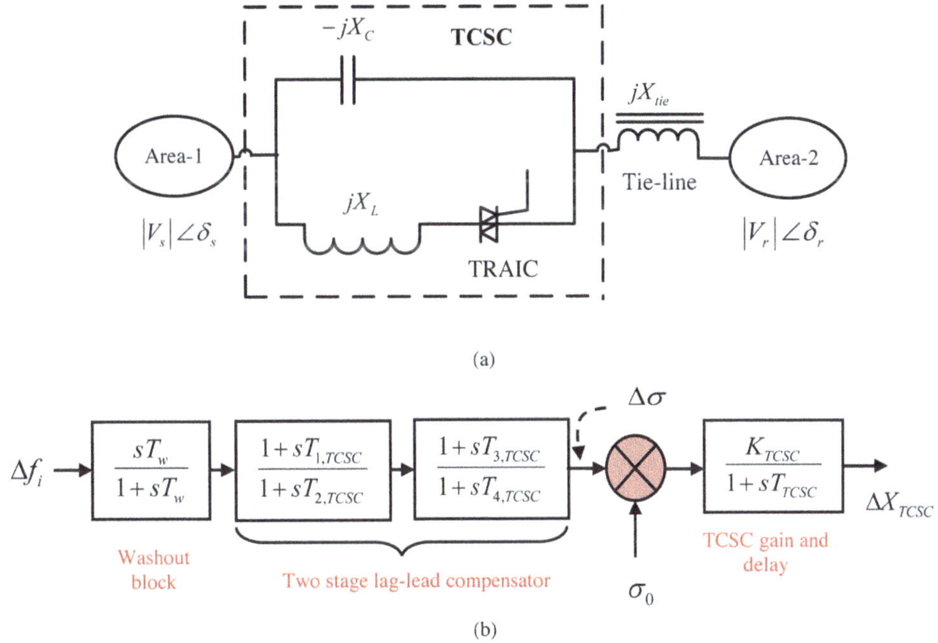

RESULTS AND COMPARATIVE STUDY

The efficacy of the computation techniques and designed control strategies under the action of wind and load perturbations has been closely studied in this section. The model of projected wind perturbation is shown in Fig. 8(a). The present study has been performed in four different phases and these are listed as

1. Initially the performance of test system is examined under the action of PI-controller.
2. Subsequently, an optimized PIDF-controller is included in the system to speed up the performance.
3. To overcome the shortcomings of PIDF-controller and to accelerate the 'speed of response' of the modeled WDG system, PIDF-controller is replaced by the CC-TID controller.
4. Finally, TCSC is coordinated with CC-TID controller for further enhancement of stability margin.

The controller parameters are optimally selected by implementing LGWO, WOA, SGA and DSO algorithms. The simulations were carried out on an Intel Core i5 processor with 4GB RAM, 2.8 GHz MATLAB 2013a platform. The simulink model of Fig. 1(c) is constructed in SIMULINK environment and optimization codes of computation algorithm are separately written in *.m-file*. The defined problem is designed as a constraint minimization problem bounded by the controller gains. It is modeled as follows

Minimize or maximize the selected objective function value (J)

Subjected to:

Figure 8. Dynamic behavior of coordinated WDG system considering optimum controller gains of Table 2 (a) profile of wind power perturbation, (b) WTG frequency error, (c) WTG output power deviation, (d) DG output power deviation

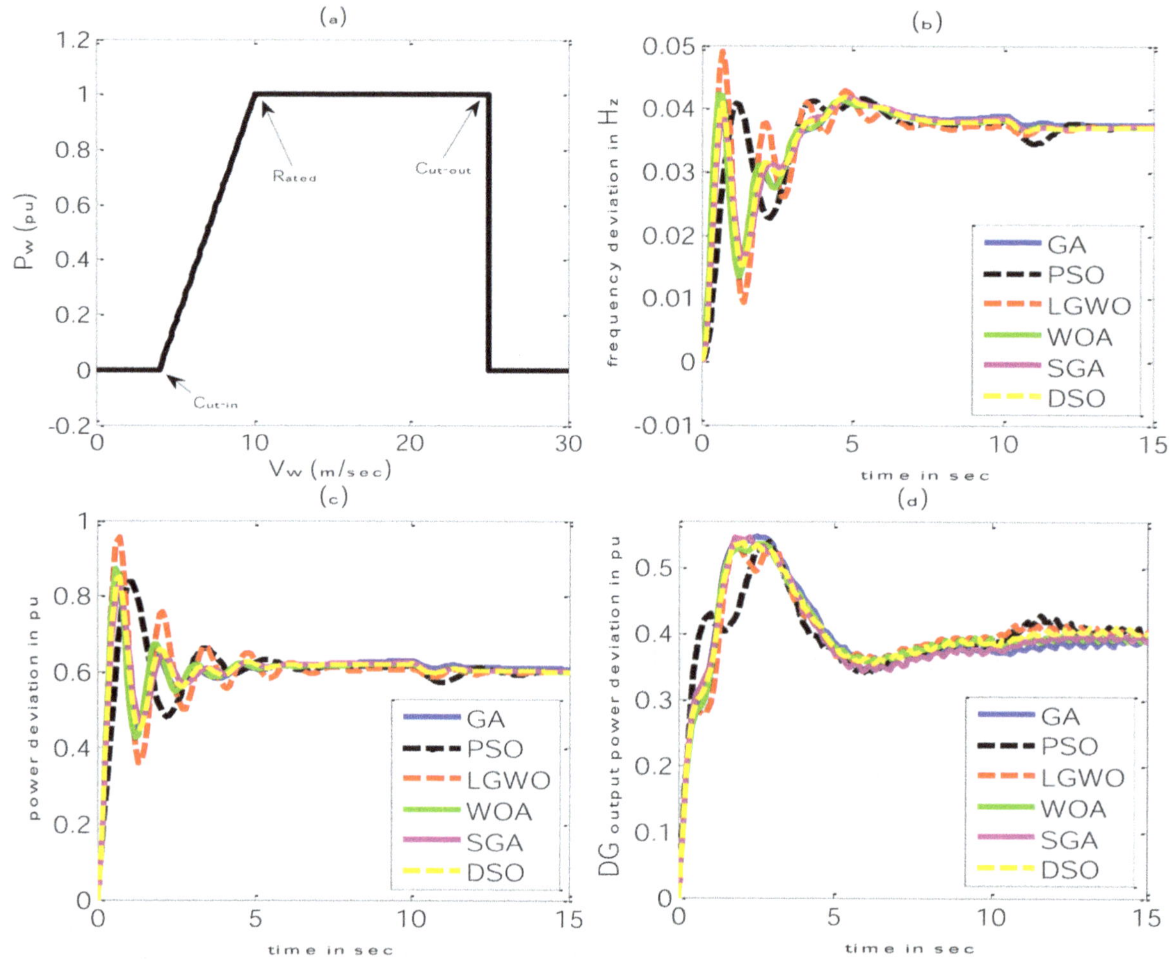

$$\left.\begin{array}{l} k_{p,lb} \le k_p \le k_{p,ub} \\ k_{i,lb} \le k_i \le k_{i,ub} \\ k_{d,lb} \le k_d \le k_{d,ub} \\ N_{lb} \le N \le N_{ub} \\ n_{lb} \le n \le n_{ub} \end{array}\right\} \left.\begin{array}{l} K_{TCSC,lb} \le K_{TCSC} \le K_{TCSC,ub} \\ T_{TCSC,lb} \le T_{TCSC} \le T_{TCSC,ub} \\ T_{1,lb} \le T_1 \le T_{1,ub} \\ T_{2,lb} \le T_2 \le T_{2,ub} \\ T_{3,lb} \le T_3 \le T_{3,ub} \\ T_{4,lb} \le T_4 \le T_{4,ub} \end{array}\right\} \tag{52}$$

where $k_{pid,lb}$ & $k_{pid,ub}$ are the lower and upper bounds of PID-controller parameters, in order; N_{ub} and N_{lb} are the upper and lower bounds of LPF constants, in order; n_{ub} and n_{lb} are the upper and lower bounds of tilt parameter, in order; $K_{TCSC,ub}$ and $K_{TCSC,lb}$ are the upper and lower bounds of TCSC gain, respectively; $T_{TCSC,ub}$ and $T_{TCSC,lb}$ are the upper and lower bounds of TCSC time constant, in order; $T_{1,2,3,4,lb}$ and $T_{1,2,3,4,ub}$ are the upper and lower bounds of the time constant of lag-lead compensator, in order.

Transient Analysis of WDG System

With PI-Controller

To assess the performance of WDG system, preliminary, the test system is conferred with a PI-controller and controller gains are selected employing LGWO, WOA, SGA, and DSO algorithms. To ensure supremacy, the performance of the designed PI-controller has been compared to the results of GA and PSO optimized PI-controller (Gampa and Das, 2015). The optimum gains of PI-controller through the maximization of J_1 and minimization of J_2 are presented in Tables 2 and 3, respectively, with the results offered by (Gampa and Das, 2015). Tables 2 and 3 illustrate the Eigen values of WDG system with PI-controller. A close observation of the fitness value of Table 2 shows that the damping factor calculated with the DSO algorithm is relatively higher than the values offered by the other algorithms as listed in Table 2. As opposed to, observation of the Table 3 reveals that the DSO algorithm provides least minimum value of J_2 compared to LGWO, WOA, SGA, GA, and PSO algorithms, which is again proven from Fig. 9(a). Fig. 9(a) demonstrates that DSO quickly attains the optimal global value and takes less iteration to reach the optimum level. Thus it may conclude that the DSO algorithm provides more satisfactory controller settings compared to other algorithms.

The dynamic responses of the test system are obtained by considering the optimum controller gains of Tables 2 and 3 and shown in Figs. 8 and 9, respectively. The settling time and peak overshoot of the system oscillation are calculated from Figs. 8 and 9 and summarized in Table 4. In Figs. 8-9, the results offered by LGWO: PI controller is suffering from large peak overshoot and takes more time to attain the final value. Moreover, DSO: PI controller efficiently attenuates frequency and power oscillation. The damping profile of the test system has been satisfactorily enhanced with the inclusion of DSO: PI controller. From this analysis, it may conclude that DSO: PI controller is best among the other controllers as shown in Figs. 8-9. This comparative analysis further helps to find that the DSO algorithm has better tuning capability than other proposed algorithms.

With PIDF-Controller

It is noticed from the results illustrated in the previous section that the system oscillation takes a long time to attain the steady-state value. Thus to increase the speed of system response, a PIDF-controller is provided to the WDG system. The computational ability of the proposed optimization algorithms is examined by an extensive comparative study with GA and PSO based controllers (Gampa and Das, 2015). The gain parameters of the PIDF-controller are selected employing LGWO, WOA, SGA, and DSO algorithms, and after optimization, results are presented in Tables 5-6 and for the fitness functions J_1 & J_2, respectively. It is elicited from Table 5 that the value of J_1 is increased with DSO: PIDF controller that exhibits further improvement of the degree of stability. The value of with DSO: PIDF controller is raised by 11.74% (SGA), 11.53% (WOA), 11.93% (LGWO), 17.18% (PSO), and 18.08% (GA).

Figure 9. Dynamic behavior of coordinated WDG system considering optimum controller gains of Table 3 (a) Convergence profile, (b) WTG frequency error, (c) WTG output power deviation, (d) DG output power deviation

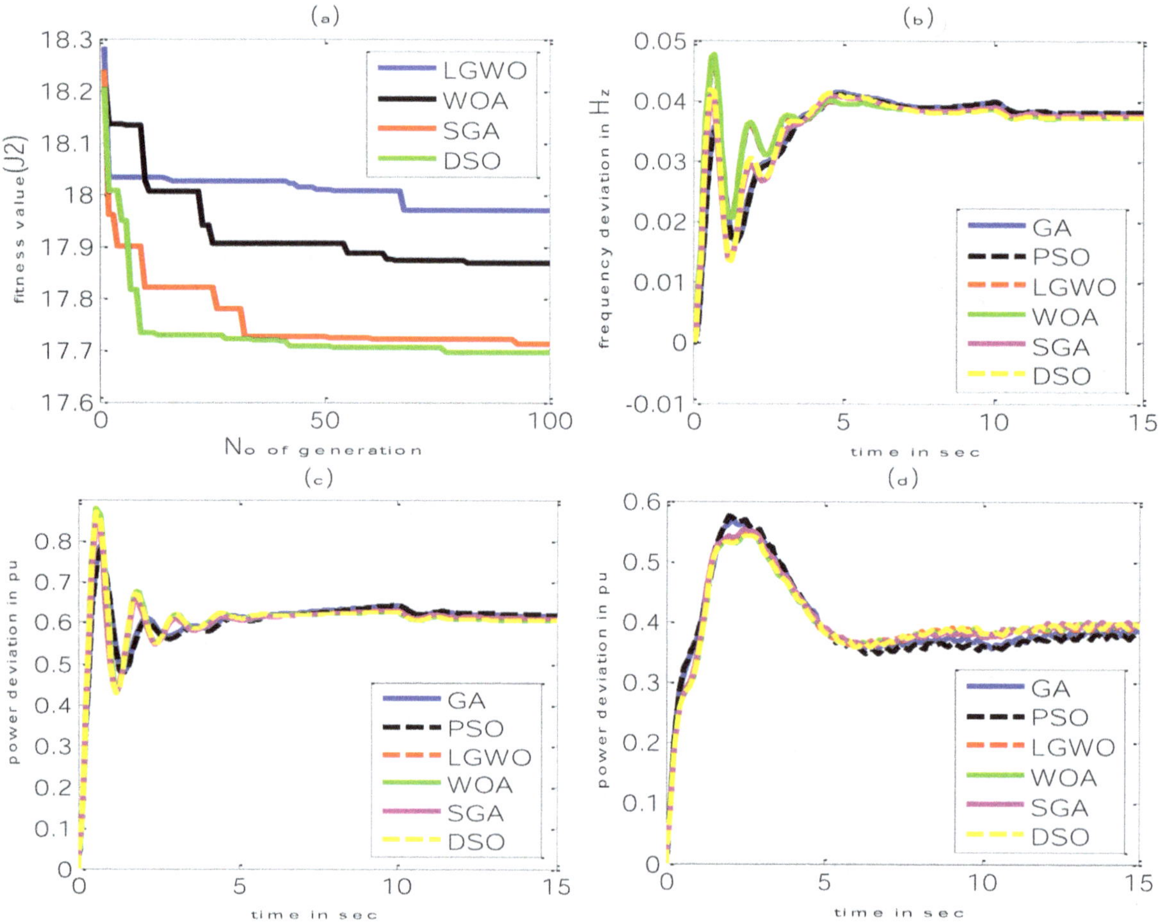

Table 2. Optimum controller gains computed via minimization of J_1

<table>
<tr><th rowspan="2">Computation Techniques</th><th rowspan="2">k_p</th><th rowspan="2">k_i</th><th rowspan="2">J_1</th><th colspan="6">Eigen Values</th></tr>
<tr><th>GA</th><th>PSO</th><th>LGWO</th><th>WOA</th><th>SGA</th><th>DSO</th></tr>
<tr><td>GA</td><td>51.18</td><td>74.12</td><td>0.4959</td><td rowspan="6">-39.028
-24.8216
-2.1863
-1.3467
-0.624±j2.6563
-0.496±j0.8912</td><td rowspan="6">-39.0281
-24.8195
-0.618±j2.649
-2.1953
-1.3512
-0.496±j0.981</td><td rowspan="6">-39.03
-25.42
-0.61±j4.59
-0.48±j0.88
-1.79
-1.20
-1</td><td rowspan="6">-39.03
-25.6
-1±j5.05
-1.70
-0.47±j0.88
-0.35
-1</td><td rowspan="6">-39.03
-25.45
-1.04±j4.68
-1.71
-0.47±j0.88
-0.4
-1</td><td rowspan="6">-39.03
-25.45
-1.04±j4.68
-1.71
-0.47±j0.88
-0.40
-1</td></tr>
<tr><td>PSO</td><td>50.93</td><td>74.23</td><td>0.4960</td></tr>
<tr><td>LGWO</td><td>126.789</td><td>52.575</td><td>0.5012</td></tr>
<tr><td>WOA</td><td>146.83</td><td>52.633</td><td>0.5071</td></tr>
<tr><td>SGA</td><td>127.585</td><td>52.713</td><td>0.5082</td></tr>
<tr><td>DSO</td><td>127.604</td><td>52.743</td><td>0.5119</td></tr>
</table>

Boldface shows best results

Table 3. Optimum controller gains computed via minimization of J_2

<table>
<tr><th rowspan="2">Computation Techniques</th><th rowspan="2">k_p</th><th rowspan="2">k_i</th><th rowspan="2">J_2</th><th colspan="6">Eigen Values</th></tr>
<tr><th>GA</th><th>PSO</th><th>LGWO</th><th>WOA</th><th>SGA</th><th>DSO</th></tr>
<tr><td>GA</td><td>140.20</td><td>25.29</td><td>20.52</td><td rowspan="6">-39.0275
-25.5599
-1.115±j4.96
-1.7011
-0.465±j0.877
-0.1741</td><td rowspan="6">-39.0281
-24.8216
-2.1863
-1.3467
-0.624±j2.6563
-0.496±j0.8912</td><td rowspan="6">-39.03
-25.63
-1.06±j5.13
-1.70
-0.47±j0.88
-0.28
-1</td><td rowspan="6">-39.03
-25.63
-1.03±j5.12
-1.70
-0.47±j0.88
-0.28
-1</td><td rowspan="6">-39.03
-25.63
-1.09±j5.14
-1.70
-0.47±j0.88
-0.15
-1.1563</td><td rowspan="6">-39.03
-25.63
-1.03±j5.12
-1.70
-0.47±j0.88
-0.27
-1.4263</td></tr>
<tr><td>PSO</td><td>139.74</td><td>11.21</td><td>20.14</td></tr>
<tr><td>LGWO</td><td>149.87</td><td>34.27</td><td>17.968</td></tr>
<tr><td>WOA</td><td>149.78</td><td>42.658</td><td>17.869</td></tr>
<tr><td>SGA</td><td>149.74</td><td>22.66</td><td>17.711</td></tr>
<tr><td>DSO</td><td>149.658</td><td>42.27</td><td>17.695</td></tr>
</table>

Boldface shows best results

Table 4. Transient specifications of coordinated WTG system with PI-controller

Control Algorithms	For J_1						For J_2					
	$\Delta\omega_1$		ΔP_{WTG}		ΔP_{DG}		$\Delta\omega_1$		ΔP_{WTG}		ΔP_{DG}	
	OS	ST	OS	ST	OS	ST	OS	ST	OS	ST	OS	ST
GA	NA	NA	NA	NA	NA	NA	NA	NA	NA	NA	NA	NA
PSO	0.017	16.92	0.224	15	0.175	17.66	0.005	17.76	0.249	15	0.143	17.06
LGWO	0.0318	10.96	0.5939	10.76	0.3548	24.33	0.0267	10.35	0.4452	11.47	0.4645	18.72
WOA	0.0138	10.72	0.4547	**10.26**	0.3723	10.67	0.0284	**10.29**	0.4589	11.39	0.4227	24.89
SGA	0.0129	10.30	0.3826	10.34	0.3663	24.72	0.0989	11.34	0.4111	12.61	0.3894	24.88
DSO	0.0127	**10.23**	0.4152	**10.26**	0.3447	**24.65**	0.0129	**10.29**	0.4449	**11.32**	0.3624	**17.75**

Boldfaces show best outputs

Similarly, a significant improvement of the value of J_2 is noticed in Table 6. It is remarkable that DSO algorithm provides least minimum fitness value compared to LGWO, WOA, SGA, GA, and PSO, which is further affirmed from the convergence characteristic (Fig. 11(a)). Thus, from the above discussion, it may conclude that the tuning competence of DSO algorithm is relatively high in comparison to LGWO, WOA, SGA, PSO, and GA.

Fig. 10 shows the variation of frequency of WTG, output power of WTG, and output power of DG considering the optimum controller gains of Table 5. Likewise, the deviation of WTG frequency, the output power of WTG, and output power of DG considering the optimum controller settings of Table 6 are displayed in Fig. 11. The transient specifications like maximum overshoot and setting time of system oscillation are computed from Figs. 10-11 and offered in Table 7. Figs. 10 and 11 illustrate that GA: PID controller offers output with high peak overshoot and more undue oscillations. As opposed to, DSO: PIDF controller provides better results in term of small peak overshoot and least settling time. However, the control signal generated from the designed controllers, as shown in Figs. 10 and 11, is nearly same. Hence it may conclude that DSO: PIDF controller outperforms other controllers as marked in Figs. 10-11. Since DSO provides the best results in comparison to other proposed algorithms, hence the rest of the study is performed with DSO-algorithm.

Figure 10. Dynamic response of coordinated WDG system considering optimum controller gains of Table 5 (a) profile of wind power, (b) WTG frequency error, (c) WTG output power deviation, (d) DG output power deviation

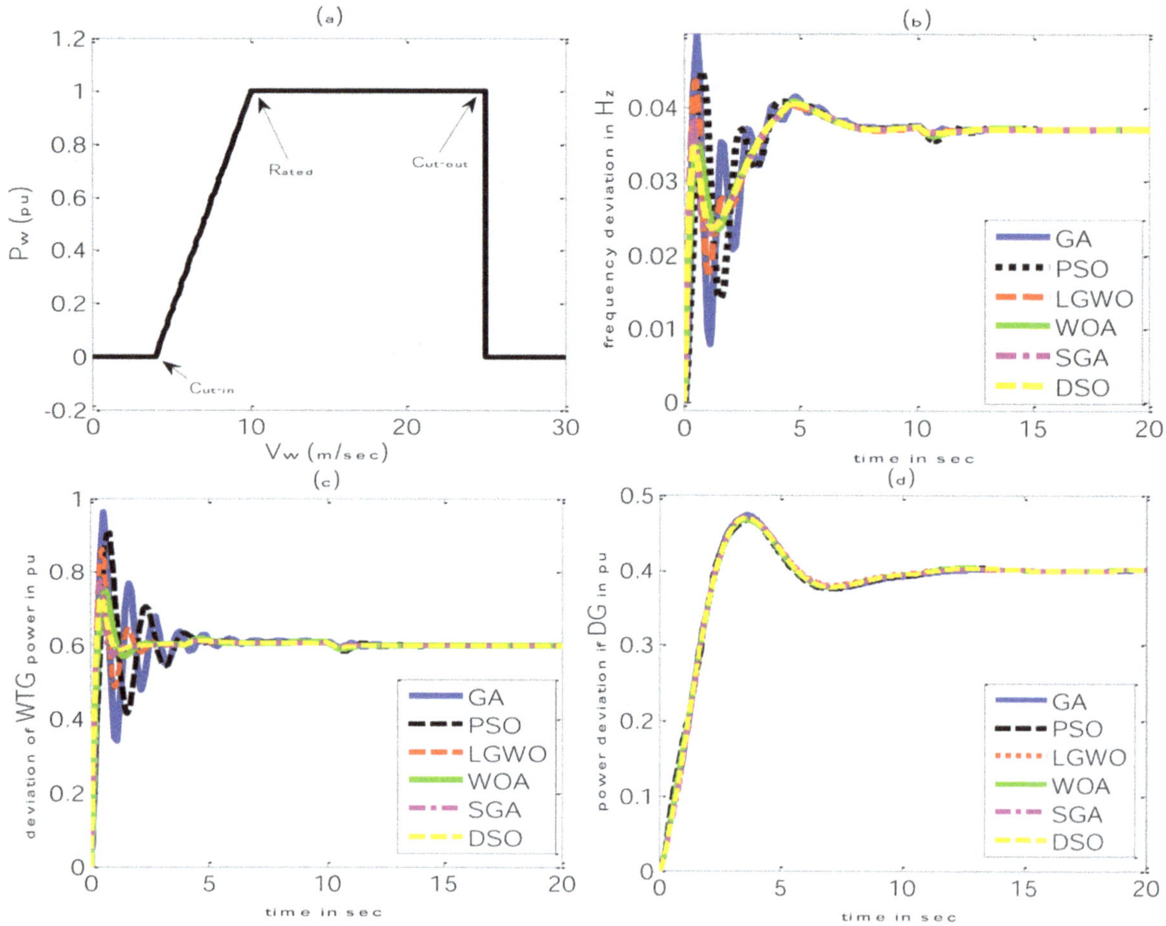

With CC-TID Controller

Under this head, small-signal stability of the modeled WDG system is evaluated providing CC-TID controller. To develop the CC-TID controller, TID-controller and PI-controller are used as slave and master controller, respectively. The DSO-algorithm is applied to tune the CC-TID gains via the minimization of J_2. The optimized parameters of CC-TID controller are listed in Table 8. To establish the supremacy of CC-TID controller, the output results are compared to that of DSO: PID controller. Fig. 12 illustrates the deviation of the frequency of WTG, deviation of WDG output power, and output power deviation of DG. Fig. 12 illustrates that DSO: CC-TID controller efficiently mitigates the system deviation to zero speedily than DSO: PID controller. The damping specifications like fitness value, peak overshoot, and settling time are computed from Fig. 12 and shown in Table 9. It is evident from Table 9 that fitness value J_2 is effectively reduced with DSO: CC-TID controller. An enhancement in the damping of system oscillations is also seen in Fig. 12. It is explicit from the above discussion that DSO: CC-TID controller outperforms the DSO: PID controller.

Figure 11. Dynamic response of coordinated WDG system considering optimum controller gains of Table 6 (a) convergence characteristic, (b) WTG frequency error, (c) WTG output power deviation, (d) DG output power deviation

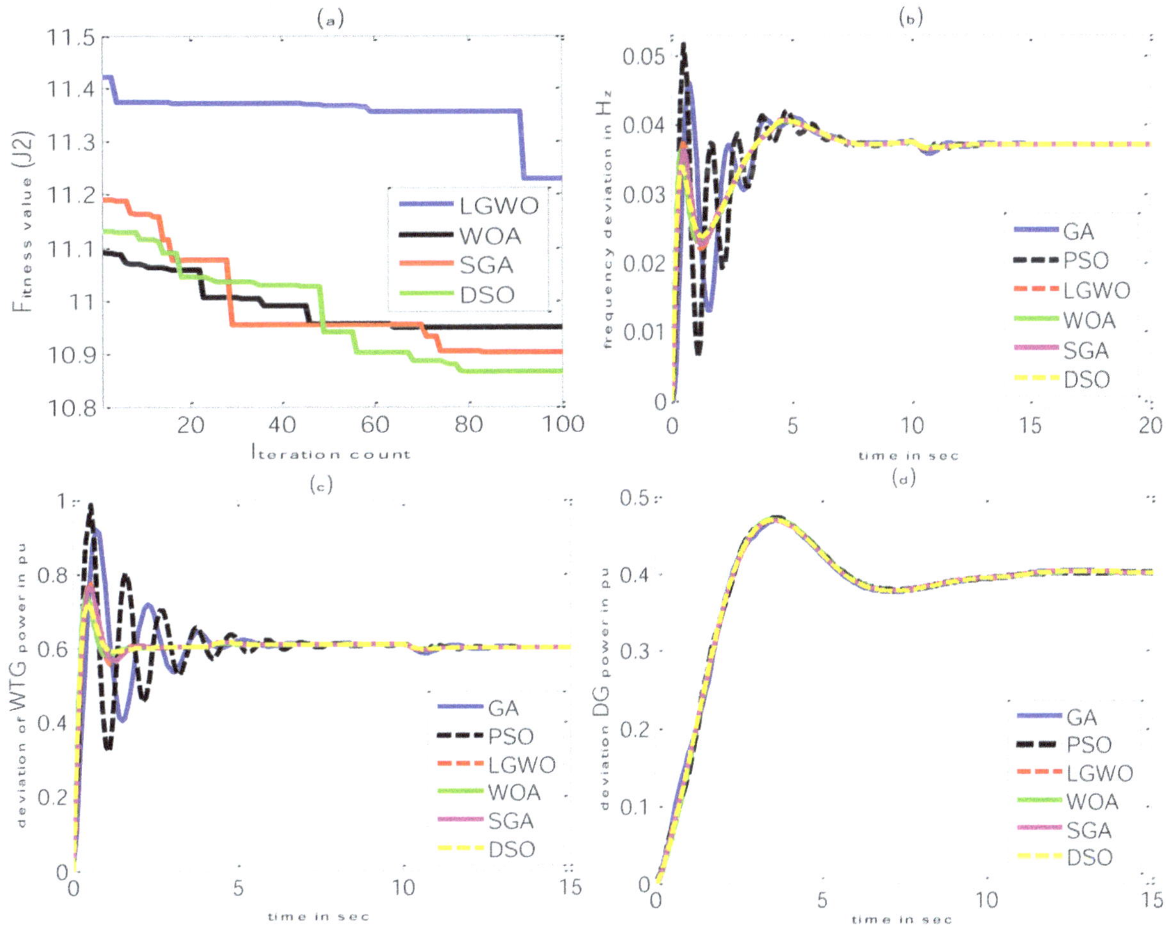

Table 5. Optimum controller gains computed via maximization of J_1

<table>
<tr><th rowspan="2">Computation Techniques</th><th rowspan="2">k_p</th><th rowspan="2">k_i</th><th rowspan="2">k_d</th><th rowspan="2">N</th><th rowspan="2">J_1</th><th colspan="6">Eigen Values</th></tr>
<tr><th>GA</th><th>PSO</th><th>LGWO</th><th>WOA</th><th>SGA</th><th>DSO</th></tr>
<tr><td>GA</td><td>103.53</td><td>124.12</td><td>73.53</td><td>-</td><td>0.5322</td><td rowspan="6">-39.0383
-13.41±j14.93
-1.6278
-0.537±j0.906
-0.532±j1.01</td><td rowspan="6">-39.0377
-13.4±j13.82
-1.6241
-0.538±j1.02
-0.538±j0.908</td><td rowspan="6">-194.39
-39.03
-23.33
-1.65 ±
j5.75
-0.48 ±
j0.88
-1.73
-0.98</td><td rowspan="6">-193.77
-39.03
-21.16
-2.26 ±
j3.57
-0.48 ±
j0.88
-2.11
-1.39</td><td rowspan="6">-49.3
-39.04
-14.64
-3.95
±
j5.76
-0.474
±
j0.877
-1.735
-0.842</td><td rowspan="6">-131.50
-39.03
-17.95
-3.82 ± j4
-0.48±j0.88
-1.83
-1.12</td></tr>
<tr><td>PSO</td><td>96.67</td><td>116.39</td><td>66.82</td><td>-</td><td>0.5381</td></tr>
<tr><td>LGWO</td><td>198.65</td><td>183.17</td><td>11.09</td><td>194.07</td><td>0.5722</td></tr>
<tr><td>WOA</td><td>121.39</td><td>146.44</td><td>15.53</td><td>193.33</td><td>0.5748</td></tr>
<tr><td>SGA</td><td>197.705</td><td>149.42</td><td>24.04</td><td>44.78</td><td>0.5734</td></tr>
<tr><td>DSO</td><td>160.66</td><td>149.64</td><td>24.26</td><td>130.416</td><td>0.6497</td></tr>
</table>

Boldface shows best results

Table 6. Optimum controller gains computed via minimization of J_2

Computation Techniques	k_p	k_i	k_d	N	J_2	Eigen Values					
						GA	PSO	LGWO	WOA	SGA	DSO
GA	197.65	108.82	50.20	-	12.85	-39.0357 -12.03±j9.16 -3.1530 -1.7735 -0.468±j0.88 -0.6554	-39.0377 -13.40±j13.82 -1.6241 -0.538±j1.017 -0.538±j0.908	-101.95 -39.03 -21.09 -2.38±j4.63 -0.48±j0.88 -1.78 -1.06	-185.80 -39.03 -21.23 -2.46±j4.28 -0.48±j0.88 -1.84 -1.18	-163.67 -39.03 -18.15 -3.83±j3.62 -0.48±j0.88 -1.86 -1.13	-171.4 -39.03 -18.51 -3.82±j4.89 -0.47±j0.88 -1.75 -0.91
PSO	198.14	147.39	29.68	-	11.28						
LGWO	150.987	142.602	15.0927	101.008	11.23						
WOA	143.633	147.205	16.1882	185.329	10.95						
SGA	151.356	140.712	24.174	162.833	10.90						
DSO	186.732	149.343	25.218	170.579	**10.86**						

Boldface shows best results

Table 7. Transient specifications of coordinated WTG system with PIDF-controller

Control Algorithms	For J_1						For J_2					
	$\Delta\omega_1$		ΔP_{WTG}		ΔP_{DG}		$\Delta\omega_1$		ΔP_{WTG}		ΔP_{DG}	
	OS	ST	OS	ST	OS	ST	OS	ST	OS	ST	OS	ST
GA	NA	NA	NA	NA	NA	NA	NA	NA	NA	NA	NA	NA
PSO	0.012	17.47	0.164	15	0.192	18.39	0.002	16.75	0.088	15	0.096	15.30
LGWO	0.969	6.89	0.2953	4.95	0.3066	8.38	0.0941	6.92	0.3285	4.81	0.312	8.42
WOA	0.970	10.70	0.2713	4.99	0.3049	8.34	0.0967	6.97	0.2826	4.93	0.3123	8.54
SGA	0.947	6.89	0.1928	5.04	0.3077	8.39	0.092	6.93	0.2345	4.79	0.3134	8.43
DSO	0.907	6.89	0.2331	4.72	0.3028	8.27	0.0926	6.89	0.2415	4.68	0.318	8.54

Table 8. Optimized values of CC-TID controller and transient characteristic

Parameters	k_p	k_i	k_d	n	N	J_2	$\Delta\omega_1$		ΔP_{WTG}		ΔP_{DG}	
							OS	ST	OS	ST	OS	ST
TID-controller	166.18	118.309	35.632	2.0593	152.687	10.37	0.0915	6.91	0.1332	4.97	0.3077	8.34
PI-controller	69.3483	25.457	-	-	-							

Table 9. A comparative studies between CC-TID controller and TCSC-CC-TID controller in terms of transient specification

Control Algorithm	$\Delta\omega_1$			ΔP_{WTG}				ΔP_{DG}	
	OS	% Improvement of OS	ST	OS	% Improvement of OS	ST	% Improvement of ST	OS	ST
CC-TID	0.0926	4.64	6.91	0.1332	32.7	4.97	94.03%	**0.3077**	8.34
TCSC-CC-TID	**0.0883**		**6.90**	**0.0896**		**0.2967**		0.3126	**8.32**

Boldfaces show best results

Figure 12. Dynamic performance of coordinated WDG system, (a) frequency deviation of WTG, (b) power deviation of WTG, and (c) power deviation of DG

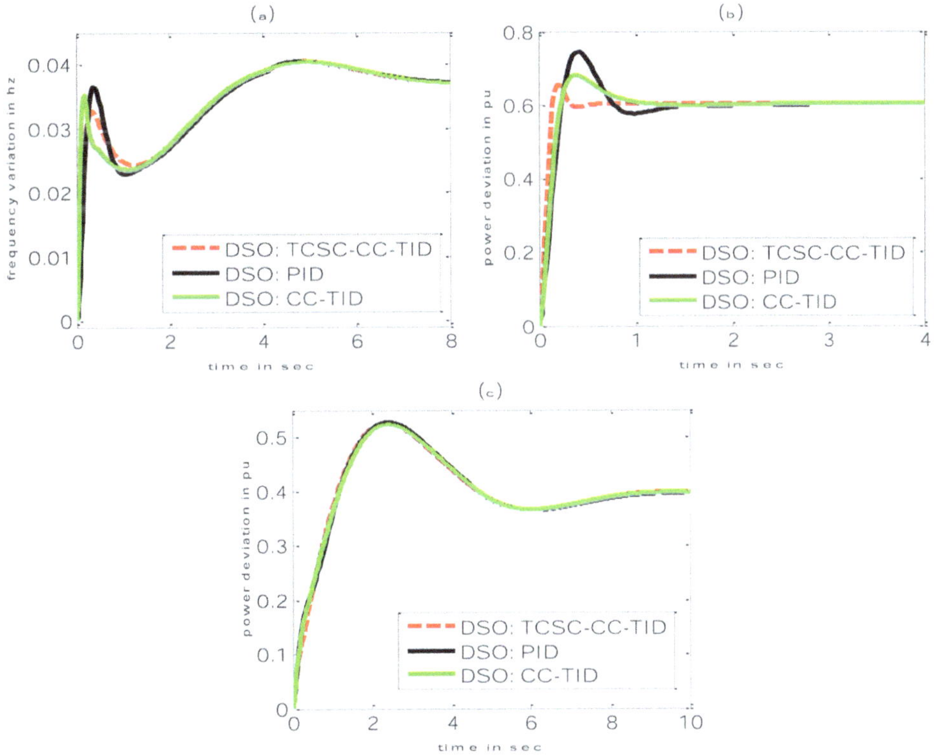

With Coordinated TCSC-CC-TID Controller

In this section, an optimized TCSC controller is developed and coordinated to work with CC-TID controller for adding extra damping to the frequency and power oscillations of WDG system. The DSO-algorithm is implemented to search the optimal gains of both TCSC and CC-TID controller parameters. In CC-TID, the master controller gains are computed as k_p =161.0045, k_i =149.4701, k_d =75.6318, n =2.4103, N =125.6874. The optimized values of slave controller parameters are k_p =41.2741, k_i =65.2030. The optimized gains TCSC controller are K_{TCSC} =0.3252, T_{TCSC} =0.8023, $T_1 = 0.5212$, T_2 =0.6355, T_3 =0.2351, T_4 =0.9598. To confirm the supremacy and effectiveness, the results of coordinated TCSC-CC-TID controller is compared to the outputs of the CC-TID controller and portrayed in Fig. 12. Table 9 shows a comparison between CC-TID and TCSC-CC-TID controller concerning maximum overshoot and settling time. It is noteworthy from Table 9 and Fig. 12 that coordinated TCSC-CC-TID controller significantly enhanced the damping of system oscillation and thereby increasing stability margin in terms of lower peak overshoot and small settling time. It is attracting from Fig. 12(b) and Table 9 that TCSC-CC-TID controller reduces the settling time of WTG output power in a superior manner. However, a minimal improvement of deviation of DG output power is attained with TCSC-CC-TID controller.

CONCLUSION

In this chapter, a novel CC-TID controller has been designed and implemented to study the dynamical behavior of WDG system. Both the load and wind power perturbations are considered for investigating the performance of WDG. Four powerful optimization algorithms, such as LGWO, SGA, WOA, and DSO are applied for searching optimum global values of CC-TID controller parameters. The computational competence of DSO algorithm is confirmed over GA, PSO, and other proposed optimization techniques concerning fitness value and speed of convergence. The dynamic performance of the WDG system under CC-TID control action is compared with PI and PID controllers. The offered results reveal the superiority of CC-TID controller over PI and PID controllers in terms of decreased peak amplitude and settling time of system oscillation. Furthermore, the study carried out with coordinated TCSC-CC-TID controller shows remarkable enhancement of WDG performance. Hence it may conclude that DSO optimized coordinated TCSC-CC-TID controller adequately improved the dynamic performance of the modeled WDG system.

FUTURE SCOPE

In this work, a linearized model of the WDG system has been considered to design and implement the proposed control strategy for frequency and power stabilization. The nonlinear effects of WTG and DG are not taken into consideration. Further, the test system is studied in an isolated mode, i.e., the dynamics of WDG is not evaluated by coordinating it with the conventional power system. Hence, in the near future authors have the aim

- to derive a more realistic model of WDG considering different physical nonlinearities and other practical constraints.
- to examine the dynamic performance of WDG by connecting it with thermal and/or hydropower systems.
- to apply oppositional-based learning for accelerating the convergence speed of proposed computation techniques.

REFERENCES

Abraham, R. J., Das, D., & Patra, A. (2007). Effect of TCPS on oscillations in tie-power and area frequencies in an interconnected hydrothermal power system. *IET Generation, Transmission & Distribution*, *1*(4), 632–639. doi:10.1049/iet-gtd:20060361

Alhelou, H., Hamedani-Golshan, M. E., Zamani, R., Heydarian-Forushani, E., & Siano, P. (2018a). Challenges and Opportunities of Load Frequency Control in Conventional, Modern and Future Smart Power Systems: A Comprehensive Review. *Energies*, *11*(10), 2497. doi:10.3390/en11102497

Alhelou, H. H. (2018c). Fault Detection and Isolation in Power Systems Using Unknown Input Observer. In Advanced Condition Monitoring and Fault Diagnosis of Electric Machines. IGI Global.

Alhelou, H. H., Golshan, M., & Fini, M. (2018). Wind Driven Optimization Algorithm Application to Load Frequency Control in Interconnected Power Systems Considering GRC and GDB Nonlinearities. *Electric Power Components and Syst.*

Alhelou, H. H., & Golshan, M. E. H. (2016, May). Hierarchical plug-in EV control based on primary frequency response in interconnected smart grid. *Proc. 2016 IEEE 24th Iranian Conference on Electrical Engineering (ICEE)*, 561-566. 10.1109/IranianCEE.2016.7585585

Alhelou, H. H., Golshan, M. H., & Askari-Marnani, J. (2018b). Robust sensor fault detection and isolation scheme for interconnected smart power systems in presence of RER and EVs using unknown input observer. Int. *J of Elect Power and Energy Syst.*, *99*, 682–694. doi:10.1016/j.ijepes.2018.02.013

Alhelou, H. H., Hamedani-Golshan, M. E., Heydarian-Forushani, E., Al-Sumaiti, A. S., & Siano, P. (2018, September). Decentralized Fractional Order Control Scheme for LFC of Deregulated Nonlinear Power Systems in Presence of EVs and RER. In *2018 International Conference on Smart Energy Systems and Technologies (SEST)* (pp. 1-6). IEEE. 10.1109/SEST.2018.8495858

Alhelou, H. S. H., Golshan, M. E. H., & Fini, M. H. (2015, December). Multi agent electric vehicle control based primary frequency support for future smart micro-grid. In *Proc. of Smart Grid Conference (SGC)*. Iran University of Science and Technology.

Alshahrestani, A., Golshan, M. E. H., & Alhelou, H. H. (2018, November). WAMS Based Online Estimation of Total Inertia Constant and Damping Coefficient for Future Smart Grid Systems. In *Smart Grid Conference (SGC)* (pp. 1-5). Academic Press.

Aziz, A., Oo, A. T., & Stojcevski, A. (2018). Analysis of frequency sensitive wind plant penetration effect on load frequency control of hybrid power system. Int. *J Elect Power Energy Syst.*, *99*, 603–617. doi:10.1016/j.ijepes.2018.01.045

Bahgaat, N. K., Ibrahim, M., Hassan, M. A. M., & Bendary, F. M. A. (2018). Load frequency Control Based on Modern Techniques in Two Areas Power Systems. In Advances in System Dynamics and Control. IGI Global. doi:10.4018/978-1-5225-4077-9.ch006

Barisal, A.K., Panigrahi, T.K., & Mishra, S. (2017). A hybrid PSO-LEVY flight algorithm based fuzzy PID controller for automatic generation control of multi-area power systems. *Int. J of Energy Optimization and Engg.*, *6*(2), 42-63.

Bhatti, T. S., Al-Ademi, A. A. F., & Bansal, N. K. (1997). Load frequency control of isolated wind-diesel-microhydro power systems (WDMHPS). *Energy*, *22*(5), 461–470. doi:10.1016/S0360-5442(96)00148-X

Chedid, R., Mrad, F., & Basma, M. (1999). Intelligent Control of a Class of Wind Energy Conversion Systems. *IEEE Transactions on Energy Conversion*, *14*(4), 1597–1604. doi:10.1109/60.815111

Chedid, R. B., Karaki, S. H., & El-Chamali, C. (2000). Adaptive Fuzzy Control for Wind-Diesel Weak Power Systems. *IEEE Transactions on Energy Conversion*, *15*(1), 71–78. doi:10.1109/60.849119

Das, D., Aditya, S. K., & Kothari, D. P. (1999). Dynamics of diesel and wind turbine generators on an isolated power system. Int. *J Elect Power Energy Syst.*, *21*(3), 183–189. doi:10.1016/S0142-0615(98)00033-7

Dash, P., Saikia, L. C., & Sinha, N. (2015). Automatic generation control of multi area thermal system using Bat algorithm optimized PD–PID cascade controller. Int. *J Elect Power Energy Syst.*, *68*, 364–372. doi:10.1016/j.ijepes.2014.12.063

Delassi, A., Arif, S., & Mokrani, L. (2018). Load frequency control problem in interconnected power systems using robust fractional PI^{λ} D controller. *Ain Shams Engg J.*, *9*(1), 77–88. doi:10.1016/j.asej.2015.10.004

Fini, M. H., Yousefi, G. R., & Alhelou, H. H. (2016). Comparative study on the performance of many-objective and single-objective optimisation algorithms in tuning load frequency controllers of multi-area power systems. *IET Generation, Transmission & Distribution*, *10*(12), 2915–2923. doi:10.1049/iet-gtd.2015.1334

Gampa, S. R., & Das, D. (2015). Real power and frequency control of a small isolated power system. Int. *J Elect Power Energy Syst.*, *64*, 221–232. doi:10.1016/j.ijepes.2014.07.037

Ganguly, S., Mahto, T., & Mukherjee, V. (2017). Integrated frequency and power control of an isolated hybrid power system considering scaling factor based fuzzy classical controller. *Swarm and Evolutionary Computation*, *32*, 184–201. doi:10.1016/j.swevo.2016.08.001

Goncalves, M. S., Lopez, R. H., & Miguel, L. F. F. (2015). Search group algorithm: A new metaheuristic method for the optimization of truss structures. *Computers & Structures*, *153*, 165–184. doi:10.1016/j.compstruc.2015.03.003

Guha, D., Roy, P. K., & Banerjee, S. (2016a). Load frequency control of interconnected power system using grey wolf optimization. *Swarm and Evolutionary Computation*, *27*, 97–115. doi:10.1016/j.swevo.2015.10.004

Guha, D., Roy, P. K., & Banerjee, S. (2016b). Oppositional biogeography-based optimisation applied to SMES and TCSC-based load frequency control with generation rate constraints and time delay. *Int J of Power and Energy Conversion*, *7*(4), 391–423. doi:10.1504/IJPEC.2016.079887

Guha, D., Roy, P. K., & Banerjee, S. (2017). Multi Verse Optimizatio: A novel method for solution of load frequency control problem in power system. *IET Generation, Transmission & Distribution*, *11*(14), 3601–3611. doi:10.1049/iet-gtd.2017.0296

Guha, D., Roy, P. K., & Banerjee, S. (2018). Robust Optimization Algorithms for Solving Automatic Generation Control of Multi-Constrained Power System. In Handbook of Research on Power and Energy System Optimization. IGI Global. doi:10.4018/978-1-5225-3935-3.ch003

Guha, D., Roy, P. K., & Banerjee, S. (2018). Optimal tuning of 3 degree-of-freedom proportional-integral-derivative controller for hybrid distributed power system using dragonfly algorithm. *Computers & Electrical Engineering*, *72*, 137–153. doi:10.1016/j.compeleceng.2018.09.003

Heidari, A. A., & Pahlavani, P. (2017). An efficient modified grey wolf optimizer with Lévy flight for optimization tasks. *Applied Soft Computing*, *60*, 115–134. doi:10.1016/j.asoc.2017.06.044

Howlader, A. M., Izumi, Y., Uehara, A., Urasaki, N., Senjyu, T., Yona, A., & Saber, A. Y. (2012). A minimal order observer based frequency control strategy for an integrated wind-battery-diesel power system. *Energy*, *46*(1), 168–178. doi:10.1016/j.energy.2012.08.039

Kaliannan, J., Baskaran, A., & Dey, N. (2017). Automatic Generation Control of Thermal-Thermal-Hydro Power Systems with PID controller using Ant Colony Optimization. In Renewable and Alternative Energy: Concepts, Methodologies, Tools, and Applications. IGI-Global. doi:10.4018/978-1-5225-1671-2.ch023

Kamwa, I. (1990). Dynamic modeling and robust regulation of a no-storage wind-Diesel hybrid power system. *Electric Power Systems Research*, *18*(3), 219–233. doi:10.1016/0378-7796(90)90054-7

Kumar, A., & Shankar, G. (2017). Quasi-oppositional harmony search algorithm based optimal dynamic load frequency control of a hybrid tidal–diesel power generation system. *IET Generation, Transmission & Distribution*, *12*(5), 1099–1108. doi:10.1049/iet-gtd.2017.1115

Kumari, S., & Shankar, G. (2018). Novel application of integral-tilt-derivative controller for performance evaluation of load frequency control of interconnected power system. *IET Generation, Transmission & Distribution*, *12*(14), 3550–3560. doi:10.1049/iet-gtd.2018.0345

Lee, D. J., & Wang, L. (2008). Small-Signal Stability Analysis of an Autonomous Hybrid Renewable Energy Power Generation/Energy Storage System Part I: Time-Domain Simulations. *IEEE Transactions on Energy Conversion*, *23*(1), 311–320. doi:10.1109/TEC.2007.914309

Lurie, B. J. (1994). *Three parameters tunable tilt-integral derivative (TID) controller*. US Patent, US5371670.

Mahto, T., & Mukherjee, V. (2017). A novel scaling factor based fuzzy logic controller for frequency control of an isolated hybrid power system. *Energy*, *130*, 339–350. doi:10.1016/j.energy.2017.04.155

Makdisie, C., Haidar, B., & Alhelou, H. H. (2018). An Optimal Photovoltaic Conversion System for Future Smart Grids. In Handbook of Research on Power and Energy System Optimization. IGI Global. doi:10.4018/978-1-5225-3935-3.ch018

Melo, V. V. D., & Banzhaf, W. (2017). *Drone Squadron Optimization: a novel self-adaptive algorithm for global numerical optimization. Neural Comput and Applic*. doi:10.100700521-017-2881-3

Mirjalili, S., & Lewis, A. (2016). The whale optimization algorithm. *Advances in Engineering Software*, *95*, 51–67. doi:10.1016/j.advengsoft.2016.01.008

Mondal, D., Chakraborty, A., & Sengupta, A. (2014). *Power System small signals stability and Control*. Academic Press.

Morsali, J., Zare, K., & Hagh, M. T. (2017). Applying fractional order PID to design TCSC-based damping controller in coordination with automatic generation control of interconnected multi-source power system. *Engineering Science and Technology, an Int. J, 20*(1), 1-17.

Nadweh, S., Hayek, G., Atieh, B., & Haes Alhelou, H. (2018). Using Four – Quadrant Chopper with Variable Speed Drive System Dc-Link to Improve the Quality of Supplied Power for Industrial Facilities. Majlesi Journal of Electrical Engineering.

Ngamroo, I. (2012). Application of electrolyzer to alleviate power fluctuation in a standalone microgrid based on an optimal fuzzy PID control. *Int. J Elect Power Energy Syst.*, 969-976.

Njenda, T. C., Golshan, M. E. H., & Alhelou, H. H. (2018, November). WAMS Based Intelligent Under Frequency Load Shedding Considering Online Disturbance Estimation. In *Smart Grid Conference (SGC)* (pp. 1-5). Academic Press.

Njenda, T. C., Golshan, M. E. H., & Alhelou, H. H. (2018, November). WAMS based Under Frequency Load Shedding Considering Minimum Frequency Prediction and Extrapolated Disturbance Magnitude. In *Smart Grid Conference (SGC)* (pp. 1-5). Academic Press.

Pan, I., & Das, S. (2015). Fractional-order load-frequency control of interconnected power systems using chaotic multi-objective optimization. *Applied Soft Computing*, *29*, 328–344. doi:10.1016/j.asoc.2014.12.032

Rahman, A., Saikia, L. C., & Sinha, N. (2015). Load frequency control of a hydro-thermal system under deregulated environment using biogeography-based optimised three degree-of-freedom integral-derivative controller. *IET Generation, Transmission & Distribution*, *9*(15), 2284–2293. doi:10.1049/iet-gtd.2015.0317

Rajbongshi, R., & Saikia, L. C. (2018). Combined voltage and frequency control of a multi-area multi-source system incorporating dish-Stirling solar thermal and HVDC link. *IET Renewable Power Generation*, *12*(3), 323–334. doi:10.1049/iet-rpg.2017.0121

Raju, M., Saikia, L. C., & Sinha, N. (2018). Maiden application of two degree of freedom cascade controller for multi-area automatic generation control. *Int Trans Electr Energ Syst.*, *e2586*, 1-20.

Semi, S., Nasri, S., Zafar, B., & Cherif, A. (2017). Multi-Input Single-Output State Space for Hybrid Power System Approach Using PEMFC: Fuel Cell and Applications. *International Journal of Energy Optimization and Engineering*, *6*(4), 35–50. doi:10.4018/IJEOE.2017100103

Senjyu, T., Nakaji, T., Uezato, T., & Funabashi, T. (2005). A Hybrid Power System Using Alternative Energy Facilities in Isolated Island. *IEEE Transactions on Energy Conversion*, *20*(2), 406–414. doi:10.1109/TEC.2004.837275

Shankar, G., & Mukjerjee, V. (2016). Load frequency control of an autonomous hybrid power system by quasi-oppositional harmony search algorithm. *Int. J Elect Power Energy Syst.*, *78*, 715–734. doi:10.1016/j.ijepes.2015.11.091

Singh, O. (2017). Automatic generation control of multi-area interconnected power systems using hybrid evolutionary algorithm. In Handbook of Research on Soft Comput and Nature-Inspired Algorithms. IGI-Global. doi:10.4018/978-1-5225-2128-0.ch010

Singh, V. P., Mohanty, S. R., Kishor, N., & Ray, P. K. (2013). Robust H-infinity load frequency control in hybrid distributed generation system. *Int. J Elect Power Energy Syst.*, *46*, 294–305. doi:10.1016/j.ijepes.2012.10.015

Tah, A., & Das, D. (2016). Operation of small hybrid autonomous power generation system in isolated, interconnected and grid connected modes. *Sustainable Energy Technologies and Assessments*, *17*, 11–25. doi:10.1016/j.seta.2016.07.001

Uhlen, K., Foss, B. A., & Gjosaeter, O. B. (1994). Robust Control and Analysis of a Wind-Diesel Hybrid Power Plant. *IEEE Transactions on Energy Conversion*, *9*(4), 710–708. doi:10.1109/60.368338

Zamani, R., Hamedani-Golshan, M. E., Haes Alhelou, H., Siano, P., & Pota, H. (2018). Islanding Detection of Synchronous Distributed Generator Based on the Active and Reactive Power Control Loops. *Energies*, *11*(10), 2819. doi:10.3390/en11102819

Zare, K., Hagh, M. T., & Morsali, J. (2015). Effective oscillation damping of an interconnected multi-source power system with automatic generation control and TCSC. Int. *J Elect Power Energy Syst.*, *65*, 220–230. doi:10.1016/j.ijepes.2014.10.009

KEY TERMS AND DEFINITIONS

Cascade Controller: Cascade controller includes secondary measurement and secondary feedback arrangement that makes it more compatible to provide good set-point tracking and better disturbance rejection ability. In a cascade controller, two feedback loops are employed, and the inner loop responds much faster than the outer loop.

Frequency Control: Frequency control is a process of maintaining the stability of a power system. In the power system, the frequency of the loop gets deviate from the steady-state value under the action of load perturbation. Load frequency controller is employed to regulate the power generation level to match the load profile to keep the area frequency at its nominal value (±2.5Hz of nominal value).

Hybrid Power System: This hybrid energy system provides centralized electrical power generation in a local area by combining renewable energy resources with some slack systems and energy storage devices. The storage system helps to avoid the energy crisis issue and provides fast active compensation to power system oscillations.

Optimization Techniques: Optimization is the process of making something better. Optimization is the selection of the best choice from among available options.

Thyristor Control Series Capacitor (TCSC): TCSC is a capacitive reactance compensator that comprises a series capacitor bank shunted with a thyristor-controlled reactor (TCR) to offer smoother variable capacitive reactance. In TCSC no high voltage interfacing transformer is required, thereby making it more economical than other FACTS devices. TCSC is usually connected in series with tie-line with an aim to improve rapid and continuous control of the transmission-line series-compensation level.

Chapter 4
Uninterrupted Power Supply to Micro-Grid During Islanding

Ruchi Chandrakar
National Institute of Technology Raipur, India

Ruchita Nale
National Institute of Technology Raipur, India

Monalisa Biswal
National Institute of Technology Raipur, India

ABSTRACT

The major purpose of uninterruptible power supply (UPS) systems is to supply regulated sinusoidal voltage at constant frequency and amplitude. UPS systems are gaining much popularity as a means of providing clean and continuous electricity to critical loads during any disturbances in main grid. Modern equipment is sensitive to power fluctuation and requires back up power supply for optimal performance. This chapter introduces a set of possible solutions so that uninterrupted power supply can be provided to emergency feeders and critical loads such as hospitals and communication systems. Different network configurations can be applied to micro-grid system for obtaining an uninterrupted power supply. Various hybrid energy and modern UPS systems for micro-grid along with their control techniques have been elucidated. A comparative assessment of all UPS technologies on the basis of cost, performance, and efficiency of the system has been presented.

GENERAL INTRODUCTION

Now-a-days the large advancement and development in the distribution system is only possible due to the integration of small local renewable generation in the distribution level. These small local renewable resources are generally installed below a couple of megawatts and can be photovoltaic, wind farms, micro-hydro turbines etc. Distributed generating resources or dispersed generation can be defined as generation resources other than conventional generating stations, which are nearby to load point usually closer to customer site (Oudalova, 2011). With the help of DG, the cost of distribution, transmission, and

DOI: 10.4018/978-1-5225-8030-0.ch004

losses can be reduced. Also, voltage profile, power quality of the distribution system can be improved. However, with the installation of DG different issues related to protection will arise and islanding is one of those issues (Redfern & Usta, 1993; Nale & Biswal, 2017).

Islanding is a condition in which distribution sector becomes electrically isolated from the main supply power but it gets continuously energized by DG connected to it (Oudalova, 2011). During any upstream side fault the main grid completely isolated from the distribution system and complete power interruption condition in distribution system will arise. In the DG connected grid this situation may not arise as under islanding condition the distribution sector will get continuous supply from DG sources. But due to main grid interruption the distribution section with local DG sources will be unsynchronised. The repeated operations of recloser connected between main grid and distribution level during disturbances is also not acceptable for the healthy operation of power system. To mitigate this issue of synchronization during islanding UPS technology is adapted. Common utility power problems are also corrected by the most UPS units such as sustained over voltage, voltage spike, noise, reduction in input voltage, harmonic distortion, the mains supply frequency instability (Dhal & Rajan, 2015).

Also, for the past many years the electric power system is facing shortage of electricity. To overcome, scheduled power cut which has resulted due to this large gap between customer demand and electricity supply, there is a need to install alternate source of energy such as Uninterruptible power supply system (UPSs) and Fossil Fuel Generators to reduce the effects of load shedding (Ali & Arsad, 2017).

Generally, an UPS system requires regulated sinusoidal output voltage with low total harmonic distortion (THD). The variations in voltage or load will not influence the operation of UPS system. For sinusoidal input current, THD is lower and power factor is unity. Higher efficiency, higher reliability, lower cost, smaller size, lesser weight etc. are the advantages of utilization of UPS technology. Generally, batteries and generators are highly used in these uninterruptible power supply systems are batteries and/ or generators (Ahmad & Kashif, 2016; Alhelou et al, 2018; Njenda et al, 2018; Makdisie et al, 2018; Fini et al, 2016; Alhelou et al, 2016; Nadweh et al, 2018).

The disadvantages associated with the utilization of battery in UPS systems are large number of charge/ discharge cycle, low power and energy density, heavy energy demand and environmental incompatibility. Also, the maintenance and cost of generator are high. Hence, alternative methods of energy storage and generation have been developed such as super capacitor, flywheel, fuel cell and their combinations (Chellappan & Enjeti, 2008).

To mitigate these issues, a new scheme is proposed in this book chapter which provides an alternative solution to be in connection with the main grid even though the main grid is disconnected through the point of common coupling (PCC). During power interruption in main grid, under fault scenario or maintenance period, the PCC breaker will switch off and distribution system gets completely disconnected from main source. This situation develops synchronization issue for DGs integrated distribution network. But with alternative connection from main grid to distribution network such an issue can be avoided. Next, with the help of flywheel we can improve the local storage in distribution network. This will help in providing emergency supply during major power failure in main grid. The proposed solutions are briefly discussed in the subsequent sections.

LITERATURE REVIEW OF UPS TECHNOLOGY

The uninterruptible power supply (UPS) is used to provide emergency power to the load. This UPS will act as an emergency or auxiliary power when utility power fails. Clean, uninterrupted and high-quality power is supplied by UPS system to critical load such as communication system, medical support system, computer etc. This section of literature review basically discusses about the architecture of UPS system, its classification, various control strategies, advantages and drawbacks of UPS technology.

ARCHITECTURE OF UPS SYSTEM FOR UNINTERRUPTED SUPPLY

The main issue with the islanding is that the power supply is disconnected from the distribution sector and the distribution system will get moved to out of synchronization condition. In modern age of power system, with the help of adaptive or advanced technology, such a condition can be avoided with the UPS system at the distribution end (Guerrero &Hang, 2008; Abusara & Guerrero, 2014). The working principle of UPS system is explained below.

The standby mode of power supply is the most commonly used topology for personal computer. The illustration of block diagram is shown in Fig. 1. The main power supply is connected to the load through a connecting switch also known as static switch. It is selected based on AC input which is the primary power source. When power supply fails, the transfer switch must be operated for transferring the load from main power to the battery or inverter to provide backup supply. Compact size, high efficiency and low cost are the main benefits of this architecture. With the addition of proper surge circuit and filter, SPS can also be used to suppress the surge and unwanted noise signal. The advantage of the standby power supply (SPS) system also includes less stress on the inverter circuit.

The drawback of this standby UPS technology includes poor voltage regulation of the mains supply and low switching time which seriously affect sensitive load. Also, the output voltage of this architecture is not completely isolated from the power supply and the filtering is required only for transient and radio-frequency signals (Brownlie, 1988; Alhelou et al, 2015).

The second mostly used configuration is of a true UPS system which is shown in Fig. 2. In this design, the inverter circuit is always connected on line to the load side. The battery charger is used for charging the battery bank and the inverter is used for converting the entire load power flow. The charger is also used for carrying the full load of the true UPS configuration in addition to supplying power to the load

Figure 1. Standby power supply configuration

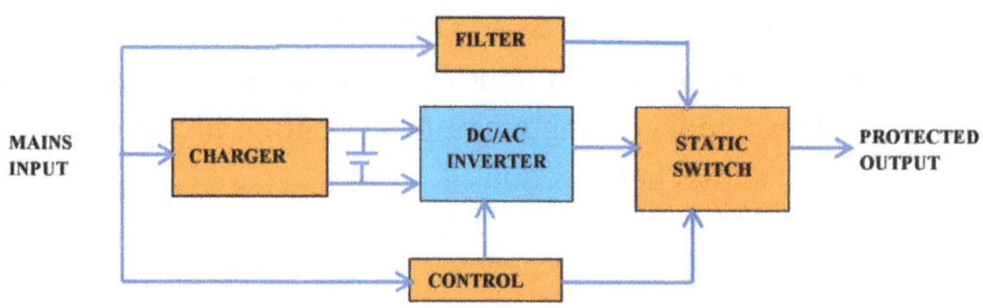

Figure 2. True UPS configuration

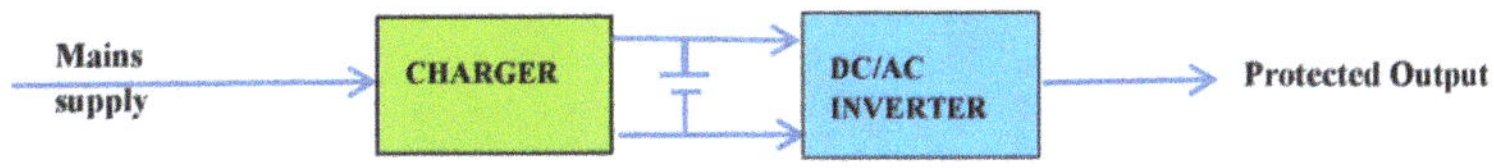

(Brownlie, 1988). The benefits of this true UPS architecture include regulated, isolated and continuous power to the load. This UPS system is mostly suitable for backup power source where good voltage regulation and filtered output power is required by the load. In event of power interruption or outage, the chance of occurrence of transient is very rare.

It also has some drawbacks like high cost and need for greater circuit reliability as both charger and inverter are fully loaded to supply continuous power supply. Also, the battery charger is drawing non – linear power from the supply. There are various classifications of UPS system whose working principle is explained in subsequent sections with the help of block diagrams.

CLASSIFICATION OF UPS TECHNOLOGY

The Conventional UPS system has been classified into various configurations such as battery powered UPS, transformer-based UPS and UPS system with hybrid energy storage topologies. All these configurations are discussed with their benefit and drawback in the subsequent section.

Battery Powered UPS Topologies

Battery powered UPS system has been classified into online UPS, offline UPS, hybrid UPS, passive standby UPS, line interactive UPS, double conversion UPS and delta conversion UPS system. These UPS systems are discussed with the help of block diagram in this section.

Online UPS System

The online UPS system basically consists of a rectifier, a static switch and an inverter as shown in Fig. 3. The function of rectifier is to charge the battery for backup power source and it also maintains constant DC link voltage during normal mode of operation. While this DC link voltage must be converted to AC by the inverter to supply power to the load. With the help of inverter, the power supply is maintained to the load. Hence, the inverter is supplying clean and regulated power to the load with the help of filter. Thus, under both operating modes the inverter can feed power to the load.

The main merit of using on-line UPS system is that it can provide better isolation to the critically sensitive load during power line disturbances. It also helps in providing voltage sag immunity or high susceptibility to the load and it has negligible switching time. But the drawbacks of the online UPS system include low efficiency, high THD, low power factor and unsuitable for protection of critically sensitive load (Solter, 2002).

Figure 3. Online UPS system architecture

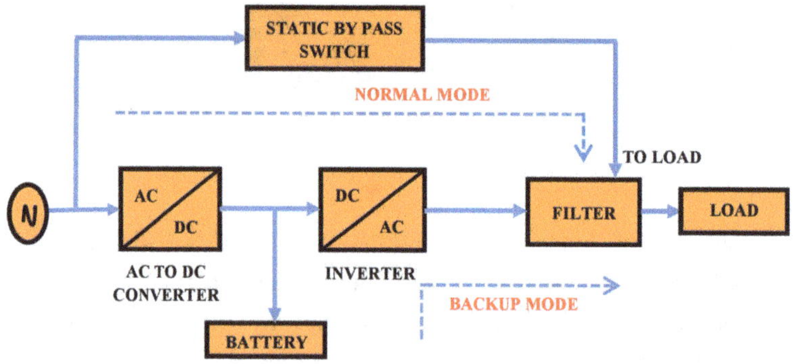

Offline UPS System

In online UPS system as shown in Fig.4, a battery charger, a static switch and inverter are used. Sometimes a surge suppressor or a filter is also included at output side for avoiding disturbance and main supply noise before directly feeding it to the sensitive load. During normal mode, the battery bank is charged by a charger or rectifier which is also used to feed power to the load since the inverter is at the standby mode. During power failure the load is connected to the inverter by the static transfer switch and the power is fed from the battery to the critical load through the inverter. The switching time of static switch is generally less which ultimately does not affect the computer load or other sensitive load connected to it.

The main advantages of the offline UPS system are low cost, small size and simple design. But the drawbacks associated are lack of real isolation from the load and poor output voltage regulation. During operation of nonlinear load, performance of the system is very poor (Niroomand & Karshenas, 2010; Karve, 2002; Bekiarov & Emadi, 2002; Alhelou et al, 2018; Njenda et al, 2018).

Hybrid UPS System

The Hybrid UPS system overcome all the drawback of online and offline UPS system. This UPS system has a voltage regulator (Ferro-resonant transformer) at the load end, for the regulation of the output voltage. The architecture of hybrid UPS system is shown in Fig.5. It also provides the advantage of voltage sag immunity or high susceptibility whenever switching take-place from normal mode to UPS mode.

Figure 4. Offline UPS system architecture

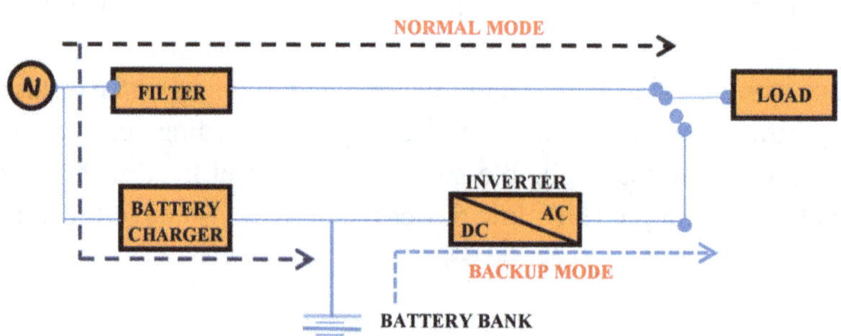

Figure 5. Hybrid UPS system architecture

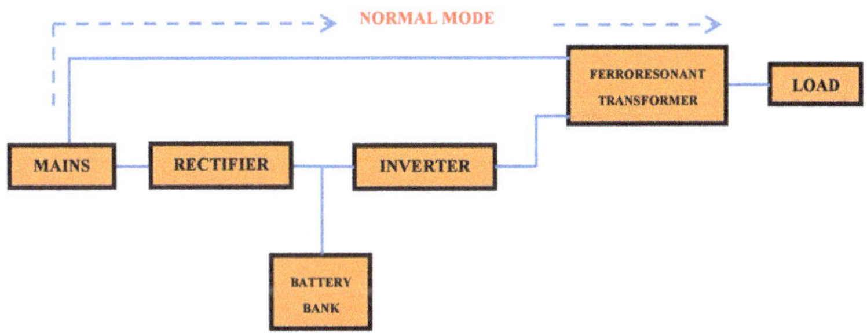

Line Interactive UPS System

In such a UPS system, the inverter is always connected in parallel at the output of UPS system as a source of backup power. The connection is shown in Fig. 6. Operating the inverter in reverse mode provides battery charging and interaction with the utility (Karve, 2002). When the input power fails, the switch is transferred from utility supply to battery so that power flows from the battery bank to the load. Basically, there are three operating modes for the line-interactive UPS topology and these are normal mode, bypass mode and stored energy mode (Karve, 2002).

- **Normal Mode:** Through the parallel connection of the AC input and UPS inverter, better filtering and adequate conditioning is obtained for the main power supply. The inverter operation provides output voltage regulation, conditioning, battery charging, and synchronization of output frequency with the input frequency.
- **Stored-Energy Mode:** The UPS system operates in stored energy mode until it is available otherwise it is returned to normal mode. During power interruption the combination of battery and inverter are used to maintain continuous power to the load side. Here, a static switch is disconnected to avoid back feeding of input supply from the inverter. Since the inverter is always connected to the load side, it provides better filtering and reduced switching transient. Thus, line interactive UPS system offers the advantage of small size, low cost, low losses and high efficiency.

Figure 6. Line interactive UPS system architecture

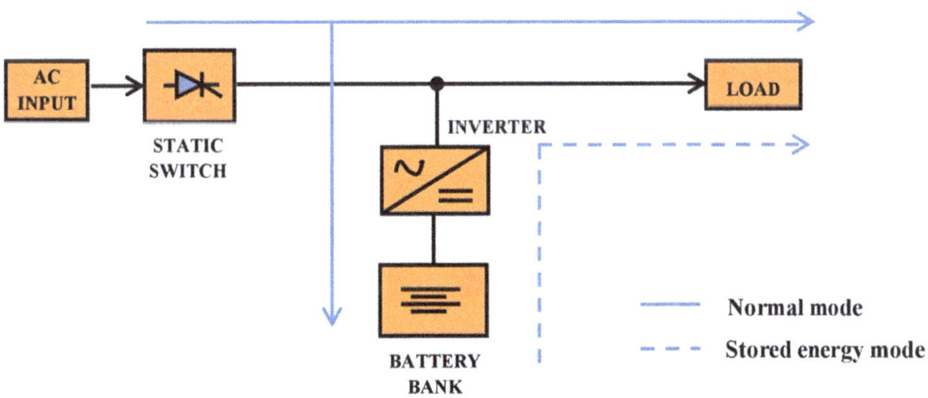

- **Bypass Mode:** In the event of internal malfunctioning of UPS, an additional maintenance bypass switch is provided. This arrangement is used to transfer load to bypass the input.

The drawback includes lack of protection from over voltage and spike, poor efficiency in case of nonlinear load operation and low output frequency regulation. It also leads to poor output voltage conditioning since the inverter is not connected in series with the supply (Karve, 2002).

Passive Standby UPS System

The operating principle of passive standby UPS system is described with the help of Fig. 7. During interruption, the inverter is connected in parallel with the supply to provide backup power. This topology has basically two operating modes: normal mode and stored energy mode.

- **Normal Mode:** In this mode of operation the inverter is at standby, since it is not performing any conversion operation. The sensitive loads are supplied from AC input through a filter or conditioner to eliminate disturbance or transient present in the supply. The filtering or conditioning devices also offer the advantage of voltage regulation.
- **Stored-Energy Mode:** During interruption or power outage, the battery and inverter work together to provide continuity of supply to the load. The switching time is very less in terms of milliseconds. In case of power failure, the switching action is directly performed with the help of electromechanical or electronic UPS switch to transfer load to the inverter.

The advantage offered by passive standby UPS system includes simple circuit design, compact in size and low cost of design (Karve, 2002). The main disadvantages are lack of complete isolation of the sensitive load from the distribution system, poor regulation of output voltage and frequency, long switching time since static switch is not utilized. It is also not suitable for frequency conversion operation.

Double Conversion Principle-Based UPS System

This UPS system is mostly used for higher power rating application. The function of Double conversion UPS system is explained with the help of Fig. 8. The inverter is used as primary power instead of normal AC input. The connection of inverter is in series with the AC input as power is continuously taken by the load from the inverter. Failure of the input power does not cause activation of static switch. Gener-

Figure 7. Passive standby UPS system architecture

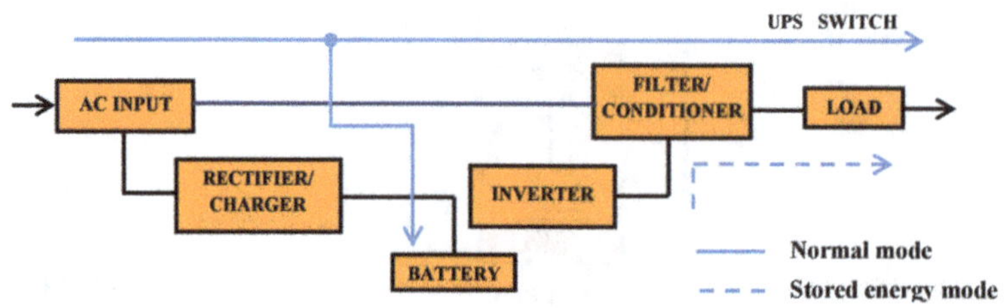

Figure 8. Double conversion UPS architecture

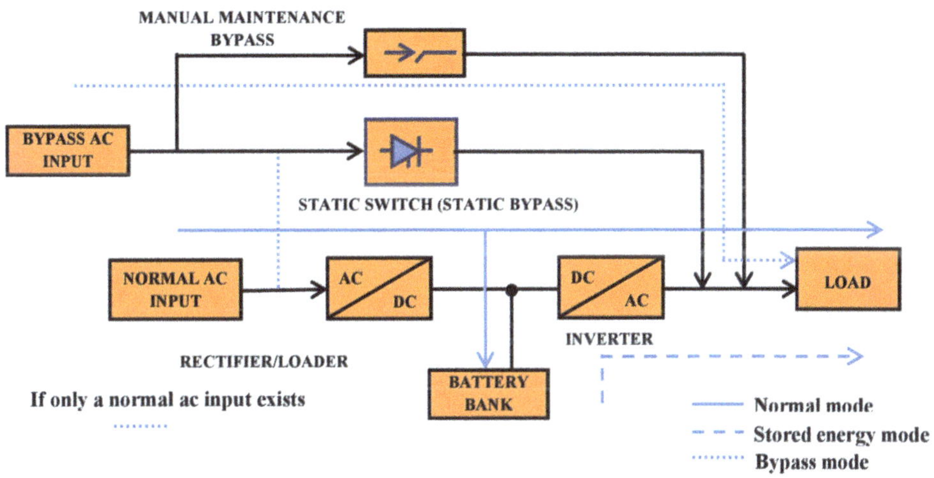

ally, three operating modes are there in double conversion UPS system: normal mode, bypass mode and stored energy mode.

- **Normal Mode Operation:** This mode includes double conversion operation from AC-DC and DC-AC through the combination of rectifier and inverter as they are continuously feeding power to the load.
- **Stored Energy Mode Operation:** During power failure stored energy is used to feed power to the load through the battery and inverter combination. This combination is used for entire power conversion resulting in reduced efficiency with increased heat losses.
- **Bypass Mode Operation:** A static switch or static bypass is commonly used to allow transfer of instantaneous amount of power to the load from the bypass AC input. This bypass mode is utilized in case of UPS internal malfunctioning, overloading, transient and at the end of battery backup time. To ensure continuity of power supply UPS system must be synchronized with the bypass AC input.

The advantages offered by double conversion UPS system include protection of the load by the inverter, isolation from the distribution system fluctuation and precise regulation of output voltage. It can also be operated as frequency converter and performance is also good during transient and steady state.

The associated drawbacks are high cost and low efficiency. It provides regulated output but also cause wearing of the components thus leading to reduced reliability. Also, the power supplied to the charger or rectifier is nonlinear in nature (Karve, 2002).

Delta Conversion Principle-Based UPS System

To overcome the drawback of double conversion UPS configuration, the delta conversion UPS topology is developed. This system has basically two inverters connected to a battery bank as shown in Fig. 9. First inverter is termed as delta inverter. This is connected with the main power supply through a transformer in a series connection. Inverter 2 is termed as main inverter which is connected to the load side. Both

Figure 9. New Delta Conversion Topology

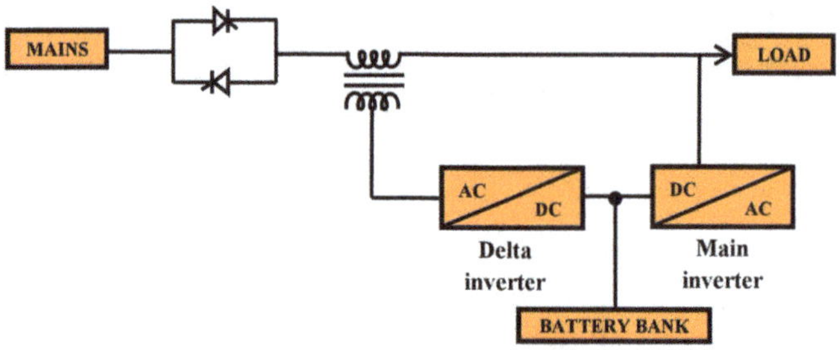

are capable of performing four quadrant operations. PWM control technique is used to maintain the output voltage with the help of main inverter. The input supply voltage is maintained sinusoidal through the choke coil hence the nature of output current is free from harmonics (Rathmann &Warner, 1996).

If any difference of output voltage occurs between the supply and the UPS system, delta inverter eliminates that difference of voltage. The delta inverter is used to maintain the unity power factor as the sinusoidal supply voltage and current are in phase with each other. It is also used to control the battery charging operation. The mains static switch provides protection against reserve supply to mains. The control scheme is based on the power balance principle with current mode control. The system has lower loss as compared to double conversion system because only the delta or difference of voltage between mains and output power is converted. This delta conversion UPS technology provides the solution for harmonics and low power factor problems. The new delta conversion topology will prove to be an extremely important step ahead in UPS technology.

TRANSFORMER BASED UPS TOPOLOGIES

A significant technological step taken towards change in power converter topologies for customer benefits and allowing the users to employ transformer free UPS design instead of transformer-based UPS designs. To achieve the need of applications, various configurations of UPS have been developed and are discussed in the subsequent sections.

Conventional Transformer-Based UPS Technology

Conventional transformer-based UPS system (Kwon et al., 2001; Alhelou et al., 2018) comprises of an inverter, rectifier, bypass switch and in some cases line frequency transformer. The battery bank is charged through rectifier to sustain the consistent DC line voltage. In order to supply sinusoidal AC voltage to loads, inverter is employed. For step up and step down of voltage two power frequency transformers are used. Such types of UPS system are applicable for high power applications, are robust and provide galvanic isolation from transients in grid system. However, owing to the operation of transformers at grid frequency, dimensions of the system is extremely increased. Also, the system is having reduced efficiency.

To improve the efficiency and reliability of the system, a single stage UPS system with single phase trapezoidal AC supply is proposed by removing power factor correction circuit and incorporating DC link capacitor. Although the current harmonics are removed by using DC link capacitor, the system power factor is noticeably low and the UPS is not good enough for high power applications. To further increase the performance and design of conventional UPS technology, a three-leg type converter based online UPS system is proposed in (Choi et al., 2005). Although the performance of such type of UPS system is increased by decrease in number of active switches, but the life of large number of batteries connected to DC bus may reduce due to continuous overcharging of the battery System.

High Frequency Transformer Isolation Based UPS System

Nowadays advent of semiconductor devices has led to development of fast switches and diodes which in turn leads to reduction in size of transformer, thus reducing the weight, size and enhancing the efficiency of the system. Various topologies with high frequency isolation transformer are proposed in (Branco et al., 2013; Nasiri et al., 2008). However, due to the usage of large number of active switches, overall cost of the system is increased and there is a need of more power conversion stages for the implementation of such a technique (Branco et al., 2013). Hence, the overall energy conversion efficiency is reduced as compared to the conventional system.

Transformer-less UPS System

To overcome the shortcomings of above mentioned topologies, transformer-less UPS systems are introduced. In this, few of the components are replaced with advanced switches and diodes. Hence the size of the UPS is reduced greatly and is more efficient as compared to other topologies. Various UPS topologies have been reported in literature, which include four legs type converter topology (Park et al., 2008). The requirement of transformer is eliminated because of the incorporation of battery charger/discharger. During power failure mode, the battery supplies the required power to the dc link capacitor. The number of batteries requirement in transformer-less UPS system is quite high, thereby enhancing the cost of batteries and reducing the reliability of the system.

Further, Z source inverter-based UPS system (Zhou et al., 2008), offline transformer-less UPS system (Marie et al., 2011) and online transformer-less UPS system (Kim et al., 2009) are introduced with an aim of improving the efficiency, reducing the size and overall cost of the system. However, the requirement of battery bank in all aforementioned techniques is excessively high. To overcome this drawback, in (Aamir & Mekhilef, 2017) a non- isolated online transformer-less UPS system with small battery bank of only 24 V is proposed but this is only applicable for low power applications.

Technical Specification and Performance Comparison Between Transformer Based and Transformer-Less UPS System

At present transformer-less UPS system is highly preferred owing to its reduced volume, high efficiency and cost-effective design for low power applications. Transformers in UPS system provide galvanic isolation, noise reduction and limit the fault current to a certain degree. Transformer-less UPS make use of all active power conversion devices to achieve identical performance whereas transformer-based UPS incorporates passive magnetics along with very few active power conversion devices which results in comparatively robust and simpler UPS.

Transformer based UPS technology is always more reliable as compared to transformer-less UPS technology and has higher mean time before failure (MTBF) owing to simple SCR operation instead of complicated control of IGBT with high frequency switching. However, transformer-less UPS can maintain unity power factor over a wide range of load with total harmonic distortion of less than 3%. Further, the cost of transformer-based UPS system is high due to the presence of input and output transformer. Also, the size of UPS system can be reduced to 50% by removing transformers. Hence, the transformer-less UPS systems are portable and can be fitted in a limited space. Also, transformer less UPS are 10% more efficient as compared to transformer-based UPS system owing to the

Next, choosing an appropriate UPS technology according to the specific requirement is a tedious task. A system designer should examine the type of UPS system which will be best utilized based on the physical and electrical distribution requirements and trade-offs.

UPS WITH HYBRID STORAGE TOPOLOGIES

During grid failure UPS system with battery as storage source generally provide backup power. These batteries used for storage purpose basically provide instant energy for backup. However, it also suffers from large charging/discharging cycle. Hence, they cannot be utilized to provide backup power for long interval of time (Ibrahim et al., 2008).

These batteries containing heavy toxic metal are hazardous for environment. Serious environmental issues are also caused by toxic metal like mercury, lead, cadmium etc. resulting dangerous for human health. Hence, some new alternative methods of storing energy has been developed such as super capacitor, fuel cell and the combination of both are also getting famous (Mekhilef et al.,2012).

Fuel Cell and Battery Based UPS System

This UPS system is based on the combination of fuel cell and battery as a source of backup power to provide continuous power supply to critical load. The fuel cell is utilized as the primary source of energy while backup power is supplied by the battery bank. Fuel cell is connected to DC bus through a DC-DC converter while sources of energy storage are connected via bi-directional converter as shown in Fig. 12. The bidirectional converter is operated as a battery charger during grid connected mode (buck operation) while they are utilized as a discharger during backup mode (boost operation). The connection of hybrid UPS system is shown in Fig. 12. The DC-Bus supply energy to the connected load through an inverter (Zhang et al., 2013).

This UPS system provides stable and continuous power to the load in case of grid failure. Whenever the power supply fails, the hydrogen gas as a fuel is supplied to the fuel cell stack. Fuel cell is unable to provide instantaneous energy for backup to the load since the time required to develop the voltage is quite long. Hence, battery bank is used to provide instantaneous energy to the external load (Zhan et al., 2015; Monfared et al., 2014). The amount of power availability is function of hydrogen supply as state of charge of battery cannot be accurately predicted. The challenges offered by this UPS system are slow response time, high cost and high sensitivity to low frequency ripples.

This approach also proves the possibility of replacement of battery bank by the fuel cell. Absence of batteries without any toxic material, clean and quiet operation makes is compatible for environment (Chiang et al., 2010; Choi et al., 2006).

Figure 10. Block diagram of hybrid energy storage UPS system

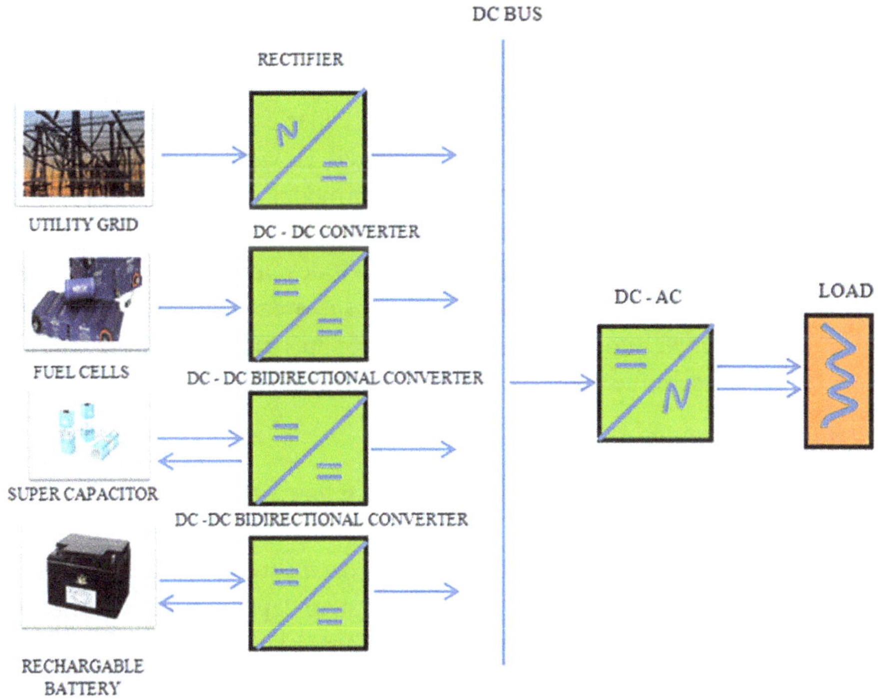

Figure 11. Hybrid UPS system with PV panel

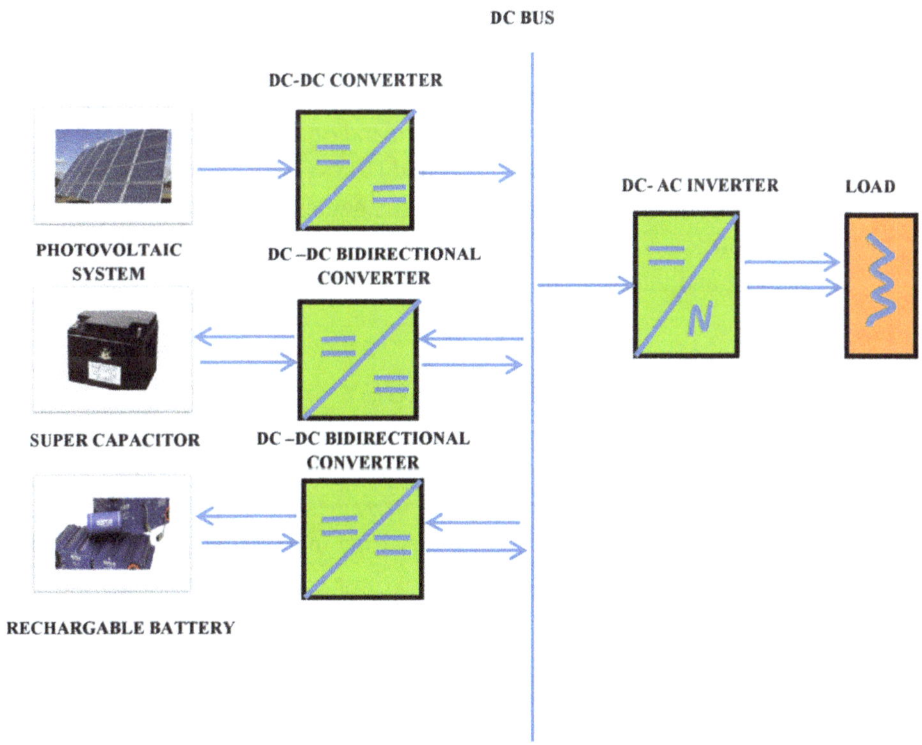

Super Capacitor and Battery Based UPS System

Super capacitor is used to provide power during transient period like starting braking and load reversal. Again, batteries are utilized to feed power during running operation. Generally super capacitor module can supply power for only few seconds as compared to the batteries thus it can only be utilized for application where long duration of backup power is not required. However, the backup power provided by the battery bank is for few minutes. Hence, super capacitor module is paralleled with the battery bank to reduce stress on it as shown in Fig. 10. (Lahyani et al., 2013).

Recently, the combination of super capacitor and battery is developed to improve reliability of power supply. This hybrid system of UPS technology consists of super capacitor module for high power and batteries to provide high energy. But the installation cost of super capacitor is very high. Reducing the overall cost of super capacitor is the main challenge in the implementation of this hybrid UPS system (Kollimalla et al., 2014).

Integrated UPS System Based on Renewable Energy

Since greenhouse effect and global warming has reached to the most critical and threatening level which is responsible for hazards related to environment and ecosystem depletion. To overcome this challenge, environmentally friendly UPS system based on renewable energy has been developed. In grid isolated area wind energy and photovoltaic panel are proposed to provide the backup power to the sensitive load (Nayar et al., 2000). The super capacitor and battery bank are linked to the DC bus via bidirectional converter and DC-DC converter is used to connect PV panel to the common DC bus as shown in Fig. 11. An inverter is also utilized for power conversion to feed power to the load. Hence good quality power and more reliability are achieved by this renewable integrated UPS system (Chauhan & Saini, 2014).

Whenever the load demand is greater than the amount of generation or during night hours, the energy stored by the super capacitor module and battery bank is utilized to fulfil the requirement of load. The super capacitor is installed to meet the transient power demand of the load and to fulfil the fast-dynamic regulation of the power (Chen et al., 2009). This integrated UPS system must be installed with photovoltaic panel to develop more efficient and reliable energy management system. It is beneficial for reducing the maximum power requirement, cost of electricity and load fluctuation. It also offers the advantage of smoothing the load transient and intermittent variation of energy requirement (Bortolini et al., 2014).

CONTROL STRATEGIES OF UPS TECHNOLOGY

The control system of UPS system is one of the major-part. The UPS control techniques offers regulated output voltage under any disturbances. Several control strategies have been devised to give regulated output voltage. These control techniques are broadly classified in to single loop control and multi loop control schemes. In the following subsection, the details of these techniques are explained and evaluated.

Single Loop Control

In such a control scheme, the regulated output voltage is provided only through output voltage feedback loop (Karshenas & Niroomand, 2005; Moriyama et al., 1998). The output voltage is regulated by comparing it with a reference signal so as to assure good performance. It makes the system simple in design and economical but the performance of the technique is unsatisfactory during non- linear and unbalanced loading condition.

Multi Loop Control

To enhance the performance of the controller, multi loop control schemes are utilized. Various control parameters like inductor/capacitor filter current or output current and voltage are utilized as feedback to the controllers which improves the performance of the control system and hence making it more flexible even under unbalanced system. But these controllers are too expensive for usage because of large number of sensors. Generally, the inner loop employs inductor/capacitor filter current as feedback signal and outer loop employs output voltage as feedback signal. An error is generated by comparing feedback signal with desired reference signal and by adopting appropriate compensators; error is compensated for attaining stabilized output. By utilizing multi loop control system various high-performance controllers have been developed which are discussed below.

Predictive Controllers

It is an emerging control technique that has numerous applications in power inverters owing to its characteristics i.e. prediction of future behaviour of the control variables by utilizing system model. Controller utilizes this data in order to obtain the optimal actuation, according to a predefined optimization criterion (Cortes et al., 2008). These controllers have fast dynamic response and can be applied to systems with different constraints and non- linearity, and are easy to perceive. The computational burden is high as compared to conventional controllers. Predictive controllers are further classified in to deadbeat control and model predictive control.

Deadbeat Control Technique

It is the most well- known technique for uninterruptible power supply systems (Mattavelli 2005; Zhang et al., 2014). In such type of control technique, computation of reference voltage is carried out at each sampling interval and at the next sampling instant, it is applied to follow the reference value. It is an interesting means for discrete time control as it decreases the state variable error to zero in limited number of sampling steps, provides rapid transient response. Apart from offering rapid dynamic response these techniques are responsive to change in parameters and model un-certainty. Furthermore, the response of these techniques are also affected by unforeseen disruptions like dead times, dc link voltage variations as there is no inherent integral action in the control system.

Model Predictive Control Technique

It is also known as receding horizon control. Model predictive control predicts the nature of control variables up to a certain time instant by utilizing system model. A particular switching state is chosen on the basis of cost function (Camacho & Bordons 2007; Goodwin et al., 2004). This controller offers high stability and better response in controlling UPS system.

It provides control flexibility, in which system constraints like switching states, voltage and current limitation and nonlinearity can be easily incorporated in designing of the controller. The complexity in implementation of model predictive controller is reduced by modelling the converter in limited number of switching states and optimization is carried out only in one-time step horizon as proposed in (Muller et al., 2005; Vargas et al., 2008) for current control of matrix converter, flying capacitor converter (Silva et al., 2007), three phase inverters (Kim et al., 2015; Rodriguez et al., 2004).

Repetitive Control Technique

To reduce the expenses caused by the necessity of high speed control for rejecting disruptions and to have reduced total harmonic distortion in output, repetitive control system is broadly used. Mostly it is adopted for removing periodic disruptions in dynamic systems (Zhang et al., 2003; Chen et al., 2013) It takes the advantage of recurring behaviour of the disruptions which employs time delay unit in multiple feedback loops resulting in removal of repetitive error effectively. However, the computational burden on the system is high and has slow response time.

The deployment of repetitive controller increases the steady state response of the controller, however due to incorporation of time delay unit between input and output, the dynamic response is unacceptable. Therefore in (Tzou et al., 1997 & Haneyoshi et al., 1988), repetitive control is applied in integration with least square error state feedback control and dead-beat control. But it increases the complexity and cost of the control system (Zhang et al., 2003). The repetitive control technique employing internal model principle (Bojoi et al., 2011) is implemented in single phase and three phase voltage source inverters (Mastromauro et al., 2009; Jiang et al., 2012) by including a time delay unit into a positive feedback loop. Generally, such type of controllers is employed for inverters with non- linear load.

Iterative Learning Technique

In this technique, with the recurrence of control task, the control command is regulated at every iteration. Hence, converges to zero tracking error. The response of the technique can be achieved even if in the absence of any information about the system. The tracking error is analysed for previous run in the result and is adjusted in the next run. The command and disruptions in iterative learning technique is considered as periodic as in repetitive controller. The implementation and designing of iterative learning controller are easier and simple to follow as compared to repetitive controller. It is particularly applicable in high performance motors (Kempf & Kobayashi, 1999) and robots (Elci et al., 2002).

Non-Linear Control Techniques

Since the inverter is purely nonlinear, nonlinear controllers are preferable over linear type controllers. A nonlinear controller provides good stability and improves the dynamic response of the system. It is more robust as compared to linear controller. Such systems are difficult to control without including feed forward action in the control structure which increases the complexity of the system. Also. the controller is insensitive to load disturbance and parameter variation. Sliding mode control technique and adaptive control technique are extensively accepted as nonlinear control techniques.

Sliding Mode Control Technique

Sliding mode control technique is widely accepted in industrial engineering and non- linear dynamic system such as high-performance robots (Wang et al., 2016), motors (Liu et al. 2014), power system, flexible air breathing hypersonic vehicle (Hu et al., 2014). The technique is also used with two level PWM inverter and having constant switching frequency and current limiter. As the technique is not affected by load disruptions, parameter variations and has good dynamic response, entailing to almost constant steady state response. Contrarily, it is difficult to find sliding surface and its response will deteriorate with definite sampling rate

Adaptive Control Technique

Adaptive control technique automatically adapts the control parameters based on uncertainties in structural and environmental conditions. It does not require previous data about uncertain parameters rather it estimates the data online based on measured signals. Several adaptive control techniques for UPS have been proposed in (Kissaoui et al., 2014; Kissaoui et al., 2016). The performance of adaptive controller is high with fast transient response, low total harmonic distortion and provides extremely good voltage regulation for non- linear and unbalanced loads, even though the computational burden is very high for adaptive controllers.

ADVANTAGES OF CONVENTIONAL UPS SYSTEM

The various advantages offered by conventional UPS system are discussed here:

Protection of Grid System from Interruption and Outages

To protect power grid system during interruption, following configuration of UPS topologies are used.

By Using Uninterruptible Power Supply (UPS)

In the event of interruption or complete blackout, UPS system is utilized to provide protection to sensitive equipment. UPS system should be used to overcome unacceptable loss of power supply resulted due to failure or down time. UPS system is also used to protect the power system from transient like noise, under-voltage surge and voltage sag depending on the application. Generally, three types of UPS are installed to provide different types of protection. These are as follows.

Three Major UPS Topologies for Protection

Online UPS System

The online UPS system offers high degree of power protection, power conditioning and power reliability. The benefit provided by this system is better protection from voltage transient and reduced time of power transfer. Good level of voltage regulation is achieved by continuous generation of ripple free sine wave at the output.

Offline or Standby UPS System

The advantage of offline UPS system includes lower cost solution for less critical applications like personal computer, peripheral and stand-alone device like PLCs (programmable logic controllers) etc. This offline UPS system act as a backup power for the load by supplying stored energy from the battery bank during transient like sag and power failure. The limited power demand is also provided to the critical load by input supply. With the help of surge suppressor or filter some amount of noise suppression is also achieved by this module.

Line – Interactive UPS System

The line-interactive UPS topology offers high level of backup power and efficient power conditioning. It is also beneficial for good regulation of output voltage without damaging the life cycle of battery bank. This unit is mostly suitable for the load where frequent voltage fluctuation usually occurs.

Improvement of Power Quality in Distribution System by Using Uninterruptible Power Supply

Whenever the grid power fails the main operation of any UPS system is to feed backup power to critical load. The capability of UPS system also includes correction of common utility issue like noise, voltage spike, sustained reduction of input voltage, harmonic distortion, sustained overvoltage, frequency instability etc. (Dhal & Rajan, 2015; Alshahrestani, 2018; Zamani, 2018). Hence, power quality is improved by using UPS system.

Protection of Computer System or Data Centre from Power Interruption

Whenever there is a sudden interruption in power supply, electronic device like computer system, data centre or logic circuit-based device are seriously affected due to hardware damage or data loss. Power interruption also affects video projector, alarm system and equipment related to data networking. In case of power unavailability, the computer system is protected from damage by installing UPS system which provides continuous power supply. When main power supply is restored, voltage surge occur which is extremely harmful for the hardware device. To protect computer system against surge, a surge protector or surge absorber can be utilized to absorb the excess amount of voltage.

Protection Against Surge by Using Offline UPS System

Offline UPS system is used to provide protection against power surge or interruption. The incoming source is directly connected to the device or equipment which is to be protected. The offline UPS internally switch-on its inverter circuit whenever the input supply is unavailable. This offline UPS system has a type of transfer switch to connect the external equipment at the output of the inverter. It is also used to reduce output voltage harmonics. The offline UPS system can also be used to provide uninterrupted power supply to sensitive load with the help of feedback system. Feedback system offers pure sinusoidal line current, regulated output voltage and low losses in distribution circuit.

DRAWBACKS OF CONVENTIONAL UPS SYSTEM

The main drawbacks of conventional UPS system are discussed in this section:

Role of UPS System in Increasing the Energy Crisis

According to the survey a large amount of energy is wasted in the utilization of UPS system. The problem with the usage of UPS system is that there is not any energy efficient solution for the wastage caused by it. The fact is that with the increase in the installation of the UPS system the shortage of energy is also increasing because of the electricity wastage. Also, the problem of load shedding is further increasing due to electricity shortage caused by UPS.

Generation of Harmonics by UPS System

The UPS system is responsible for injecting current containing harmonics to protect critically sensitive load. The increased level of THDs in supply current are also responsible for over stressing of capacitor used for power factor correction, heating of transformer and unwanted tripping of connected circuit breaker in power system. High value of THD causes generation of harmonic on the low voltage circuit of transformer which sometime also cause noise in induction motor, commercial and domestic electrical appliance. Thus, reducing the life cycle of equipment and damaging the critically sensitive load.

Reduction of Power Quality and Increase in Losses by Using the UPS System

By calculating the increase in losses caused by the utilization of UPS system its efficiency can also be monitored. The issues related to UPS system are low energy efficiency and poor power quality. The non-linear nature of UPS system is responsible for increase in THD, voltage dip and poor voltage regulation. The constant losses have more impact on the efficiency of UPS system. The power quality issue like voltage unbalance, voltage sag, voltage and current harmonic also have significant impact on the distribution power network.

PROPOSED FUTURE GENERATION UPS TECHNOLOGY

In these proposed system batteries are replaced by the flywheel energy storage system, combination of super capacitor bank and fuel cell stack to overcome the drawback of battery powered UPS system. To provide more reliable and flexible method for accommodation of DG sources, an intelligent method of storing energy has been developed. Also, a scheme is proposed consisting of main and reserve bus to mitigate the problem of power interruption during fault, maintenance and overloading. These proposed systems are discussed below.

Battery-less UPS System Powered by Flywheel

In this battery-less UPS topology, batteries are replaced by Flywheel energy storage system to supply backup power to the critical load during power interruption. Flywheel is especially designed to supply maximum rate of power for few second only. In contrast, batteries are used to supply maximum power to the load for about few minutes. While combination of diesel generator and flywheel energy storage system are also capable of providing full rated power for few minutes. Hence, battery-based UPS system

Figure 12. Battery less UPS system with battery and/ or flywheel panel

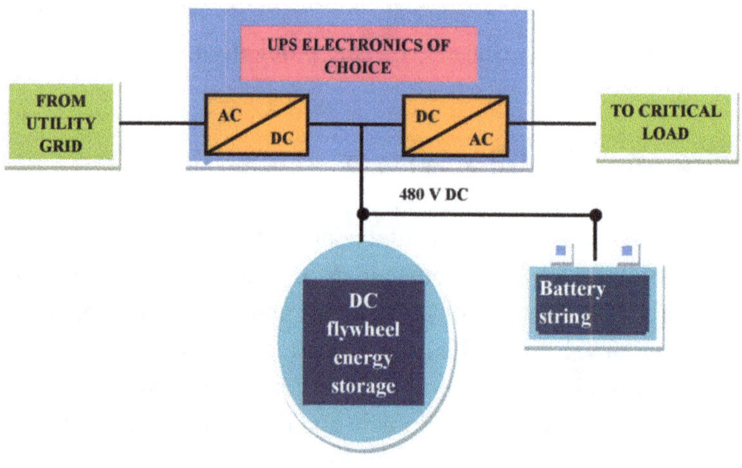

can be replaced by the DC flywheel energy storage system in combination with diesel-fired generator. This flywheel storage system has high cost of installation, reduced maintenance, better reliability, long life, small footprint on environment as compared to battery-based UPS system.

This Flywheel energy storage system can also be paralleled with battery bank as shown in Fig. 12, to increase reliability of power. Frequent cycle of charging and discharging are extremely harmful for life of batteries. DC flywheel system is capable of handling power disturbance or transient. Thus, life cycle of batteries is improved and energy of battery bank is saved for power outage. This proposed technology is capable of feeding power for longer duration outage and are extremely insensitive to frequent cycling as compared to batteries. This proposed flywheel energy storage system can also be used alone to provide uninterrupted supply for some application where longer duration back-up power is not necessary (Brown & Chvala, 2009).

Battery-less UPS System Powered by Fuel Cell and Super Capacitor

This battery less UPS topology is powered by the combination of fuel cell stack and super capacitor bank as shown in Fig. 13. In this proposed system, batteries are replaced by the super capacitor bank which is used to provide transient power during starting, load reversal and braking while steady state power is supplied by the fuel cell stack. In case of power failure or interruption, the super capacitor bank is installed in combination with fuel cell stack to reduce its starting time. Until the fuel cell is ramped to provide full load power super capacitor is used to feed power to the critical load.

Batteries and standby generator are used for storage purpose to provide required backup power. But the battery bank system has their own drawback like low energy and power density, environmentally incompatible, large number of charge/discharge cycle, generation of heat and pressure due to the high energy requirement. Also, standby generators have challenge of high installation and maintenance cost. To overcome all these drawbacks alternative energy storage system such as super capacitor, fuel cell and also their combination have been researched for generation and storage purpose.

Recently fuel cell is extremely used due to long duration operation, low maintenance requirement, low cost, high power capacity and environmental compatibility. The combination of fuel cell stack and super capacitor module also offer certain advantage such as supply of peak current, suppression of second order harmonic, removal of ripple by filtering and smoothing of glitches present in the power supply. It also help in improvement of power quality and economy of fuel (Chellappan & Enjeti, 2008).

Figure 13. Battery-less UPS system with Flywheel and super capacitor bank panel

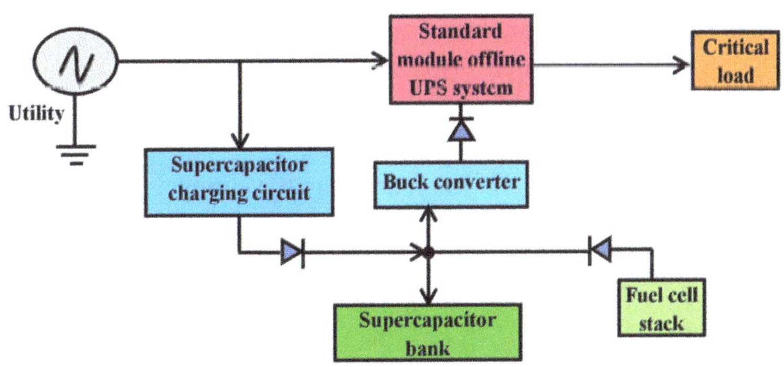

Future Generation Uninterruptible Power Supply System

Recently researcher are more focusing on the development and improvement of smart grid and it is going to be the future generation of conventional power system. The sustainable method of generating electricity such as photovoltaic panel is environmentally compatible and it can be operated in parallel with the smart grid system. With the help of this proposed distributed generation system load is shared among the generating sources during maximum power demand. In case of interruption, the passive energy storage system such as super-capacitor, battery bank can be used to provide backup power to the load. Thus, reliable power is provided to the sensitive load and the overall cost of generation and distribution is reduced (Colak et al., 2015; Farhangi, 2010).

The concept of smart distributed generation is also included in the category of the uninterruptible power supply (UPS) system. For accommodation of DG sources an intelligent method of energy storage system has been developed for energy saving and providing reliable power supply. The operation of future generation UPS system has been explained with the help of block diagram as shown in Fig. 14.

The block diagram includes high frequency and bidirectional converter which basically allow the parallel operation of battery bank with other distributed generation system (DG) in smart grid. Hence, this proposed system facilitate the cyclic utilization of electrical power between power grid and improved energy storage system. It can be efficiently utilized for driving motor and supplying the auxiliary component of hybrid electric vehicle in Distributed Generation system (Chiang et al., 2010; Zhao et al., 2013).

This intelligent UPS system has been developed to improve the economy, reliability and efficiency of the smart grid. When the tariffs rates are profitable, power can be transferred to the micro-grid. Multiple energy source, improved power converter and energy storage system can work together to supply uninterrupted power to the load. Instead of large development this proposed UPS system need further research work to efficiently utilize the concept of micro –grid and smart grid system. (Abusara et al., 2014; Xu et al., 2007).

Proposed Back-Up Connection Based Uninterrupted Power Supply

In this back-up line-based UPS system, the utilization of reserve bus is encouraged to provide back-up synchronized supply from main bus of the supply side main grid. The basis architectural diagram for back-up connection-based UPS system is shown in Fig. 17. The main purpose of reserve bus in grid system is to provide temporary supply during the maintenance or failure of main bus. So, in the proposed UPS system, the back-up supply from main bus can be extended to distribution end through reserve bus during emergency period so that any failure of system synchronization can be avoided. Also, during the maintenance of point of common coupling bus breakers, the proposed arrangement helps in avoiding the islanding condition of microgrid system. Hence, the uninterrupted power supply can be provided to load and DG with the implementation of the new adaptive scheme. The main operational diagram of the proposed UPS system is shown in Fig. 15.

As shown in Fig. 17, if any abnormal condition arises in the connecting line between main grid and microgrid with the availability of main grid supply, the main circuit breaker (CB) will trip to isolate the two connected grids. It may cause DGs to move out of synchronization state. Under the development of such a condition, the main CB signal will be transferred to the bus coupler through communication channel to charge the reserve bus so that power can be extended from reserve bus. Next, with the help

Figure 14. Block diagram of proposed intelligent UPS system

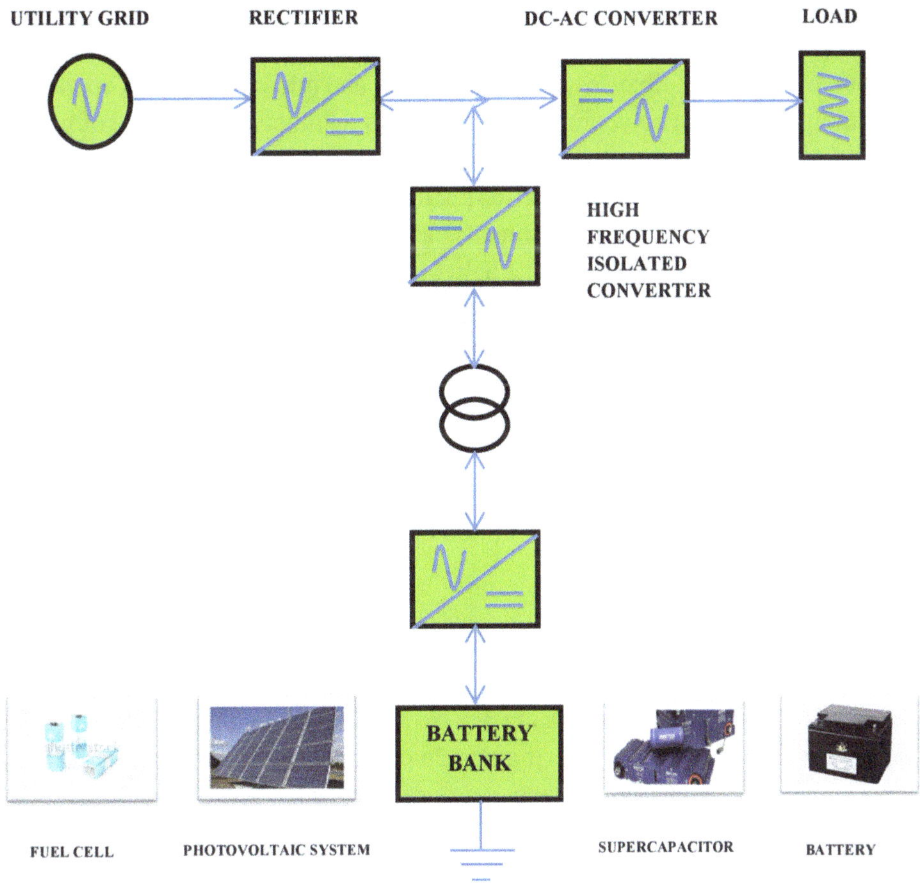

Figure 15. Proposed scheme for uninterrupted supply

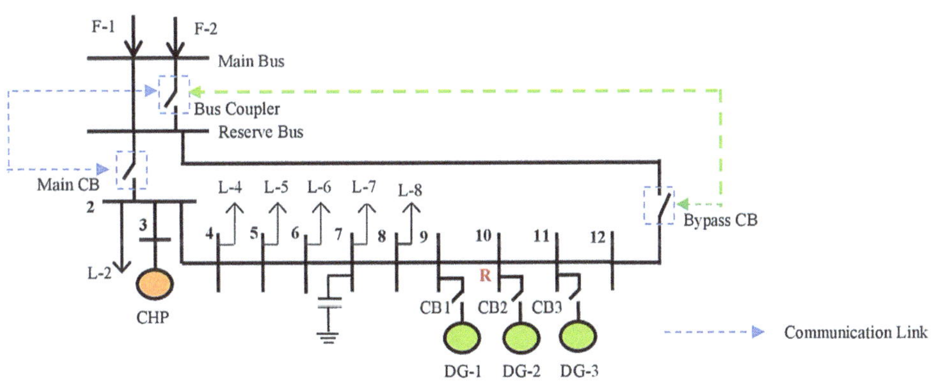

of another feeder, power from reserve bus can be supplied to the other end of microgrid system and the arrangement is shown in dotted green lines. In this scheme, the bus coupler open/close information is translated to the bypass CB. Both bus coupler and bypass CB close with the opening of main CB. Most of the cases, due to main grid disturbances main CB pull-out and microgrid system operate under islanding mode even through with zero power mismatch condition. Unintentional failure of main CB leads to in-synchronization of microgrid system which further causes power interruption at distribution level. But the proposed scheme helps in improving grid functionality as well as provides uninterrupted power supply to distribution grid penetrated through renewable sources. Also, during fault on distribution side, continuity of supply can be maintained with the help of such an arrangement. In this proposed scheme, the bus coupler can also be utilized for transferring the load from main bus to reserve bus in case of overloading. Under maintenance of main bus also synchronization with microgrid can be maintained using reserve bus using this arrangement.

Thus, this proposed system is capable of maintaining continuity of the supply in case of power failure. The proposed scheme offers only initial cost of communication and additional bypass line between reserve bus to dead end of microgrid system. Although this proposed scheme has higher initial cost but once installed is able to mitigate 50-60% interruption of power supply at distribution level.

CONCLUSION

The popularity of microgrid system is increasing day-by-day so that continuous, clean and cost-effective electricity can be supplied to end users. During certain disturbances or islanding condition, the power supply is disconnected from the distributed generation system, as the voltage and frequency may not be maintained within prescribed limits. It may cause DGs to move out of synchronization state. With the implementation of adaptive and advanced technology, such a condition can be avoided so that uninterrupted power supply can be maintained in the microgrid system. In this chapter, a detail literature review on UPS technology has been provided including architecture, classification and control strategies. A comparison between different UPS topologies has also been provided. Performance of different UPS technologies on the basis of their efficiency, merits and demerits have been analysed and mentioned. The comparative assessments of various UPS system and their control technique have been presented to provide effective information for selection of UPS technology for specific application. Also, for nonlinear load different nonlinear control techniques can be implemented. But such control techniques are complicated in nature. For the control of the inverter model predictive control can be performed excellently. UPS system based on hybrid energy sources and their application in micro and smart grid system shows new path for technological development and research in this area. Depending on the requirement of power rating and backup time a suitable UPS topology can be selected.

To mitigate the drawbacks of conventional UPS system such as reduction of power quality and generation of harmonics, new battery-less UPS systems have been proposed in this chapter. This proposed system has flywheel as an alternative to batteries and also the combination of super capacitor and fuelcell has replaced the conventional energy storage technology to provide more reliable power for longer duration. A new scheme has been proposed consisting of main and reserve bus to feed power to sensitive load during outage, interruption and over-loading. The proposed scheme helps in reducing the power outage rate to 30-40% at distribution level. This chapter will be very helpful for reference in the field of research, design and manufacturing of UPS system.

REFERENCES

Aamir, M., & Mekhilef, S. (2017). An online transformerless uninterruptible power supply system with a smaller battery bank for low power applications. *IEEE Transactions on Power Electronics, 32*(1), 233–247.

Abusara, M. A., Guerrero, J. M., & Sharkh, S. M. (2014). Line-interactive UPS for micro-grids. *IEEE Transactions on Industrial Electronics, 61*(3), 1292–1300.

Ahmad, A., Saqib, A. S., & Kashif, A. R. (2016). Impact of wide-spread use of uninterruptible power supplies on Pakistan's power system. *Energy Policy, 98*, 629–636.

Ahmed, O., & Bleijs, J. (2015). An overview of DC–DC converter topologies for fuel cellultracapacitor hybrid distribution system. *Renewable & Sustainable Energy Reviews, 42*, 609–626.

Alhelou, H., Hamedani-Golshan, M. E., Zamani, R., Heydarian-Forushani, E., & Siano, P. (2018). Challenges and Opportunities of Load Frequency Control in Conventional, Modern and Future Smart Power Systems: A Comprehensive Review. *Energies, 11*(10), 2497.

Alhelou, H. H. (2018). *Fault Detection and Isolation in Power Systems Using Unknown Input Observer. In Advanced Condition Monitoring and Fault Diagnosis of Electric Machines* (p. 38). Hershey, PA: IGI Global.

Alhelou, H. H., Golshan, M., & Fini, M. (2018). Wind Driven Optimization Algorithm Application to Load Frequency Control in Interconnected Power Systems Considering GRC and GDB Nonlinearities. *Electric Power Components and Syst.*

Alhelou, H. H., & Golshan, M. E. H. (2016, May). Hierarchical plug-in EV control based on primary frequency response in interconnected smart grid. In *Electrical Engineering (ICEE), 2016 24th Iranian Conference on* (pp. 561-566). IEEE.

Alhelou, H. H., Golshan, M. H., & Askari-Marnani, J. (2018). Robust sensor fault detection and isolation scheme for interconnected smart power systems in presence of RER and EVs using unknown input observer. *International Journal of Electrical Power & Energy Systems, 99*, 682–694.

Alhelou, H. H., Hamedani-Golshan, M. E., Heydarian-Forushani, E., Al-Sumaiti, A. S., & Siano, P. (2018, September). Decentralized Fractional Order Control Scheme for LFC of Deregulated Nonlinear Power Systems in Presence of EVs and RER. In *2018 International Conference on Smart Energy Systems and Technologies (SEST)* (pp. 1-6). IEEE.

Alhelou, H. S. H., Golshan, M. E. H., & Fini, M. H. (2015, December). Multi agent electric vehicle control based primary frequency support for future smart micro-grid. In *Smart Grid Conference (SGC)* (pp. 22-27). Academic Press.

Alshahrestani, A., Golshan, M. E. H., & Alhelou, H. H. (2018, November). WAMS Based Online Estimation of Total Inertia Constant and Damping Coefficient for Future Smart Grid Systems. In *Smart Grid Conference (SGC)* (pp. 1-5). Academic Press.

Arsad, N., & Ali, U. (2017). An analysis of the effects of residential uninterruptable power supply systems on Pakistan's power sector. *Energy for Sustainable Development, 36*, 16–21.

Bekiarov, S. B., & Emadi, A. (2002).Uninterruptible power supplies: classification, operation, dynamics, and control. *APEC Seventeenth Annual IEEE Applied Power Electronics Conference and Exposition*, 597-604.

Bojoi, R. I., Limongi, L. R., Roiu, D., & Tenconi, A. (2011). Enhanced power quality control strategy for single-phase inverters in distributed generation systems. *IEEE Transactions on Power Electronics*, *26*(3), 789–806.

Bortolini, M., Gamberi, M., & Graziani, A. (2014). Technical and economic design of photovoltaic and battery energy storage system. *Energy Conversion and Management*, *86*, 81–92.

Branco, C. G. C., Torrico-Bascope, R. P., Cruz, C. M. T., & Lima, F. K. de A. (2013). Proposal of three-phase high-frequency transformer isolation UPS topologies for distributed generation applications. *IEEE Transactions on Industrial Electronics*, *60*(4), 1520–1531.

Brown, D. R., & Chvala, W. D. (2009). Flywheel Energy Storage: An Alternative to Batteries for UPS Systems. *Energy Engineering*, *102*(5), 7–26.

Brownlie, d. (1988). Battery requirements for uninterruptible power-supply applications. *Journal of Power Sources*, *23*, 211–220.

Camacho, E. F., & Bordons, C. (2007). *Model Predictive Control*. New York: Springer-Verlag.

Chauhan, A., & Saini, R. (2014). A review on integrated renewable energy system based power generation for stand-alone applications: Configurations, storage options, sizing methodologies and control. *Renewable & Sustainable Energy Reviews*, *38*, 99–120.

Chellappan, M. V., Todorovic, M. H., & Enjeti, P. N. (2008). Fuel Cell Based Battery-less UPS System. *IEEE Industry Applications society Annual meeting*, 1-8.

Chen, D., Zhang, J., & Qian, Z. (2013). An improved repetitive control scheme for grid-connected inverter with frequency-adaptive capability. *IEEE Transactions on Industrial Electronics*, *60*(2), 814–823.

Chen, X., Fu, Q., & Wang, D. (2009). Performance analysis of PV grid-connected power conditioning system with UPS. *4th IEEE Conference on Industrial Electronics and Applications (ICIEA)*, 2172-2176.

Chiang, H., Ma, T., Cheng, Y., Chang, J., & Chang, W. (2010). Design and implementation of a hybrid regenerative power system combining grid–tie and uninterruptible power supply functions. *IET Renewable Power Generation*, *4*, 85–99.

Choi, J. H., Kwon, J. M., Jung, J. H., & Kwon, B. H. (2005). High performance online UPS using three-leg-type converter. *IEEE Transactions on Industrial Electronics*, *52*(3), 889–897.

Choi, W., Howze, J. W., & Enjeti, P. (2006). Fuel-cell powered uninterruptible power supply systems: Design considerations. *Journal of Power Sources*, *157*, 311–317.

Colak, I., Kabalci, E., Fulli, G., & Lazarou, S. (2015). A survey on the contributions of power electronics to smart grid systems. *Renewable & Sustainable Energy Reviews*, *47*, 562–579.

Cortes, P., Kazmierkowski, M. P., Kennel, R. M., Quevedo, D. E., & Rodriguez, J. (2008). Predictive Control in Power Electronics and Drives. *IEEE Transactions on Industrial Electronics*, *55*(12), 4312–4324.

Dhal, P. K., & Rajan, C. A. (2015). Power Quality Improvement Based on Uninterruptible Power Supply (UPS) in Distribution System. *IEEE Sponsored 2nd International Conference on Electronics and Communication System,* 286-290.

Elci, H., Longman, R. W., Phan, M. Q., Juang, J., & Ugoletti, R. (2002). Simple learning control made practical by zero-phase filtering: Applications to robotics. *IEEE Transactions on Circuits and Systems. I, Fundamental Theory and Applications*, *49*(6), 753–767.

Farhangi, H. (2010). The path of the smart grid. *Power Energy and Magazine. IEEE*, *8*, 18–28.

Fini, M. H., Yousefi, G. R., & Alhelou, H. H. (2016). Comparative study on the performance of many-objective and single-objective optimisation algorithms in tuning load frequency controllers of multi-area power systems. *IET Generation, Transmission & Distribution*, *10*(12), 2915–2923.

Fu, Q., Montoya, L. F., Solanki, A., Nasiri, A., Bhavaraju, V., Abdallah, T., & … . (2012). Microgrid generation capacity design with renewables and energy storage addressing power qualityand surety. *IEEE Transactions on Smart Grid*, *3*, 2019–2027.

Goodwin, G. C., Seron, M. M., & Dona, J. D. (2004). *Constrained Control and Estimation*. New York: Springer-Verlag.

Guerrero, J. M., Hang, L., & Uceda, J. (2008). Control of distribution uninterruptible power supply systems. *IEEE Transactions on Industrial Electronics*, *55*(8), 2845–2859.

Hadjipaschalis, I., Poullikkas, A., & Efthimiou, V. (2009). Overview of current and future energy storage technologies for electric power applications. *Renewable & Sustainable Energy Reviews*, *13*, 1513–1522.

Haneyoshi, T., Kawamura, A., & Hoft, R. G. (1988). Waveform compensation of PWM inverter with cyclic fluctuating loads. *IEEE Transactions on Industry Applications*, *24*(4), 582–589.

Hu, X., Karimi, H. R., Wu, L., & Guo, Y. (2014). Model predictive control-based non-linear fault tolerant control for air-breathing hypersonic vehicles. *IET Control Theory & Applications*, *8*(13), 1147–1153.

Ibrahim, H., Ilinca, A., & Perron, J. (2008). Energy storage systems—characteristics and comparisons. *Renewable & Sustainable Energy Reviews*, *12*, 1221–1250.

Jiang, S., Cao, D., Li, Y., Liu, J., & Peng, F. Z. (2012). Low-THD, fast-transient, and cost-effective synchronous-frame repetitive controller for three-phase UPS inverters. *IEEE Transactions on Power Electronics*, *27*(6), 2994–3005.

Karshenas, H. R., & Niroomand, M. (2005). Design and implementation of a single phase inverter with sine wave tracking method for emergency power supply with high performance. *International conference on electrical machines and systems*, 1232-1237.

Karve, S. (2002). Three of a kind. *IEE Review*, *46*(2), 27–31.

Kempf, C. J., & Kobayashi, S. (1999). Disturbance observer and feed forward design for a high-speed direct-drive positioning table. *IEEE Transactions on Control Systems Technology*, *7*(5), 513–526.

Kim, E.-H., Kwon, J.-M., & Kwon, B.-H. (2009). Transformerless three-phase on-line UPS with high performance. *IET Power Electronics*, *2*(2), 103–112.

Kim, S. K., Park, C. R., Yoon, T. W., & Lee, Y. I. (2015). Disturbance observer based model predictive control for output voltage regulation of three-phase inverter for uninterruptible power supply applications. *European Journal of Control*, *23*, 71–83.

Kirubakaran, A., Jain, S., & Nema, R. (2009). A review on fuel cell technologies and power electronic interface. *Renewable & Sustainable Energy Reviews*, *13*, 2430–2440.

Kissaoui, M., Al Tahir, A. A. R., Abouloifa, A., Chaoui, F. Z., Abouelmahjoub, Y., & Giri, F. (2016). Output-Feedback Nonlinear Adaptive Control Strategy of Three-phase AC/DC Boost Power Converter for On-line UPS Systems. *IFAC-Papers On Line*, *49*(13), 324–329.

Kissaoui, M., Chaoui, F. Z., Abouloifa, A., Giri, F., & Abouelmahjoub, Y. (2014). Adaptive control of uninterruptible power supply based on AC/AC Power Converter. *2014 International Conference on Multimedia Computing and Systems (ICMCS)*, 1557-1562.

Kollimalla, S. K., Mishra, M. K., & Narasamma, N. L. (2014). Design and Analysis of Novel Control Strategy for Battery and Super-capacitor Storage System. *IEEE Transaction on Sustainable Energy*, *5*, 1137–1144.

Koohi-Kamali, S., Tyagi, V., Rahim, N., Panwar, N., & Mokhlis, H. (2013). Emergence of energy storage technologies as the solution for reliable operation of smart power systems: A review. *Renewable & Sustainable Energy Reviews*, *25*, 135–165.

Kwon, B. H., Choi, J. H., & Kim, T. W. (2001). Improved single-phase line-interactive UPS. *IEEE Transactions on Industrial Electronics*, *48*(4), 804–811.

Lahyani, A., Venet, P., Guermazi, A., & Troudi, A. (2013). Battery/super-capacitors combination in uninterruptible power supply (UPS). *IEEE Transactions on Power Electronics*, *28*, 1509–1522.

Lasseter, R. H. (2011). Smart distribution: Coupled micro-grids. *Proceedings of the IEEE*, *99*(6), 1074–1082.

Liu, J., Laghrouche, S., Ahmed, F. S., & Wack, M. (2014). PEM fuel cell air-feed system observer design for automotive applications: An adaptive numerical differentiation approach. *International Journal of Hydrogen Energy*, *39*(30), 17210–17221.

Liu, J., Laghrouche, S., & Wack, M. (2014). Observer-based higher order sliding mode control of power factor in three-phase ac/dc converter for hybrid electric vehicle applications. *International Journal of Control*, *87*(6), 1117–1130.

Liu, S., Zhuang, S., Xu, Q., & Xiao, J. (2016). Improved voltage shift islanding detection method for multi-inverter grid-connected photovoltaic systems. *IET Generation, Transmission & Distribution*, *10*(13), 3163–3169.

Maciejowski, J. M. (2001). *Predictive Control with Constraints*. Englewood Cliffs, NJ: Prentice-Hall.

Makdisie, C., Haidar, B., & Alhelou, H. H. (2018). An Optimal Photovoltaic Conversion System for Future Smart Grids. In *Handbook of Research on Power and Energy System Optimization* (pp. 601–657). IGI Global.

Marei, M. I., Abdallah, I., & Ashour, H. (2011). Transformerless Uninterruptible Power Supply with Reduced Power Device Count. *Electric Power Components and Systems*, *39*(11), 1097–1116.

Mastromauro, R. A., Liserre, M., Kerekes, T., & Dell'Aquila, A. (2009). A single-phase voltage-controlled grid-connected photovoltaic system with power quality conditioner functionality. *IEEE Transactions on Industrial Electronics*, *56*(11), 4436–4444.

Mattavelli, P. (2005). An improved deadbeat control for UPS using disturbance observers. *IEEE Transactions on Industrial Electronics*, *52*, 206–212.

Mekhilef, S., Saidur, R., & Safari, A. (2012). Comparative study of different fuel cell technologies. *Renewable & Sustainable Energy Reviews*, *16*, 981–989.

Monfared, M., Golestan, S., & Guerrero, J. M. (2014). Analysis, design, and experimental verification of a synchronous reference frame voltage control for single-phase inverters. *IEEE Transactions on Industrial Electronics*, *61*, 258–269.

Moriyama, A., Ando, I., & Takahashi, I. (1998). Sinusoidal voltage control of a single phase uninterruptible power supply by a high gain PI circuit. In *Proceedings of 24th Annual Conference of Industrial Electronics Society*. IEEE.

Muller, S., Ammann, U., & Rees, S. (2005). New time-discrete modulation scheme for matrix converters. *IEEE Transactions on Industrial Electronics*, *52*(6), 1607–1615.

Nadweh, S., Hayek, G., Atieh, B., & Haes Alhelou, H. (2018). Using Four – Quadrant Chopper with Variable Speed Drive System Dc-Link to Improve the Quality of Supplied Power for Industrial Facilities. Majlesi Journal of Electrical Engineering.

Nale, R., & Biswal, M. (2017).Comparative assessment of passive islanding detection techniques for microgrid. *International Conference on Innovations in Information, Embedded and Communication System*, 1-5.

Nasiri, A. (2007). Digital control of three-phase series-parallel uninterruptible power supply systems. *IEEE Transactions on Power Electronics*, *22*, 1116–1127.

Nasiri, A., Nie, Z., Bekiarov, S. B., & Emadi, A. (2008). An on-line UPS system with power factor correction and electric isolation using BIFRED converter. *IEEE Transactions on Industrial Electronics*, *55*(2), 722–730.

Nayar, C. V., Ashari, M., & Keerthipala, W. (2000). A grid-interactive photovoltaic uninterruptible power supply system using battery storage and a back -up diesel generator. *IEEE Transactions on Energy Conversion*, *15*, 348–353.

Niroomand, M., & Karshenas, H. (2010). Review and comparison of control methods for uninterruptible power supplies. IEEE Power Electronic & Drive Systems & Technologies Conference (PEDSTC), 18-23.

Njenda, T. C., Golshan, M. E. H., & Alhelou, H. H. (2018, November). WAMS Based Intelligent Under Frequency Load Shedding Considering Online Disturbance Estimation. In *Smart Grid Conference (SGC)* (pp. 1-5). Academic Press.

Njenda, T. C., Golshan, M. E. H., & Alhelou, H. H. (2018, November). WAMS based Under Frequency Load Shedding Considering Minimum Frequency Prediction and Extrapolated Disturbance Magnitude. In *Smart Grid Conference (SGC)* (pp. 1-5). Academic Press.

Oudalova, A. (2011). *Advanced architectures and control concepts for more micro-grids*. Retrieved from www.microgrids.eu/documents/654.pdf

Park, J. K., Kwon, J. M., Kim, E. H., & Kwon, B. H. (2008). High-performance transformerless online UPS. *IEEE Transactions on Industrial Electronics*, *55*(8), 2943–2953.

Racine, M. S., Parham, J. D., & Rashid, M. (2005). An overview of uninterruptible power supplies. *Proceedings of the 37th Annual North American Power Symposium*, 159-64.

Rathmann, S., & Warner, H. A. (1996). New generation UPS technology, the delta conversion principle. *IAS '96. Conference Record of the 1996 IEEE Industry Applications Conference Thirty-First IAS Annual Meeting*.

Redfern, M. A., Usta, O., & Fielding, G. (1993). Protection against loss of utility grid supply for a dispersed storage and generation unit. *IEEE Transactions on Power Delivery*, *8*(3), 948–954.

Ren, G., Ma, G., & Cong, N. (2015). Review of electrical energy storage system for vehicular applications. *Renewable & Sustainable Energy Reviews*, *41*, 225–236.

Rodríguez, J., Pontt, J., Silva, C., Salgado, M., Rees, S., Ammann, U., ... Cortes, P. (2004). Predictive control of a three-phase inverter. *Electronics Letters*, *40*(9), 561–562.

Shi, K., Li, H., Hu, C., & Xu, D. (2015). Topology of super uninterruptible power supply with multiple energy sources. *2015 9th International Conference on Power Electronics and ECCE Asia (ICPE-ECCE Asia)*, 1742-1749.

Silva, E. I., McGrath, B. P., Quevedo, D. E., & Goodwin, G. C. (2007). Predictive control of a flying capacitor converter. *Proceedings of American Control Conference*, 3763-3768.

Solter, W. (2002). A new international UPS classification by IEC 62040-3. *IEEE INTELEC 24th Annual International Telecommunications Energy Conference*, 541-555.

Tzou, Y. Y., Ou, R. S., Jung, S. L., & Chang, M. Y. (1997). High-performance programmable ac power source with low-harmonic distortion using DSP-based repetitive control technique. *IEEE Transactions on Power Electronics*, *2*(4), 715–725.

Vargas, R., Rodriguez, J., Ammann, U., & Wheeler, P. W. (2008). Predictive current control of an induction machine fed by a matrix converter with reactive power control. *IEEE Transactions on Industrial Electronics*, *55*(12), 4362–4371.

Wang, W., Kliber, J., Zhang, G., Xu, W., Howell, B., & Palladino, T. (2007). A power line signaling based scheme for anti-islanding protection of distributed generators-part II: Field test results. *IEEE Transactions on Power Delivery*, *22*(3), 1767–1772.

Wang, X., Li, X., Wang, J., Fang, X., & Zhu, X. (2016). Data-driven model-free adaptive sliding mode control for the multi degree-of-freedom robotic exoskeleton. *Information Sciences*, *327*, 246–257.

Xu, W., Zhang, G. C., Wang, W., Wang, G., & Kliber, J. (2007). A power line signaling based technique for anti-islanding protection of distributed generators-part 1: Scheme and analysis. *IEEE Transactions on Power Delivery*, *22*(3), 1758–1766.

Yafaoui, A., Wu, B., & Kouro, S. (2012). Improved active frequency drift anti-islanding detection method for grid connected photovoltaic systems. *IEEE Transactions on Power Electronics*, *27*(5), 2367–2375.

Zamani, R., Hamedani-Golshan, M. E., Haes Alhelou, H., Siano, P., & Pota, R, H. (2018). Islanding Detection of Synchronous Distributed Generator Based on the Active and Reactive Power Control Loops. *Energies*, *11*(10), 2819.

Zhan, Y., Guo, Y., Zhu, J., & Li, Li. (2015). Performance comparison of input current ripple reduction methods in UPS applications with hybrid PEM fuel cell/super-capacitor power sources. *International Journal of Electrical Power & Energy Systems*, *64*, 96–103.

Zhan, Y., Guo, Y., Zhu, J., & Li, Li. (2015). Power and energy management of grid/PEMFC/ battery/ super-capacitor hybrid power sources for UPS applications. *International Journal of Electrical Power & Energy Systems*, *67*, 598–612.

Zhan, Y., Guo, Y., Zhu, J., & Wang, H. (2008). Intelligent uninterruptible power supply system with back-up fuel cell/battery hybrid power source. *Journal of Power Sources*, *179*, 745–753.

Zhang, K., Kang, Y., Xiong, J., & Chen, J. (2003). Direct repetitive control of SPWM inverter for UPS purpose. *IEEE Transactions on Power Electronics*, *18*(3), 784–792.

Zhang, W., Xu, D., Li, X., Xie, R., & Li, H. (2013). Seamless transfer control strategy for fuel cell uninterruptible power supply system. *IEEE Transactions on Power Electronics*, *28*, 717–729.

Zhao, B., Song, Q., Liu, W., & Xiao, Y. (2013). Next-generation multi-functional modular intelligent UPS system for smart grid. *IEEE Transactions on Industrial Electronics*, *60*, 3602–3618.

Zhou, Z. J., Zhang, X., Xu, P., & Shen, W. X. (2008). Single-phase uninterruptible power supply based on Z-source inverter. *IEEE Transactions on Industrial Electronics*, *55*(8), 2997–3004.

KEY TERMS AND DEFINITIONS

Distributed Energy Resources (DERs): Distributed Energy Resources (DERs) are electricity-producing resources or controllable loads that are directly connected to a local distribution system or connected to a host facility within the local distribution system.

Distributed/Decentralised Generation (DG): Distributed/Decentralised generation (DG) is electrical generation and storage performed by a variety of small, grid-connected devices referred to as distributed energy resources (DER).

Islanding: Islanding is a condition in which distribution sector becomes electrically isolated from the main supply power, but it gets continuously energized by DG connected to it.

Microgrid: Microgrid is a group of interconnected loads and distributed energy resources within clearly defined electrical boundaries that acts as a single controllable entity with respect to the grid. A microgrid can connect and disconnect from the grid to enable it to operate in both grid-connected or island-mode.

Point of Common Coupling: The point where the microgrid is connected with the main grid through a breaker mechanism.

Uninterrupted Power Supply (UPS): Uninterrupted Power Supply (UPS) is a kind of an electrical system which is used to provide backup power supply to critical loads and emergency feeders.

Chapter 5
Home Load-Side Management in Smart Grids Using Global Optimization

Abdelmadjid Recioui
University of Boumerdes, Algeria

ABSTRACT

Demand-side management (DSM) is a strategy enabling the power supplying companies to effectively manage the increasing demand for electricity and the quality of the supplied power. The main objectives of DSM programs are to improve the financial performance and customer relations. The idea is to encourage the consumer to use less energy during peak hours, or to move the time of energy use to off-peak times. The DSM controls the match between the demand and supply of electricity. Another objective of DSM is to maintain the power quality in order to level the load curves. In this chapter, a genetic algorithm is used in conjunction with demand-side management techniques to find the optimal scheduling of energy consumption inside N buildings in a neighborhood. The issue is formulated as multi-objective optimization problem aiming at reducing the peak load as well as minimizing the energy cost. The simulations reveal that the adopted strategy is able to plan the daily energy consumptions of a great number of electrical devices with good performance in terms of computational cost.

INTRODUCTION

Traditional power grids face some challenges that limit their efficiency. Among these deficiencies are the non-optimal dimensioning and usage of grid resources. In order to meet customers' demand of electric energy, power capacity must be able to meet the worst case scenario that is the peak of demand. Moreover, additional capacity must be available to deal with the uncertainty in generation and consumption. Based on that, the grids resources are, for most of the time, underutilized (Barbato and Capone, 2014).

Another proliferating issue in the current power grids is the integration of medium- to small-sized renewable energy source (RES) plants. The tremendous employment of RESs is motivated by the enormous socio-economic benefits obtainable with these sources. These include: reduction of greenhouse gas emissions and air pollution, diversification of energy supply, reduced dependence on imported fuels,

DOI: 10.4018/978-1-5225-8030-0.ch005

economic development and jobs in manufacturing, installation and management of RESs plants (Ortega et al., 2013; Pfeiffer and Mulder, 2013; Chellali et al., 2011). Despite these benefits, industry and consumers themselves, there are barriers that can limit renewable sources' integration (Richards et al., 2012).

Solutions to the previous dilemma have been found in the revolutionizing aspect of Smart grids (SGs). SGs can make grids more efficient and smarter by means of facilitating the deployment of renewable energy sources, decreasing oil consumption by reducing the need for inefficient generation during peak usage periods, optimizing resources utilization and construction of back-up (peak load) power plants and enabling the integration of plug-in electric vehicles(PEVs) and energy storage systems (ESSs) (NIST, 2010). The energy produced is dispatched through the transmission and distribution sectors, which are controlled by the operation domain. The balance between supply and the demand is guaranteed by the market domain, which consists of suppliers of bulk electricity, retailers who supply electricity to users and traders who buy electricity from suppliers and sell it to retailers and aggregators of distributed small-scale power plants. The service provider manages services for utilities companies and end-users, like billing and consumers' account management. Finally, customers consume energy, but can also generate and store electricity locally. This domain includes residential, commercial and industrial customers, who can actively contribute to the efficiency of the grid (Barbato and Capone, 2014).

Demand management mechanisms can be classified into two main categories: demand-response (DR) and demand-side management (DSM). DR methods are reactive solutions designed to encourage consumers to dynamically change their electricity demand in the short term, according to signals provided by the grid/utilities, such as prices or emergency condition requests. Typically, these techniques are used to reduce the peak demand or to avoid system emergencies, such as blackouts. On the other hand, DSM is a proactive approach aimed at making consumers energy-efficient in the long-term. In the literature, demand-response and demand-side management terms are often used interchangeably. Thus, in some works on DSM, proposed solutions are called DR methods and vice-versa. Actually, demand-response and demand-side management are two different methodologies, which can also be used in conjunction with each other (Barbato and Capone, 2014)..

Demand management mechanisms can be designed to control the electric resource of individual users. However, this approach may have some undesirable effects (Strbac, 2008). In fact, consumers are characterized by diversity in terms of appliance usage. This feature is fully exploited by the power system to optimize its efficiency in generating and distributing energy. Demand management mechanisms for individual users may actually disturb this diversity. As an example, in the case of systems for consumers' payment reduction, all users would shift their loads to periods of the day where the electricity prices are low. Unfortunately, this would determine large peaks of demand during such low-cost periods and, possibly, service interruptions (i.e., blackout or brownouts) (Barbato and Capone, 2014).. To contain these unwanted side effects, management mechanisms can be designed to control the community of users, thus managing their resources based on a system-wide perspective.

Two different approaches are proposed in the literature to define these methods: optimization and game theory. In the first case, all consumers are supposed to cooperate in managing their resources, and optimization models are used to minimize a shared utility function. However, these solutions do not incorporate conflicts among users. In the case of real-time tariffs, for example, energy prices depend on the overall users' demand, and one consumer's actions directly affect the others in terms of costs. As a consequence, minimizing the overall bill may be unfair in terms of payment sharing among customers. In order to address conflicts, game theory is usually used, since it can model complex interactions among the independent rational players of the power grid (Saad et al., 2012).

In this chapter, the concept of intelligent home energy management in a neighborhood is introduced with an energy consumption planning system of the daily tasks of a set of household users being developed. The planning strategy aims at reducing the peak load as well as minimizing the energy cost. The maximization of the peak-to-average ratio of the total energy demand is considered as a desired objective function for the utility and the minimization of the energy cost is considered for consumer. The optimal solution of this multi-objective planning problem is found using a Genetic Algorithm (GA).

BACKGROUND

In power grids, generation capacity is required to meet peak-hour load demand plus a security margin. However, according to recent studies, the average utilization of the generation capacity is below 55% (US energy information administration, 2014). This leads to inefficient operation of power grids because a portion of generation plants is largely unused or underutilized but must still be maintained and supervised to guarantee its reliability. On the other hand, as energy demand and peak load demand continue to increase, additional generation capacity will be needed to accommodate future load demand, which requires a large investment and might lead to even lower utilization.

Recently, the smart grid (SG) has been proposed as a new type of electrical grid to modernize current power grids to efficiently deliver reliable, economic, and sustainable electricity services (US energy information administration, 2014). One of the key features of the SG is the replacement of conventional mechanical meters with smart meters to enable two-way communications between users and grid operators. Using the communication infrastructure of the SG, it is possible to shape load demand curves of the users by means of demand side management (DSM) programs (Goudarzi et al., 2011; Barbato et al., 2011; Agnetis et al., 2013). Commonly, demand side management is a term used by the electric companies to describe programs developed for the sake of influencing the electricity usage patterns of customers to control the energy consumption at the consumer/meter side.

The DSM is an opportunity to cancel or delay the need to construct new generating capacity by a reduction or a shift in the consumers' energy. Also, for domestic or industrial consumer, DSM can be considered an opportunity to save money by reducing their electricity bill taking the advantage of financial incentive provided by utility. In the global energy scenario, the demand management is an important function of the smart grid, which ensures the grid sustainability and reliability. Demand management is not entirely new for the electric grid, but it is moving towards a customer driven activity in the future (Agnetis et al., 2011; Zhao etal., 2013; Guo et al., 2012).

Demand management mechanisms can be designed to control the electric resources of individual users (Strbac, 2008). Two different approaches are proposed literature to address the DSM problem: optimization and game theory (Barbato and Capone, 2014). Game theory is practically used since it can model complex interactions among the independent rational players of the power grid (Saad et al., 2012). The extension of demand management mechanisms for communities of users is represented by techniques designed for micro-grids, which are small-scale versions of the electricity systems that locally generate and distribute electricity to consumers. These grids constitute an ideal way to integrate renewable resources at the community level and allow for customer participation in the electricity market (Liang and Zhuang, 2014; Prabaakaran et al., 2013; Ravichandran et al., 2013; Fini et al, 2016; Alhelou et al, 2018; Alhelou et al 2015; Njenda et al, 2018; Haes Alhelou et al, 2018; Zmani et al, 2018).

Many approaches are introduced to solve Demand side management as an optimization problem with proper objective functions (Barbato t al., 2013). Optimization methods for demand-side management can be classified based on three main Characteristics (Barbato and Capone, 2014; Makdisie et al, 2018; Alhelou et al., 2018; Alhelou et al., 2016; Nadweh et al., 2018; Alhelou et al., 2015): First, DSM systems can be designed to optimize the usage of electric resources of either individual users or a community of cooperative consumers. Users are individually managed, while in the second case, consumers collaborate in defining their operating plans and DSM methods are used to optimize a shared utility function. A further classification can be obtained based on whether deterministic or stochastic techniques are utilized to design the demand management mechanism. Finally, DSM systems can be classified based on the time scale used to manage the resources of customers: day-ahead and real-time. In the day-ahead stage, the operating plan of electric resources of users is defined over the next 24-h time period (or a different time horizon).

Various dynamic and effective schemes for autonomous DSM in smart girds have been proposed in literature. Examples of pioneering works include the one of (Mohsenian-Rad et al., 2010; Alhelou et al., 2018; Alhelou et al., 2015; Alhelou et al., 2016; Njenda et al., 2018) who proposed an autonomous load scheduling algorithm based on cooperative game theory, where each user is a player and their load schedules are the strategies. The authors in (Agarwal and Cui, 2011) proposed a load scheduling no cooperative game among users that can be reduced to a congestion game. In both studies, the single optimization objective is to minimize the electric bill of the users, while the reduction of the peak-hour consumption is considered as a desirable secondary effect. (Samadi et al., 2011) proposed an auction based scheme where users provide their utility functions and energy constraints to the utility company, who then replies with a set of prices that maximizes the utility functions of users. A similar auction scheme is also proposed by (Li et al., 2011). We can notice that previous studies mostly aim at a single objective, e.g., to minimize the cost of the users.

THE SMART GRID

A smart grid is an electricity network that uses digital and other advanced technologies to monitor and manage the transport of electricity from all generation sources to meet the varying electricity demands of end-users. It offers a lot of valuable technologies that are already in use today or that can be used within the near future. Smart grids co-ordinate the needs and capabilities of all generators, grid operators, end-users and electricity market stakeholders to operate all parts of the system as efficiently as possible i.e. minimizing costs and environmental impacts while maximizing system reliability, and stability. Smart Grid includes electric network, digital control appliance, and intelligent monitoring system (Zhao et al., 2013). All of these can:

- Deliver electricity from producers to consumer.
- Control energy flow.
- Reduce the loss of energy.
- Make the performance of the electric network more reliable and controllable.

In the short term, a smarter grid will function more efficiently, enabling it to deliver a specific level of service at an expected less costs while offering great societal benefits such as less impact on the environment. We can expect the Smart Grid to support the kind of transformation that the internet has already brought to our daily life (European Smart Grids Technology Platform, 2006).

Significance and Goals for Smart Grid

As the demand in energy climbed quickly, there has been serious shortage in energy transmission and distribution. With reference to US, since 2000 (U.S. Energy Information Administration, 2014), only 668 additional miles of interstate electric energy transmission line have been created, compared to the hundreds of thousands of high-voltage transmission lines. Consequently, the outage and power quality costs American Electric Utilities a lot. People need an optimal and efficient way to 'broadcast' the power flows from a few of central power generators to a large amount of consumers, then, Smart Grid came out.

Smart Grid can offer a lot of potential economic and environmental benefits and significance such as:

- Improving reliability of power quality and transmission
- Increasing power distribution efficiency and conservation
- Reducing costs for electric utilities
- Reducing expenditures on electricity by households and businesses
- Lower Greenhouse Gas (GHG) and other gas emissions

Improving the Reliability of Power Quality and Transmission

Because of the requirement of the increasing power, we can figure a lot of troubles such as; "the slow response time of mechanical switches", "a lack of automatic analytics", "more and more blackouts and brownouts happen". Taking US as an example, in the past 40 years, there have been 5 massive blackouts, and three of them occurred in the past 9 years. However, Smart Grid can solve these problems today.

As technology evolves, people can make power more controllable and planned centrally. Now, with Smart Grid's help, we avoid this kind of risks before they happen (U.S. Energy Information Administration, 2014).

Smart Grid Technologies

There are many smart grid technology areas (each consisting of sets of individual technologies) spanning the entire grid, from generation through transmission and distribution to various types of electricity consumers. Some of the technologies are actively being deployed and are considered mature in both their development and application, while others require further development and demonstration. A fully optimized electricity system will deploy all the technology areas in Figure 1. However, not all technology areas need to be installed to increase the "smartness" of the grid (Technology Roadmap Smart Grids, 2011).

1. **Wide-Area Monitoring and Control:** Real-time monitoring and display of power system components and performance across interconnections and over large geographic areas help system operators to understand and optimize power system components, behavior and performance. Advanced system operation tools avoid blackouts and facilitate the integration of variable renewable energy resources (Technology Roadmap Smart Grids, 2011).
2. **Information and Communications Technology Integration:** The underlying communication infrastructure, whether using private utility communication networks (radio networks, meter mesh networks) or public carriers and networks (Internet, cellular, cable or telephone), supports data transmission for deferred and real-time operation, and during outages (Technology Roadmap Smart Grids, 2011).

Figure 1. Smart Grid Technology Areas

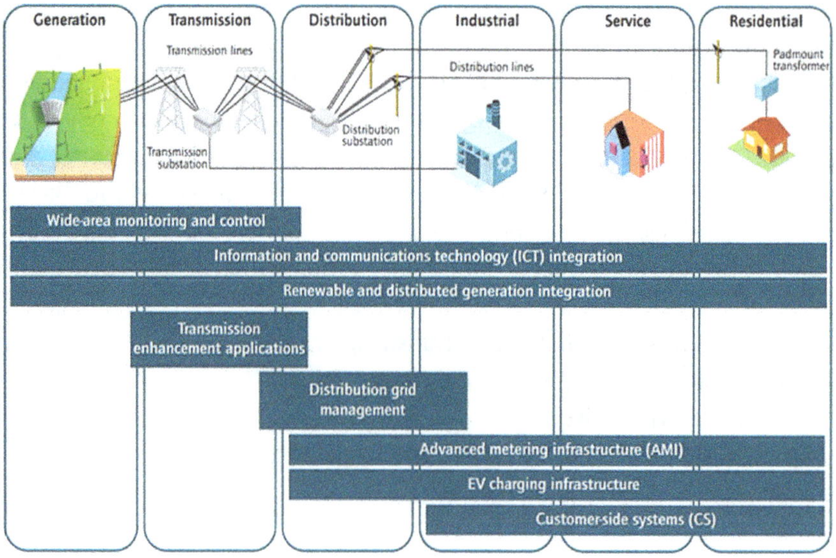

3. **Renewable and Distributed Generation Integration:** Integration of renewable and distributed energy resources can present challenges for the dispatch ability and controllability of these resources and for operation of the electricity system. Energy storage systems can alleviate such problems by decoupling the production and delivery of energy. Smart grids can help through automation of control of generation and to ensure supply and demand balance (Technology Roadmap Smart Grids, 2011).
4. **Transmission Enhancement Applications:** There is a number of technologies and applications for the transmission system:
    a. Flexible AC transmission systems (FACTS) are used to enhance the controllability of transmission networks and maximize power transfer capability.
    b. High voltage DC (HVDC) technologies are used to connect wind and solar farms to large power areas, with decreased system losses and enhanced system controllability.
    c. Dynamic line rating (DLR), which uses sensors to identify the current carrying capability of a section of network in real time, can optimize utilization of existing transmission assets, without the risk of causing overloads.
    d. High-temperature superconductors (HTS) can significantly reduce transmission losses and enable economical fault-current limiting with higher performance (Technology Roadmap Smart Grids, 2011).
5. **Distribution Grid Management:** Distribution and sub-station sensing and automation can reduce outage and repair time, maintain voltage level and improve asset management. Advanced distribution automation processes real-time information from sensors and meters for fault location, automatic reconfiguration of feeders, voltage and reactive power optimization, or to control distributed generation (Technology Roadmap Smart Grids, 2011).
6. **Advanced Metering Infrastructure:** Advanced metering infrastructure (AMI) involves the deployment of a number of technologies providing customers and utilities with data on electricity price and consumption, including the time and amount of electricity consumed (Technology Roadmap Smart Grids, 2011). AMI will provide a wide range of functionalities:

   a. Remote consumer price signals, which can provide time-of-use pricing information.
   b. Ability to collect, store and report customer energy consumption data for any required time intervals or near real time.
   c. Improved energy diagnostics from more detailed load profiles.
   d. Ability to identify location and extent of outages remotely via a metering function that sends a signal when the meter goes out and when power is restored.
   e. Remote connection and disconnection.
   f. Losses and theft detection.
7. **Electric Vehicle Charging Infrastructure:** Electric vehicle charging infrastructure handles billing, scheduling and other intelligent features for smart charging during low energy demand.
8. **Customer-Side Systems:** Customer-side systems, which are used to help manage electricity consumption at the industrial service and residential levels, include energy management systems, energy storage devices, smart appliances and distributed generation.

Functions of Smart Grid

The government and utilities funding development of grid modernization have defined the functions required for smart grids. According to the United States Department of Energy's Modern Grid Initiative report, a modern smart grid must have:

1. **Self-Healing From Power Disturbance Events:** The management of the smart grid requires digital control, automated analysis of problems, and automatic switching capabilities. This is defined as an intelligent grid control system which must become automated because decision speeds are increasingly becoming too fast for humans to manage.
   Operators or managers can use the real-time information to automatically avoid or mitigate power outages (U.S. Department of Energy, 2008), blackout, power quality problems, and system collision. Smart Grid will likely have a control system that can analyze its performance while it might be used to control electronic switches that are tied to multiple substations with varying costs of generation and reliability (Balls et al., 2008).
2. **Enabling Active Consumers Participation and Operating Resiliently Against Attack:** Smart Grid allows consumers to change their behaviors around variable electric rates. It incorporates consumer equipment and behavior in grid design, operation, and communication system. The connection between energy management systems enables consumers to manage energy better, and help them access to real-time pricing.
   Smart grid can identify and respond to hacker attack or natural disruptions better because real-time information enables both grid operators and managers to isolate affected areas and redirect power flows around damaged facilities.
   The smart monitoring of power grids can control and manage smart grids to avoid the system disruptions like blackouts (Balls et al., 2008).
3. **Providing Power Quality and Optimizing Assets:** According to recent data, losses of outages and power quality issues in US businesses are more than $100 billion on average annually. More stable power provided by smart grid technologies will reduce downtime and prevent such high cost.

Smart Grid can optimize capital assets by minimizing operations and maintaining lower costs. Optimizing power flows can make full use of lowest-cost generation resources and reduce waste which can ultimately save consumers money.

4. **Accommodating All Generation and Enabling New Products, Services, and Markets:** As smart grids support traditional power loads all time, integration of small-scale, localized, or on-site power generation allows residential, commercial, and industrial customers to self-generate and sell excess power to the grid with minimal technical or regulatory barriers. This also improves reliability and power quality, reduces electricity costs, and offers more customer choice. Intelligence in distribution grids will enable small producers to generate and sell electricity at the local level using many alternative sources (U.S. Department of Energy, 2008).

Features of Smart Grid Implementations

Existing and planned implementations of smart grids provide a wide range of features to perform the required functions.

1. **Load Reduction:** Usually, responding time to a rapid increase in power consumption should be longer than the start-up time of a large generator. If there is a smart grid, it may restrict all individual devices, or another larger customer, to reduce the load. With mathematical prediction algorithms' help, it is possible to figure out how many standby generators need to be used to reach a certain failure rate. In the traditional grid, the failure rate can only be reduced at the cost of more standby generators.
   In a smart grid, the load reduction by even a small portion of the clients may eliminate the problem (U.S. Department of Energy, 2008).
2. **Elimination of the Demand Fraction:** Normally, information only flows from the users and the loads they control back to the utilities. The utilities attempt to supply the demand and may succeed or fail to varying degrees. Eliminating the fraction of demand avoids the cost of adding reserve generators and allows users to cut their energy bills by allowing low priority devices to use energy only when it is cheapest (Energy future coalition, 2010)).
3. **Distribution of Power Generation:** Generation Distribution allows individual consumers to create power by themselves. This allows individual consumers to manage their generation directly to their load by which they can avoid power failure. If a local sub-network generates more power than it is consuming, the reverse flow can raise safety and reliability issues. A smart grid can manage these situations (Energy future coalition, 2010).

Energy Management

The term Energy Management has many meanings. It can be considered as the wisdom and effective use of energy to maximize profits (minimize costs) and enhance competitive position (Barney et al., 2006). Energy management is also the strategy of adjusting and optimizing energy using systems and procedures to reduce energy requirements per unit of output while holding constant or reducing total costs of producing the output from these systems. In fact, we are mainly concerned with the one that relates to saving energy businesses, public-sector, government organizations, and home (Kahlenborn et al., 2010).

1. **Meaning of Energy-Saving:** Energy management is the process of monitoring, controlling, and conserving energy in a building or an organization. Typically this involves the following steps:
    a. Metering our energy consumption and collecting the data.
    b. Finding opportunities to save energy, and estimating how much energy each opportunity could save.
    c. Taking action to target the opportunities to save energy. Typically we start with the best opportunities first.
    d. Tracking our progress by analyzing our meter data to see how well our energy-saving efforts have worked (Kahlenborn et al., 2010).
2. **Objectives of Energy Management:** A whole system viewpoint to energy management is required to ensure that many important activities will be examined and optimized. Some desirable sub-objectives of energy management programs include:
    a. Improving energy efficiency and reducing energy use, thereby reducing costs.
    b. Cultivating good communications on energy matters.
    c. Developing and maintaining effective monitoring, reporting and management strategies for wise energy usage.
    d. Finding new and better ways to increase returns from energy investments through research and development.
    e. Reducing the impacts of curtailments, brownouts, or any interruption in energy supplies (Barney et al., 2006).
3. **Importance of Energy Management:** Energy management is the key to saving energy. Much of the importance of energy saving stems from the global need to save energy - this global need affects energy prices, emissions targets, and legislation; all of which lead to several compelling reasons why we should save energy at our society specifically. Globally we need to save energy in order to:
    a. Reduce the damage that we are doing to our planet, Earth.
    b. Reduce our dependence on the fossil fuels that are becoming increasingly limited in supply.

    Energy management is the means to control and reduce energy consumption of our organization, and this is important because it enables us to:
    a. Reduce costs and this is becoming increasingly important as energy costs rise.
    b. Reduce carbon emissions and the environmental damage that they cause as well as the cost-related implications of carbon
    c. Reduce risk, the more energy we consume, the greater risk will have because the energy price increases or supply shortages could seriously affect our profitability, or even make it impossible for the business/organization to continue.

    With energy management, we can reduce this risk by reducing our demand for energy and by controlling it so as to make it more predictable (Kahlenborn et al., 2010).
4. **Managing Energy Consumption:** Four steps to the energy-management process above have been identified (rules of thumb):
    a. **Metering Energy Consumption and Data Collection:** The old approach to energy data collection is to manually read meters once a week or once a month. This is quite a unpleasant task, and weekly or monthly data is not nearly as good. The data that comes easily and automatically form the modern approach.

        The modern approach to energy data collection is to fit interval-metering systems that automatically measure and record energy consumption at short, regular intervals such as every

15 minutes or half an hour. Detailed interval energy consumption data makes it possible to visualize patterns of energy waste that would be impossible to see otherwise. For example, there is simply way that weekly or monthly meter readings can show us how much energy we are using at different times of the day, or on different days of the week and seeing these patterns, it makes it much easier to find the routine waste in our building (Kahlenborn et al., 2010).. So, the more data we can get, the more detailed information we get.

b. **Finding and Quantifying Opportunities to Save Energy:** The detailed meter data that we are collecting will be invaluable for helping us to find and quantify energy-saving opportunities. The easiest and most cost-effective energy-saving opportunities typically require little or no capital investment.

For example, a number of buildings have advanced control systems that could and should be controlling well the HVAC (**H**eating, **V**entilation and **A**ir **C**onditioning) but unknown to the facilities-management staff, are faulty or mis-configured, and consequently committing such errors such as heating or cooling an empty building every night and every weekend.

One of the simplest ways to save a significant amount of energy is to encourage staff to switch equipment off at the end of each working day. Looking at detailed interval energy data is the ideal way to find routine energy waste. We can check whether staff and timers are switching things off without having to patrol the building day and night, and; with a little detective work, we can usually figure out who or what is causing the energy wastage that we will inevitably find. Using our detailed interval data, it is usually easy to make reasonable estimates of how much energy is being wasted at different times. For example, if we have identified that a lot of energy is being wasted by equipment left on over the weekends, we can:

- Use our interval data to calculate how much energy (in kWh) is being used each weekend.
- Estimate the proportion of that energy that is being wasted (by equipment that should be switched off).
- Using the figures from "a" and "b", we can calculate an estimate of the total kWh that are wasted each weekend.

Alternatively, if we have no idea of the proportion of energy that is being wasted by equipment left on unnecessarily, we could:

- Walk the building one evening to ensure that everything that should be switched off is switched off.
- Look back at the data for that evening to see how many (kW) were being used after we switched everything off.
- Subtract the target (kW) from the typical (kW) for weekends to estimate the potential savings in (kW).
- Multiply the (kW) savings by the number of hours over the weekend to get the total potential (kWh) energy savings for a weekend.

Also, most buildings have opened a variety of equipment or building-fabric-related energy-saving opportunities, most of which require a more significant capital investment. We are probably aware of many of these, such as upgrading insulation or replacing lighting equipment.

Although our detailed meter data will not necessarily help us to find these equipment- or building-fabric-related opportunities, it will be useful for helping us to quantify the potential savings that each opportunity could bring. It is much more reliable to base our savings estimates on real metered data than on rules of thumb alone. It is critically important to quantify the expected savings for any opportunity that we are considering investing a lot of time or money into. It is the only way we can figure out how to invest on the biggest, easiest energy savings first (Kahlenborn et al., 2010).

Supply Side Management

Supply-side management (SSM) refers to actions taken to ensure the generation, transmission and distribution of energy are conducted efficiently. This has become especially important with the deregulation of the electricity industry in many countries, where the efficient use of available energy sources becomes essential to remain competitive. The generated electricity should be utilized efficiently to meet the demand of countries. This improves the reliability of the power supply system. SSM is used primarily with reference to electricity but it can also be applied to actions concerning the supply of other energy resources such as fossil fuels and renewable resources. Energy users will normally focus their efforts on demand-side management methods (DSM) but some will consider the supply side too.

For example, they may look at on-site generation alternatives (including cogeneration) or consider diversifying to alternative fuel sources such as natural gas, solar, wind and biomass (Barney et al., 2006).

Importance of SSM

For an electricity system, effective SSM will increase the efficiency with which the end-users are supplied, allowing the utility company to defer major capital expenditure, which might be required for increasing their capacity in growing markets. SSM makes installed generating capacity able to provide electricity at lower cost and reduce environmental emissions; SSM can also contribute to improving the reliability of a supply system. With the current trend of deregulating the supply industry, it is becoming more important to embark on supply-side management where the supplier, user and the environment are all winning (Barney et al., 2006).

In brief, an electrical utility may embark on SSM to:

- Ensure reliable availability of energy at the minimum economic cost
- Provide maximum value to its customers by reducing energy prices
- Meet increasing electricity demand without incurring in unnecessary major
- capital investments in new generating capacity
- Minimize environmental impact.

Techniques Used in SSM

- **Direct load control (DLC):** This is the program designed to direct control the utility power supply to individual appliances on a consumer premises. The control usually involves residential consumers. The cost benefit of DLC includes:
    - Power system production cost savings
    - Power system generating capacity cost savings
    - Power system loss reduction.

The direct load control options are:

- ◦ Direct load control: the utility can switch off the load directly when required.
- ◦ Interruptible load control: the utility provides advance notice to the customer for switching off their loads.
- ◦ TOU (time-of-use) tariffs: where utility rate structure is designed according to the time (Chiu et al., 2013).
- ◦ Load management by time dependent tariffs

In this method, load management (LM) is carried out by the influence of tariffs setting. The total cost of generating and delivering of electricity to consumers was being divided into four fundamental categories of services:

- ◦ Customer services.
- ◦ Distribution services.
- ◦ Transmission services.
- ◦ Generation services (Barney et al., 2006).

Integrated utilities in regulated states set the rates to cover the costs of all services.

The electric consumers are billed as: flat rate tariffs two part tariff, time of use tariff (TOU), and Spot price.

In a flat rate tariff, a customer pays the same amount for electricity at any time of day. (Chamberlin & Herman, 1996) as well as (Ashok & Benergee, 2001) presented load management by TOU rates. In this method, the utility provides transparent information on the electricity price at different periods to the customers to encourage off peak and discourage peak period consumption by varying price of electricity. (Babu, 1995) developed discriminatory time of use tariffs in which the price corresponds to the marginal cost of supply. He identified that the price of electricity has a significant contribution to the LM scheme. An econometric model for electricity demand in the domestic sector was developed. Linear programming (LP) was used to set electricity tariffs (Barney et al., 2006). Another type of tariff setting for LM is spot price. The message is sent to customers to indicate the price of electricity for an instant of time. A spot price scheme is appreciable if electricity price fluctuation is high and if the consumer can anticipate the price behavior as well as being able to respond quickly when the electricity price is high or low (Barney et al., 2006).

- **Dispatch Load Management (DLM):** In this method, the utility make agreements with the customers on the reduction of their load for a certain time. In exchange, customers receive discounts on their electric bills. These programs are primarily oriented towards large commercial and industrial customers. The number of participants is generally low, but load reductions per customer can be significant and overall load savings is substantial (Barney et al., 2006).
- **Rebate Program:** Rebates are a type of financial incentive offered by electric utilities over the past decade. In the residential sector, rebates have been commonly offered for the purchase of efficient appliances and compact fluorescent bulbs. In the C&I sectors, lighting rebate programs are most common, followed by air conditioning and motor rebate programs. Rebate levels vary widely, from approximately 20 –100% of the cost depending on the price (Barney et al., 2006).
- **Thermal Energy Storage:** It is the use of thermal energy storage to store heat for space or water heating during off peak period and use at the peak time. This system is also widely used in air conditioners. The on- off switching of the storage elements was accomplished by communication technology and control as voluntary load control (Barney et al., 2006).

Demand Side Management

Energy demand management, also known as demand side management (DSM), is the modification of consumer demand for energy through various methods such as financial incentives (Chiu et al., 2013) and education. Usually, the goal of demand side management is to encourage the consumer to use less energy during peak hours, or to move the time of energy use to off-peak times such as nighttime and weekends (Office of Energy, 2010). Peak demand management does not necessarily decrease total energy consumption, but could be expected to reduce the need for investments in networks and/or power plants for meeting peak demands. An example is the use of energy storage units to store energy during off-peak hours and discharge them during peak hours (Chiu et al., 2012).

The term DSM was coined following the time of the 1973 energy crisis and 1979 energy crisis. Demand Side Management was introduced publicly by Electric Power Research Institute (EPRI) in the 1980s (Balijepalli, 2011).

Types of Energy Demand Management

- **Energy Efficiency**: Using less power to perform the same tasks.
- **Demand Response:** Any reactive or preventative method to reduce, flatten or shift peak demand. Demand Response includes all intentional modifications to consumption patterns of electricity of end-user customers that are intended to alter the timing, level of instantaneous demand, or the total electricity consumption (Albadi and El-Saadany, 2007). Demand Response refers to a wide range of actions which can be taken at the customer side of the electricity meter in response to particular conditions within the electricity system (such as peak period network congestion or high prices)(Torriti et al., 2010).
- **Dynamic Demand:** Advance or delay appliance operating cycles by a few seconds to increase the diversity factor of the set of loads.

The concept is that by monitoring the power factor of the power grid, as well as their own control parameters, individual, intermittent loads would switch on or off at optimal moments to balance the overall system load with generation, reducing critical power mismatches.

As this switching would only advance or delay the appliance operating cycle by a few seconds, it would be unnoticeable to the end user. In the United States, in 1982, a patent for this idea was issued to power systems engineer Fred Schweppes.

Problems With DSM

Some people argue that demand-side management has been ineffective because it has often resulted in higher utility costs for consumers and less profit for utilities (Myron, 2010).

One of the main goals of demand side management is to be able to charge the consumer based on the true price of the utilities at that time. If consumers could be charged less for using electricity during off-peak hours, and more during peak hours, then supply and demand would theoretically encourage the consumer to use less electricity during peak hours, thus achieving the main goal of demand side management. Another problem of DSM is privacy: The consumers have to provide some information about their usage of electricity to their electricity company. This is less of a problem now as people are used to suppliers noting purchasing patterns through mechanisms such as "loyalty cards".

Mathematical Formulation of the Demand Side Management (DSM) Problem and Its Optimal Solution

The scope of the DSM programs is the planning, development and implementing of programs whose objective is to shape actively the daily and seasonal electric load profiles of customers to realize or to achieve better overall system utilization. DSM activity has grown and matured over the past decades. Many utilities have implemented DSM programs on a routine basis and more utilities are considering DSM as a part of their resource planning process.

The benefits from applying DSM programs are mutual for both the customer and the utility; utilities will have better utilization of the available system capacity. For customers, the amount of monthly electric bill will be decreased besides the improvement in the electrical service quality.

At the heart of the DSM programs, there is a series of measures intended to encourage specific groups of customers to modify their energy usage patterns in a manner consistent with the utility's DSM objectives while maintaining or enhancing customer satisfaction.

Different utilities have different programs to be applied on their customers. These programs are different according to the number of participated customers in the program, nature of the targeted load type (commercial, industrial or residential), the revenue from each program and the level of customer's satisfaction or reaction towards similar applied programs. These programs can be augmented in five steps: DSM targets, financial and feasibility study, designing of effective programs, program implementation & monitoring and program evaluation.

DSM Programs

In this section, the different DSM programs will be briefly explained. This explanation will include the main objective of the program and the changes made on a typical load curve after applying this program. Firstly, a DSM program is a program used to control the load profile indirectly in order to achieve the utility objectives (Attia, 2010).

These objectives are:

- To have the load factor as close as possible to 1.0
- To have the peak load within the proper margin.

By achieving the previous objectives, the utility would get the maximum possible energy from the installed units, thus maximizing the total profit and minimizing the average cost per KWh (Elsobki, 1996) has listed these programs as follows:

- **Valley Filling:** In this program, the main objective is to increase the demand during the off peak periods while having the same load peak (Fig. 2). This could be achieved by encouraging the consumers to increase their demand (Attia, 2010).

- **Load Shifting:** In this program, it is required to shift part of the demand at the peak period to the off peak periods (Fig. 3). This program could be used in case that the installed capacity is not enough during the peak load (Attia, 2010).
- **Peak Clipping:** This program is used to decrease the demand during the peak load periods. Also, these loads can't be shifted to the off peak periods (Fig. 4). This could be due to lack of installed capacity during these periods. This program could be achieved be indirectly forcing the consumers to decrease their loads by the use of miniatures on their supply points (Attia, 2010).
- **Energy Conservation:** This program is used when it is required to decrease the energy consumption all over load period (Fig. 5). This could be achieved by using high efficiency components (Attia, 2010).

Figure 2. Valley filling program effect

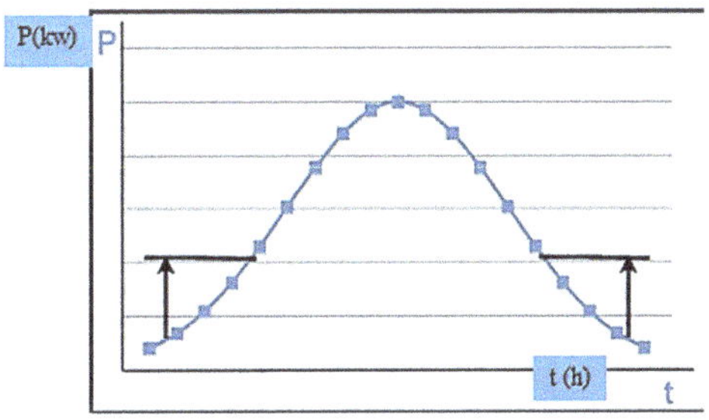

Figure 3. Load shifting program effect

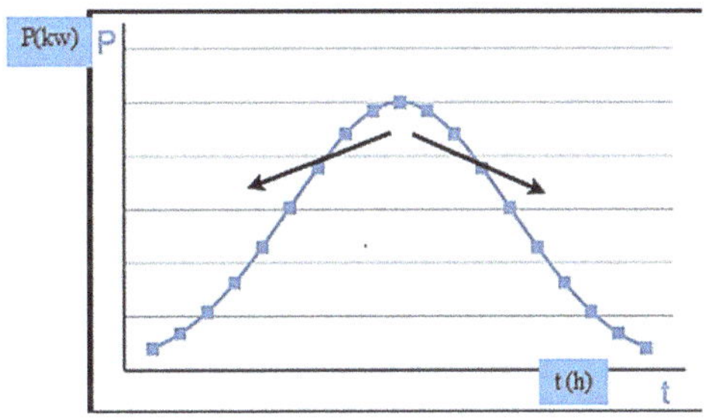

Figure 4. Peak clipping program effect

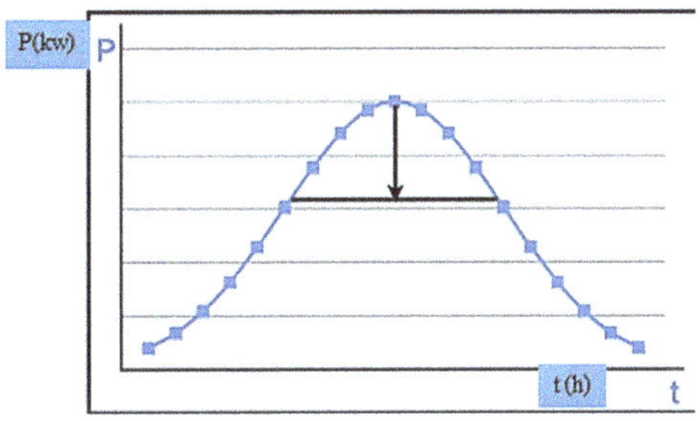

Figure 5. Energy conservation program effect

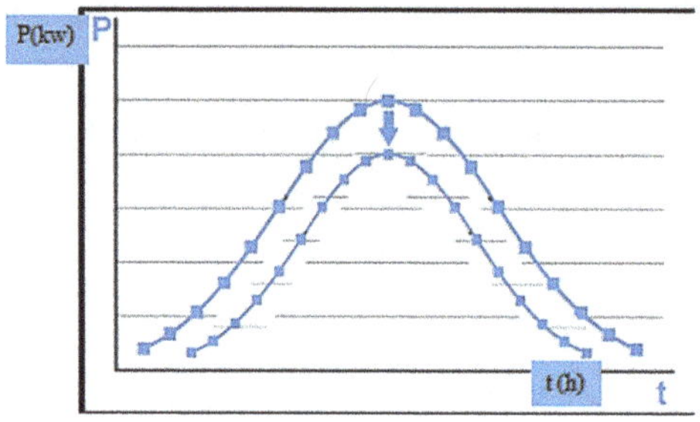

Figure 6. Load building program effect

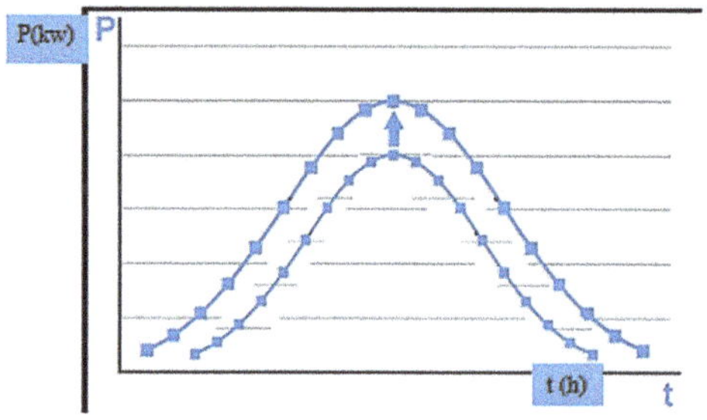

- **Load Building:** This program is used when it is required to increase the energy consumption. This could be very beneficial in case of surplus capacity. This is because the average cost per KWh will decrease (Attia, 2010).

It may be noted that peak clipping, load shifting and energy conservation techniques are considered as new resources that can help the utility to meet the increasing demand of its customers. Loads building and valley filling are economic efficiency options for power systems with long term exceptions of surplus power (Attia, 2010; Alshahrestani et al, 2018; Alhelou et al., 2018).

Means of Implementing DSM Techniques

References (El Sobki, 1997; EPRI, 1993) have introduced the prospective means of implementing DSM techniques that can be categorized into:

1. **Direct Load Control:** It is an obligation way by which the utility can modify customers load pattern. It can be applied by switching off the power supply on specific category of customers at specific time interval, or force the customers not to use a specific type of electrical load at a specific time interval.
2. **Indirect Load Control:** It is an optimal way by which the utility can change the customers load pattern by using special methods such as:
   -Time of use rates
   -Thermal energy storage

-Efficient end use technologies
-Electric tariff system
-Electrification technologies.
The more commonly used methods are the electric tariff system and time of use rates.

FORMULATION OF DSM PROGRAMS AS AN OPTIMIZATION PROBLEM

DSM has a major role of utility planning and operation. In this section, an optimal based formulation is developed to simulate the implementation process of the DSM program to assess its technical and financial impacts for both utility and users. The objective function is formulated either to control the use of the supply side resources subject to end user demand for power and energy without loss of production or comfort, or to improve system performance by increasing load factor and enhance the customer service quality. (Gellings and Chemberlin, 1993 and Elsobki and Wahdan, 1999) have offered the questions that confront the demand side management planner to construct the DSM optimization model. The mathematical construction model for any optimization problem is generally determined by clarifying the following questions:

- What does the model seek to determine?
- What are the objectives (goals) needed to be achieved to determine the best solution?
- What are the variables of the problem?
- What constraints must be imposed on variables to simulate properly actual variables?

The mathematical formulation of the DSM techniques as an optimization problem is given. Two sorts for the objective function are contributed, either to maximize the system load factor for the utility, or to minimize the total cost of the bill for the customer. While there are two sorts for the objective function for the five DSM techniques, the imposed constraints on the demand type at different time intervals (control variables) differ from a technique to another and depend, also, on the load peculiarities and the power system.

DSM programs seek to optimize either of the following two objective functions:

$$\text{Max.L.F.} = \left[\left[\sum_{i=1}^{N}\sum_{j=1}^{J} P(i,j).t(j)\right] / \sum_{j=1}^{J} t(j)\right] / \sum_{i=1}^{N} p(i,k) \tag{1}$$

$$\text{Max.L.F.} = \left[\left[\sum_{j=1}^{J} \mathbf{PTO}(\boldsymbol{j}).t(j)\right] / \sum_{j=1}^{J} t(j)\right] / \boldsymbol{PTO}(\boldsymbol{k}) \tag{2}$$

$$\text{Min.C} = \left[\sum_{i=1}^{N}\sum_{j=1}^{J} P(i,j).t(j).ce(i,j)\right] + \left[\sum_{i=1}^{N}\sum_{j=1}^{J} P(i,j).cd(i,j)\right] \tag{3}$$

$$Max.\,Revenue = \left[\sum_{i=1}^{N}\sum_{j=1}^{J} P(i,j)\cdot t(j)\cdot ce(i,j)\right] + \left[\sum_{i=1}^{N}\sum_{j=1}^{J} P(i,j)\cdot cd(i,j)\right] \tag{4}$$

where:

- L.F.: is the system load factor.
- P (i,j): is the demand of load type i at time interval number j.
- N: is the total number of load demand types.
- J: is the total number of time intervals.
- $P_{TO(j)}$: is the total demand for all the loads types from j=1 to j=J over the time interval number j.
- k: is the number of time interval at which the maximum demand for all the load types numbers from i=1, N over all the time duration from j=1, J occurs.
- C: is the total cost of the electrical demand and energy consumption.
- ce(i,j): is the cost of energy for load type i at time interval number j.
- cd(i,j): is the cost of demand for load type i at time interval number j.

In the following, the different DSM techniques including the related objective function and constraints as an optimization problem are presented.

The description of the method and effect on load shape in addition to means of implementation are also given (Attia, 2010).

Valley Filling

Program description and effect on load shape:It entails building of off-peak loads. This is often the case when there is under-utilized capacity that can operate on low cost fuels. The net effect is an increase in total energy consumption, while the peak demand is kept fixed (Figure 7). Consequently, the load factor will be improved.

- **Means of Implementation:** This can be achieved by creation of new off-peak electric loads such as charging of electric cars and thermal energy storage.
- **Objective Function:** The objective function is formulated to maximize the system load factor using rather equation (1) or equation (2) subject to:

Equality constraint:

$$P_{new}(i,j) = P_{old}(i,j)\ \forall tk \rightarrow th P_{new}(i,j) \geq P_{old}(i,j)\ \forall to \rightarrow tk, th \rightarrow TD \tag{5}$$

Inequality constraints:

$$P_{new}(i,j) \leq P(value)\ \forall to \rightarrow tk, th \rightarrow TD \tag{6}$$

where

- P_{new}(i,j): is the demand of load type i at time interval j after applying DSM technique.
- P_{old}(i,j): is the demand of load type i at time interval j before applying DSM techniquc.
- The P_{new}(i, j): is not permitted to increase the P(value) which is an extreme limiting value given by the planner.

Figure 7. Load demand control to achieve valley filling DSM program

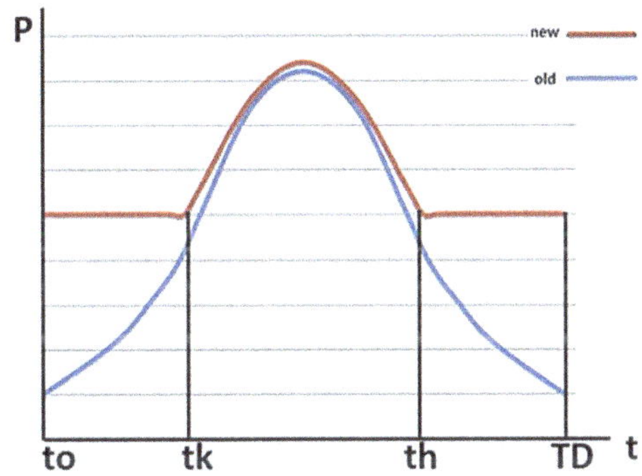

- P (value): is an extreme limiting value given by the planner for load demand after applying DSM program

Load Shifting

Program description and effect on load shape: It involves shifting loads from on-peak to off-peak periods (Figure 8). The net effect is a decrease in peak demand, but no change in the total energy consumption. This effectively, improves the system load factor and decreases the cost of the electricity bill (Attia, 2010).

- **Means of Implementation:** This can be achieved by time of use rates and/or use of storage devices that shift the timing of conventional electric appliances operation.
- **Objective Function:** The objective function is formulated either to maximize the system load factor or to minimize the customer electricity bill using equation (2) and equation (3) subject to:

Equality constraint:

$$\sum_{i=1}^{N}\sum_{j=1}^{J} P_{new}(i,j) * t(j) = \sum_{i=1}^{N}\sum_{j=1}^{J} P_{old}(i,j) * t(j)$$

$$P_{new}(i,j) = P(value)\ \forall tk \rightarrow th \tag{7}$$

Inequality constraints:

$$P_{new}(i) \geq P_{old}(i)\ \forall to \rightarrow tk,\ th \rightarrow TD$$

$$P_{new}(i) \leq P(value)\ \forall to \rightarrow tk, th \rightarrow TD \tag{8}$$

Figure 8. Load demand control to achieve load shifting DSM program

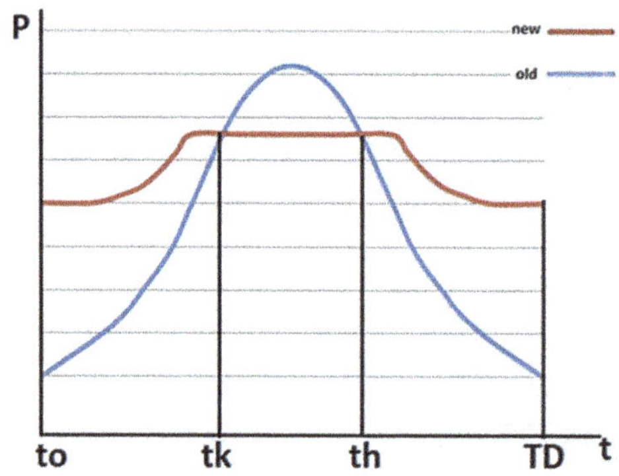

Peak Clipping

Program description and effect on load shape: Peak clipping refers to reduction of utility loads during peak demand periods (Figure 9). The net effect is a reduction in both demand and total energy consumption. Therefore, the system load factor is improved and, also, the customer electricity bill is decreased (Attia, 2010).

- **Means of Implementation:** Direct utility control on customer appliances or end-use equipment can be carried out to reduce peak demand periods.
- **Objective Function:** The objective function is formulated either to maximize the system load factor using equation (2) subject to:

Equality constraint:

$$P_{new}(i) = P_{old}(i)\ \forall to \rightarrow tk,\ th \rightarrow TD \tag{9}$$

Inequality constraints:

$$P_{new}(i) \leq P(value1)\ \forall tk \rightarrow th;\ P_{new}(i) \geq P(value2)\ \forall tk \rightarrow th;\ P(value2) \leq P(value1)) \tag{10}$$

Figure 9. Load demand control to achieve peak clipping DSM program

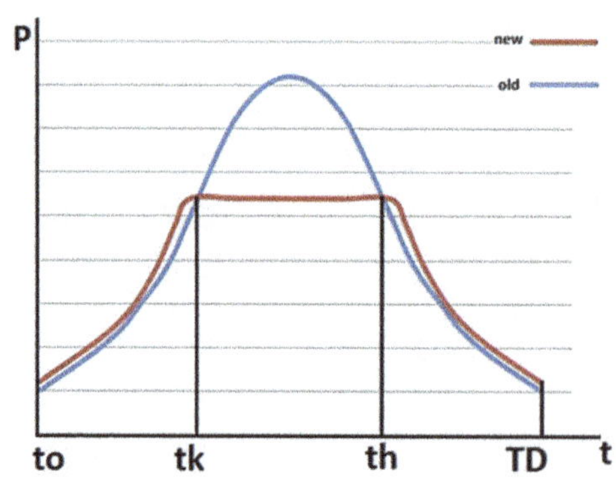

Figure 10. Load demand control to achieve energy conservation DSM program

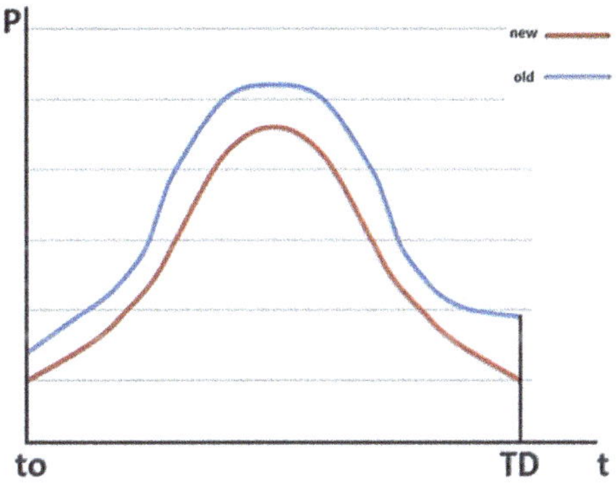

P(*value1*), P(*value2*) are limiting values given by the planner, that depends on the nature of the load and user activity, for load demand after applying DSM program.

Energy Conservation

Program description and effect on load shape: References (Gellings, 1985 and Attia et al., 2006) have clarified the energy conservation technique as an effective mean for reducing the end-users consumptions. In such a method, both peak demand and total energy consumption are reduced (Figure 10).

- **Means of Implementation:** Appliances efficiency improvement and weatherization are some examples for energy conservation.
- **Objective Function:** The objective function is formulated to minimize the cost of the customer electricity bill using equation (3) subject to:

Inequality constraints:

$$P_{new}(i) \leq P_{old}(i)\ \forall to \rightarrow TD \qquad (11)$$

Load Building

Program description and effect on load shape: It refers to an increase in overall sales. The net effect is an increase in both peak demand and total energy consumption (Figure 11).

- **Means of Implementation:** Load building involves increased market share of loads that can use electric energy instead of fuel. Electric vehicles, industrial heating and electrification may be, also, effective means for load building (Attia, 2010).
- **Objective Function:** The objective function is formulated to maximize utility revenue referred to equation (4) subject to:

Figure 11. Load demand control to achieve load building DSM program

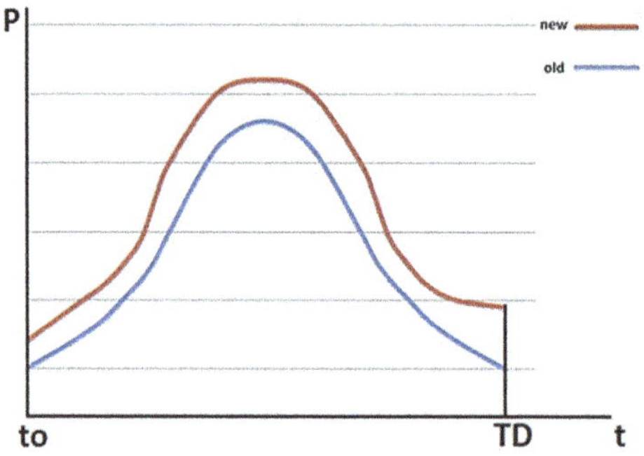

Inequality constraints:

$$P_{new}(i) \geq P_{old}(i) \; \forall to \rightarrow TD \qquad (12)$$

GENETIC ALGORITHMS

Genetic algorithms are a part of evolutionary computing, which is a rapidly growing area of artificial intelligence. Genetic algorithms are inspired by Darwin's theory about evolution i.e. solution to a problem solved by genetic algorithms is evolved Genetic Algorithms (GAs) were invented by John Holland in 1975 (Haupt, 2004 and Coley, 1999).

Evolution in the natural world displays a remarkable problem solving ability, as demonstrated by the fact that a myriad species have evolved on Earth, demonstrating a diverse range of survival strategies. It would therefore not be unreasonable to deduce that a problem solving strategy, inspired by the mechanics of natural selection and genetics, may prove highly effective in solving certain classes of problems

Genetic algorithms emulate the mechanics of natural selection by a process of randomized data exchange. In this way they are able to solve of range of difficult problems which cannot be tackled by other approaches. The fact that they are able to search in a randomized, yet directed manner, allows them to reproduce some of the innovative capabilities of natural systems

Because genetic algorithms were inspired by the behavior of natural systems, the terminology used to describe them is a mix from both biological and computer fields. A genetic algorithm manipulates strings of information, usually called chromosomes. These encode potential solutions to a given problem. Chromosomes are evaluated and assigned a score (fitness value) in terms of how well they solve the given problem according to criteria defined by the programmer. These fitness values are used as a probability of survival during a round of reproduction. New chromosomes are produced by combining two (or more) parent chromosomes. This process is designed to lead to a succession of fitter offspring, each encoding better solutions, until an acceptably good solution is found.

A genetic algorithm (GA) is an iterative optimization process that imitates the adaptation and evolution of a single species of organism. Using a chromosomal mapping system, the GA starts with a large number of potential design configurations. The range of possible configurations is determined by the constraints of the problem and the method of encoding all configuration information into the chromosome (Haupt, 2004 and Coley, 1999).

In genetic algorithms, evolution towards a global optimum occurs as a result of pressure exerted by a fitness-weighted selection process and exploration of the solution space is accomplished through combination and mutation of existing characteristics present in the current population. Other optimization techniques (such as gradient descent methods) search a region of the solution space around an initial guess for the best local solution (Griffiths et al., 1993).

Genetic algorithms belong to the class of global optimizers as opposed to the familiar traditional local optimizers, such as conjugate gradient and the quasi-Newtonian methods. The distinction between local and global search of optimization techniques is that the local techniques produce results that are highly dependent on the starting point or initial guess, while the global methods are highly independent of the initial conditions. Though they possess the characteristic of being fast in convergence, local techniques have a direct dependence on the existence of at least the first derivative. Furthermore, they place constraints on the solution space such as differentiability and continuity, conditions that are hard or even impossible to satisfy in practice. Conjugate gradient techniques depend either explicitly or implicitly on a derivative in the form of the gradient (Stace, 1997;Weile and Michielssen, 1997; Johnson and Rahmat-Smaii, 1997). The global techniques, on the other hand, are largely independent of and place few constraints on the solution space (Recioui, 2012).

GA Operators

A general flow chart of a genetic algorithm is shown in fig. 12. The tasks that a genetic algorithm must perform lead to the existence of three phases in the genetic algorithm optimization (Recioui, 2012).

- **Initiation:** Means filling the initial population with encoded, usually randomly created parameter strings or chromosomes. The coding is a mapping from the parameter space to the chromosome space. Often, binary coding is utilized. In some cases, as in the real coded GA, the parameters are mapped to themselves. These two coding schemes have in common some operators but some differences as well.
- **Reproduction:** Consists in three main operators: selection, crossover and mutation.The smallest unit of a genetic algorithm is called a *gene*, which represents a unit of information in the problem domain. A series of genes, known as a *chromosome*, represents one possible solution to the problem. Each gene in the chromosome represents one component of the solution pattern. The most common form of representing a solution as a chromosome is a string of binary digits. Each bit in this string is a gene. The process of converting the solution from its original form into the bit string is known as *coding*. The specific coding scheme used is application dependent. The solution bit strings are decoded to enable their evaluation using a fitness measure.
- **Selection:** In biological evolution, only the fittest survive and their gene pool contributes to the creation of the next generation. Selection in GA is also based on a similar process. In a common form of selection, known as *fitness proportional selection*, each chromosome's likelihood of being selected as a good one is proportional to its fitness value.
- **Alteration to Improve Good Solutions:** The alteration step in the genetic algorithm refines the good solution from the current generation to produce the next generation of candidate solutions. It is carried out by performing crossover and mutation.
- **Crossover:** May be regarded as artificial mating in which chromosomes from two individuals are combined to create the chromosome for the next generation. This is done by splicing two chro-

mosomes from two different solutions at a crossover point and swapping the spliced parts. The idea is that some genes with good characteristics from one chromosome may as a result combine with some good genes in the other chromosome to create a better solution represented by the new chromosome.

- **Mutation:** Is a random adjustment in the genetic composition. It is useful for introducing new characteristics in a population (something not achieved through crossover alone). Crossover only rearranges existing characteristics to give new combinations. For example, if the first bit in every chromosome of a generation happens to be a 1, any new chromosome created through crossover will also have 1 as the first bit. The mutation operator changes the current value of a gene to a different one. For bit string chromosome this change amounts to flipping a 0 bit to a 1 or vice versa. Although useful for introducing new traits in the solution pool, mutations can be counterproductive, and applied only infrequently and randomly.

Components of Genetic Algorithms

The most important components in a GA consist of:

1. Representation (definition of individuals)
2. Evaluation function (or fitness function)
3. Population
4. Parent selection mechanism
5. Variation operators (crossover and mutation)
6. Survivor selection mechanism (replacement)

Figure 12. General flow chart of the genetic algorithm

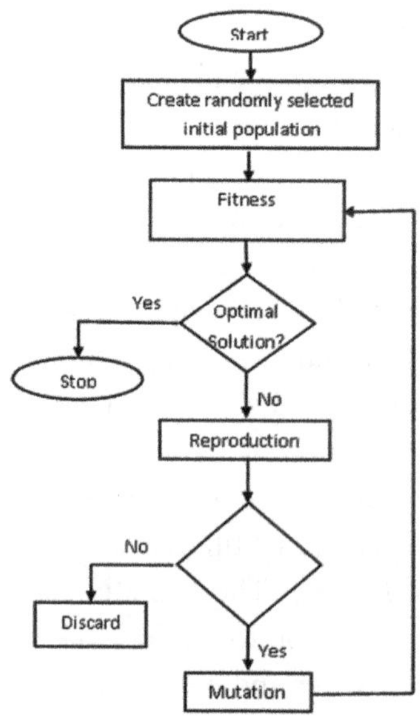

Representation

Objects forming possible solution within original problem context are called *phenotypes*, their encoding, the individuals within the GA, are called *genotypes*.

The representation step specifies the mapping from the phenotypes onto a set of genotypes.

Candidate solution, *phenotype* and *individual* are used to denotes points of the space of possible solutions. This space is called *phenotype space*.

Chromosome, and *individual* can be used for points in the *genotye space*.

Elements of a chromosome are called *genes*. A value of a gene is called an *allele*.

Variation Operators

The role of variation operators is to create new individuals from old ones. Variation operators form the implementation of the elementary steps with the search space.

Mutation Operator

A unary variation operator is called *mutation*. It is applied to one genotype and delivers a modified *mutant*, the *child* or *offspring* of it.

In general, mutation is supposed to cause a random unbiased change. Mutation has a theoretical role: it can guarantee that the space is connected.

Crossover Operator

A binary variation operator is called *recombination* or *crossover*. This operator merges information from two parent genotypes into one or two offspring genotypes.

Similar to mutation, crossover is a stochastic operator: the choice of what parts of each parent are combined, and the way these parts are combined, depend on random drawings.

The principle behind crossover is simple: by mating two individuals with different but desirable features, we can produce an offspring which combines both of those features.

Figure 13. Example of the crossover and mutation operators in the binary GA

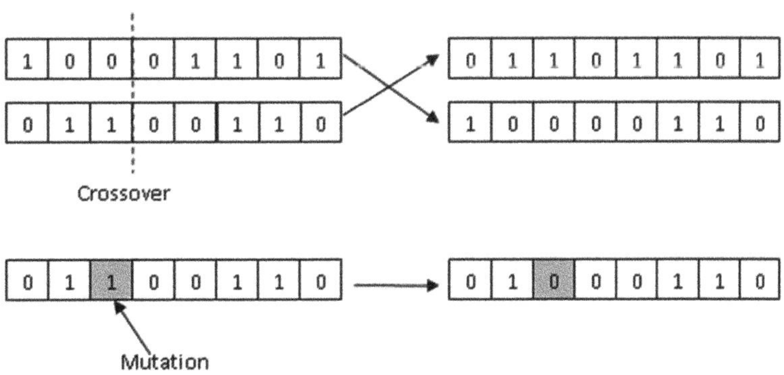

Parent Selection Mechanism

The role of *parent selection* (*mating selection*) is to distinguish among individuals based on their quality to allow the better individuals to become parents of the next generation.

Parent selection is *probabilistic*. Thus, high quality individuals get a higher chance to become parents than those with low quality. Nevertheless, low quality individuals are often given a small, but positive chance, otherwise the whole search could become too greedy and get stuck in a local optimum.

Survivor Selection Mechanism

The role of survivor selection is to distinguish among individuals based on their quality. In GA, the population size is (almost always) constant, thus a choice has to be made on which individuals will be allowed in the next generation. This decision is based on their fitness values, favoring those with higher quality.

As opposed to parent selection which is stochastic, survivor selection is often *deterministic*, for instance, ranking the unified multiset of parents and offspring and selecting the top segment (fitness biased), or selection only from the offspring (age-biased).

Initialization

Initialization is kept simple in most GA applications. Whether this step is worth the extra computational effort or not is very much depending on the application at hand.

Termination Condition

It is to notice that GA is stochastic and mostly there are no guarantees to reach an optimum.

Commonly used conditions for terminations are the following:

1. the maximally allowed CPU times elapses
2. The total number of fitness evaluations reaches a given limit
3. for a given period of time, the fitness improvement remains under a threshold value
4. The population diversity drops under a given threshold.

Population

The role of the population is to hold possible solutions. A *population* is a multiset of genotypes. In almost all GA applications, the population size is constant, not changing during the evolutional search.

Elitism

When creating a new population through crossover and mutation, there is a big chance that we will lose the best chromosome. Elitism is a method, which copies the best chromosome (or a few best chromosomes) to insert it (them) to the new population if the newly created children are less fit then the individuals in the old population. As such, Elitism can increase the performance of GA very rapidly, because it prevents losing the best found solution.

CASE STUDY RESULTS AND DISCUSSIONS

This section considers a dynamic demand management for the residential sector and formulates the energy consumption-scheduling problem as a multi- objective optimization problem, addressed with a heuristic approach. The adopted planning strategy aims at reducing the peak load as well as minimizing the energy cost. It has to be noted that the considered optimization objectives are mostly conflicting and non-commensurable. Therefore the optimal solution of this multi-objective planning problem is found using a Multi-Objective Genetic Algorithm.

Problem Description

The increasing number of automation system and electrical appliances in residential sector makes the enhancing of electrical efficiency of commercial and domestic buildings highly desirable. On the other hand, comfort of user and quality of life must be preserved. This part of project addresses this challenging issue as a constrained multi-objective optimization problem. The aim is the balancing of the energy consumption in a residential district to avoid the concentration of simultaneous electricity request on the same time. This has to be done by saving the cost and by shifting loads from on-peak to off-peak periods. We considered a neighborhood of N buildings, whose **U** users program the set of daily tasks to be done (basic electrical appliances). It is assumed that each electrical appliance of the buildings is equipped with a terminal unit controller (TUC), which collects and transmit consumption to a building controller, connected to the energy consumption-planning system (ECPS) of the residential area. The TUC turn on and off the appliances according to the scheduling pattern planned by ECPS. The ECPS schedule the tasks at times multiple of t in a discrete time setting.

The consumption scheduling system shifts in time the execution of the tasks according to the minimization of the two objective functions, associated with the electrical load profile and the energy cost. The first function is a measure of the maximum load factor to reduce the peak load power consumption by the following equation (1). The second objective function takes into account the energy price as in equation (3).

Results and Discussions

To demonstrate the effectiveness of the proposed heuristic strategy for optimal planning of daily electrical consumptions, we consider a case study of a residential area with 100 smart homes that have several basic electrical appliances. We supposed that each domestic unit has at least five electrical devices (fridge, washing machine, dishwasher, electrical car and interior lighting) and up to twelve appliances. So, there are at least 1000 electrical devices over the given residential area with different consumption patterns specified in (table .1) (Recioui et al., 2016).

The planning system schedules the tasks according to the given earliest starting times, the latest finishing times, the durations, the power requirements and the start times specified by the user. Taking a single house as a first step, we started the work with the electricity consumption and work for each task with the earliest possible starting time specified by the consumer to construct the total electrical load profile of that home with the earliest starting time specified by the same consumer. Finally, the total power consumption of that home is shown in Fig. 14. The total power is not uniformly consumed; which causes a peak load.

Figure 14. Daily home load profile

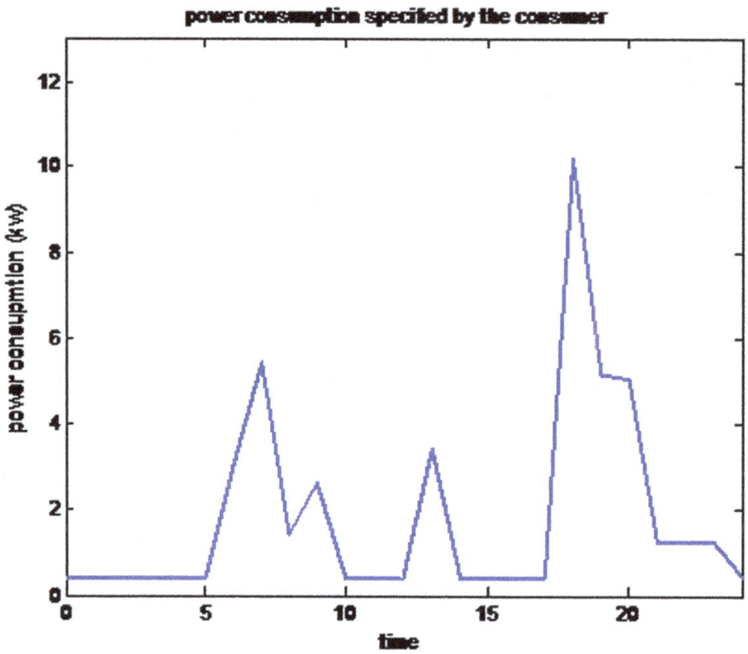

Table 1. Simulated home load scenario

Tasks	Earliest Starting Time	Latest Finishing Time	Duration (hours)	Power (kw/H)
Dish washer	7	19	3	1
Washing machine	6	24	2	1
Spin dryer	13	18	1	3
Cooker hob	7	9	1	3
Cooker oven	18	21	1	5
Cooker microwave	6	9	1	1.7
Interior Lighting	18	24	6	0.85
Laptop	18	24	2	0.1
Desktop	18	24	3	0.3
Vacuum cleaner	9	17	1	1.2
Fridge	0	24	24	0.3
Electrical car	18	8	3	3.5
Other tasks	0	24	24	0.1

So, The TCU will fix that by turning on and off the appliances according to the scheduling pattern planned by ECPS. The ECPS schedule the tasks at times multiple of t in a discrete time setting. (Table. 2) indicates the new electrical load profile of that home with the best starting times.

Table 2. New scheduled load profile with best starting times

Time(h)	1	2	3	4	5	6
power (kw)	1.45	1.55	1.65	1.575	1.45	1.25
Time(h)	13	14	15	16	17	18
power (kw)	1.05	1.1	1.25	1.275	1.225	0.75
power (kw)	1.4	3.275	3.05	0.85	1.025	1.0
Time(h)	19	20	21	22	23	24
power (kw)	2.3	3.425	3.2	2.85	2.825	2.525

Figure 15. Old and new home power consumption

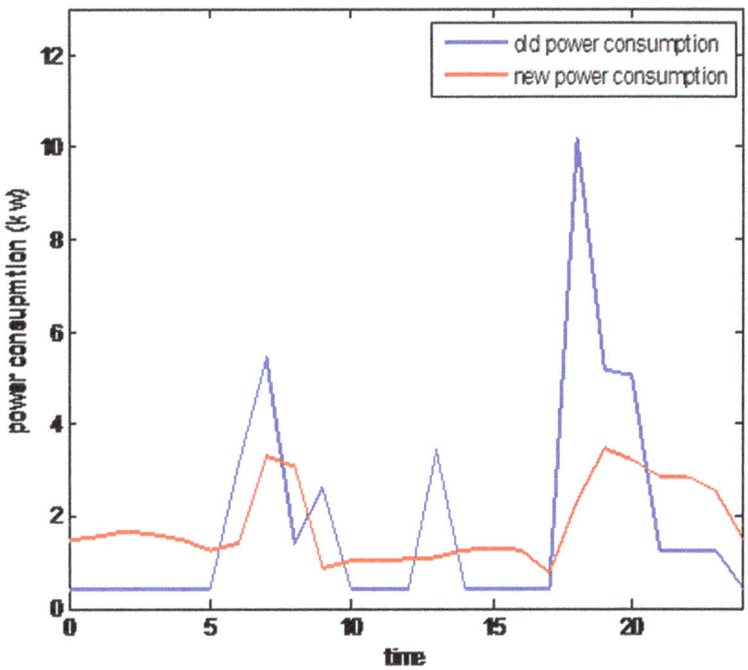

It is clear that the overall power consumption of the newly scheduled activities is less than the old profile (Fig. 15). This can be accomplished by selectively shifting some loads (as suggested by Genetic algorithms). As for the user, the shifting can be done either manually (user implication) or automatically through the smart meter. One main characteristic of the scheduled activities is that the main (high consuming power) activities are shifted to be executed in the off-peak periods. This work describes an efficient multi-objective planning system to manage the electricity demand in smart residential area. The proposed strategy is based on a heuristic approach using the Genetic Optimization Algorithm. The obtained results show that the scheduler not only decreases the pick load and reduces the utility bills but also preserves the user satisfaction.

CONCLUSION AND FUTURE WORKS

This chapter highlighted the benefits of smart grids and energy management in managing the electricity consumption, the utility bills and preserving the user satisfaction. The demand side management was modeled as an optimization task where the optimal usage times were found using Genetic Algorithms. We succeeded in achieving the best optimal solution of rescheduling the total power consumption of the selected residential area by clipping the total peak demand (a consistent reduction of approximately 20% of peak load) and minimizing the cost. Moreover, the first objective of minimizing the total peak- was tested through three DSM programs which are: peak clipping, valley filling and load shifting (each one subjected to specific constrains). Finally, it can be concluded that load shift technique has more accuracy in minimizing the total peak demand and maximizing the system's load factor without affecting the total energy consumption (same amount of energy consumption before and after rescheduling).

As a future direction, the work can be improved by having a real time monitoring and control system where load control will be done on a real time basis. This is because genetic algorithms are known to be slow in convergence and hence they are not adequate to online optimization. The classical techniques are; on the other hand, very adequate and fast but care must be taken regarding tier convergence and starting solutions.

REFERENCES

Agarwal, T., & Cui, S. (2011). *Noncooperative games for autonomous con-sumer load balancing over smart grid.* Available at: http://arxiv.org/abs/1104.3802

Agnetis, A., de Pascale, G., Detti, P., & Vicino, A. (2013). Load Scheduling for Household Energy Consumption Optimization. *IEEE Transactions on Smart Grid*, *4*(4), 2364–2373. doi:10.1109/TSG.2013.2254506

Agnetis, A., Dellino, G., Detti, P., Innocenti, G., de Pascale, G., & Vicino, A. (2011), Appliance operation scheduling for electricity consumption optimization. *Proceedings of the 50th IEEE Conference on Decision and Control and European Control Conference (CDC-ECC)*, 5899–5904. 10.1109/CDC.2011.6160450

Albadi, M. H. E. F. E.-S. (2007). *Demand Response in Electricity Markets: An Overview*. IEEE.

Alhelou, H., Hamedani-Golshan, M. E., Zamani, R., Heydarian-Forushani, E., & Siano, P. (2018). Challenges and Opportunities of Load Frequency Control in Conventional, Modern and Future Smart Power Systems: A Comprehensive Review. *Energies*, *11*(10), 2497. doi:10.3390/en11102497

Alhelou, H. H. (2018). *Fault Detection and Isolation in Power Systems Using Unknown Input Observer. In Advanced Condition Monitoring and Fault Diagnosis of Electric Machines* (p. 38). Hershey, PA: IGI Global.

Alhelou, H. H., Golshan, M., & Fini, M. (2018). Wind Driven Optimization Algorithm Application to Load Frequency Control in Interconnected Power Systems Considering GRC and GDB Nonlinearities. *Electric Power Components and Syst.*

Alhelou, H. H., & Golshan, M. E. H. (2016, May). Hierarchical plug-in EV control based on primary frequency response in interconnected smart grid. In *Electrical Engineering (ICEE), 2016 24th Iranian Conference on* (pp. 561-566). IEEE. 10.1109/IranianCEE.2016.7585585

Alhelou, H. H., Golshan, M. H., & Askari-Marnani, J. (2018). Robust sensor fault detection and isolation scheme for interconnected smart power systems in presence of RER and EVs using unknown input observer. *International Journal of Electrical Power & Energy Systems*, *99*, 682–694. doi:10.1016/j.ijepes.2018.02.013

Alhelou, H. H., Hamedani-Golshan, M. E., Heydarian-Forushani, E., Al-Sumaiti, A. S., & Siano, P. (2018, September). Decentralized Fractional Order Control Scheme for LFC of Deregulated Nonlinear Power Systems in Presence of EVs and RER. In *2018 International Conference on Smart Energy Systems and Technologies (SEST)* (pp. 1-6). IEEE. 10.1109/SEST.2018.8495858

Alshahrestani, A., Golshan, M. E. H., & Alhelou, H. H. (2018, November). WAMS Based Online Estimation of Total Inertia Constant and Damping Coefficient for Future Smart Grid Systems. In *Smart Grid Conference (SGC)* (pp. 1-5). Academic Press.

Antony, S. (1997). *Use of genetic algorithms in electromagnetics* (Bachelor thesis). University of Queensland.

Attia, H., El-Sobki, M., & Wahadan, S. (2006). *Priority ranking of Industrial Loads and Application of Demand Side Management (DSM) Technique*. The Eleventh International Middle East Power System Conference, ElMinia, Egypt.

Balijepalli, M., & Pradhan, K. (2011). *Review of Demand Response under Smart Grid Paradigm*. IEEE PES Innovative Smart Grid Technologies. doi:10.1109/ISET-India.2011.6145388

Balls, C., Battaglini, A., Haas, A., & Lilliestam, J. (2008). *The SuperSmart Grid*. Retrieved from http://www.supersmartgrid.net/wp-content/uploads/2008/06/battaglini-lilliestam-2008- supersmart-grid-tallberg1.pdf

Barbato, A., & Capone, A. (2014). Optimization Models and Methods for Demand-Side Management of Residential Users: A Survey. *Energies*, *7*(9), 5787–5824. doi:10.3390/en7095787

Barbato, A., & Capone, A. (2014). Optimization Models and Methods for Demand-Side Management of Residential Users: A Survey. *Energies*, *7*(9), 5787–5824. doi:10.3390/en7095787

Capehart, B. L., Turner, W. C., & Kennedy, W. J. (2006). *Guide to Energy Management* (5th ed.). The Fairmont Press.

Chellali, F., Khellaf, A., Belouchrani, A., & Recioui, A. (2011). A Contribution to the actualization of the wind map of Algeria. *Renewable and Sustainable Energy Reviews, Elsevier.*, *15*(2), 993–1002. doi:10.1016/j.rser.2010.11.025

Chiu, W.-Y., Sun, H., & Poor, H. V. (2012). Demand-side energy storage system management in smart grid. *2012 IEEE Third International Conference on Smart Grid Communications (Smart Grid Comm.)*, 73,. 10.1109/SmartGridComm.2012.6485962

Chiu, W.-Y., Sun, H., & Poor, H. V. (2013). Energy Imbalance Management Using a Robust Pricing Scheme. *IEEE Transactions on Smart Grid, 4*(2), 896–904. doi:10.1109/TSG.2012.2216554

Coley David, A. (1999). *An introduction to genetic algorithms for scientists and engineers*. World Scientific. doi:10.1142/3904

El-Sobki, M. S., Jr. (1997). Tariff as a Demand Side Management Tool-its Design and Impact in View of Demand Reduction and Energy Savings technologies. Academic Press.

Electric Power Research Institute (EPRI). (1993). *Principles and Practice of DSM*. TR-102556, Project # 2342-16. EPRI.

Elsobki, M. S. (1996). DSM Strategy Options – An Optimal Based formulation. *Proceedings AUPT-DECIRED International Symposium*, 58-65.

Elsobki, M. S., & Wahadan, S. (1999). *A Demand Side Management (DSM) Priority Selection Technique – its Design and Implementation*. CIRED, International Conference & Exhibition on Electricity Distribution, Nice, France.

Fini, M. H., Yousefi, G. R., & Alhelou, H. H. (2016). Comparative study on the performance of many-objective and single-objective optimisation algorithms in tuning load frequency controllers of multi-area power systems. *IET Generation, Transmission & Distribution, 10*(12), 2915-2923.

Gellings, C. W. (1985). The concept of Demand Side Management for Electric Utilities. *IEEE, 73*(10).

Gellings, C. W., & Chamberlin, J. H. (1993). *DSM Concepts and Methods. The Fairmont Press.*

Goudarzi, H., Hatami, S., & Pedram, M. (2011), Demand-side load scheduling incentivized by dynamic energy prices. *Proceedings of the IEEE International Conference on Smart Grid Communications (SmartGridComm)*, 351–356. 10.1109/SmartGridComm.2011.6102346

Griffiths Anthony, J. (1993). *An introduction to genetic analysis*. New York: W. H. Freeman and Company.

Guo, Y., Pan, M., & Fang, Y. (2012). Optimal power management of residential customers in the smart grid. *IEEE Transactions on Parallel and Distributed Systems, 23*(9), 1593–1606. doi:10.1109/TPDS.2012.25

Haupt, R. L., & Ellen, H. S. (2004). *Practical genetic algorithms* (2nd ed.). John Wiley & Sons, Inc.

Huh. (2017). *Smart Grid Test Bed Using OPNET and Power Line Communication: Emerging Research and Opportunities*. IGI-Global.

Hussein, A. A. (2010). Optimal Based Demand Side Management DSM Formulation. In *Proceedings of the 14th International Middle East Power Systems Conference (MEPCON'10)*. Cairo University.

Johnson, J. M., & Rahmat-Samii, Y. (1997). Genetic algorithms in engineering electromagnetics. *IEEE Antennas & Propagation Magazine, 39*(4), 7–21. doi:10.1109/74.632992

Katz, M. (2010). Demand Side Management: Reflections of an Irreverent Regulator.. Oregon Public Utility Commission.

Li, D., Jayaweera, S., & Naseri, A. (2011). Auctioning game based demand-response scheduling in smart grid. *Online Conf. Green Commun. (GreenCom)*. 10.1109/GreenCom.2011.6082508

Liang, H., & Zhuang, W. (2014). Stochastic modeling and optimization in a microgrid: A survey. *Energies, 7*(4), 2027–2050. doi:10.3390/en7042027

Litos Strategic Communication. (2008). *The Smart Grid: An introduction.* U.S. Department of Energy.

Makdisie, C., Haidar, B., & Alhelou, H. H. (2018). An Optimal Photovoltaic Conversion System for Future Smart Grids. In *Handbook of Research on Power and Energy System Optimization* (pp. 601–657). IGI Global. doi:10.4018/978-1-5225-3935-3.ch018

Mohsenian-Rad, A., Wong, V., Jatskevich, J., Schober, R., & Leon-Garcia, A. (2010). Autonomous demand-side management based on game-theo-retic energy consumption scheduling for the future smart grid. *IEEE Transactions on Smart Grid, 1*(3), 320–331. doi:10.1109/TSG.2010.2089069

Nadweh, S., Hayek, G., Atieh, B., & Haes Alhelou, H. (2018). Using Four – Quadrant Chopper with Variable Speed Drive System Dc-Link to Improve the Quality of Supplied Power for Industrial Facilities. Majlesi Journal of Electrical Engineering.

Njenda, T. C., Golshan, M. E. H., & Alhelou, H. H. (2018, November). WAMS Based Intelligent Under Frequency Load Shedding Considering Online Disturbance Estimation. In *Smart Grid Conference (SGC)* (pp. 1-5). Academic Press.

Njenda, T. C., Golshan, M. E. H., & Alhelou, H. H. (2018, November). WAMS based Under Frequency Load Shedding Considering Minimum Frequency Prediction and Extrapolated Disturbance Magnitude. In *Smart Grid Conference (SGC)* (pp. 1-5). Academic Press.

Ortega, M., del Río, P., & Montero, E. A. (2013). Assessing the benefits and costs of renewable electricity. The Spanish case. *Renewable & Sustainable Energy Reviews, 27*, 294–304. doi:10.1016/j.rser.2013.06.012

Pfeiffer, B., & Mulder, P. (2013). Explaining the diffusion of renewable energy technology in developing countries. *Energy Econ., 40*, 285–296. doi:10.1016/j.eneco.2013.07.005

Prabaakaran, K., Chitra, N., & Kumar, A. S. (2013). Power quality enhancement in microgrid—A survey. *Proceedings of the IEEE International Conference on Circuits, Power and Computing Technologies (ICCPCT)*, 126–131.

Ravichandran, A., Malysz, P., Sirouspour, S., & Emadi, A. (2013). The critical role of microgrids in transition to a smarter grid: A technical review. *Proceedings of the IEEE Transportation Electrification Conference and Expo (ITEC)*, 1–7. 10.1109/ITEC.2013.6573507

Recioui, A. (2012). *Mutiple antenna systems over MIMO channels: Multiple antenna communication systems.* LAP Lambert Academic Publishing.

Recioui, A., Djehaiche, M., & Boumezrag, A. (2016). *Load Side Management in Smart Grids using a Global Optimizer.* Presented at the 8th International Exergy, Energy and Environment Symposium, Antalya, Turkey.

Richards, G., Noble, B., & Belcher, K. (2012). Barriers to renewable energy development: A case study oflarge-scale wind energy in Saskatchewan, Canada. *Energy Policy, 42*, 691–698. doi:10.1016/j.enpol.2011.12.049

Saad, W., Han, Z., Poor, H. V., & Basar, T. (2012). Game-theoretic methods for the smart grid: An overview of microgrid systems, demand-side management, and smart grid communications. *IEEE Signal Processing Magazine*, *29*(5), 86–105. doi:10.1109/MSP.2012.2186410

Saad, W., Han, Z., Poor, H. V., & Basar, T. (2012). Game-theoretic methods for the smart grid: An overview of microgrid systems, demand-side management, and smart grid communications. *IEEE Signal Processing Magazine*, *29*(5), 86–105. doi:10.1109/MSP.2012.2186410

Samadi, P., Schober, R., & Wong, V. (2011). Optimal energy consumption scheduling using mechanism design for the future smart grid. *Proc. IEEE Int. Conf. Smart Grid Commun. (Smart Grid Comm.)*. 10.1109/SmartGridComm.2011.6102349

Shandilya, S. (2016). *Handbook of Research on Emerging Technologies for Electrical Power Planning, Analysis and Optimization*. IGI Global. doi:10.4018/978-1-4666-9911-3

Strbac, G. (2008). Demand side management: Benefits and challenges. *Energy Policy*, *36*(12), 4419–4426. doi:10.1016/j.enpol.2008.09.030

Strbac, G. (2008). Demand side management:Benefits and challenges. *Energy Policy*, *36*(12), 4419–4426. doi:10.1016/j.enpol.2008.09.030

Torriti, Hassan, M. G., & Leach, M. (2010). Demand response experience in Europe: Policies, programmes and implementation. *Energy*, *35*(4), 1575–1583. doi:10.1016/j.energy.2009.05.021

U.S. Energy Information Administration. (2014). *Independent Statistics and Analysis*. U.S. Energy Information Administration.

Walter, K., Sibylle, K., Johanna, K., Ina, R., & Silas, S. (2010). *DIN EN 16001: Energy Management Systems in Practice: A Guide for Companies and Organisations. Federal Ministry for the Environment, Nature Conservation and Nuclear Safety*. BMU.

Weile, D. S., & Eric, M. (1997). Genetic algorithm optimization applied to electromagnetics; A review. IEEE Transactions on Antenna and Propagation, 45(3). doi:10.1109/8.558650

Zamani, R., Hamedani-Golshan, M. E., Haes Alhelou, H., Siano, P., & Pota, R, H. (. (2018). Islanding Detection of Synchronous Distributed Generator Based on the Active and Reactive Power Control Loops. *Energies*, *11*(10), 2819. doi:10.3390/en11102819

Zaminga, A. (2011). House energy demand optimization in single and multi-user scenarios. *Proceedings of the IEEE International Conference on Smart Grid Communications (SmartGridComm)*, 345–350.

Zhao, Z., Lee, W. C., Shin, Y., & Song, K. B. (2013). An optimal power scheduling method for demand response in home energy management system. *IEEE Transactions on Smart Grid*, *4*(3), 1391–1400. doi:10.1109/TSG.2013.2251018

KEY TERMS AND DEFINITIONS

Advanced Metering Infrastructure (AMI): It is architecture for automated two-way communication between a smart utility meter and a utility company.

Demand Response: In demand response, consumers play a significant role in the operation of the electric grid by reducing or shifting their electricity usage during peak periods.

Energy Management: Energy management includes planning and operation of energy production and energy consumption for resource conservation, climate protection and cost savings, while maintaining the users to have permanent access to the energy.

Genetic Algorithms: A metaheuristic inspired from the process of natural selection and are used to produce high-quality solutions to optimization and search problems by relying on bio-inspired operators such as mutation, crossover, and selection.

Load Side Management: Is the modification of consumer demand for energy through various methods such as financial incentives and behavioral change through education to encourage them to use less energy during peak hours, or to move the time of energy use to off-peak times such as nighttime and weekends.

Optimization: It is an act, process, or methodology to make something as perfect, functional, or effective as possible.

Peak Load Reduction: It is a process of reducing the energy consumption at high demand by shifting or switching off some users.

Smart Grids: It is a mix between the traditional power grid and the modern information and communication technologies to react to changes in usage.

Smart Meters: A smart meter is an electronic device that records consumption of electric energy and communicates the information to the electricity utility for monitoring and billing.

Chapter 6
Reliable Electricity Generation in RES-Based Microgrids

Valeriy Kharchenko
Federal Scientific Agroengineering Centre VIM, Russia

Valentin Gusarov
Federal Scientific Agroengineering Centre VIM, Russia

Vadim Bolshev
https://orcid.org/0000-0002-5787-8581
Federal Scientific Agroengineering Centre VIM, Russia

ABSTRACT

Using microgrid generation technologies is proposed in order to organize reliable power supply to rural areas. The concept of microgrid based on RES is considered as one of the realization forms of the distributed energy paradigm. In this chapter, there are the principles of generating complex formation in any given microgrid considering the specifics of the region, consumption patterns, and the potential of renewable energy sources in a given area. The algorithm for meeting the challenges of forming the structure of the microgrid generating structure is shown. The criteria for selection of power generation sources when solving the issue of their inclusion in the microgrid is proposed. The chapter also suggests the design of the micro gas turbine that is able to operate on biogas.

INTRODUCTION

Among the many directions and ways of development of modern energy, it can be singled out two tendencies, which have become especially active recently. This is the formation of a global energy system, on the one hand, and the all-round development of distributed energy production, on the other. The realization of both these tendencies is hard to imagine without a large-scale use of renewable energy sources (RES), which step by step consistently occupy more and more significant positions in the world energy balance.

The formation of a global energy system based on RES occurs consistently through building large solar and wind power stations which are gradually moving to large regional and interregional systems

DOI: 10.4018/978-1-5225-8030-0.ch006

(Strebkov & Kharchenko, 2011). There were published proposals for the creation of a global solar power plant that can work 24 hours a day and cover the global electricity demand all year round. The work of Strebkov & Irodionov (2004) proposes a model for the future development of world energy based on direct conversion of solar energy at solar power plants. The plant consists of three large solar power plants connected by special power transmission lines. The work of Strebkov et al. (2018) consideres the transfer of transcontinental energy flows by means of resonant waveguide technologies invented by N. Tesla.

No less important direction of the development of world energy is distributed energy based on the universal construction of individual sources of energy generation needed to supply facilities at the place of production. This direction of energy development is the most interesting for the agricultural sector (Camblon et al., 2009; Alhelou et al., 2018; Alhelou et al., 2016; Alhelou et al, 2015). Power supply of agricultural facilities has a number of specific features: the dispersal of consumers, small unit capacity, the large length of electric, thermal and gas networks, the presence of large sparsely populated areas where agricultural production is conducted and centralized electric and heat supply are absent. Ensuring reliable energy supply of these territories is a task of primary importance.

The reliability of power networks, minimization of energy losses and high economic efficiency of these networks play a significant role (Vinogradov et al., 2018; Vinogradov et al., 2019; Njenda et al., 2018; Zamani et al., 2018). The use of a microgrid formed mainly on the basis of renewable energy sources (RES) is a modern form of realization of the concept of distributed energy and is of great importance for solving the problem of sustainable power supply to agricultural producers.

It is generally accepted that the term "Microgrid" was first introduced by Professor Robert Lesseter in the University of Wisconsin. According to R. Lasseter the key feature of the microgrid is its ability to separate and isolate itself from the centralized power grid in the event of problems with power supply in order to ensure uninterrupted power supply inside the microgrid without power outage. This should be done in accordance with the concept of the network for CERTS (Consortium for Electric Reliability Technology Solutions) with minimal impact on the power within the microgrid. When the state of the network is normalized the microgrid has to automatically reconnects to the centralized network without interruptions in power supply.

The concept of a microgrid was formed quite a long time ago. Most experts agree that the microgrid, in the most general sense, is a set of generation sources with the energy accumulation systems distributed in a certain territory as well as the final consumers of electric power united in a single network. Nevertheless, there is still no consensus on the parameters characterizing microgrids such as the aggregate power of the generation sources attached to the microgrid and the size of the area where the network is realized. In the case of microgrid operation in parallel with the centralized network it is also necessary to solve the question of the connection quantity between them (there might be one connection or several anchor points). In the most common case, the definition of a microgrid looks like "A microgrid is an integrated low-power energy system with generators of electricity and consumers of energy distributed along with whole system".

One of the strategic tasks of rural energy today is to reduce the energy intensity of agricultural production based on the widespread use of new advanced technologies for the production and consumption of energy resources (Liu & Su, 2008; Alshahrestani et al., 2018; Alhelou et al., 2018). A promising way to solve the problem is to expand the scale of use of renewable energy sources. In addition to saving traditional energy resources, the use of renewable energy sources opens up the possibility of providing electricity for remote, non-centralized networks, primarily agricultural facilities.

However, the main sources of renewable energy like the sun and the wind depend on impermanent external conditions and they often can cause unpredictable power supply interruptions. This requires the creation of powerful storage systems and the creation of backup sources of energy supply, which significantly increases the cost of the created systems (Alhelou et al., 2015; Alhelou & Golshan, 2016).

Therefore, combined systems based on the application of two or more types of renewable energy sources can be used for autonomous power supply to rural consumers. They can successfully complement each other with a subsequent reduction in the need for the energy accumulation and the frequency of backup power supply use. Successful autonomous systems can be organized if there is comprehensive information on the potential of various RES in the considered area, especially at the facility (Daus et al., 2018).

However, it is often difficult at one facility, especially at a small peasant house, to realize combined power systems on the basis of all possible sources. In addition, the ratio of power loads and electricity generation in most cases is difficult to balance since both generation sources and consumers are few.

It will be much easier to eliminate all these problems if both the number of generation sources and the number of electricity consumers are much larger and they are more diverse. These conditions are easy to implement if a small local micro network (so called "microgrid") is created (Laria, 2009; Fini et al., 2016).

GENERAL DESCRIPTION OF MICROGRID

As can be seen from the above, the microgrid is an integrated low-power energy system with distributed generators and energy consumers. In the microgrid there is possible a wide integration of local non-fuel renewable energy sources such as solar, wind energy and small hydro power plants. A small power station on biogas with a microturbine can be also used as a generating source. This will significantly improve the reliability of the entire microgrid since the generation of electricity by such systems is much more predictable.

There are many variants of microgrids. They can work not only autonomously, but also in parallel with the central network (Mariam, Basu & Conlon, 2013; Nadweh et al., 2018; Njenda et al., 2018; Makdisie et al., 2108). The innovations that have occurred in energy, electronics, management technologies, informatics and communications create favorable conditions for the development and improvement of microgrids and, above all, their optimal control with the maintenance of standard and stable power parameters, despite the use of unstable energy sources such as wind power plants (WPP) and solar power plants (SPP).

Unlike centralized power systems, microgrids make it easier to balance power and get a good balance between generating capacity and the amount of generated energy. This can be achieved through dynamic reserve capacity and efficient energy storage while large power systems require the need to maintain an expensive and bulky backup capacity (Luo et al., 2015).

Microgrids in addition do not have a big impact on the operation of the power system in the case of connection to a central power grid since most of the energy is produced and consumed within the microgrid. It also allows eliminating the losses arising from the transmission of electricity through power networks. There are many opportunities to set the price for electricity below the market, because microgrids do not have a huge infrastructure, a large number of maintenance personnel and high energy costs (Adomavicius et al., 2013).

To be an owner of microgrids as well as to be their exploiter can be owners of residential buildings, businesses, companies, villages, villages etc. Here, at the same time energy consumers can be its producers, exploiting their micro power generation plant and / or energy storage units. The integration of WPP, SPP and other power plants based on RES into a microgrid meets significantly fewer bureaucratic obstacles than connection to the centralized power grid.

Good prospects for the construction of microgrids are available in rural areas, where access to local primary renewable energy sources is less restricted than in urbanized areas. In these cases.

All generation schedules (daily, seasonal, monthly, annual) are plotted for the development of all sources considering the potential of RES in the locations where they are deployed (where their locations are selected taking into account the provision of maximum capacity). The method of superposition of consumption and energy production schedules for different compositions of generation sources determines their combination that fully ensures the load coverage of all consumers with minimal capacity of accumulation and redundancy, that is, with minimal capital and operating costs.

When a microgrid works in parallel with the centralized network, the need for energy storage and backup can be minimized. At the same time, working with a centralized network can use two agreed schemes. In the first case the microgrid receives electricity from the network to cover peak loads and behaves like an ordinary consumer. In the second case the exchange of energy with the network takes place in two directions, that is, the microgrid covers its deficit through receiving energy from the network and in case of the presence of excess power the it transfers electricity back to the grid. The main amount of electricity produced and consumed at the same time remains in the microgrid, so this connection does not greatly affect the network.

Despite the seeming simplicity the formation of a microgrid is a rather difficult task in practice. Here it is necessary to solve the problem with the composition of participants in micro networks (subscribers), the number and type of electricity generators, the correct organization of the microgrid control system and so on.

A particularly important work is the procedure of forming a generating set of a microgrid, that is, a set of technical means that generates electricity to the grid, what will ensure its optimal structure. And for this it is important to determine the criteria by which this or that technical tool will be selected for inclusion in the structure of the created micro network.

Many works are devoted to the theory and practice of constructing and exploiting microgrid. Here there are some of the arguments on this issue as well as the presentation of some developed propositions.

First, at the first stage of work on the creation of a microgrid it is necessary to find answers to the following questions:

1. How large should this system be?
2. How much will the system cost?
3. Will the system make a profit in the foreseeable future?

Answers to these questions make it possible to understand how the microgrid will be acceptable for investment.

The configuration of a microgrid is primarily determined by factors such as:

1. The composition of consumers and their categorization;
2. Availability of consumers requiring uninterrupted power supply;
3. The potential of various types of renewable energy sources in the entire microgrid, especially in the locations of consumer facilities (that makes it convenient to manage the operation of the generating source).

TWO LEVELS OF TASKS FOR MICROGRIDS CREATING

The deployment of an autonomous microgrid is especially relevant in regions with high energy potential of RES and without the possibility of joining a centralized network. At the same time, the power storage capacity should be calculated considering the length of probable time intervals without wind and solar radiation. Developing projects for the microgrid creation in a given area is supposed to be solved problems at two levels:

The first level involves considering issues related to the microgrid as a whole (the highest level). The following issues at this level are considered:

1. The development of the schematic diagram of the microgrid including the determination of the layout of consumers, generators, storage systems and standby power supplies (SPS);
2. The determination of the category of electricity consumers, their consumption (daily, seasonal, annual), the degree of the need to ensure uninterrupted power supply and the assessment of the consequences of power network disturbances;
3. The assessment of RES potential in various locations of the territory covered by the micro-network, including the potential of renewable energy sources in consumer locations;
4. The Determination of RES type and the development of a schematic diagram of the coupling of generation sources;
5. The development of possible modes of microgrid operation under the condition of reliable energy supply to consumers;
6. The assessment of the possibility of microgrid developing in terms of including new consumers and new generation sources for those microgrids which are already in operation. It is necessary to take into account that in the microgrid energy consumers can be also energy producers at the same time;
7. The evaluation of the acceptable form of ownership of the microgrid and its organizational structure. It should be taken into account the fact that owners of residential buildings (associations of homeowners), enterprises, CJSC, municipalities etc. can own and operate the microgrid. In carrying out this work it is important to consider which consumers are planned for inclusion in the microgrid (already existing or newly connected) and how they will function as part of a microgrid;
8. The evaluation of the presence in the immediate vicinity of a centralized power grid and the possibility of technological connection to it;

9. Consideration of electricity supply reliability of consumers in the microgrid. The reliability of supplying electricity in various emergency cases within the microgrid is much easier to provide than in large centralized power systems. Power consumers in the microgrid may participate in the process of balancing power by regulating their own loads, generating and accumulating electricity;
10. The consideration of the economics of the microgrid and tariffs for the generated and supplied electricity as well as the payback period of the project for the creation of the microgrid;
11. Ensuring the optimal choice of the system for controlling the operation of the microgrid implementation.

An approximate microgrid scheme for power supply of a cottage community connected to a central network is shown in Figure 1.

Since this scheme provides for connection to a centralized power network, the requirements for storage and backup power systems are significantly reduced. In this version, the microgrid is designed to work in parallel with the network, but the electricity is produced and consumed mainly within the microgrid itself. The exchange of electric power flows is carried out in the case of transmission of the surplus electricity produced in a microgrid to the centralized network or getting electricity from the centralized power grid to cover peak loads inside the microgrids.

The second level provides for consideration of issues related to the selection of the microgrid elements. At this level issues such as the type and power of the generators, storage and SPS systems, the design of the interface devices for transmitting electricity from generators to a microgrid and a number of other issues related to the microgrid components are discussed.

Figure 1. Simplified scheme of a microgrid for power supply of a cottage community connected to a central network. 1 is a consumer; 2 is an electric energy storage battery; 3 is a solar array; 4 is a wind power installation; 5, 6 is a hydropower plant; 7 is a steam power plants with wooden fuel.

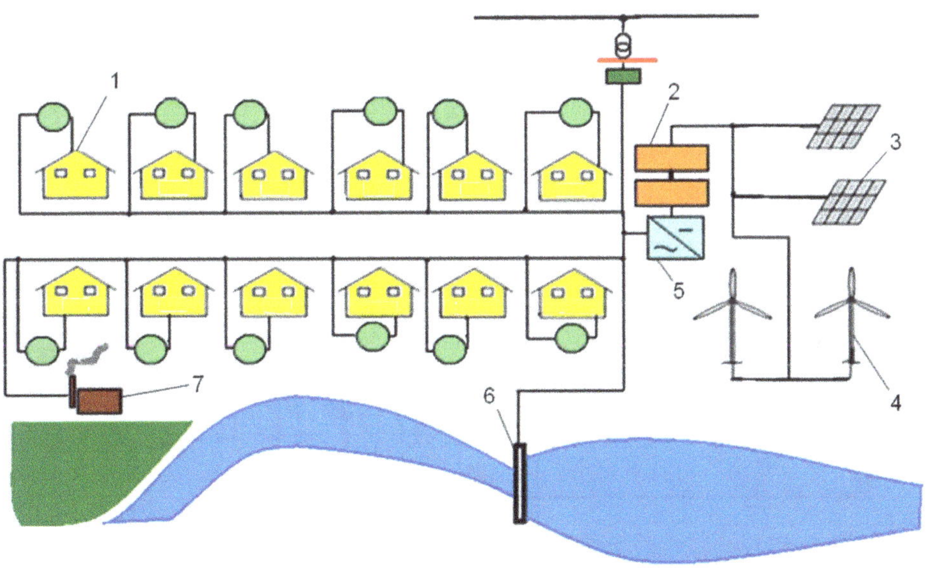

SELECTION OF ELECTRICITY GENERATION SOURCES

The correct selection of the microgrid elements is one of the first steps formed in terms of ensuring the reliability of its functioning, the payback period, the cost of the produced and supplied electricity. Therefore, this stage of work looks to be extremely important.

The composition of a complex of generation sources in such a microgrid is determined by a number of factors, the most important of which is the composition of consumers and their loads as well as the potential of renewable energy sources of various types available for use in the territory where the microgrid is implemented. A schematic diagram of the microgrid for power supply to rural consumers is given in Figure 2.

Figure 2. One of the possible options for a local rural microgrid

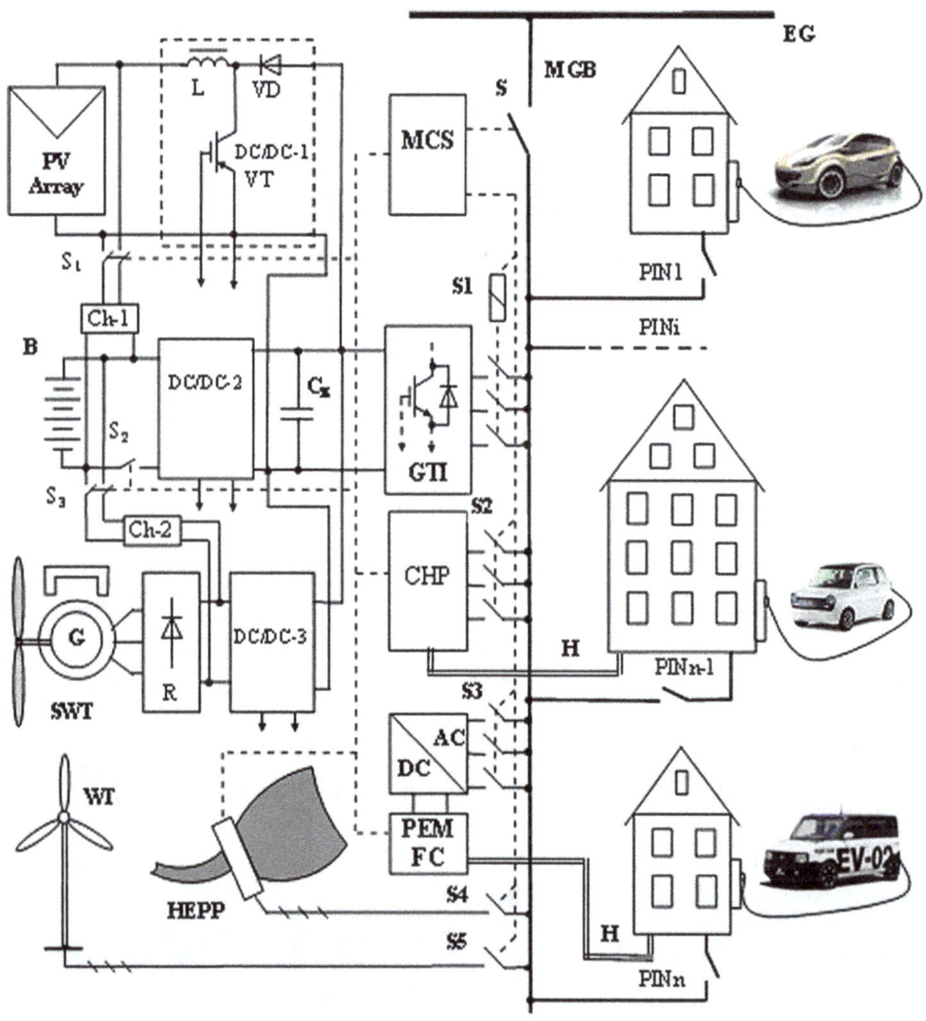

Criteria for Selecting Sources of Electricity Generation

The existence of such criteria creates favorable conditions for expanding the volumes and capacity of microgrids implemented on the basis of renewable energy sources. Their use in practice will provide an effective tool for selecting sources of electricity generation. In order to select and substantiate such criteria it is extremely important to evaluate the effectiveness of all sources of generation which can be both traditional and non-traditional, that is renewable energy sources and different variants of their combination in the created microgrids. This requires the use of a unified system approach to the analysis of both individual energy conversion technologies and combined systems as well as the consideration of many related factors.

Significantly, the renewable energy sources are most closely corresponding to the technical conditions of microgrid functioning, which in turn are the most effective form of their implementation in practice.

The choice of effective options for power supply to consumers united into a microgrid is a complex problem of system analysis since the power supply systems differ in many ways according parameters as below:

1. The type of used energy production technology;
2. The degree of interchangeability and complementarity of energy sources;
3. The operating modes of the microgrid;
4. The planned lifetime of the created microgrid;
5. The assumed degree of substitution of energy generated by energy generating devices;
6. The presence of the energy potential at a given location for a given specific type of RES;
7. The operation modes with respect to the centralized power networks of a higher level;
8. The possibility of autonomous operation (the ability to work in the "island" mode);
9. A number of economic indicators;
10. The presence of consumers in the network that require uninterrupted power supply and their relationship with consumers of other categories;
11. The degree of permanent generation of electricity by this generator and the possibility of working in a "basic" mode.

As it was mentioned above the main sources of renewable energy, that is solar radiation and wind, are impermanent and often have unpredictable interruptions. Due to that the provision of powerful storage systems and backup sources of energy supply are required what leads to significantly increase in the cost of such power generation systems. At the same time the joint use of two or more types of RES in one energy system, which can complement each other, reduces the need for energy accumulation and use of standby power supplies. Especially effective systems are autonomous ones when comprehensive information on the potential of various renewable energy sources is known and used in order to create these systems. That is why combined systems based on the use of two or more renewable energy sources are more preferable for providing autonomous power supply to remote consumers.

In general, power generation sources can be classified according to the following enlarged criteria:

1. Weather and climatic conditions (the potential of the given source);

2. Technical characteristics (a complex of parameters determined by the type of installation, its characteristics etc.);
3. Economic indicators (unit cost of installed capacity, operating costs, maintenance costs etc.).

Characteristics of the Most Common Sources of Generation

The generating complex of any microgrid consists of two types of elements:

1. The power generating devices (Solar Power Plants, Wind Power Plants, Small Hydro Power plants, Diesel-generators, Small Gas Turbine Installations etc.);
2. The power accumulators (batteries) which are designed to store energy in the case of energy production more than its consumption in the microgrid and to give it back to the microgrid in the event of an energy shortage.

The power generating devices in turn can subdivided into 2 types:

1. Based on traditional sources of generation (diesel generators, gas generators etc.)
2. Based on non-traditional sources of generation, that is, renewable energy sources (wind power plants, solar power plants, small hydroelectric power plants etc.)

Among power generation sources used in a microgrid majority belongs to renewable energy sources. There are however two types of traditional sources of power generation which are the most common and of greatest practical importance. Among the traditional sources of generation, the most common ones in practice are diesel generators. All their advantages and disadvantages are well known and do not need special discussion. Therefore, we consider gas microturbines as traditional sources of generation in the present work. The use of such microturbines in agriculture has a number of advantages. For more effective use of micro-gas turbines under our management conditions we have developed a new gas microturbine design with several advantages. A brief description of such a turbine will be given below.

Solar Power Plants

Technical and economic efficiency of the solar power plants (SPP) in the first place depends on the efficiency of photovoltaic modules (Makdisie et al., 2018; Kharchenko et al., 2010). The installed capacity of SPP grew at an extremely slow rate at the first stage of their use due to the high cost of solar cells and panels based on them. In recent years prices for solar cells began to fall sharply against the backdrop of growing prices for traditional fuel. Today, the cost of electricity generated at SES almost corresponds to the cost of electricity obtained by traditional methods (Figure 3). In the future, a widening gap in favor of solar panels is expected.

Currently, there is a large number of solar power plants of various types from small-scale laboratory solar installations (Figure 4) to more powerful ones including SPP located on the roofs of buildings (Figure 5) and powerful solar stations ready to work in large microgrids (Figure 6) (Belenov et al., 2016; Daus et al., 2016).

Figure 3. Cost of electricity produced by conventional power plants (1) and by the solar power plants (2)

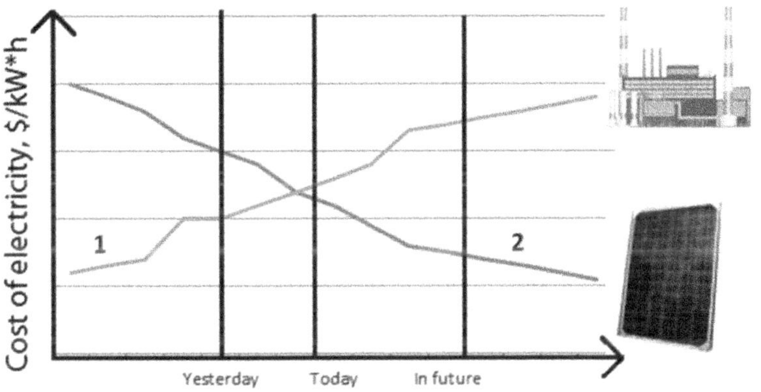

Figure 4. Solar power plant with a power of 1 kW designed to study the operation in parallel with the network in Federal Scientific Agroengineering Center VIM, Russia

Figure 5. Solar power plant with a power of 73 kW on the roof of a commercial building in Kamyshin, Volgograd region, Russia

Wind Power Plants

Wind-driven power plants are also widespread electric power generation systems, which are second only to solar energy systems in terms of prevalence (Ramonas, 2009). Wind power plants are usually divided into three types of installations:

1. System WPP;
2. Wind diesel WPP;
3. Charging WPP designed to charge batteries.

A small WPP is a wind power plant with a power of up to 50 kW. They have as a rule a swept surface area of up to 200 mF, an AC output voltage of up to 1000 V and a DC output voltage of up to 1500 V. Along with horizontally axial systems, which have the most widespread today, vertically-axial systems increasingly began to be used in practice.

The possibility of using a modern and efficient generator is of great importance for promoting the use of wind power stations. The work of Gribkov (2017) presents information on wind-powered vertically axial installations and systems of guaranteed energy supply of low power based on them (Figure 7) as well as the appearance of new generators designed for use in wind power systems (Figure 8).

Figure 6. Solar station with a power of 1 MW in Shymkent, Republic of Kazakhstan

Figure 7. Aerodynamic scheme of a vertical-axis turbine-type wind power plant with a directing device (a); modular vertical axis wind turbine with guide device (b); wind-solar-diesel complex of guaranteed power supply with a power of 16 kW (c)

Figure 8. Generators of various types for use in wind power plants

a is VG-05(12)/650-56-0306, VG-05(12)/650-56-0406, VG-1(12)/450-57-0206, VG-1(12)/650-57-0206; б is VG-5(28)/300-114-02B; в is VG-5(28)/300-114-02G

Development of Gas Turbine Plants for Use in Power Supply Microgrids of Agricultural Facilities

Micro gas turbine stationary power plants (MGTPP) with a power of up to 30 kW have never been produced in Russia except for special purpose products. The practice of world production of MGTPP shows that only a narrow range of companies are able to produce such products. The following Micro gas turbine installations are represented in the market: Capstone Turbine Corporation, Ingersoll-Rand, Calnetix Power Solutions, Turbec AB, Ingersoll-Rand Energy Systems, Honeywell Power Systems, Bowman Power System, UTC Power, Wilson Turbo Power, Toyota Turbine System etc. These MGTPP work in cogeneration mode. They differ from each other in terms of designs, applied know-how, operating modes, packaging options. The power of the proposed installations is from 30 to 350 kW. Microturbines Capstone and Calnetix can work both independently and with the transition to parallel operation with clusters of several similar installations.

Gas-turbine plants produce heat energy approximately twice as much as electric one. Agriculture is a very serious consumer of thermal energy (Lachuga & Strebkov, 2009). In the balance of consumption of energy resources by agriculture the share of thermal energy accounts for about two-thirds. Therefore, the use of gas turbine plants (in case of the use of generated heat) is the most promising in the agricultural sector. The possibility to obtain a large amount of thermal energy from gas turbine plants implies a faster payback of the project. In addition, unlike other sectors of the economy the agricultural sector has enough raw material for the production of biogas, which in principle could become a promising raw material for gas turbine systems.

The high efficiency of using micro gas turbine power plants is provided by a wide operating range for varying electrical loads from 1 to 115%.

The main advantage of using MGTPP is the low content of harmful emissions (the range is 9-25 ppm) what allows them to be placed close to the places where people live. This parameter for gas turbine systems is better than for piston power plants (their closest competitors) as well as for diesel-generator systems.

The use of micro gas turbine power plants in comparison with gas-piston units allows reducing costs for exhaust gas purification systems. Gas turbine plants have little vibration and noises within 65-75 dB (which corresponds to the noise level of a household vacuum cleaner). Special sound insulation for noise reduction as a rule is not applied.

However, as the analysis showed, the micro gas turbine systems present in the market do not fully meet our requirements, and therefore the need to develop a new-generation MGTPP arose (Nikitin, 2012).

The goal set for the development and research of micro gas turbine units for autonomous power supply of agricultural facilities is a complex task due to the following reasons:

1. A small number of publications on the research and design work devoted to the MGTPP of low power;
2. Small applicability of gas turbine power plant parameters of high power to MGTPP parameters of low power;
3. The need to create a methodology for calculating various parameters of the MGTPP using an ICE (internal combustion engine) turbo compressor;
4. The need to conduct a large number of tests of experimental MGTPP;
5. The need to select the various nodes of MGTPP when designing.

In the works of Gusarov & Gusarova (2017) and Gusarov & Kharchenko (2018) a micro gas turbine power plant was created and the following tasks were accomplished:

1. The analysis of existing power supply systems on the basis of serial MGTPP and the determination of the ways of reducing their prime cost, which are applied in the developed installation;
2. The development of the methods for calculating MGTPP with an ICE turbo compressor using various types of fuel have been developed;
3. The production of MGTPP with the ability to work on different kinds of gaseous fuel;
4. The test of the created MGTPP in accordance with modern test requirements and analyzing the effectiveness of this micro gas turbine installation;
5. Carrying out the feasibility study of the possibility of applying MGTPP in agriculture.

The micro gas turbine cogeneration power generator consists of a micro gas turbine engine (MGTE) with a periphery, a power turbine required to select mechanical power, a high-speed electric generator, an electronic power conversion system, an exhaust gas thermal energy recovery system and an automatic control system.

Based on the analysis of the existing design solutions of gas turbine installations for MGTPP the radial centripetal turbine is recognized as the most suitable for the following reasons:

1. The turbo compressor of this type is particularly suitable for small gas turbine engines since it is possible to make the design more compact by means of constructive connection of the turbine with the impeller of the compressor with the same external diameter;
2. The rotor of a radial centripetal turbine, in contrast to the rotor of an axial turbine, consists of a disk and individual blades and can be manufactured in a cheaper way by both forging or precision casting;
3. A radial turbine of small dimensions can theoretically be even more efficient than a corresponding axial turbine due to the much smaller influence of the Reynolds number on its characteristics and therefore the scale;
4. A radial turbine has higher strength and reliability in operation compared to an axial turbine
5. The blades of the radial turbine are practically insensitive to the action of small solid particles trapped in the combustion gases, while the ingress of solid particles into the axial turbine can cause serious erosion of the blades;
6. The radial turbine has higher pressure differential than the turbine of the axial type. A two or three-stage radial turbine can be used for pressure differential of the 3 or 4 order;
7. A radial turbine with adjustable nozzle blades can maintain its calculated (maximum) efficiency in a relatively wide power range and thus have significantly better performance at partial loads than an axial-type turbine;
8. Like the axial type turbine, the radial one has the same characteristics of maximum torque when starting at low speed.

The above reasons show that inward flow radial turbines have great potential and the problem of creating design methods of MGTPP based on such turbines is important and actual. These design methods undoubtedly have to meet the operating conditions, the technical equipment of enterprises and the consumer financial capabilities.

The existing designs of turbo compressors manufactured in Russia are produced on the basis of inward flow radial turbines which are widely available in the market and relatively cheap. This allows them to be used for the production of low-cost MGTPP.

The appearance of the micro gas turbine power plant is shown in Figure 9.

MGTPP was developed with the approach of maximizing the use of standardized units and parts manufactured by the Russian industry. So, a serial turbocharger of an internal combustion engine was used as a turbine and a compressor of MGTPP. However, a special combustion chamber has been developed for this purpose. The operation of the turbo compressor is provided by a lubrication system on an automotive oil which includes an oil tank, an oil pump, an oil filter and an oil cooling radiator. The pressure in the system is 2.5 - 5.0 kg/cm^2 provided by a standard oil pump used in the Russian industry. The operating temperature is 80 - 95 °C and is maintained by means of an oil cooling radiator.

A step-by-step description of all the theoretical and experimental work carried out to create a micro gas turbine with the necessary parameters and all its innovative components is a separate description. We note only that the developed turbine is very promising for use in the conditions of agricultural production and can be the subject of close cooperation with interested organizations.

Figure 9. Micro gas turbine power plant GTE-10C

POWER GENERATION MANAGEMENT IN A DISTRIBUTED NETWORK WITH THE GENERATION SOURCES OF VARIOUS NATURE.

The use of local generating capacity not connected with centralized grids and operating to meet the electricity needs of small areas is currently a worldwide trend. The advantage of using low-power generators operating in local low-voltage microgrids is the possibility of reducing the cost of generating electricity by reducing transmission losses, increasing the efficiency of generating capacity, reducing equipment downtime and reducing maintenance costs.

The use of generating capacities using renewable energy sources (wind power plants, micro and mini hydro power plants, solar electrical installations, gas piston plants using biogas as fuel) is one way to reduce the cost of electricity generation in microgrids.

The basis for optimizing the operation of a distributed system using generation sources based on different physical principles is a control system that ensures the optimal input and output of generating capacities depending on the specific conditions at each moment of time. For example, the condition for the use of wind power plants in the system is the availability of a sufficient amount of wind flow capable of providing load requirements and providing the consumer with electricity of acceptable quality. When the power of the wind flow decreases, it is necessary to carry out a gradual withdrawal from the generation of a wind power installation of the system with simultaneous generation on the basis of another energy source in a volume depending on the electricity demand of the consumer at a given time. Such operational management of electricity generation depending on the load schedules will significantly reduce the cost of generating electricity by maximizing the use of renewable energy sources and performing the operational control of generation via a distributed control system.

The structural scheme of the distributed microgrid is shown in Figure 10.

In this scheme the generating capacities "Generator 1", "Generator 2" and "Generator n" with control capability are united into a single system providing electricity to the consumers "Consumer 1", "Consumer 2" and "Consumer n". The distributed control system provides the control of generating capacities in such a way that at each specific moment of time the energy generation is carried out with a minimum cost of electricity generation as possible. A distinctive feature of the system under consideration is the lack of a common computing center and the need to organize information channels for the exchange of control information in the system. To a certain extent, at the same time it is possible to organize electricity exchange with medium-voltage networks and to connect the information distributed control system to a single dispatch center.

A distinctive feature of using the method of organizing information exchange and using a distributed generation management system is the ability to scale the microgrids into power systems operating under a single algorithm for managing the system as a whole. (Lapshin, 2013)

The use of information exchange over the electric network provides the possibility of implementing an electricity generation system using both renewable and non-renewable energy sources when working together in a single low voltage electrical network. This allows managing the electricity generation depending on the conditions of generation and consumption of electrical energy on the basis of feedback sensors, thus reducing the cost of generated electricity and improving the overall system efficiency. It also enables the exchange of electricity with higher-level power grids.

The structural scheme of the system realizing the method of generation control is shown in Figure 12. This method lies in analyzing environmental data (wind direction and velocity, solar radiation, water flow velocity, volume and pressure of biogas in the main line), load power as well as voltage and fre-

Figure 10. The structural scheme of the distributed microgrid

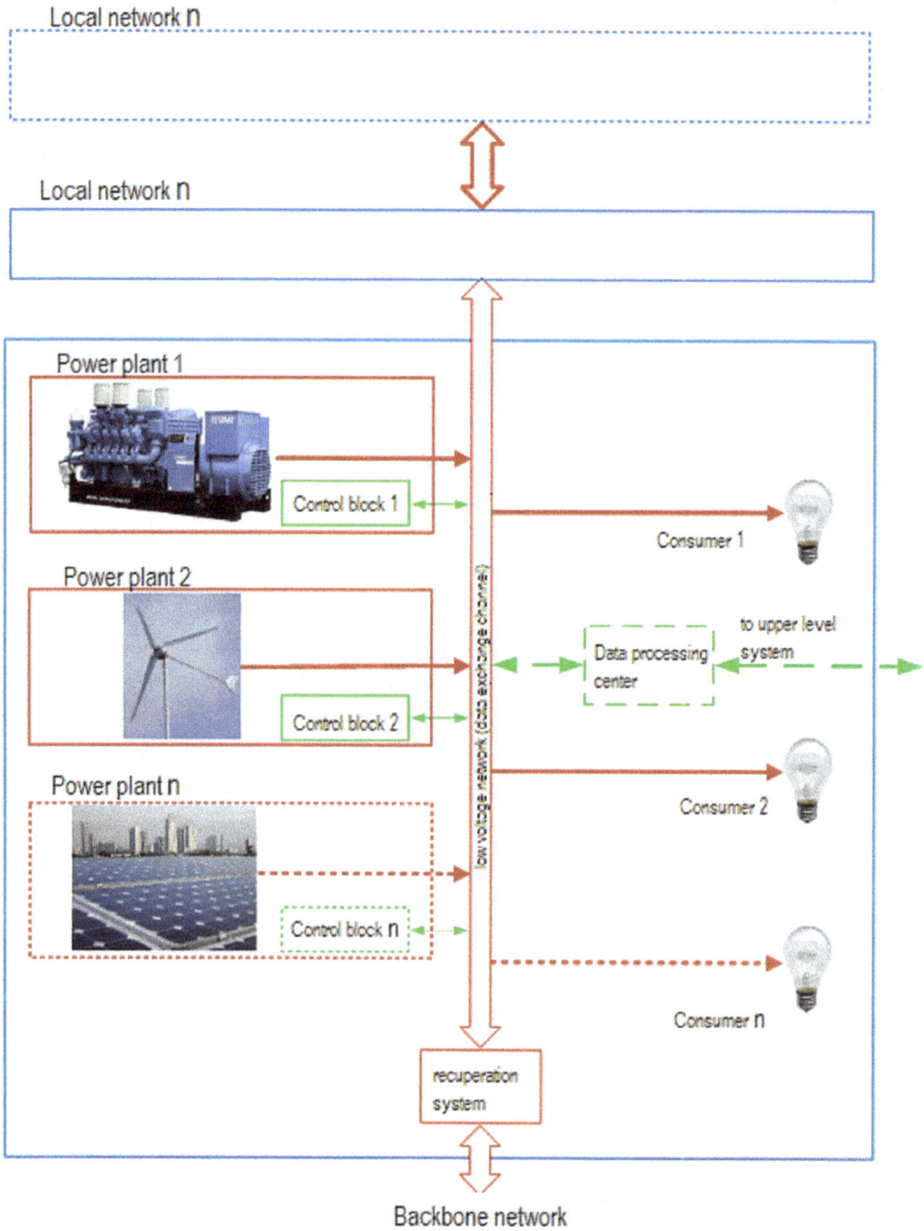

quency at the input of the main power line. On the basis of the obtained data an algorithm for choosing the most optimal generation source is implemented in order to provide the consumer with electricity of acceptable quality corresponding to existing standards and with the lowest cost of generating electricity. When the state of the environment changes and (or) electricity consumption, the source of generation is changed as well.

The system shown in Figure 11 includes a stationary control system 1, local control modules 2 that are installed directly on generation sources using non-renewable energy A1 ... An (diesel fuel, main

Figure 11. The structural scheme of the system realizing the method of generation control

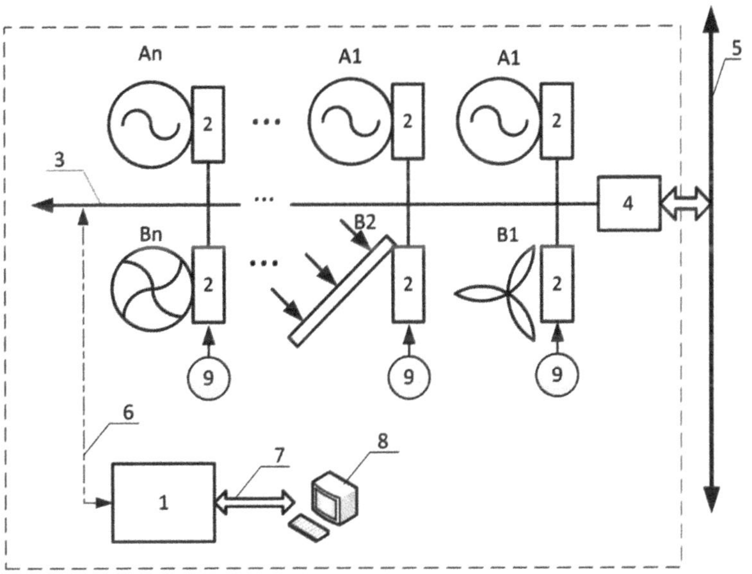

gas, boiler fuel) as well as generation sources using renewable energy B1 ... Bn (wind turbines, mini-hydropower plants, solar panels, biogas generators). The local control modules installed on renewable energy generation sources transmit data from environmental sensors 9 to the stationary control system 1 through an information channel 6 using the local low voltage network 3 as the data transmission medium. All sources of generation are combined into the local low voltage network 3, delivering electricity directly to the consumer. The local low voltage network 3 uses the system of electric energy exchange 4 with the networks of medium or high voltage 5. The stationary control system 1 has an information channel 7 for exchanging data with the upper layer system 8.

The system that implements the above method works as follows. Electricity from the backbone network 5 through the electric energy exchange system 4 is received to consumers. The stationary control system 1 analyzes the load in the network and the state of the environment by means of the sensors 9. If there is sufficient wind flow, solar radiation or water pressure, the energy produced by generation sources is introduced in the local network 3 through local control modules 2 mounted on each generation sources. The energy received from the backbone network 5 at the same time is limited in such a way that the main sources of generation in the micro network are renewable energy sources B1 ... Bn. In case of excess capacity from local renewable energy sources B1 ... Bn, it is received back to through the system of electricity exchange 4 into the backbone network 5. If the generation power from renewable energy sources B1 ... Bn is not enough (for example, at peak load on the grid) and also there is the absence of connection to the backbone network 5, the generation from non-renewable energy sources A1 ... An is managed in such a way that the sources with a cheaper type fuel have primacy (the use of gas generation has the advantage over diesel one while diesel one does over gasoline one). Each such local system can be embedded as an element of a larger network and operated under the control of the top-level system 8 via an information exchange channel 7.

Managing the Electricity Generation in a Microgrid

The concept of the microgrid assumes a wide variety of electrical schemes of supply networks. The network can have AC and DC sections of different voltages. The type of current in the network is determined by the type of generators and the type of load. In the networks where there are diesel power plants (DPP), gas turbine power plants (GTPP), biogas power plants (BGPP) or hydropower plants (HPP) along with solar (SPP) and wind power (WPP) power stations the ac schemes are generally accepted (Husarov, Kharchenko & Lapshin, 2013).

The problem of AC power systems with heterogeneous energy sources is the adaptation of generators to the mains voltage and frequency compatible with the load of consumers. The structure of the microgrid can be linear (consisting of several lines) and ring one. The network can provide consumers with electricity of both single-phase and three-phase voltage. An approximate electrical block diagram of the microgrid is shown in Figure 12.

According to Figures 12: 1 are solar power plants SPP1 and SPP2; 2 is the wind power plant for charging of storage batteries WPP1; 3 is the wind power plant of mains voltage WPP2; 4 is a micro gas turbine power plant; 5 is a hydroelectric power station HPP; 6 is a storage battery (SB); 7 is a network controller NC-1; 8 is a charge controller of SPP1 and SPP2; 9 is a block of parallel operation of the wind farm; 10 is a charge controller of WPP; 11 is a network controller NC-2; 12 are rural houses; 13 is street lighting; 14 is a water supply system; 15 is an entrance barrier.

A linear circuit showed in Figure 12 consists of three electrical lines. The "A" line is used for the connection of residential houses of a rural settlement. The "B" line is used for the connection of a water supply system and a security system. The "C" line is used for the connection of a street lighting system. The stable operation of the electrical network requires the creation of conditions under which the power of the load is completely provided by the power of the generators. Since the power supply of the microgrid is carried out by fuel generators and stochastically operating generators (solar and wind power plants), the

Figure 12. An approximate electrical block diagram of a linear microgrid (a) and a ring microgrid (b) with generation centers of various types

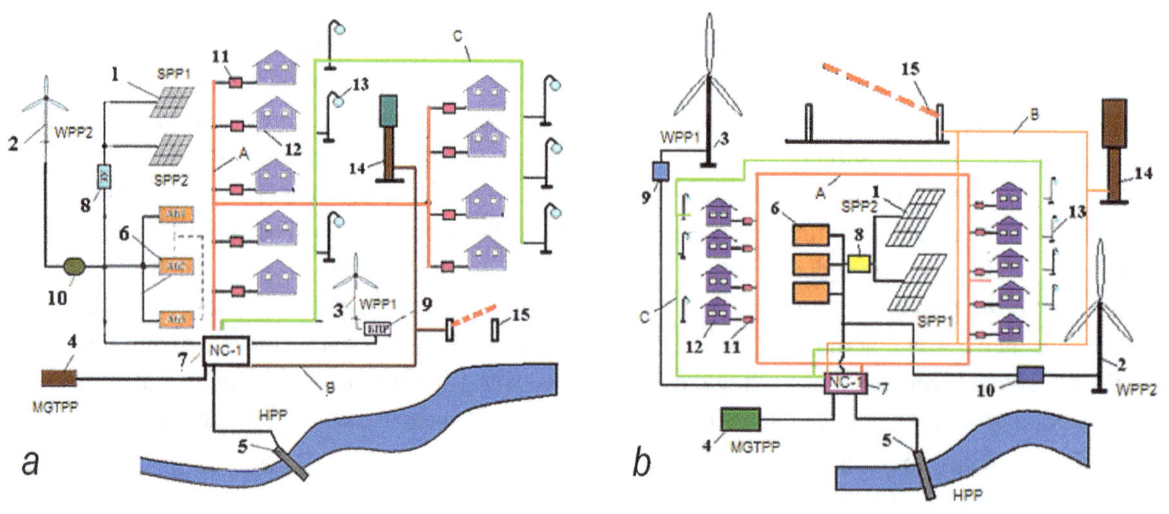

network includes electric energy accumulators (storage batteries or supercapacitors) allowing the store of peak loads. The distribution of the load on the electric lines of the microgrid is made according to the level of responsibility of the power supply. The most important load is the "A" line, residential buildings are connected to. After the «A" line there is the "B" line, the water supply system and the security system are connected to. The last "C" line with the outdoor lighting system will be disconnected first in case of capacity shortage. The "A" line should be disconnected only as a last resort. The management of electricity distribution through the lines of the microgrid is made by the network controller of the first level (NC-1), the management of electricity distribution directly to consumers (houses, farms, shops, workshops etc.) is made by the network controller of the second level (NC-2).

The network controller NC-2 is installed directly at the consumer. Since the selected scheme of the microgrid assumes the presence of energy storage devices directly in the network, the power supply of the load depends on the voltage level of the batteries. All the load available to consumers, i.e. an internal network of home ownership, a small farm, a shop, workshops etc., is the load of the second network layer. All the load of the second level is connected to the NC-2, which has three output ports. Each load group is connected to an appropriate port. A refrigerator, a water pump, a fire alarm system, emergency lighting etc. is the main load connected to the first port. A secondary load is connected to the second port (lighting, TV, computer, kitchen equipment etc.) A less important load is connected to the third port (construction tools, automatic door opening, electric heating devices, outdoor lighting etc.).

An autonomous microgrid is a risky power supply system and the task of a developer is to minimize the risks of power failure. The effect of reliability is achieved by including fuel generators (diesel power plants, gas turbine power plants, biogas power plants) and hydropower plants when conditions permit.

SPP, WPP, BGPP and HPP are the main sources of energy. Their total power should provide the total load power of the microgrid and the power of charging the energy storage devices.

$$\sum P_{main.generation} \geq P_{load} + P_{charge} \tag{1}$$

where $P_{main.generation}$ is total power produced by the main sources of generation, kW

P_{load} is total load power of the microgrid, kW

P_{charge} is total power required for of charging the energy storage devices, kW

The fuel generators are in turn the reserve sources of energy. Their power should also be higher than the load in the network by the amount of the charging power of storage batteries.

$$\sum P_{fuel.generation} \geq P_{load} + P_{charge} \tag{2}$$

where $P_{fuel.generation}$ is total power produced by the fuel sources of generation, kW

The main tasks of the microgrid are to ensure uninterrupted power supply to consumers and to generate electricity with a low cost. The algorithm of power supply system operation has been developed to save fuel from standby generators. According to the algorithm the start of fuel generators is managed after the voltage level of storage batteries drops to a critical level. Fuel generators provide the work of consumers and charging storage batteries during RES downtime.

The NC-1 with a power of up to 100 kW is designed to control the electric lines of a microgrid of low power. The controller provides the specified algorithm of microgrid operation and disconnects the electric lines in a predetermined order to provide power supply to the main consumers. Its task is to analyze the balance of generation power and load. One of the possible algorithms for the operation of a small power microgrid is given below.

If an electricity generation level provides the entire load of the rural settlement including the charge of the storage batteries, the voltage level in the network is 230 V which is simultaneously a signal for the network controllers of the second level (NC-2) about sufficient power supply. If the level of generation in the network decreases, the control system enters the mode of deficit covering from the storage batteries. If the voltage in the storage batteries reaches the set level "α", the system reduces the network voltage to the level of 225 V. When the network controller NC-2 determines the voltage decline to the level of 225 V, it switch off the "C" line, that is the third group of the least important load.

With the further use of accumulated energy and the reduction of the voltage in the storage batteries to the level "β", the voltage in the network is reduced to 220 V. The NC-2 determines this fact and switch off the line «β», that is the second group of consumers.

When the voltage drops to the critical low level on the storage batteries, the NC-1 signals the inclusion of standby generators, diesel or gasoline power stations, which must provide the load of the rural settlement with lack of energy and charge the battery. When the standby generators are started, the NC-2 will switch on all the lines and load groups.

The voltage levels "α" and "β" are set by programming and determined by the importance of the load at each facility specifically. Figure 13 shows a block diagram of the first level network controller.

As shown in the electrical block diagram (Figure 13), solar power plants SPP1 and SPP2 and wind power plant SPP2 are connected to the section of the DC inverter circuit while simultaneously charging the batteries. The DC circuit has a voltage of 320 V and is connected to the common bus of the inverter 6 via port 1. GTPP via port 2 and AC/DC converter, DPP via port 3 and AC/DC converter, WPP1 through port 4 and AC/DC converter are connected to the same bus. The hydroelectric power station is a constantly operating generator, that is the source of the reference voltage. So, the output parameters of the voltage and frequency of the HPP are determinative for the generators of the entire system and the inverter converting the DC voltage to AC. As we know, solar power stations are the most affordable sources of energy in the market for use in a microgrid. Not only because their cost is constantly decreasing, but also because their installation does not require special preparation. Photovoltaic modules can

Figure 13. Block diagram of the layout of the ports NC-1

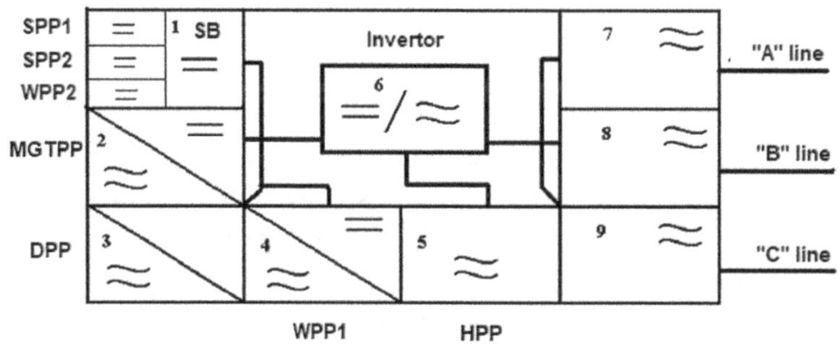

be configured for a given voltage and their current-voltage characteristic does not varies much when a change in the level of solar radiation. Therefore, the SPP is almost always workable in the daytime. Very effective SPP are installations with concentrators and a solar tracking system (Strebkov et al., 2002). The most noticeable change is the level of current when the significant change of solar radiation. The voltage level changes in insignificant limits in this case (Strebkov & Irodionov, 2005). In the middle zone of Russia solar power plants are most effective in the summer while wind power plants are the most effective in the winter.

FUTURE RESEARCH DIRECTIONS

Despite the great work done in the field of building microgrids by scientists from all over the world, there are many problems in this theme. Here It is possible to indicate the problems in the management of the micro-network, consisting of sources of generation of different nature. Both the technical and organizational-legal aspects of the operation of the microgrid with a centralized power network are not completely resolved, especially in cases where the electric power is transferred from the microgrid to a centralized network.

CONCLUSION

The scale of the use of microgrid is continuously increasing all over the world. The realization of microgrid concept is most active in North America. The power of the microgrids introduced into operation there exceeds the indicators in Europe, the Asia-Pacific region and the rest of the world combined together. Regardless. this direction is actively developing in many countries including Russia, China, etc. Therefore, the further efforts to create solid scientific and methodological foundations of the technology of microgrid is not losing its relevance.

In this chapter an attempt is made to create the prerequisites for the formation of provisions and the principles of generating complex formation in any given microgrid taking into account the specifics of the region, consumption patterns and the potential of renewable energy sources in a given area. The proposed algorithm for meeting the challenges of forming the structure of the microgrid generating structure as well as the proposed criteria for selection of power generation sources when solving the issue of their inclusion in the micro-network can effectively form the necessary elements of the microgrid.

The developed design of the micro gas turbine is a promising source of electric power generation as it has a number of advantages over other sources operating on fuel and can be successfully used as a basic source of generation of a rural microgrid. The design features make it possible to ensure its operation on biogas which is accessible in rural areas. Prospects for successful application of the developed micro gas turbine for power supply of rural objects are due to the huge reserves of bioresources in the countryside which can easily be processed into organic fuels according to the technologies available on the market.

REFERENCES

Adomavicius, V., Kharchenko, V., Valickas, J., & Gusarov, V. (2013). RES-based microgrids for environmentally friendly energy supply in agriculture. *Proceeding of 5th International Conference Trends in Agriculture Engineering*, 51-55.

Alhelou, H., Hamedani-Golshan, M. E., Zamani, R., Heydarian-Forushani, E., & Siano, P. (2018). Challenges and Opportunities of Load Frequency Control in Conventional, Modern and Future Smart Power Systems: A Comprehensive Review. *Energies*, *11*(10), 2497. doi:10.3390/en11102497

Alhelou, H. H. (2018). *Fault Detection and Isolation in Power Systems Using Unknown Input Observer. In Advanced Condition Monitoring and Fault Diagnosis of Electric Machines* (p. 38). Hershey, PA: IGI Global.

Alhelou, H. H., Golshan, M., & Fini, M. (2018). Wind Driven Optimization Algorithm Application to Load Frequency Control in Interconnected Power Systems Considering GRC and GDB Nonlinearities. *Electric Power Components and Syst.*

Alhelou, H. H., & Golshan, M. E. H. (2016, May). Hierarchical plug-in EV control based on primary frequency response in interconnected smart grid. In *Electrical Engineering (ICEE), 2016 24th Iranian Conference on* (pp. 561-566). IEEE. 10.1109/IranianCEE.2016.7585585

Alhelou, H. H., Golshan, M. H., & Askari-Marnani, J. (2018). Robust sensor fault detection and isolation scheme for interconnected smart power systems in presence of RER and EVs using unknown input observer. *International Journal of Electrical Power & Energy Systems*, *99*, 682–694. doi:10.1016/j.ijepes.2018.02.013

Alhelou, H. H., Hamedani-Golshan, M. E., Heydarian-Forushani, E., Al-Sumaiti, A. S., & Siano, P. (2018, September). Decentralized Fractional Order Control Scheme for LFC of Deregulated Nonlinear Power Systems in Presence of EVs and RER. In *2018 International Conference on Smart Energy Systems and Technologies (SEST)* (pp. 1-6). IEEE. 10.1109/SEST.2018.8495858

Alhelou, H. S. H., Golshan, M. E. H., & Fini, M. H. (2015, December). Multi agent electric vehicle control based primary frequency support for future smart micro-grid. In *Smart Grid Conference (SGC)* (pp. 22-27). Academic Press.

Alshahrestani, A., Golshan, M. E. H., & Alhelou, H. H. (2018, November). WAMS Based Online Estimation of Total Inertia Constant and Damping Coefficient for Future Smart Grid Systems. In *Smart Grid Conference (SGC)* (pp. 1-5). Academic Press.

Belenov, A. T., Daus, Yu. V., Rakitov, S. A., Yudaev, I. V., & Kharchenko, V. V. (2016). The Experience of Operation of the Solar Power Plant on the Roof of the Administrative Building in the Town of Kamyshin, Volgograd Oblast. *Applied Solar Energy*, *52*(2), 151–156. doi:10.3103/S0003701X16020092

Camblong, H., Sarrb, J., Niangc, A. T., Curea, O., Alzola, J. A., Sylla, E. H., & Santos, M. (2009). Micro-grids project, Part 1: Analysis of rural electrification with high content of renewable energy sources in Senegal. *Renewable Energy*, *34*(10), 2141–2150. doi:10.1016/j.renene.2009.01.015

Daus, Yu. V., Kharchenko, V. V., & Yudaev, I. V. (2016). Evaluation of Solar Radiation Intensity for the Territory of the Southern Federal District of Russia when Designing Microgrids Based on Renewable Energy Sources. *Applied Solar Energy, 52*(2), 124–129. doi:10.3103/S0003701X16020109

Daus, Yu. V., Kharchenko, V. V., & Yudaev, I. V. (2018). Solar Radiation Intensity Data as Basis for Predicting Functioning Modes of Solar Power Plants. In V. Kharchenko & P. Vasant (Ed.), Handbook of Research on Renewable Energy and Electric Resources for Sustainable Rural Development (pp. 283-310). Academic Press. doi:10.4018/978-1-5225-3867-7.ch012

Fini, M. H., Yousefi, G. R., & Alhelou, H. H. (2016). Comparative study on the performance of many-objective and single-objective optimisation algorithms in tuning load frequency controllers of multi-area power systems. *IET Generation, Transmission & Distribution, 10*(12), 2915–2923. doi:10.1049/iet-gtd.2015.1334

Gribkov, S. V. (2017). Vetroehnergeticheskoe oborudovanie kompleksy garantirovannogo ehlektrosnabzheniya maloj moshchnosti nauchno-inzhenernogo centra «Vindek» [Wind power equipment complexes of guaranteed power supply of low power of the scientific and engineering center "Vindek"]. *Ehnergetika i ehlektrooborudovanie, 6*(44), 24-27.

Gusarov, V. A., & Gusarova, E. V. (2017). Mikrogazoturbinnaya ustanovka dlya avtonomnogo ehnergosnabzheniya [Micro-gas turbine installation for autonomous power supply]. *Gazoturbinnye tekhnologii, 6*(149), 8-12.

Gusarov, V. A., & Kharchenko, V. V. (2018). Mikrogazoturbinnaya ustanovka GTEH-10S [Micro-Gas Turbine Installation GTE-10S]. *Vestnik VIEHSKH, 1*(30), 49–55.

Gusarov, V. A., Kharchenko, V. V., & Lapshin, S. A. (2013). Al'ternativnaya ehnergetika i ehkologiya [Alternative energy and ecology]. *Nauchno-tekhnicheskij centr. TATA, 7*, 15–18.

Kharchenko, V. V., Nikitin, B. A., & Tikhonov, P. V. (2010). Estimation and forecasting of PV cells and modules parameters on the basis of the analysis of interaction of a sunlight with a solar cell material. *Proceedings of 4th International Conference TAE 2010, Trends in Agricultural Engineering,* 307-310.

Lachuga, Y. U. F., & Strebkov, D. S. (2009). *EHnergeticheskaya strategiya sel'skogo hozyajstva Rossii na period do 2020 g* [Energy strategy of agriculture in Russia for the period until 2020]. Moscow, Russia: VIEHSKH.

Lapshin, S. A. (2013). Sistema upravleniya generaciej i raspredeleniem ehnergii v lokal'nyh setyah nizkogo napryazheniya s ispol'zovaniem vozobnovlyaemyh i ne vozobnovlyaemyh vidov ehnergii [Control system for generation and distribution of energy in local low-voltage networks using renewable and non-renewable types of energy]. Materialy X Mezhdunarodnoj ezhegodnoj konferencii "Vozobnovlyaemaya i malaya ehnergetika 2013".

Laria, A. (2009). Survey on microgrids: analysis of technical limitations to carry out new solutions. *Proceeding of the 13th European Conference on Power Electronics and Applications 2009 (EPE'09),* 1-8.

Liu, X., & Su, B. (2008). Microgrids – An Integration of RE Technologies. *Proceeding of China International Conference on Electricity Distribution,* 1-7.

Luo, X., Wang, J., Dooner, M., & Clarke, J. (2015). Overview of Current Development in Electric Energy Storage Technologies and the Application Potential in Power Systems Operation. *Applied Energy, 137*, 511–536. doi:10.1016/j.apenergy.2014.09.081

Makdisie, C., Haidar, B., & Alhelou, H. H. (2018). An Optimal Photovoltaic Conversion System for Future Smart Grids. In P. Kumar, S. Singh, I. Ali, & T. Ustun (Eds.), *Handbook of Research on Power and Energy System Optimization* (pp. 601–657). Hershey, PA: IGI Global. doi:10.4018/978-1-5225-3935-3.ch018

Mariam, L., Basu, M., & Conlon, M. F. (2013). A Review of Existing Microgrid Architectures. *Journal of Engineering*. doi:10.1155/2013/937614

Nadweh, S., Hayek, G., Atieh, B., & Haes Alhelou, H. (2018). Using Four – Quadrant Chopper with Variable Speed Drive System Dc-Link to Improve the Quality of Supplied Power for Industrial Facilities. Majlesi Journal of Electrical Engineering

Nikitin, O. (2012). Mikroturbiny v bor'be za potrebitelya [Microturbines in the struggle for the consumer]. *Ehkologicheskie sistemy, 11.*

Njenda, T. C., Golshan, M. E. H., & Alhelou, H. H. (2018, November). WAMS Based Intelligent Under Frequency Load Shedding Considering Online Disturbance Estimation. In *Smart Grid Conference (SGC)* (pp. 1-5). Academic Press.

Njenda, T. C., Golshan, M. E. H., & Alhelou, H. H. (2018, November). WAMS based Under Frequency Load Shedding Considering Minimum Frequency Prediction and Extrapolated Disturbance Magnitude. In *Smart Grid Conference (SGC)* (pp. 1-5). Academic Press.

Ramonas, C., Adomavicius, V., & Kepalas, V. (2009). Research of the power conversion processes in the system of power supply from a number of wind turbines over the one grid-tied inverter. *Proceedings of International Conference Electrical and Control Technologies*, 368-373.

Strebkov, D., Nekrasov, A., Trubnikov, V., & Nekrasov, An. (2018). Single-Wire Resonant Electric Power Systems for Renewable-Based Electric Grid. In V. Kharchenko & P. Vasant (Eds.), Handbook of Research on Renewable Energy and Electric Resources for Sustainable Rural Development (pp. 449–474). Academic Press. doi:10.4018/978-1-5225-3867-7.ch019

Strebkov, D. S., & Irodionov, A. E. (2004). Global Solar Power System. In *Eurosun 2004. 14 Intern. Sonnen Forum, Frebing, Germany. Vol. 3, PV systems and PV Cells* (pp. 336 – 343). Academic Press.

Strebkov, D. S., & Irodionov, A. E. (2005). PV research and technological development in Russia. In *Proceedings of the 2005 Solar World Congress.* American Solar Energy Society, International Solar Energy Society CD, ISES.

Strebkov, D.S., & Kharchenko, V.V. (2011). Rol' i mesto VIEH v razvitii global'noj ehnergetiki [The role and place of renewable energy in the development of global energy]. *ZH. Malaya ehnergetika, 3*(3), 3-12.

Strebkov, D. S., Tver'yanovich, EH. V., Tyuhov, I. I., & Irodionov, A. E., & Yarcev, N.V. (2002). Solnechnye koncentratornye tekhnologii dlya ehnergoobespecheniya zdanij [Solar concentrator technologies for power supply of buildings]. *Geliotekhnika, 3*, 64–68.

Vinogradov, A., Bolshev, V., Vinogradova, A., Kudinova, T., Borodin, M., Selesneva, A., & Sorokin, N. (2019). A System for Monitoring the Number and Duration of Power Outages and Power Quality in 0.38 kV Electrical Networks. In P. Vasant, I. Zelinka, & G. W. Weber (Eds.), *Intelligent Computing & Optimization. ICO 2018. Advances in Intelligent Systems and Computing* (Vol. 866, pp. 1–10). Cham: Springer; doi:10.1007/978-3-030-00979-3_1

Vinogradov, A., Vasiliev, A., Bolshev, V., Semenov, A., & Borodin, M. (2018). Time Factor for Determination of Power Supply System Efficiency of Rural Consumers. In V. Kharchenko & P. Vasant (Eds.), Handbook of Research on Renewable Energy and Electric Resources for Sustainable Rural Development (pp. 394–420).Academic Press. doi:10.4018/978-1-5225-3867-7.ch017

Zamani, R., Hamedani-Golshan, M. E., Haes Alhelou, H., Siano, P., & Pota, H. (2018). Islanding Detection of Synchronous Distributed Generator Based on the Active and Reactive Power Control Loops. *Energies*, *11*(10), 2819. doi:10.3390/en11102819

KEY TERMS AND DEFINITIONS

Distributed Generation: Electrical generation and storage performed by a variety of small, grid-connected devices referred to as distributed energy resources.

Hydroelectric Power Plant: A power plant used the energy of falling water or fast running water.

Microgrid: A localized group of electricity sources and loads that can operates both connected to the centralized power network and function autonomously as physical or economic conditions dictate.

Power Plant: An industrial facility for the generation of electric power.

Power Supply Consumer: A legal entity or a private person exercising the use of electric energy (capacity) on the basis of a concluded contract.

Renewable Energy Sources: The sources of power generation worked on renewable energy which is naturally replenished on a human timescale, such as sunlight, wind, rain, tides, waves, and geothermal heat.

Solar Power Plants: A power plant used the conversion of light into electricity using semiconducting materials that exhibit the photovoltaic effect.

Wind Power Plant: A power plant used air flow through wind turbines providing the mechanical power to turn electric generators.

Chapter 7
Under Frequency Load Shedding Techniques for Future Smart Power Systems

H. H. Alhelou
Tishreen University, Syria

ABSTRACT

It is critical for today's power system to remain in a state of equilibrium under normal conditions and severe disturbances. Power imbalance between the load and the generation can severely affect system stability. Therefore, it is necessary that these imbalance conditions be addressed in the minimum time possible. It is well known that power system frequency is directly proportional to the speed of rotation of synchronous machines and is also a function of the active power demand. As a consequence, when active power demand is greater than the generation, synchronous generators tends to slow down and the frequency decreases to even below threshold if not quickly addressed. One of the most common methods of restoring frequency is the use of under frequency load shedding (UFLS) techniques. In this chapter, load shedding techniques are presented in general but with special focus on UFLS.

LOAD SHEDDING TECHNIQUES

Figure 1 shows the most common types of load shedding techniques and their sub-categories. Generally, load shedding techniques are divided into three main categories which are conventional, adaptive, and computational intelligence-based techniques.

COMPUTATIONAL INTELLIGENCE TECHNIQUES

Computational intelligence includes techniques such as artificial neural networks (ANN), genetic algorithms (GA), fuzzy logic control (FLC), adaptive neuro-fuzzy inference system (ANF), and particle swarm optimization (PSO).

DOI: 10.4018/978-1-5225-8030-0.ch007

Figure 1. Load shedding techniques

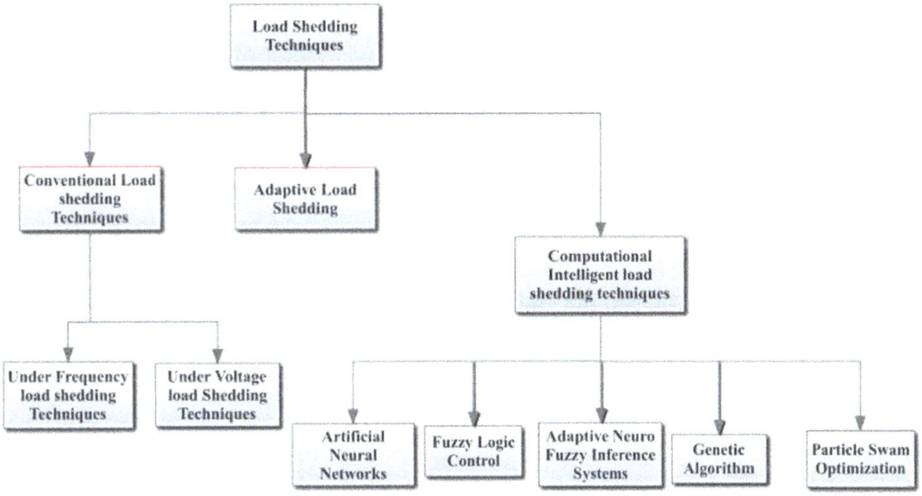

Advantages:

They are robust and flexible in dealing with complex non-linear systems. Though further research is still in progress they have been implemented in different scenarios of load shedding in power systems. Their advantages can be summarized as below based on cases they were implemented.

- ANN can guarantee an optimum amount of load shedding
- FLC can be used for load shedding application on a power system of any size.
- FLC parameters are optimized by using ANN, which may lead to accurate load shedding
- GA is a global optimization technique for solving non-linear, multi-objective problems. GA ensures a minimum amount of load shedding.
- PSO computation is simple and has the ability to find the optimum value.

Limitations:

- ANN can provide satisfactory results for known cases only and may fail to predict accurate results for unknown or varying cases.
- The membership parameters of FLC require prior system knowledge. Otherwise, it may fail to provide optimum load shedding.
- ANN can only work with Sugeno-type systems.
- GAs take a long time to determine the load shedding amount. This relative slowness limits their usage for online application.
- PSO is easily interrupted by partial optimization

Conventional Load Shedding Techniques

Conventional load shedding techniques fall in two categories which are traditional-UFLS and under voltage load shedding UVLS. Fig.2 shows the generalized process of traditional-UFLS and UVLS.

Traditional Under Frequency Load Shedding

Traditional UFLS is considered the most common and the basis of all other UFLS techniques. In this method relays makes use of locally measured frequency which is constantly compared against a certain threshold as the input. The frequency thresholds may differ from each power system and a violation of each threshold triggers relay action.

Advantages.

- They are simple and easy to implement.
- have proven to be efficient though not optimum

limitations

- A steep frequency gradient (df/dt) and a gradual df/dt are usually treated equally and may result in over or under-shedding.
- Inability to provide optimum load shedding. They simply follow a preset rule in which a fixed amount of load is shed when frequency deviates from the nominal value.
- The main disadvantage of this method is that it does not estimate the actual amount of the power imbalance which results in either over-shedding, which affects power quality, or under-shedding, which leads to tripping of electricity service (Tang et al, 2013 & Njenda et al, 2018).

Under Voltage Load Shedding (UVLS) Techniques

As indicated earlier the primary focus of this report is on UFLS, but it is of necessity to indicate other issues that results in load shedding. The majority of severe power blackouts that have been recorded around the world have their root causes in voltage instability problems (El-Sadek et al, 1998& Alhelou et al, 2018). UVLS techniques are implemented by power companies to protect the power system voltage instability and restore voltage to its nominal value thereby protecting the system from total voltage collapse. Voltage instability is usually a result of either forced outage of the generator or the line, or overloading. When overloading or forced outage occurs, the reactive power demand in transmission lines varies severely and may result in a blackout if not recovered in time.

Under Frequency Load Shedding (UFLS) Techniques

The aim of an Under Frequency Load Shedding (UFLS) scheme is to protect the Power System against major generation losses or total system collapse. UFLS scheme accomplish this task through planned and controlled load tripping until load levels match remaining generation capacity. For the power system to remain stable the total demand must match with the total generation, if demand is higher than the generation the system frequency begins to drop until a certain threshold where the system collapses if no action is taken to address the deficit. UFLS come as a tool to prevent system collapse. It should be noted however that while some load is lost, shedding a small amount of load will prevent continued and uncontrolled loss of generation that, if allowed, could result in system black out or interconnection-wide blackouts.

Figure 2. The flow chart for under frequency and under voltage load shedding techniques
(Zin et al 2004)

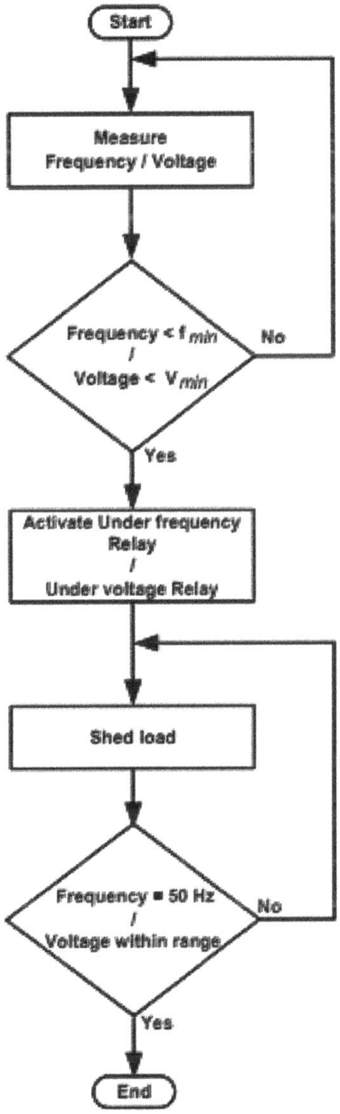

The UFLS scheme was developed in response to system-wide disturbances that have been recorded in literature. For example, the Western Electricity Coordinating Council (WECC) developed an Off-Nominal Frequency Load Shedding Plan (WECC Plan) in response to three system-wide disturbances that occurred in 1996. The plan was updated in 2011 which indicates the importance of managing frequency to avoid total system collapse. Though over frequency load shading can be encountered, in this report the primary focus is on UFLS.

Some of the primary objectives of the UFLS scheme are to:

- Minimize the risk of total system collapse.
- Protect generating equipment and transmission facilities against damage.

- Provide effective load shedding within the power system to arrest frequency decline.
- Improve overall system reliability.
- Match overall generation to overall load.
- Coordinate load shedding with under frequency protection of generating units.
- Coordinate load shedding with any other actions that can be expected to occur under conditions of frequency decline.
- Base load shedding on studies of system dynamic performance, using the latest state-of-the-art computer analytical techniques.
- Minimize the risk of further separation, loss of generation, or excessive load shedding which is accompanied by excessive over-frequency conditions.
- Address load controlled by customer-owned relays where the load is counted toward meeting minimum load shedding requirements.

During the design of an UFLS scheme the following must be taken into account.

- Sufficient load must be dropped by UFLS scheme to keep the system frequency within the continuous operating range of the generating units.
- Minimum and maximum permissible dynamic frequency during a disturbance to be specified.
- Load shedding blocks in case of the traditional scheme with a minimum separation between steps of 0.1 Hz or other agreed value to be specified.
- Specify the Under frequency relays maximum operating time for the high speed trip.

Blackouts that are being faced around the world poses a question on the reliability of conventional and adaptive load shedding techniques in avoiding such power outages. To overcome this problem, it is required that reliable ULFS techniques, provide quick and accurate load shedding to prevent power system collapse. Various methods can be used in UFLS and each method has its advantages and disadvantages.

Under frequency load shedding is commonly applied in the case of a generation loss where a fast decrease in frequency is experienced. The loss of generation can be due to technical system faults or human operation error. According to Institute of Electrical and Electronics Engineers (IEEE) standards, ''under frequency load shedding must be performed quickly to arrest power system frequency decline by decreasing power system load to match available generating capacity'' (IEEE standard, 2007). To keep in adherence to the standard frequency threshold values are set to implement the under frequency load shedding once a power imbalance is recognized. The minimum acceptable frequency varies from system to system since it depends on the generator type, its auxiliary devices, and the turbine (Delfino et al, 2001& Alhelou et al, 2016). In order to prevent total system collapse, the UFLS relay is initialized to shed a certain amount of load in predefined steps when frequency falls below a certain predefined threshold (Tang et al, 2013 & Alhelou et al, 2018). As an example, the European Network of Transmission System Operators for Electricity (ENTSOE) has recommended the following steps for under frequency load shedding (Saadat 1999):

The first stage of automatic load shedding should be initiated at 49 Hz, where at least 5% of total consumption should be shed.

- A stepwise 50% of the nominal load should be disconnected by using under frequency relays in the frequency range of 49.0–48.0 Hz.

- In each step of load shedding, a disconnection of not greater than 10% of the load is advised.
- The maximum disconnection delay should be 350ms (according to WECS) including breakers' operating time.

Recommendations for power plants.

The following recommendations are necessary for the safe operation of power plants it is recommended in (Saadat 1999& Njenda et al, 2018) that when frequency decrease to 49.8 Hz, 'quick-start' plants should be connected to the grid. The minimum allowable operating frequencies are 47.5 and 57.5 for 50Hz and 60Hz systems respectively in order to protect the generator and its auxiliaries. Extended generator operation at 47.5 Hz or below damage the turbine blades and reduce its lifespan.

Before developing a load shedding program, it is necessary to have a certain criterion as developed in (Tofis et al, 2013 & Alhelou et al, 2015).

- The steam has priority over the electrical system: The steam system must be able to recover or both will fail.
- The electrical load shedding should coordinate with the steam system by quickly shedding the load as soon.
- Essential or critical facilities from a safety standpoint should not be shed.
- Non-essential and easy to restore facilities must be shed before those which are difficult to restore.
- The shedding process has to be quick so that the frequency drop is stopped before it's too late.
- Avoid unnecessary actions and the protection system has to be liable and redundant.
- Amount of load to be shed must be as minimum as possible but sufficient enough to restore grid security.
- Load shedding scheme must strike a balance between maximizing system protection and minimizing amount to be shed. The design decisions, which must be considered separately are as follows.
    - Maximize the anticipated overload.
    - Calculate the relay settings.
    - Select the number of load shedding stages.
    - Determine the amount of load to be shed at each stage.
    - Select which loads should be shed at each stage.

GENERAL OVERVIEW ON FREQUENCY CONTROL IN A POWER SYSTEM

For continuous operation of a power system it is necessary that frequency and voltage be constant or at least remain within acceptable limits. In practical power systems load is constantly changing and therefore frequency is not stable. To maintain equilibrium in the power system, generation must be always equal to demand since system frequency decreases if load exceeds generation and increases when power generation is greater than load demand (Hsu et al, 2011). It necessary to maintain system frequency since it is directly proportional to the generator speed. Hence, by controlling the speed of the generator we can control the frequency. Generally, generators are equipped with governors to control the frequency. During severe faults such as three phase short circuits or load variations a three stage control action can be implemented in interconnected power systems to recover system frequency before a black out occurs. Fig 3 shows the three stage process to recover the frequency to a set point of 50Hz (Saadat 1999). In the event that a power deviation is sensed, during the first 15seconds the primary control tries to clear the power imbalance between load and generation. When 15seconds lapses the secondary control is activated to restore the power and frequency deviation from the primary control to their nominal values.

The secondary control remains in action up to a few minutes before the tertiary control finally takes over. The tertiary control reserve is connected either manually or automatically to establish new operating points. The problem with modern power systems is that they operate very close to steady state stability limit due to high demand for energy. When a transient disturbance such as generator or transmission line fault occurs, a series of events happen which can result in total system collapse. In such cases frequency declines very fast and goes below the specific threshold value, therefore it is necessary to implement UFLS technique to recover the system frequency and avoid a complete power system blackout (Delfino et al, 2001& Alhelou et al, 2018).

SEMI-ADAPTIVE UFLS SCHEMES

While traditional methods only make use of frequency threshold, the semi-adaptive also makes use of the rate of change of frequency df/dt. With the advancement in computational technology, it is now possible to use micro-processor-based protection relays which have the capability of calculating time-derivatives of measured quantities. The system frequency has a direct relationship with the speed of synchronous generators and the summarize swing equation can be used to show the relationship.

$$2H_i \cdot \dot{w}_{i,pu} = \Delta P_{i,pu} \tag{1}$$

where *Hi* is the inertia constant of the *i-th* generating unit in seconds, for a generator with rated apparent power *Sr*, ωi,pu is the angular mechanical speed of the *i-th* machine in p.u. based on the rated frequency ωr and Δ*Pi*,pu the power imbalance in p.u. on the *Sr* base as seen by the generator.

Figure 3. Frequency control in power system

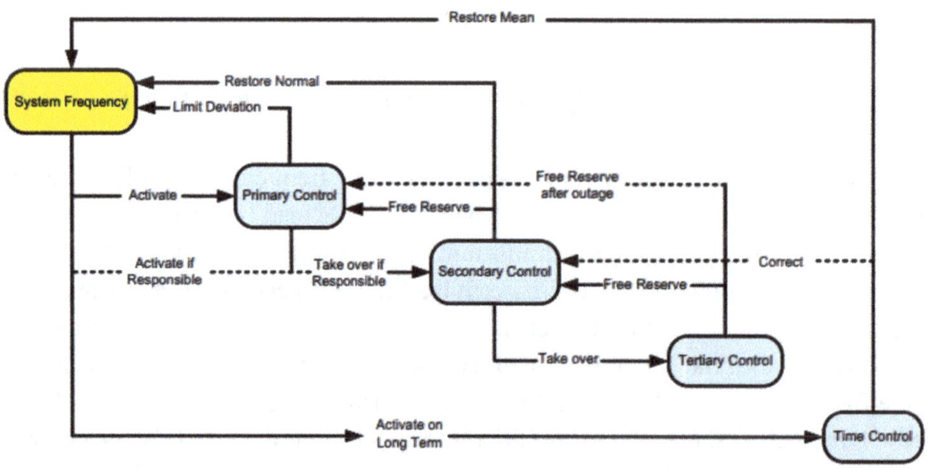

ADAPTIVE LOAD SHEDDING TECHNIQUES

Adaptive load shedding techniques uses the power swing equation to shed the required amount of load however, can be improved by also considering voltage stability. Whenever, the system suffers a disturbance (fault or islanding), a variation in frequency is known as rate of change of frequency (ROCOF). ROCOF relay is the most common example of the adaptive load shedding technique. to stabilize the frequency a certain amount of load has to shed.

The power imbalance within the system can be obtained by using equation:

$$\Delta P = \frac{2H}{f} \times \frac{df}{dt} \tag{2}$$

This equation can be applied to an isolated power system having only a single generator as well as to an interconnected power system. By putting the values in equation (2) power imbalance can be estimated. The flow chart for adaptive load shedding techniques is shown in Fig. 4.

Advantages:

- Adaptive load shedding techniques are more reliable than conventional load shedding.
- The use of rate of frequency change (df/dt) is used which has the following advantages.
  - Improved response time.
  - Reduced frequency swing.
  - One can begin to trip load blocks without waiting up until frequency drops to critical levels
  - Load shedding steps can trip simultaneously instead of sequentially.
  - Flexible, and can be tailored to different level of imports

Limitations:

- adaptive load shedding techniques also suffer from un-optimum load shedding due to variations in df/dt behavior. The df/dt value has been found to depend upon the operating load than required will result in excessive power outage (Mahat et al, 2010).
- the optimal load shedding technique is a nonlinear optimization problem dealing with multiple constraints. It is difficult to come up with the exact load to shed.

Adaptive UFLS Schemes with Power Imbalance Estimation

As compare to semi-adaptive methods adaptive UFLS schemes are fully adaptable are capable of modifying protective actions. Power imbalance estimation is one of the ways to achieve the adaptability feature. Considering a system with more than one machine a simplified swing equation with ignored damping can be used (Tofis et al, 2013).

$$2H_{COI} \cdot \dot{w}_{COI,pu} = \Delta P_{COI,pu} \tag{3}$$

Figure 4. Flow chart of adaptive load shedding techniques

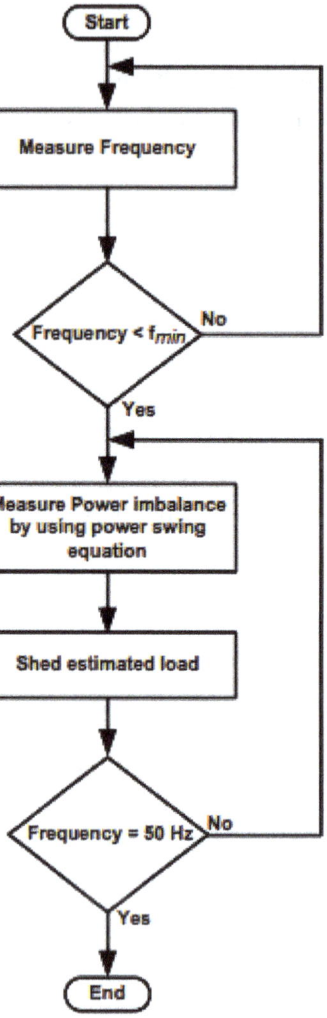

where

$$w_{COI,pu} = \left(\sum_{i=1}^{n} w_{i,pu} \cdot H_{i,sys} \right) / \left(\sum_{i=1}^{n} H_{i,sys} \right)$$

$$H_{COI} = \sum_{i=1}^{n} H_{i,sys}$$

For an EPS with n generating units, Center of Inertia (COI) represents an imaginary substitute generation unit for all units in the system where ωCOI,pu is inertia-averaged frequency response of all units included in the EPS (ωi,pu where i = 1, 2, …, n). HCOI and ΔPCOI, pu are average COI inertia in seconds and the system active-power imbalance in pu. respectively. Equation (3) is only valid immediately after the imbalance appearance.

$$\Delta P_{COI,pu} = \Delta P_{mech}\left(P_{prim}\right) - \Delta P_{ele}(P_{dist}, P_{load}\left(U, f\right) \quad (4)$$

As can be observed in equation P is a function of voltage and frequency as well and this has an effect in the proper functioning of adaptive power imbalance UFLS. As a result, HCOI estimation yields errors which might impair power system stability.

Limitations:

- There probability that consequential voltage drop is high is very significant and it occurs much faster than the change in active power.
- It is difficult to do frequency measurements just in time to capture the immediate power imbalance moment thus power imbalance might be underestimated.
- Another challenge of the power-imbalance estimation approach is knowing HCOI since no information is provided on how to select the correct measurement location for frequency and power in order to apply (2) (Wang et al, 2012).
- Sudden large power imbalance triggers electro-mechanical oscillatory modes in which generators begin to oscillate, but the COI frequency experiences only slight oscillations.

Adaptive UFLS Schemes With Prediction Capabilities

Predictive UFLS schemes introduces a new concept where it is not necessary to find the root cause for frequency variations such as load-generation imbalance. In this scheme a forecast in frequency behavior is made is made a few seconds before if prediction indicates the violation of allowable frequency tolerances load shedding is triggered.

Limitations:

- Predictive UFLS always have inherent prediction uncertainty and as time increase the prediction becomes even less reliable.

Advantages

- Frequency prediction can be performed in real-time, based on WAMS frequency measurements where prediction is constantly being updated.

WAMS Based Under Frequency Load Shedding

Modern power systems have become sophisticated and required more advanced methods for constant monitoring. As a result, the importance of Wide Area Monitoring System (WAMS) is inevitable and has significantly grown over the past few years. Due to the invention of the first Phasor Measurement Unit

(PMU) at the Center of Energy Engineering Lab of Virginia Tech, the WAMS became a reality. Before then, power systems operation and control only relied on local measurements. Since electrical grids cover large geographical areas it was really difficult to quickly pick the root cause of system collapse. The introduction of PMUs brought important improvements to power systems measurement and relaying. The measurement data were time tagged with GPS synchronization as a result data samples could be arranged to give the real time status of the power system. Data became easily available and delays were significantly reduced. The objective of this section is to introduce the use of WAMS to enhance under frequency load shedding (UFLS) to minimize the risk of ultimate system collapse due to extremely low frequency. The use of WAMS in power systems monitoring and control allows predictive UFLS which has been proven to be highly superior to traditional UFLS. Traditional based Under-frequency load shedding (UFLS) relays have fixed bands to shed and thus are not adaptive to system operating conditions (Wang et al, 2012 & Makdisie et al, 2018). They only rely on frequency as the input and are not reliable and can result in over or under shedding. With the use of WAMS global adaptive UFLS scheme can be implemented.

Figure 5. WAMS-based predictive UFLS concept representation
(Rudez et al, 2016)

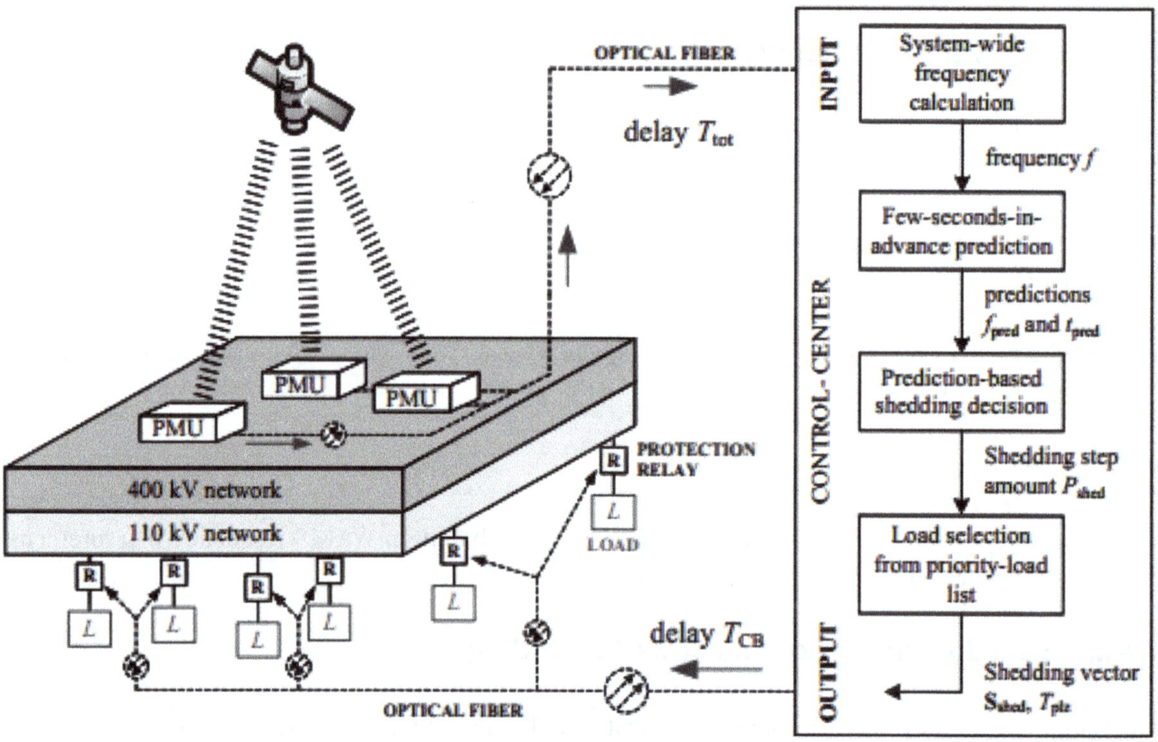

CONCEPT OVERVIEW

Practical real time operation of an actual EPS has the following limitations, noise, measurement-gathering time delays and communication time delay. Fig. 5 graphically presents the basic concept of WAMS based predictive under frequency load shedding. PMU devices which are usually installed on the high-voltage EPS buses are used to measure the local frequency and angles (Tsai 2005 & Alshahrestani et al, 2018& Alhelou et al, 2018) The frequency-measurement data from each PMU is then transferred to transmission system operator control-center through an optical-fiber network which has some delays. System wide frequency approximation is done and prediction is done to see if the limits ae not violated. Decision to shed is done based on the prediction, to minimize the effects of over shedding the results are processed in real time where data is constantly being updated. Each of the partial steps of the entire procedure is separately discussed in the following subsections.

CONCLUSION

Power systems blackouts have been a subject of great concern in electrical engineering studies. In this report load shedding techniques have been presented with primary focus on UFLS to set out a foundation for future work. The relative merits and demerits of each against the others have been outline as applied in power systems load shedding. Adaptive methods of UFLS have been highlighted to be more superior than traditional ones where the risk of over shedding is high. An example of predictive UFLS has been outlined which uses a few PMU frequency measurements provided by WAMS. The superiority of WAMS in modern power systems protection and control has been proven over the years without doubt. As a continuation for project work the use of WAMS in power systems UFLS will be presented in detail with reference to case studies.

REFERENCES

Alhelou, H., Hamedani-Golshan, M. E., Zamani, R., Heydarian-Forushani, E., & Siano, P. (2018). Challenges and Opportunities of Load Frequency Control in Conventional, Modern and Future Smart Power Systems: A Comprehensive Review. *Energies*, *11*(10), 2497. doi:10.3390/en11102497

Alhelou, H. H. (2018). *Fault Detection and Isolation in Power Systems Using Unknown Input Observer. In Advanced Condition Monitoring and Fault Diagnosis of Electric Machines* (p. 38). Hershey, PA: IGI Global.

Alhelou, H. H., Golshan, M., & Fini, M. (2018). Wind Driven Optimization Algorithm Application to Load Frequency Control in Interconnected Power Systems Considering GRC and GDB Nonlinearities. *Electric Power Components and Syst*.

Alhelou, H. H., & Golshan, M. E. H. (2016, May). Hierarchical plug-in EV control based on primary frequency response in interconnected smart grid. In *Electrical Engineering (ICEE), 2016 24th Iranian Conference on* (pp. 561-566). IEEE. 10.1109/IranianCEE.2016.7585585

Alhelou, H. H., Golshan, M. H., & Askari-Marnani, J. (2018). Robust sensor fault detection and isolation scheme for interconnected smart power systems in presence of RER and EVs using unknown input observer. *International Journal of Electrical Power & Energy Systems*, *99*, 682–694. doi:10.1016/j.ijepes.2018.02.013

Alhelou, H. H., Hamedani-Golshan, M. E., Heydarian-Forushani, E., Al-Sumaiti, A. S., & Siano, P. (2018, September). Decentralized Fractional Order Control Scheme for LFC of Deregulated Nonlinear Power Systems in Presence of EVs and RER. In *2018 International Conference on Smart Energy Systems and Technologies (SEST)* (pp. 1-6). IEEE. 10.1109/SEST.2018.8495858

Alhelou, H. S. H., Golshan, M. E. H., & Fini, M. H. (2015, December). Multi agent electric vehicle control based primary frequency support for future smart micro-grid. In *Smart Grid Conference (SGC)* (pp. 22-27). Academic Press.

Alshahrestani, A., Golshan, M. E. H., & Alhelou, H. H. (2018, November). WAMS Based Online Estimation of Total Inertia Constant and Damping Coefficient for Future Smart Grid Systems. In *Smart Grid Conference (SGC)* (pp. 1-5). Academic Press.

Delfino, B., Massucco, S., Morini, A., Scalera, P., & Silvestro, F. (2001, July). Implementation and comparison of different under frequency load-shedding schemes. In *2001 Power Engineering Society Summer Meeting. Conference Proceedings (Cat. No. 01CH37262)* (Vol. 1, pp. 307-312). IEEE. 10.1109/PESS.2001.970031

El-Sadek, M. Z. (1998). Preventive measures for voltage collapses and voltage failures in the Egyptian power system. *Electric Power Systems Research*, *44*(3), 203–211. doi:10.1016/S0378-7796(97)01200-5

Fini, M. H., Yousefi, G. R., & Alhelou, H. H. (2016). Comparative study on the performance of many-objective and single-objective optimisation algorithms in tuning load frequency controllers of multi-area power systems. *IET Generation, Transmission & Distribution*, *10*(12), 2915–2923. doi:10.1049/iet-gtd.2015.1334

Hsu, C. T., Chuang, H. J., & Chen, C. S. (2011). Adaptive load shedding for an industrial petroleum cogeneration system. *Expert Systems with Applications*, *38*(11), 13967–13974.

Mahat, P., Chen, Z., & Bak-Jensen, B. (2010). Underfrequency load shedding for an islanded distribution system with distributed generators. *IEEE Transactions on Power Delivery*, *25*(2), 911–918. doi:10.1109/TPWRD.2009.2032327

Makdisie, C., Haidar, B., & Alhelou, H. H. (2018). An Optimal Photovoltaic Conversion System for Future Smart Grids. In *Handbook of Research on Power and Energy System Optimization* (pp. 601–657). IGI Global. doi:10.4018/978-1-5225-3935-3.ch018

Nadweh, S., Hayek, G., Atieh, B., & Haes Alhelou, H. (2018). Using Four – Quadrant Chopper with Variable Speed Drive System Dc-Link to Improve the Quality of Supplied Power for Industrial Facilities. Majlesi Journal of Electrical Engineering.

Njenda, T. C., Golshan, M. E. H., & Alhelou, H. H. (2018, November). WAMS Based Intelligent Under Frequency Load Shedding Considering Online Disturbance Estimation. In *Smart Grid Conference (SGC)* (pp. 1-5). Academic Press.

Njenda, T. C., Golshan, M. E. H., & Alhelou, H. H. (2018, November). WAMS based Under Frequency Load Shedding Considering Minimum Frequency Prediction and Extrapolated Disturbance Magnitude. In *Smart Grid Conference (SGC)* (pp. 1-5). Academic Press.

Power Systems Relaying Committee. (2007). IEEE Guide for the Application of Protective Relays Used for Abnormal Frequency Load Shedding and Restoration. *IEEE Std C*, *37*, c1–c43.

Rudez, U., & Mihalic, R. (2016). WAMS-based underfrequency load shedding with short-term frequency prediction. *IEEE Transactions on Power Delivery*, *31*(4), 1912–1920. doi:10.1109/TPWRD.2015.2503734

Saadat, H. (1999). *Power System Analysis McGraw-Hill Series in Electrical Computer Engineering*. Academic Press.

Tang, J., Liu, J., Ponci, F., & Monti, A. (2013). Adaptive load shedding based on combined frequency and voltage stability assessment using synchrophasor measurements. *IEEE Transactions on Power Systems*, *28*(2), 2035–2047. doi:10.1109/TPWRS.2013.2241794

Tang, J., Liu, J., Ponci, F., & Monti, A. (2013). Adaptive load shedding based on combined frequency and voltage stability assessment using synchrophasor measurements. *IEEE Transactions on Power Systems*, *28*(2), 2035–2047. doi:10.1109/TPWRS.2013.2241794

Tofis, Y., Yiasemi, Y., & Kyriakides, E. (2013, September). A plug and play, approximation-based, selective load shedding mechanism for the future electrical grid. In *International Workshop on Critical Information Infrastructures Security* (pp. 74-83). Springer. 10.1007/978-3-319-03964-0_7

Wang, G., Xin, H., Gan, D., Li, N., & Wang, Z. (2012, July). An investigation into WAMS-based Under-frequency load shedding. In *Power and Energy Society General Meeting, 2012 IEEE* (pp. 1-7). IEEE.

Zin, A. M., Hafiz, H. M., & Aziz, M. S. (2004, November). A review of under-frequency load shedding scheme on TNB system. In *Power and Energy Conference, 2004. PECon 2004. Proceedings. National* (pp. 170-174). IEEE. 10.1109/PECON.2004.1461637

Zin, A. M., Hafiz, H. M., & Wong, W. K. (2004, November). Static and dynamic under-frequency load shedding: a comparison. In *Power System Technology, 2004. PowerCon 2004. 2004 International Conference on* (Vol. 1, pp. 941-945). IEEE. 10.1109/ICPST.2004.1460129

KEY TERMS AND DEFINITIONS

Electric Frequency.: AC frequency is the number of cycles per second in an alternating current (ac) sine wave. Said another way, frequency is the rate at which current changes direction per second. It is measured in hertz (Hz), an international unit of measure where 1 hertz is equal to1 cycle per second.

Power Oscillations: Power System Oscillations deals with the analysis and control of low frequency oscillations in the 0.2-3 Hz range, which are a characteristic of interconnected power systems.

Power Systems: An electric power system is a network of electrical components deployed to supply, transfer, and use electric power. An example of an electric power system is the grid that provides power to an extended area.

Under-Frequency Load Shedding: Under-frequency load shedding (UFLS) is a common technique to maintain power system stability by removing the overload in some part of the system.

Voltage Quality: Is used to refer to all disturbances in the supply of electricity, excluding interruptions that are covered.

Chapter 8
Transition From Traditional Grid to Smart One

Haitham Daghrour
Tishreen University, Syria

Razan Mohammad Al-Rhia
Tishreen University, Syria

ABSTRACT

Smart grids have become an urgent need to overcome the challenges of the 21st century. To transit the traditional grid to smart one, there must be a well thought out plan, called road map, which is also being carefully developed by organizations according to standards for deploying smart networks. Most studies focused on modernizing distribution networks because it was passive and technologically poor. Two approaches to developing distribution networks were presented. The smart grid modernization was also presented from social and psychological perspectives.

INTRODUCTION

As we know the electrical system consists of various components: different generating stations, substations, transmission lines and distribution network. Besides that, it involves monitoring and control centers that monitor the operation of the system components and the substation centers near the consumers. In addition, the measurement devices that measure the amount of energy between the system and users.

Good electrical networks may be characterized by four attributes:

1. **Flexibility:** The ability to respond to the rapidly increasing demand for electrical energy and challenges related to the future.
2. The ability to connect electricity between all producers and consumers.
3. **Reliability:** The ability to adapt to unexpected events so as to ensure the continued delivery and maintenance the quality required of electrical energy.
4. **Economic Operation:** The ability to be operated economically.

DOI: 10.4018/978-1-5225-8030-0.ch008

Today's grids are facing many of principal problems, which are growing in severity.

- Global electricity demand is rising faster than demand for any other final energy source. In addition, the electrification of the economy intensifies end-user demand around peak hours, stressing grids and making rapid expansion a necessity.
- Aging infrastructure tends to compromise reliability of power supply and exacerbate energy losses to the detriment of economies undergoing rapid electrification.
- The power grid will need to become more flexible to match supply and demand in real time due to the share of variable renewable energy (VRE) in the energy mix grows
- Issues relating to power quality and bi-directional electricity flows arise that cannot be properly managed by traditional grids, due to the penetration of distributed generation (DG) rises to very high levels in some areas (Lajoie, Debarre, Fayet, &Dreyfus, 2015; Alhelou, Hamedani-Golshan, Zamani, Heydarian-Forushani, & Siano,2018).

The major driver for the evolution of the power system is the need to meet rising demand for electricity while reducing carbon emissions to avoid irreversible changes to the earth's environment. All of this must be achieved in parallel with the reliability of electricity supplies on which the world's economies are increasingly dependent.

Between 2000 and 2007, global electricity demand rose, on average, by 2.5 per cent annually, and is still on the increase. By 2030, global electricity demand is expected to double to 30,000 terawatthours (TWh) annually. If this level of demand is to be met, we will need to build one gigawatt (GW) power station and its associated infrastructure every week for the next 20 years - a daunting prospect. Where global carbon emissions from power generation will increase significantly, with the knowledge that they are currently 40%, so just increasing today's operations to meet the increase in demand will not be acceptable. So, how can we meet demand and keep carbon dioxide emissions in check?

According to the International Energy Agency (IEA), which has proposed a number of scenarios for the future of global carbon emissions, annual emissions in 2030 could be reduced from the current prediction of over 40 Gt (gigatons) CO2 to just over 26 Gt by the implementation of a carefully designed set of policies. These policies aim to limit global warming to 2°C above preindustrial levels, which should limit the effects of climate change to an acceptable economic, social and environmental cost. Where the scenario indicates that the savings come from the implementation of energy efficiency measures, and from the increase in renewable energy generation.

The development of more intelligent power systems will directly support these two objectives.

In a smart grid, advanced technologies improve energy efficiency by managing demand so that it matches the availability of electricity, and they feed renewable energy into the network without letting changes in weather patterns affect the stability or reliability of the supply.

At the same time, using satellite, wireless and real-time communication, advanced technologies will enable utilities to pinpoint problems in the grid faster than they are able to today ("An Introduction to Smart Grids ", n.d.).

What Are the Smart Grids?

Until now is difficult to find a standard definition to smart grid because it is new and its components are increasing, so each academic and research center define the smart grid in particular (shabanzadeh & Moghaddam, 2013).

After reviewing all published definitions, we can say: A smart grid is an electrical network that uses digital and other advanced technologies to monitor and optimally manage the generation, transmission, distribution, consumption and business to meet the varying electricity demands of end-users. Smart grids co-ordinate the needs and capabilities of all generators, grid operators, end-users and electricity market stakeholders to operate all parts of the system as efficiently as possible, minimizing costs and environmental impacts while maximizing system reliability, resilience and stability.

While many regions have already begun to "smarten" their electricity system, all regions will require significant additional investment and planning to achieve a smarter grid. Smart grids are an evolving set of technologies that will be deployed at different rates in a variety of settings around the world, depending on local commercial attractiveness, compatibility with existing technologies, regulatory developments and investment frameworks. Figure 1 demonstrates how the smart grids evaluated (International Energy Agency [IEA], 2011).

In summary smart grids may defined as following:

Electrical Networks (low level) + Information Systems (high level)= Smart Grid

Figure 1. Smarter electricity systems

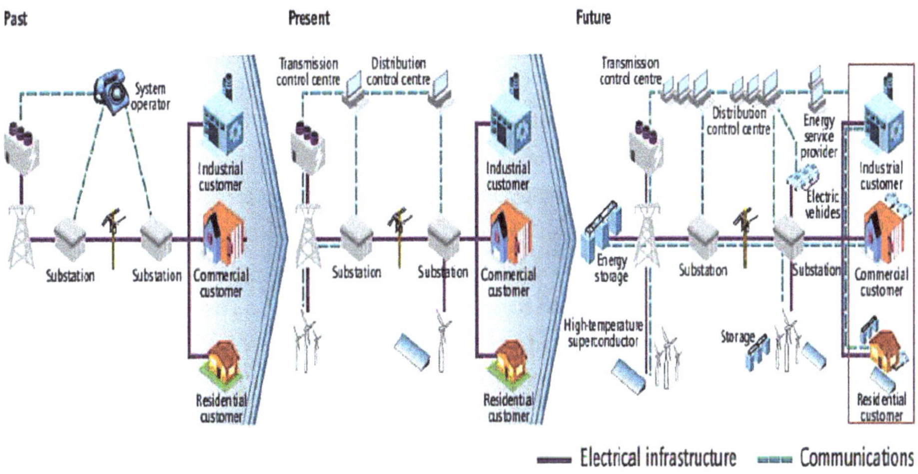

Source: Unless otherwise indicated, all material derives from IEA data and analysis.

A smart grid refers to a modernized electricity network that monitors, protects, and optimizes the operation of its interconnected elements.

The notion of grid modernization differs from country to country, depending on the smartness of the existing system. However, notwithstanding such differences, smart grids are generally characterized by the use of digital information and communications technologies to manage both the bi-directional flow of data between end users and system operators, and the bi-directional flow of power between centralized and decentralized generation (Lajoie, Debarre, Fayet, &Dreyfus, 2015).

Benefits of Smart Grid

The benefits will result from improvements in the following six key value areas (Hamilton, Miller, & Renz, 2010):

- **Reliability:** by reducing the cost of interruptions and power quality disturbances and reducing the probability and consequences of widespread blackouts
- **Economics:** by keeping downward prices on electricity prices, reducing the amount paid by consumers as compared to the "business as usual" (BAU) grid, creating new jobs, and stimulating the U.S. gross domestic product (GDP).
- **Efficiency:** by reducing the cost to produce, deliver, and consume electricity
- **Environmental:** by reducing emissions when compared to BAU by enabling a larger penetration of renewables and improving efficiency of generation, delivery, and consumption
- **Security:** by reducing the probability and consequences of manmade attacks and natural disasters
- **Safety:** by reducing injuries and loss of life from grid-related events

The Principal Characteristics of the Smart Grid

The seven principal characteristics of the smart grid are:

- **Enables Active Participation by Consumers:** Consumer choices and increased interaction with the grid bring tangible benefits to both the grid and the environment, while reducing the cost of delivered electricity.
- **Accommodates All Generation and Storage Options:** Diverse resources with "plug-and-play" connections multiply the options for electrical generation and storage, including new opportunities for more efficient, cleaner power production.
- **Enables New Products, Services, and Markets:** The grid's open-access market reveals waste and inefficiency and helps drive them out of the system while offering new consumer choices such as green power products and a new generation of electric vehicles. Reduced transmission congestion also leads to more efficient electricity markets.
- **Provides Power Quality for the Digital Economy:** Digital-grade power quality for those who need it avoids production and productivity losses, especially in digital-device environments.
- **Optimizes Asset Utilization and Operates Efficiently:** Desired functionality at minimum cost guides operations and allows fuller utilization of assets. More targeted and efficient grid maintenance programs result in fewer equipment failures and safer operations.

- **Anticipates and Responds to System Disturbances (Self-Heals):** The smart grid will perform continuous self-assessments to detect, analyze, respond to, and as needed, restore grid components or network sections.
- **Operates Resiliently Against Attack and Natural Disaster:** The grid deters or withstands physical or cyber-attack and improves public safety.

The deployment of technology solutions that achieve these characteristics will improve how the smart grid is planned, designed, operated, and maintained. These improvements—in each of the key value areas presented above—lead to specific benefits that are enjoyed by all.

The following technology solutions are generally considered when a smart grid implementation plan is developed:

- Advanced Metering Infrastructure (AMI)
- Customer Side Systems (CS)
- Demand Response (DR)
- Distribution Management System/Distribution Automation (DMS)
- Transmission Enhancement Applications (TA)
- Asset/System Optimization (AO)
- Distributed Energy Resources (DER)
- Information and Communications Integration (ICT)

The deployment of these technology solutions is expected to create improvements in the six key value areas—reliability, economics, efficiency, environmental, safety and security (Hamilton, Miller & Renz, 2010).

Figure 2. Technology solution – key value area relationships

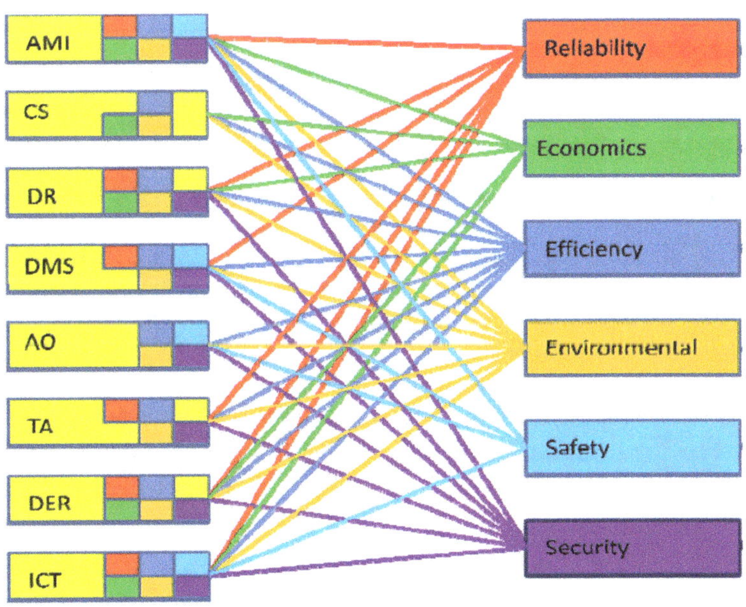

Figure 2 identifies the relationships among these technology solutions and the key value areas. This "many-to-many" relationship illustrates the synergy of smart grid solutions, an affect that must be considered when the scope of the smart grid is planned to ensure the benefits are optimized.

The high level of the smart grid consists of four major layers (Hard infrastructure, Telecommunication, Data, Applications) each layer is illustrated in figure 3 (Madrigal, Uluski, &Gaba, 2017).

Smart Planning for a Successful Smart Grid Deployment

To achieve a successful smart grid, we need a successful plan to deployment it, and this plan should address the following topics (finnigan, 2014):

- **Strategic Purpose:** Determine the objectives for deploying a smart grid and the guiding principles which will govern the project
- **Road Map:** The plan should provide a step-by-step overview for each phase of the deployment plan, in chronological order.

Figure 3. Visualization of smart grid by Pacific gas and electric

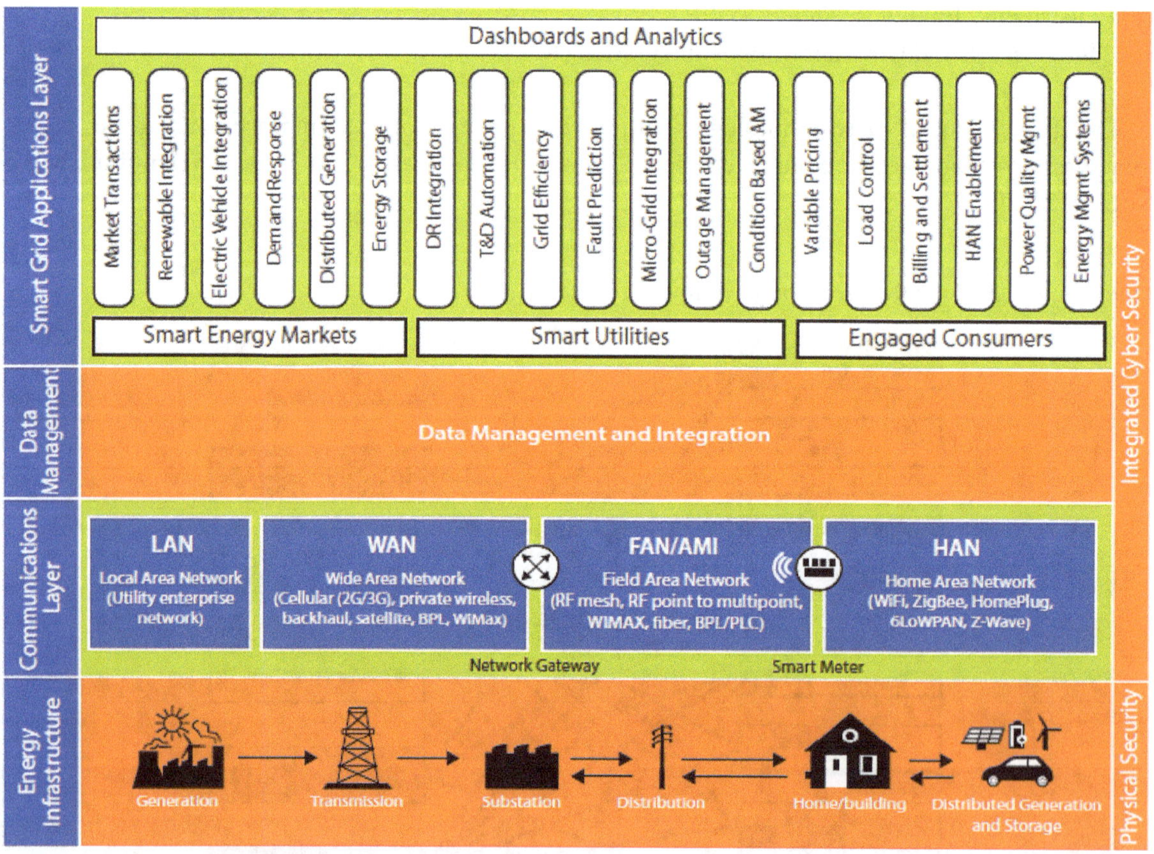

Source: PG&E 2011.
Note: HAN = Home Area Network.

- **Technologies:** Describe the technologies selected by the utility and explain how these technologies will function together as a unified system.
- **Implementation:** Explain how the utility will manage the project and coordinate the activities of the different departments involved in the deployment.
- **Customer Impacts:** Determine the changes that the customer will see by the smart grid deployment as the meters, meter data, billing, collection, connection/disconnection of electricity service, and customer service.
- **New Services:** Include the new products and services that the utility will provide after installing the smart meters, including new rate plans.
- **Customer Education:** How will the utility educate customers about the smart grid plan? What channels with the utility use to communicate with customers and how often will these communications occur?
- **Cybersecurity and Data Privacy:** How will the utility keep the customers' usage information secure? How can customers provide information to other providers of energy products and services? What types of information will be available to these third parties?

It can be seen that all the above topics are very important, but this chapter will focus only on roadmaps.

What Is a Roadmap?

Like the term "Smart Grid" the expression "Roadmap" is still without deterministic definition and hasn't a clear purpose. There are different types of roadmaps for different purposes. Also the best design of roadmap that give a set of components that describe the stages and layers of implementation plans required to transform a utility from its current state to become a Smart one. The potential components of a full business roadmap include:

- Strategy/Vision
- Business Needs/Functional Requirements
- Technology Deployment Requirements
- Capital Investment
- Implementation Plan/Phasing Plan/Release Schedule
- Organizational Change and Business Readiness Plan

In this way, each component provides insight into the basis and prioritization of the final roadmap. Each phase is built upon the phase of other inter-dependent components. Roadmaps should include the priorities, and the specific time period over which each phase of the roadmap is achieved (Buxton, 2011).

Why Do You Need a Roadmap?

A clear, well-designed roadmap is needed for utilities for the following reasons (Buxton, 2011):

- To develop consensus about their business drivers; and the technologies required to meet the business needs they prompt.
- To provide a clear path forward.

- It can be useful in both internally and externally marketing tool.
- Envision a future state to desired operation.
- Make the utility understands their future business needs (both customer and operation needs).
- Provide clarity and confidence for employees, stakeholders, regulators, shareholders, vendors and customers.

Why We Need a Standards

Good standards are needed because in the absence of standards:

- There is a risk that smart grid technologies will become prematurely obsolete or be implemented without adequate security measures.
- The future innovation will be impeded, that make the visions of smart grid difficult.

The standards are seen as a means of guiding the development of emerging "green markets" such as electric vehicles, renewable energy and smart grid whose developing nature poses significant risks to government, industry and society as a whole. The standards can therefore be used to provide a guiding framework that can be used to manage key consumer risks and minimize early commercialization risks for business and industry.

The standards also enable economies of scale and scope that help create competitive markets that encourage faster deployment of smart grid technologies and customer benefits (Berker & throndsen, 2017; Alhelou et al., 2018; Alhelou et al., 2016; Njenda et al., 2018).

Who Is Writing Smart Grid Road Maps?

Road maps are exclusively authored by actors embedded in existing governance structures. Governments are involved above all, but so are technological and legal regulatory bodies, and interest and industry organizations. In addition, road maps are written with involvement from and cooperation with research and university communities (Berker & Throndsen, 2016; Fini et al., 2016; Alhelou et al., 2018; Zamani et al., 2018; Alhelou et al., 2015; Njenda et al., 2018; Haes Alhelou et al., 2018)).

The major Smart Grid-related organizations are:

- **Standards Developing Organizations (SDOs):** Which are the organizations that develop, revise, coordinate, and amend technical standards. also it deal with different types of standards to address applications or sets of applications, and deal with specifications that lead to formal standards which are approved by law. SDOs are classified according to their roles, positions, and domains of applications. SDOs can be local, regional, or international organizations, and might be governmental, semi-governmental, or non-governmental entities.
- **Regulatory Organizations, Technical Consortia, Forums and Panels, and Marketing/ Advocacy Organizations:** Which are also actively involved in developing or evaluating Smart Grid-related technical specifications and cooperating with SDOs in promoting the standardization process.

1-Standards Developing Organizations (SDOs) which Dealing with the Smart Grid:

- International Electrotechnical Commission (IEC) (IEC,2010)
- International Organization for Standardization (ISO)
- International Telecommunication Union (ITU)
- Society of Automotive Engineers (SAE) International
- Institute of Electrical and Electronics Engineers (IEEE)
- European Committee for Standardization (CEN)
- European Committee for Electrotechnical Standardization (CENELEC)
- Telecommunications Industry Association (TIA)
- Internet Engineering Task Force (IETF)
- Alliance for Telecommunications Industry Solutions (ATIS)

2-Technical Consortia, Forums, and Panels Dealing with the Smart Grid

- Wi-Fi Alliance
- ZigBee Alliance
- WiMAX Forum
- UCA International Users Group
- National Electrical Manufactures Association (NEMA)
- Organization for the Advancement of Structured Information Standards (OASIS)
- HomePlug Power line Alliance
- HomeGrid Forum (HGF)
- GridWise Architecture Council (GWAC)

3-Other Political, Market, and Trade Organizations, Forums, and Alliances

- International Energy Agency (IEA)
- Clean Energy Ministerial (CEM)
- Demand Response and Smart Grid Coalition (DRSG)
- China Electricity Council (CEC)
- Global Smart Grid Federation
- National Institute of Science and Technology and Smart Grid Interoperability Panel (NIST, 2010).

The standards development and mapping will continue for decades, as lessons learned are picked up along the long road of real-world smart grid deployment (Sato et al., 2015).

it is obvious that some standards will be indispensable to efficiently operate smart grids and to meet the upcoming integration and interoperability problems. Furthermore, the possibility to compare smart grid projects and roadmaps will be an additional benefit because it enables the measurability of their success (spcht,Rohjans, Trefke, Uslar,& Gonzalez, 2013).

Recommended Core Standards

If we take a quick look at the road maps for a number of countries (see (spcht et al, 2 013); (Akhter & Biswas, 2016); (Bichler, 2013); (Navigant, 2015); (NYSSGC, 2010); (Du et al, 2012), we will find that the scope of smart grids differs from country to country. Where organizations and vendors came up with their own roadmaps regarding the topic of smart grid standardization. This differences between roadmaps are due to their vision and what are the requirements of each country, some countries mainly focus on reducing non-technical losses through smart metering or improving the outage management for radial feeder systems. While reliability may be a scope for some countries to provide the security of supply for the digital economy, focusing on markets and economic benefits for the country other regions have a more sustainable energy vision in mind trying to reduce carbon dioxide emissions and to cope with distributed, renewable generation like Micro Combined Heat and Power (CHP), photo-voltaic, fuel cells or electric vehicles in the distribution grid.

According to the number of recommendations the IEC TC 5 standards are of highest importance from the perspective of most experts: Since national perspectives differ, organizational standards e.g. like IEEE ones have less impact on worldwide scale.

- **IEC TR 62357:** Reference Architecture
- **IEC 61968/61970:** Common Information Model for EMS and DMS
- **IEC 61850:** Intelligent Electronic Device (IED) Communications at Substation level and DER
- **IEC 62351:** Vertical security for the TR 62357
- **IEC 60870:** Telecontrol protocols
- **IEC 62541:** OPC UA – OPC Unified Architecture, Automation Standard
- **IEC 62325:** Market Communications using CIM

For more details see ((spcht et al, 2013) and its references).

In the European context, they depended on the National Institute of Science and Technology roadmap, as were those of the states of California, Kentucky and New York. In the European context, the road maps of the UK, Germany, Denmark and Ireland were selected, along with the road map of the European Electricity Grid Initiative (EEGI).

On the international level, the road maps of the International Energy Agency (IEA) and the International Electrotechnical Commission (IEC) were selected (Berker & Throndsen, 2016).

In conclusion, the Smart Grid Roadmap aims to:

- Increase understanding among a range of stakeholders of the nature, function, costs and benefits of smart grids.
- Identify the most important actions required to develop smart grid technologies and policies that help to attain global energy and climate goals.
- Develop pathways to follow and milestones to target based on regional conditions (IEA, 2011).

Figure 4. Key technology areas

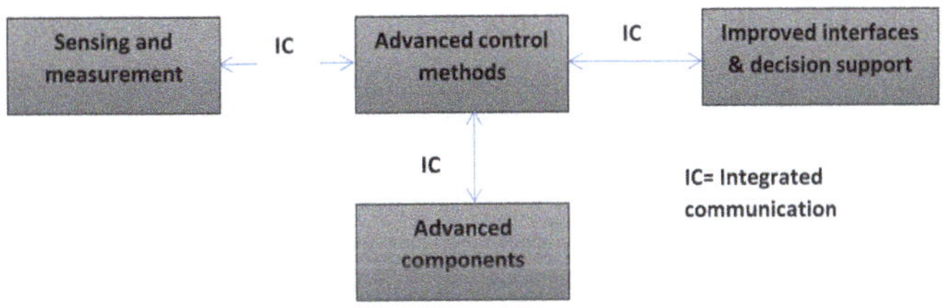

Smart Grid Roadmapping

The concept of a smart grid includes the entire electric power delivery system, beginning at the output of all generation sources to transmission system to distribution to the final use of the delivered energy.

If a smart roadmap is developed, this should include both the transmission and distribution systems, but the studies and researchers have focused more on the distribution system because it was passive and technologically poor.

This does not mean that the transmission grid – either itself or the related stakeholders – should be ignored. The interface between transmission system and distribution system operation is a significant challenge. This should be addressed by coordinating efforts on all levels in terms of planning, road mapping for smart grids and for other energy or infrastructure technologies, and operation. Transmission system stakeholders should be consulted during the road mapping process for smart grids in distribution networks to consider and co-ordinate appropriately the impacts from investments into and modification of distribution networks (IEA, 2015).

Making Transmission Bigger and Smarter

Transmission expansion is clearly an important aspect of grid modernization, this can be done by applying advanced technology to enhance the existing grid, one that deploys the concepts of a smart grid.

The transmission system have had a technical advances in his history, with advances in monitoring, protection, analysis and control, accompanied by periodic breakthroughs in transmission capacity. Power electronics has also played an important role, by enabling DC transmission and a variety of Flexible AC Transmission Systems (FACTS) enhancements. This continuous technical progress that has perhaps placed transmission on the smart grid backburner; it is already pretty smart.

While it is true that today's transmission is more advanced than distribution, the transition to a smart grid requires much more transmission capability and now is the time to make the required investment.

While recognizing that transmission challenges are large, the first requirement for any transmission system remains the same—it must be extremely reliable (NETL, 2009).

We need to make the grid:

- **Bigger:** Bolster our transmission system to bring renewable energy resources to the population centers that need them.

- **Smarter:** Broadly deploy "smart grid" technologies to make the grid more reliable, resilient, and secure, and enable much greater energy efficiency for consumers and businesses.
- **Better Integrated:** Elevate planning, siting, and cost-allocation processes from the local level to a regional level with input from all stakeholders

To address most transmission concerns, (National Energy Technology Laboratory [NETL], 2009) proposed five key technology areas (KTAs), as shown in the following diagram (figure 4), which serve as smart grid enablers.

Integrated Communications

The key to a smarter transmission system, as with the distribution system, is a reliable, high-speed integrated communications (IC) platform. The ability to rapidly move information between transmission stations, and from these stations to system control centers, provides the basis for virtually all advanced applications.

And when IC blankets both transmission and distribution, new opportunities are created for each to support the other. For example, demand response, distributed generation, distributed storage, and voltage dispatch can all help an RTO (Regional Transmission Organizations) ensure a reliable transmission grid, but this IC must be designed with the future in mind. Capacity, security, and performance must be sufficient to accommodate not only the applications of today, but also those that will be conceived tomorrow.

Sensing and Measurements

Recognizing that you can only manage what you measure, there is a clear need for more and better measurement tools.

- **Dynamic Line and Equipment Rating:** Includes measurement of temperature, wind speed, incident sunlight, the potential output of renewable generation and Sensors that measure transmission line and substation health parameters.
- **Synchophasor Monitoring:** Includes phasor measurement units
- **Reliability Assessment:** Operators are asked to make critical decisions within seconds, as opposed to minutes or hours in years past.
- **AMI:** Advanced metering infrastructure

Advanced Control Methods

A more complex transmission system will require more sophisticated controls.

Advanced Protection

- **Special Protective Systems:** Beyond advanced line protection, system integrity protection systems (SIPS) and Remedial Action Schemes (RAS) could protect large regions rather than individual elements.

- **Coordination of Renewable Generation and Storage:** Storage can provide a useful, and in some cases critical, buffer by absorbing renewable generation when load is low and supplementing it when load is high. But storage can take many forms and be spread across a large geographic area. And it can be located at any voltage level, even including the distribution secondary.
- Centralized Flow Control

Advanced Components

These components provide tools that help shape the character of tomorrow's grid.

- **Advanced Flow Control Devices:** The list includes various FACTS devices, Variable Frequency Transformers (VFTs), solid state transformers, superconducting condensers, and HVDC. Another advanced component under development, the Thin AC Converter (TACC), could allow extended dynamic voltage and power flow control. A TACC located between the station bus and an existing conventional asset such as a transformer or capacitor bank could provide dynamic reactive power insertion and controlled flow of real power.
- **Fault Current Limiters:** An electronic dynamic Short Circuit Current Limiter is today available at 500 kV; it has near zero impedance at steady state, and during a fault it electronically switches in milliseconds to a current limiting reactor. Other approaches to fault current limiters can employ the inherent characteristics of superconductivity.
- High Temperature Superconducting (Hts) Cable And High Capacity (High Temperature) Conductors
- Advanced Storage

Decision Support

The complex world of transmission has made the operator's job extremely challenging, but new tools can make it a bit easier.

- **Data Mining:** Some data that is available from devices currently deployed across the transmission system is not being collected, or does not have adequate communications to be transmitted, or is not used because it cannot be processed efficiently. It is important to take advantage of data and technology that is already available as part of the transformation to a smarter grid.
- **Fast Simulation:** Fast Simulation & Modeling (FSM) is designed to provide the mathematical underpinning and look-ahead capability for a self-healing grid—one capable of automatically anticipating and responding to power system disturbances, while continually optimizing its own performance. It will provide a tool to aid in decision-making by permitting an operator to get an accurate estimation of the state of the grid in real time. This will allow the operator to optimize grid operations as well as predict grid behavior based upon historical and real time data.
- **Advanced Visualization:** Advanced system analysis and visualization are essential technologies that must be implemented if grid operators and managers are to have the tools and training they will need to operate tomorrow's more complex grid. These technologies convert masses of power-system data into information that can be understood by human operators at a glance. Animation,

color contouring, virtual reality, and other data display techniques will prevent "data overload" and help operators identify, analyze, and act on emerging problems in a timely manner. In many situations, the time available for operators to make decisions has now shortened from hours to minutes, sometimes even seconds.

These are the five KTAs (illustrated very well in (NETL, 2009) that will be needed to make the transmission smart grid a reality. Other elements will surely emerge, but they will likely fall into one of these fundamental areas (Alshahrestani et al., 2018; Makdisie et al., 2018; Alhelou et al., 2018; Alhelou et al., 2016; Nadweh et al., 2018).

Modernization Distribution Systems

Grid modernization impacts electric distribution systems more than any other part of the electric power grid, where Distribution networks make up a very large percentage of the total electricity system network length in addition most of renewable generation is connected to the distribution networks, and distribution network investments will make up more than 80% of all the network investments to 2050 due to size and complexity of most distribution networks (IEA, 2015).

A targeted examination of the distribution network will moderate the size of the roadmap effort and provide the necessary focus to enable practical decisions that can be made to yield benefits in this much needed area. Figure 5 shows distribution system of future.

Figure 5. Distribution system of future
(Madrigal et al, 2017)

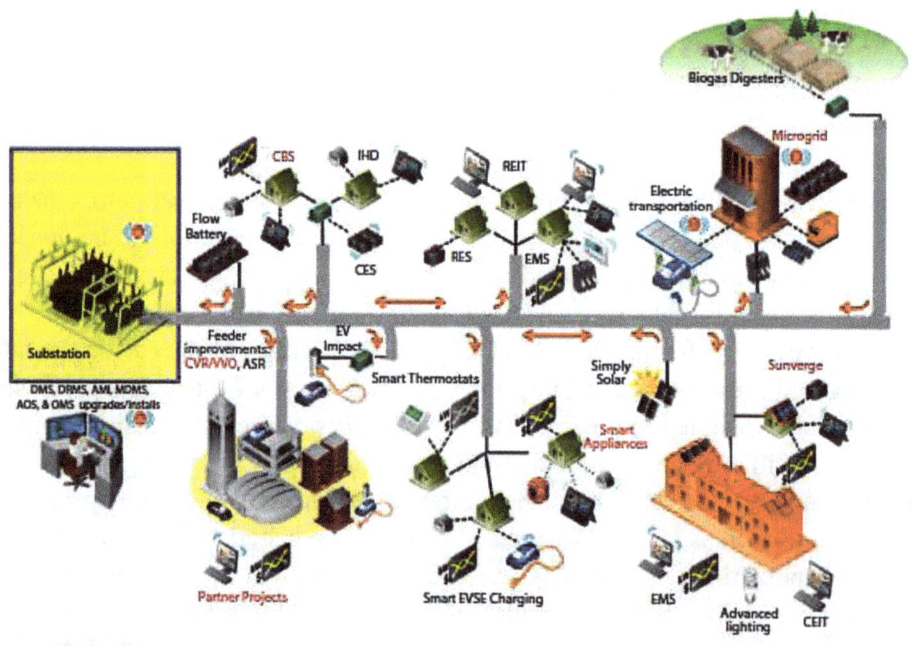

Source: World Bank.

When roadmapping the smart grid, most of papers and international agencies focused on distribution system as mentioned above, all of them had the same strategy that is many of stages or phases to develop roadmap from the planning to their wanted target, see ((Akhter & Biswas,2016); (Seyedfarshi &Pourmostadam,2013); (Heuberger,2015); (Dirkman, 2014)). The smart grid road maps of different countries and their goals are reviewed in table (1) according to (Seyedfarshi &Pourmostadam, 2013).

Selection criteria for the target countries are their high investments in smart grid technologies, the existence of a roadmap, their having a leading role in related research and standardization, their large scale smart grid implementation projects and the accessibility to their information.

As one can see phase 1 generally encompasses planning, pilot projects and infrastructure development. In phase 2 standard developments as well as the participation of customers are carried out. In phase 3 the migration to a smart grid is gradually completed.

The authors chose the technology road map for distribution system according to (IEA,2014;IEA,2015) With the assurance that the writing of the road map is not enough the true measure of success is whether or not the roadmap is implemented and achieves the organization's desired outcome. Such progress can be tracked with proper monitoring indicators (IEA, 2015).

The Roadmap Development Process

Figure 6 represents the roadmap development process. The process includes two types of activities (expert judgement and consensus, and data and analysis) and four phases (planning and preparation, visioning, roadmap development, and roadmap implementation and revision). After a roadmap is completed, implementation and updating ensure the complete realisation of the vision and goals.

The success of a roadmap is based on early planning and foresight, establishing a commonly "owned" vision, a full understanding of the national challenges and opportunities, the importance of "champions",

Table 1. Smart grid road map in sample countries

	Phase 1	Phase 2	Phase 3
Japan	Technology development of renewable energies and grid-connected system development	Technology demonstration in real environment "smart community"	
China	Planning, pilot projects and standard development	Roll-out construction	System improvements
Australia	Foundations to monitor and manage data across the network business	Analytics to improve business decisions and operations	Enabling technology to automate function
South Korea	Smart pilot city	Wide area extension of pilot project (consumer intelligence)	National smart grid completion
U.S.A (New York state)	Smart grid efforts on providing a solid smart grid foundation	Smart grid related technologies	
U.S.A (South California Edison)	Foundation, information, Technology and automation	Interactive (Grid 1.0 evolution to Grid 2.0)	Intuitive & Transactive grid
Canada (Toronto hydro)	Establish Toronto's smart community, demonstration projects	Expansion of demonstrated initiatives pilot projects results	Complete integration of technologies and services, collaboration between the utility and customers

commitment to outcomes by both public and private stakeholders, and ongoing evaluation and reports on the progress. Ideally, a roadmap is a dynamic document that incorporates metrics to facilitate the monitoring of progress towards its stated goals, with the flexibility to be updated as the market evolves (IEA, 2014; IEA, 2015).

Roadmapping Initiatives (Expert Judgement and Consensus, and Data and Analysis)

it rely upon sound data and analysis in combination with expert workshops to build consensus, to establish current baseline conditions, so that milestones and performance targets can be set and technology pathways defined to achieve the roadmap goals, i.e gathering the information needed for the roadmap while also building awareness and support throughout the process.

Where Expert judgement and consensus consist of experts in technology, policy, economics, finance, social sciences and other disciplines to formulate roadmap goals and milestones, identify gaps, determine priorities and assign tasks. Expert judgement is also often needed to make choices among possible scenarios or options revealed by data and analysis activities.

Phase 1: Planning and Preparation

The first phase of the roadmap process looks at essential planning and preparation that should be considered when starting a roadmap exercise where the organization undertaking the roadmapping initiative needs to know several considerations as:

- boundaries of the roadmapping effort

Figure 6. Road map process outline

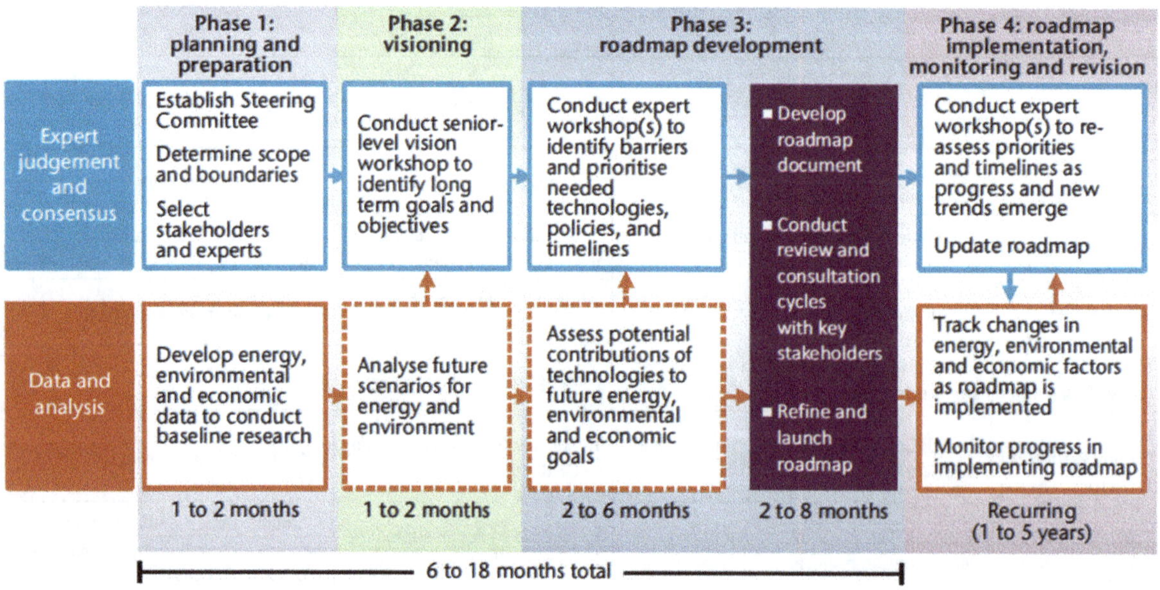

- technology areas or classes that the roadmap will consider
- energy sources or end-use sectors will be considered
- time frame for the roadmapping effort
- current state of the technology under consideration
- method of implementation and usage the resulting road map
- Private sector participation
- national government decision making
- help of other roadmaps
- What is existing tools or analysis, such as other roadmaps, can be used to influence scoping decisions.

Phase 2: Visioning

Setting a vision is the process of defining the desired pathway for a technology's deployment within a given timeframe, this will serve as the mission statement for the roadmap, framing what the roadmap will aim to do in broad terms. This visioning phase builds on the knowledge gained from the first phase.

Successful roadmapping processes often include senior-level workshops to identify long-term smart grid goals and objectives. Typical vision workshop participants include government leaders, senior industry representatives and leading researchers. At these workshops, participants consider the trends environmental, technological and other consideration.

At vision workshops, if the results of data analysis was available, participants can use it to consider alternative scenarios and projections, otherwise vision workshops will rely on the collective expert judgment of the participants.

Phase 3: Roadmap Development

The third phase of roadmap development concerns the preparation and review of the roadmap document itself. After a vision is established, the roadmap development phase begins, drawing on analysis and expert judgment to define the activities, priorities and timelines required to reach the desired vision.

As roadmap implementation usually involves a wide range of stakeholders from the public and private sectors, it is crucial to involve these actors when drafting the roadmap document to secure buy-in and support for their identified actions.

Phase 4: Implementation, Monitoring, and Revision

A crucial fourth phase in the life of a technology roadmap is the actual implementation, monitoring and revision of the roadmap document itself, by monitor key energy, environmental and economic indicators to track progress, in addition to conduct regular roadmap revision workshops to adapt roadmap goals and priorities to changing circumstances.

The road map process are explained in detail in (IEA, 2014; IEA 2015)

Another view to transition distribution system to smart one you can find it in (Madrigal et al., 2017). Here, it will be illustrated briefly strategies for distribution grid modernization: these strategies are recommended according to ((Madrigal et al., 2017)).

Each utility will have a specific strategy for grid modernization according to its technology starting point, level of available resources, and vision.

Also, strategies for distribution grid modernization in developed countries will differ from strategies in developing countries, because electric distribution utilities in developing countries may lack some of the basic building blocks and resources needed for grid modernization.

The information of the grid needed to define the amount of grid modernization are availability of key equipment including controllable power apparatus (line switches, capacitor banks, voltage regulators, and distributed energy resources [DERs]), the number and locations of distribution sensors, and the availability of reliable, effective telecommunication facilities and the availability of financial and technical resources.

Grid Modernization Levels

There are four levels of grid modernization needed to accomplish the utility's vision for grid modernization —that is, the levels of automation that currently exist at a given electric distribution utility and progressively more sophisticated levels of grid modernization that may be needed. Each level is characterized by the incremental use of more advanced applications to progress toward grid objectives such as improved efficiency, reduced losses, improved reliability, and the integration of renewable energy resources. The levels are listed below:

Level 0: Manual control and local automation define a situation in which most processes are performed manually with little or no automation. This is a situation that exists at many utilities in developing countries.

Level 1: Substation automation and remote control build on level 0 by adding IEDs and data communication facilities to achieve greater monitoring and control capabilities at HV/MV substations.

Level 2: Feeder automation and remote control build on level 1 by extending remote monitoring and advanced control to the feeders themselves (outside the substation fence).This level also includes information from communicating meters at some large customers for improved control and decision making.

Level 3: DER integration and control and demand response—the highest level of grid modernization—add energy storage, static VAR sources, and advanced communication and control facilities to effectively integrate and manage high penetrations of DERs on the distribution feeders. This level of grid modernization also includes deployment of AMI to enable on-demand reading of customer meters along with DR capabilities.

It is not necessary to have a gradual transition between this levels. That is, a utility may leapfrog one or more levels of grid modernization to achieve some of the benefits offered by the highest levels of grid modernization.

After defining the four smart grid modernization levels relevant to the distribution sector, a smart grid investment plan must be defined.

utilities must determine in any level they are, to identify the smart grid applications they need, that could help modernize their grid and achieve strategic objectives and smart grid applications.

A utility needs to clearly assess the cost, benefits, and potential risks of implementing new applications to define a sensible investment plan. Creating an investment plan is needed to include the specific list of projects to be implemented, their cost, and time frames for their implementation. Creating an investment plan is key to ensuring that the overall budgeting process considers the needs of modernization.

The broad steps toward defining the investment plan are described in the Figure 7.

Figure 7. Flowchart for creating an investment plan

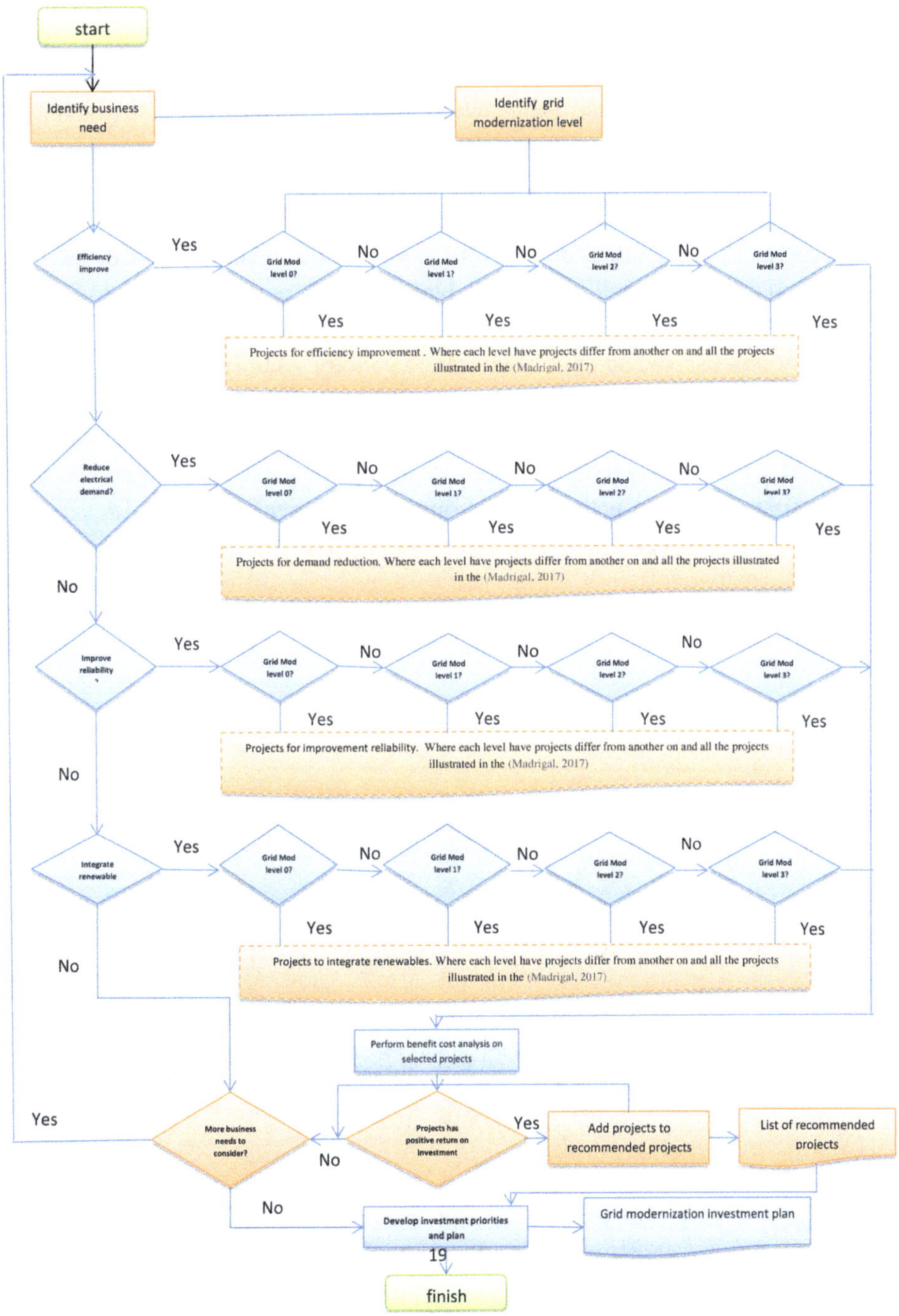

Social and Business Dimension of Smart Grid Developments

It can be considered that Smart Grid will have been as Panacea for the Environment and the infrastructure of electric system by introducing intermittent low carbon energy sources, introducing information and communication technology, technological and economic operation of the electricity grid and efficient operation of the grid and at the same time it offers economic advantages for all stakeholders.

The expectation is that the role of users will change from a passive end user into a more active one.

More general, on the role of users in smart grids, the main lesson is that user roles should be taken more seriously in relation to smart grids: experts should no longer regard users exclusively and/or simply as potential barriers to smart grid innovation but also as important stakeholders and potential participants in the innovation process.

The concept of the smart grid provides a solution to all the problems and a great economic opportunity for the major countries and industries. From the point of (verbong, Verkade, Verhees, Huijben & hoffken, 2016), these smart grid visions are economic technological visions that they focus too much on technological fixes and pay little attention to social dynamics and contexts, relating to beliefs, decisions, struggles and interactions between various actors and social groups.

Users are not only purchasers and consumers of a technology but they can be involved in various degrees in the production process (e.g. through providing input to designers) or even act as a co-producer and add value themselves.

(verbong et al., 2016) create a typology of roles by juxtaposing them according to two dichotomies (see Figure 8). The first dichotomy is that of constraining or barrier-like user roles versus enabling or empowering ones: users can either help a transition to smart grids or block it. The second dichotomy is that of passive versus active roles (i.e. is the positive or negative influence the result of strategic behavior or not?).

Figure 8. A topology of user roles in sustainable innovation

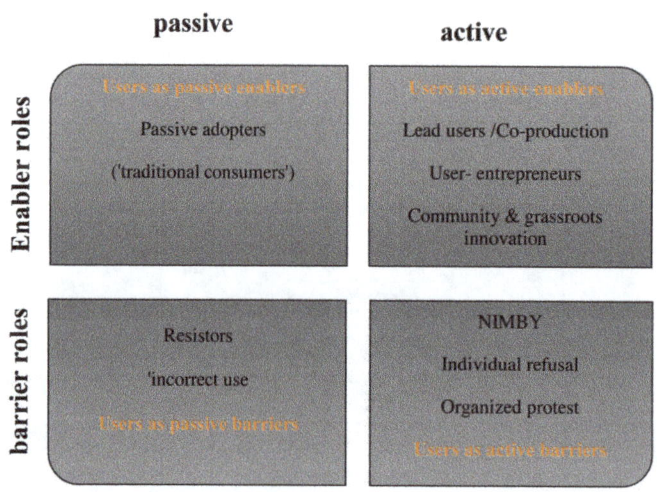

Passive Barrier Roles

we can find it in the lower left quadrant and contain the users who can't participate in using the available options of the smart energy system due to individual preferences and/or collective practices, or 'incorrect' use due to lack of knowledge, the solutions often take the form of educating (or 'domesticating') users to move them to the upper left quadrant.

Passive Enabler Roles

In that upper left quadrant, we find the 'traditional consumer': a passive adopter of an innovation (in this case: a participant in smart energy systems).

Active Barrier Roles

We find these active barrier-roles in the lower right quadrant: active resistance by

- Individual households to a smart grid innovation (e.g. refusing to install smart meters or to give access to data) and the so-called NIMBY ('Not in My Back Yard')-phenomenon
- Large-scale social movements actively resisting innovations through organized protests and political pressure (e.g. against nuclear power)
- Local, yet highly organised resistance.

Active Enabler Roles

Active enabler roles reside in the upper right quadrant. the users can become a 'lead users' that act as a key sources of information and ideas that lead to innovations which are then marketed by firms (e.g. households actively engaging in smart grid projects and providing feedback to suppliers, DNOs and utilities).

Other active enabling roles are

- Individual households as small decentralised renewable energy producers or even 'user entrepreneurs' who convert sustainable solutions to a problem they experience into a business.
- Collective user roles that actively enable sustainable innovation are captured by concepts such as 'collaborative consumption' (e.g. co-housing, car sharing),
- autonomous associations of users who cooperate for mutual benefit (e.g. collective purchasing of PV panels to bring down prices;
- 'crowdfunding' (wherein collective users are a source of capital for technological innovations.
- 'cooperatives' (groups of users that do not own their own land or roofs but collectively rent plots or roof space and install relatively large capacities of collectively purchased wind turbines or solar panels and in doing so, effectively become small, collective energy producers.
- 'community innovation': groups of users collective users that act as initiators, designers and maintainers of technological projects in their own locality (e.g. street, neighborhood, village).

And these are just the roles that 'actual' users can play, all these possible user roles have their own complex social dynamics and generative mechanisms. Research exists on all of these roles: individual or collective, constraining or enabling, passive or active.

The main lesson here is that user roles should be taken more seriously in relation to smart grids: experts should no longer regard users exclusively and/or simply as potential barriers to smart grid innovation but also as important stakeholders and potential participants in the innovation process.

The Main Elements of a Social Science Research Agenda are:

- Developing more socially embedded visions on smart grids and the services it will provide; this should not be left to the 'experts', but include all relevant actors
- A shift in the focus on developing smart grids components and systems towards the services it will deliver, taking energy consuming practices as a focal starting point
- Development and testing of innovative user-centered business models and ecosystems; there are pilot projects that experiment with new business models, but often still too much technology driven
- More general more attention to the innovative role users can have in smart grid development, and broader in sustainable innovations

Transition to Smart Grids: A Psychological Perspective

Smart grids involve substantial changes in energy infrastructures, technologies and user behaviour, and to develop a reliable and affordable smart grid, we need the acceptance of people to smart grid and changing their behaviour accordingly.

The changes in behaviours-needed in smart grid- includes the adoption of sustainable energy sources and technologies, the adoption of smart grid technologies and energy-efficient appliances, and changes in user behaviour. These behaviours will be referred as smart energy behaviours.

The authors in (van der Werff, Perlaviciute, &steg, 2016) reviewed psychological studies aimed at understanding and promoting behaviour in smart grids by individuals and households. They proposed a general framework, comprising four key issues:

1. Identification and measurement of behaviours that need to be changed to promote smart grids,
2. examination of factors underlying smart energy behaviours, including the adoption of sustainable energy resources, energy-efficient technologies, and automated control technology; investments in energy efficiency measures in buildings such as insulation; and user behaviour.
3. Designing and testing interventions to change behaviour needed to optimise smart grids, including information, financial incentives, regulations and technological changes,
4. Studying factors underlying public acceptability of energy policies and changes in energy systems aimed to promote smart grids.

The four key issues will be illustrated from the view of (van der Werff et al., 2016)

Which Behaviour Changes Are Needed to Promote Smart Grids?

The smart grid relies on renewable sources of energy. These sources are intermittent in its nature, In order to ensure a stable and efficient network, energy supply and demand must be identical. To achieve this, consumers can either accept or adopt technologies for storing renewable energy (e.g. electric cars and batteries) or shift energy use to times when renewable energy sources are abundant (e.g., when the sun is shining). Shifting energy use to times when renewable energy is available can either be done autonomously or by installing technologies that automatically switch specific appliances on or off on the basis of the available energy supply.

As we have seen, smart grids require changes in a wide range of energy behaviors. Thus, the important question is how these different types of behaviors are related, and how changes can be realized in a wide range of smart energy behaviors.

Some studies have found negative effects on the effect of some kind of energy behavior on other behaviors. In this case, participation in smart energy behavior reduces the likelihood of subsequent intelligent energy behavior. For example, people were likely to increase their energy consumption after reducing their water consumption.

Other studies have found that initial smart energy behavior makes people more likely to participate in other smart energy behaviors. For example, individuals who recycled were more likely to buy organic food and use environmentally-friendly modes of transport one and two years later ... etc. Thus, when people realize that they are engaging in smart energy behaviors (or pro-environmental behaviors in general), they are likely to see themselves as an environmentalist, prompting them to engage in environmental behaviors or energy savings in later cases.

Factors Underlying Behaviour in Smart Grids

People may engage in intelligent energy behaviors when they believe that these behaviors will benefit them greatly (prices, time and convenience) for lower costs, resulting in overall positive assessments of the relevant actions.

People also look at emotional and social costs and benefits (for example, people participate in pro-environment behaviors when they expect to enjoy such behavior. People may also engage in smart energy behavior when they expect others to agree)

Many people care about the environment and take environmental considerations into account when making decisions that motivate people to see themselves as true moralists, which may encourage smart energy behaviors

People consider that values are an important factor to take into account the individual and collective results of behavior in the field of smart energy,

Values reflect life goals or ideals that determine what is important to people and what outcomes they seek in their lives in general. Values can affect a wide range of assessments, beliefs and actions. Four types of values have been found to be relevant for people's evaluations and behaviour in smart grids: hedonic values that make people focus on pleasure and comfort, egoistic values that make people focus on safeguarding and promoting one's personal resources (i.e., money, status), altruistic values that make people focus on the well-being of other people, and biospheric values that make people focus on consequences for nature and the environment. Values affect how important people find different consequences of smart grids, and how they evaluate these consequences. More specially, people focus on consequences of smart grids that have positive or negative implications for their important values. For example, the stronger their biospheric values, the more positively people evaluated renewable energy sources, which are generally seen as having positive implications for one's biospheric values, such as reducing CO2 emissions.

Value-based acceptance judgments may affect the assessment of other characteristics of smart networks, which may be less important to people based on their values. In other words, people are more likely to evaluate smart networks in a way that is too positive or negative in line with their values-based judgments. Thus, a careful understanding of the values that actually lie in people's assessments and admissions assessments is critical to developing effective strategies for intervention and communication.

Many smart energy behaviours have positive collective consequences, and negative individual consequences. In line with this, research revealed that in general, people have more favourable evaluations of and are more likely to engage in smart energy behaviours if they strongly endorse biospheric and, to a lesser extent, altruistic values, while they are less likely do so if they strongly endorse egoistic and/or hedonic values Strong biospheric values also strengthen environmental self-identity, which in turn increases the likelihood of positive spillover effects.

So, people need to be motivated to engage in the relevant behaviours, and they need to be able to do so.

Interventions to Promote a Transition to Smart Grids

To encourage smart energy behaviour, two kind of strategies must be discussed:

1. Structural strategies that aim to enhance people's ability and motivation to engage in smart energy actions by making such actions more attractive via incentives.
2. Psychological strategies that aim to increase people's ability and motivation to engage in smart energy actions without actually changing their costs and benefits.

Acceptability of Smart Grids

Smart grids can be promoted via different energy policies and changes in energy systems, which should be acceptable to the public. Where, public acceptability depends on how and by whom a transition to smart grids is developed and implemented. The authors describe three factors that play a crucial role in this respect, namely distributive fairness, trust in involved parties, and public engagement and participation.

Distributive Fairness

Means how to distribute the benefits, costs and risks of the smart grid among the groups involved. Intelligent network solutions will be seen as unfair if some groups in the community experience most of the costs, while other groups enjoy benefits; this may reduce their acceptance.

Fair distribution of costs and benefits can be pursued in multiple ways, which are not mutually exclusive.

First, the risks and costs of smart networks can be minimized to enhance fairness and secure public acceptability. For example, renewable energy costs can be reduced by subsidies.

Second, Develop a strategy to provide additional benefits to those exposed to most costs and risks. For example, individuals can be compensated financially.

Thirdly, Offering reductions on energy prices

Trust in Involved Parties and Acceptability

People don't know the complex energy technology needed in transition to smart grid, so they need to rely on other parties such as government and energy companies.

Trust in involved parties will especially affect evaluations and perceptions when people have little knowledge about new systems or solutions, which is the case for smart grids.

If people don't trust in involved parties this will effect in acceptability of smart grids.

Public Involvement

In smart grid, the role of consumers will change from passive recipients to active producers of energy, so-called prosumers. Smart grids may also require more active public user involvement in planning, developing, and implementing smart grids. Public involvement comprises different dimensions, varying from one-way communication from developers to consumers to active public involvement in decision-making processes, which can have important implications for public acceptability of smart grids. Acceptability of smart grids is likely to be higher if people have been actively involved in the decisions-making process, as this enhances

CONCLUSION

To achieve a successful smart grid, we need a successful plan to deployment it, and this plan should have a road map "which is a plan should provide a step-by-step overview for each phase of the deployment plan, in chronological order". Road maps are exclusively authored by organizations which could be Standards Developing Organizations (SDOs) or regulatory organizations, technical consortia, forums and panels, and marketing/advocacy organizations. According to the number of recommendations the IEC TC 5 standards are of highest importance from the perspective of most experts.Smart grid road mapping means roadmapping transmission and distribution grids. Most of studies focused on distribution grid, this chapter illustrated the distribution modernization in two viewpoints: 1-roadmap by IEA which consist of four phases 2- and the Grid Modernization Levels which consists of four levels and smart grid investment. With the emphasis that each country has its own road map, according to its vision and its existing technologies. Smart grids involve substantial changes in energy infrastructures, technologies and user behaviour, and to develop a reliable and affordable smart grid, we need the acceptance of people to smart grid and changing their behaviour accordingly.

REFERENCES

Akhter, S. R., & Biswas, M. M. (2016). Roadmap of smart grid for Bangladesh based on functions and technological challenges. *Electric Power Components and Systems*, *44*(8), 864–872. doi:10.1080/15325008.2016.1138159

Alhelou, H., Hamedani-Golshan, M. E., Zamani, R., Heydarian-Forushani, E., & Siano, P. (2018). Challenges and Opportunities of Load Frequency Control in Conventional, Modern and Future Smart Power Systems: A Comprehensive Review. *Energies, 11*(10), 2497. doi:10.3390/en11102497

Alhelou, H. H. (2018). *Fault Detection and Isolation in Power Systems Using Unknown Input Observer. In Advanced Condition Monitoring and Fault Diagnosis of Electric Machines* (p. 38). Hershey, PA: IGI Global.

Alhelou, H. H., Golshan, M., & Fini, M. (2018). Wind Driven Optimization Algorithm Application to Load Frequency Control in Interconnected Power Systems Considering GRC and GDB Nonlinearities. *Electric Power Components and Syst.*

Alhelou, H. H., & Golshan, M. E. H. (2016, May). Hierarchical plug-in EV control based on primary frequency response in interconnected smart grid. In *Electrical Engineering (ICEE), 2016 24th Iranian Conference on* (pp. 561-566). IEEE. 10.1109/IranianCEE.2016.7585585

Alhelou, H. H., Golshan, M. H., & Askari-Marnani, J. (2018). Robust sensor fault detection and isolation scheme for interconnected smart power systems in presence of RER and EVs using unknown input observer. *International Journal of Electrical Power & Energy Systems, 99*, 682–694. doi:10.1016/j.ijepes.2018.02.013

Alhelou, H. H., Hamedani-Golshan, M. E., Heydarian-Forushani, E., Al-Sumaiti, A. S., & Siano, P. (2018, September). Decentralized Fractional Order Control Scheme for LFC of Deregulated Nonlinear Power Systems in Presence of EVs and RER. In *2018 International Conference on Smart Energy Systems and Technologies (SEST)* (pp. 1-6). IEEE. 10.1109/SEST.2018.8495858

Alhelou, H. S. H., Golshan, M. E. H., & Fini, M. H. (2015, December). Multi agent electric vehicle control based primary frequency support for future smart micro-grid. In *Smart Grid Conference (SGC)* (pp. 22-27). Academic Press.

Alshahrestani, A., Golshan, M. E. H., & Alhelou, H. H. (2018, November). WAMS Based Online Estimation of Total Inertia Constant and Damping Coefficient for Future Smart Grid Systems. In *Smart Grid Conference (SGC)* (pp. 1-5). Academic Press.

An Introduction to Smart Grids. (n.d.). *ABB Back Ground information: Smart Grids.* Retrieved from: https://mycourses.aalto.fi/pluginfile.php/365574/mod_folder/content/0/Smart_grids_introduction.pdf?forcedownload=1

Berker, T., & Throndsen, W. (2017). Planning story lines in smart grid road maps (2010–2014): Three types of maps for coordinated time travel. *Journal of Environmental Policy and Planning, 19*(2), 214–228. doi:10.1080/1523908X.2016.1207159

Bichler, M. (2013). *Smart Grids and the Energy Transformation: Mapping Smart Grid Activities in Germany*. Heinrich Böll Foundation.

Buxton, J. (2011). *Creating a smart grid roadmap. Black & Veatch building a world of difference.* Retrieved from https: // www.bv.com/docs/articles/creating–a–smart–grid-roadmap.pdf

Dirkman, J. (2014). *Developing a Roadmap to a Smarter Utility.* Retrieved from https://www.schneider-electric.com.vn/en/download/document/998-2095-10-28-14AR0_EN/

Du, Y., Turner, M., Liao, Y., Sunkara, M., DeRouen, J., Greenwell, A., . . . Gardner, J. (2012, January). Kentucky smart grid roadmap initiative. In Innovative Smart Grid Technologies (ISGT), 2012 IEEE PES (pp. 1-6). IEEE.doi:10.1109/ISGT.2012.6175628

Fini, M. H., Yousefi, G. R., & Alhelou, H. H. (2016). Comparative study on the performance of many-objective and single-objective optimisation algorithms in tuning load frequency controllers of multi-area power systems. *IET Generation, Transmission & Distribution, 10*(12), 2915–2923. doi:10.1049/iet-gtd.2015.1334

Finnigan, J. (2014). *Smart Planning for a Successful Smart Grid Roll-Out. Environmental Defense Fund-EFD.* Retrieved from http://blogs.edf.org/energyexchange/2014/01/30/smart-planning-for-a-successful-smart-grid-roll-out/

Hamilton, B. A., Miller, J., & Renz, B. (2010). *Understanding the benefits of the smart grid-smart grid implementation strategy.* United States Department of Energy's National Energy Technology Laboratory.

Heuberger, D. (2015, November). Roadmap for Smart Grids: Four Steps to an intelligent electrical distribution grid. In *International ETG Congress 2015; Die Energiewende-Blueprints for the new energy age; Proceedings of* (pp. 1-6). VDE.

IEA. (2014a). Energy Technology Roadmaps: A Guide to Development and Implementation. OECD/IEA.

International Electrotechnical Commission. (2010). *IEC smart grid standardization roadmap.* Author.

International Energy Agency. (2015). *How 2 Guide for Smart Grids in Distribution Networks Roadmap Development and Implementation.* Available at https://www.iea.org/publications/freepublications/publication/TechnologyRoadmapHow2GuideforSmartGridsinDistributionNetworks.pdf

Lajoie, B., Debarre, R., Fayet, Y., & Dreyfus, G. (2015). *Beyond smart meters Introduction to Smart Grids.* A.T. Kearney Energy Transition Institute. Retrieved from: http://www.energy-transition-institute.com/_/media/Files/ETI/Summary%20Introduction%20to%20Smart%20Grids.pdf

Madrigal, M., Uluski, R., & Gaba, M. K. (2017). Practical Guidance for Defining a Smart Grid Modernization Strategy: The Case of Distribution (Revised Edition). The World Bank.

Makdisie, C., Haidar, B., & Alhelou, H. H. (2018). An Optimal Photovoltaic Conversion System for Future Smart Grids. In *Handbook of Research on Power and Energy System Optimization* (pp. 601–657). IGI Global. doi:10.4018/978-1-5225-3935-3.ch018

Mckenzie, M. (n.d.). *Developing a standards road map for smart grid Deployment in Australia.* Retrieved from http://www.standards.org.au/OurOrganisation/Events/Documents/Smart%20Grid%20Forum%20-%20Presentation%204% 20-%20Mark% 20 McKenzie.pdf

Nadweh, S., Hayek, G., Atieh, B., & Haes Alhelou, H. (2018). Using Four – Quadrant Chopper with Variable Speed Drive System Dc-Link to Improve the Quality of Supplied Power for Industrial Facilities. Majlesi Journal of Electrical Engineering.

National Energy Technology Laboratory. (2009). *The Transmission Smart Grid Imperative.* Retrieved from https://www.netl.doe.gov/File%20Library/research/energy%20efficiency/smart%20grid/whitepapers/The-Transmission-Smart-Grid-Imperative_2009_09_29.pdf

Navigant. (2015). *Ontario Smart Grid Assessment and Roadmap*. Report of Navigant. Retrieved from https://www.ontarioenergyreport.ca/pdfs/Navigant-Smart-Grid-Assessment-and-Roadmap-Final-Report-.pdf

New York State Smart Grid Consortium. (2010). *Smart Grid Roadmap for the State of New York*. Report to New York State Smart Grid Consortium. Retrieved from https://www.energy.gov/sites/prod/files/oeprod/DocumentsandMedia/NYSSGC_Attachment.pdf

NIST-Office of the National Coordinator for Smart Grid Interoperability & Office of the National Coordinator for Smart Grid Interoperability. (2018). *NIST framework and roadmap for smart grid interoperability standards, release 1.0*. US Department of Commerce, National Institute of Standards and Technology.

Njenda, T. C., Golshan, M. E. H., & Alhelou, H. H. (2018, November). WAMS Based Intelligent Under Frequency Load Shedding Considering Online Disturbance Estimation. In *Smart Grid Conference (SGC)* (pp. 1-5). Academic Press.

Njenda, T. C., Golshan, M. E. H., & Alhelou, H. H. (2018, November). WAMS based Under Frequency Load Shedding Considering Minimum Frequency Prediction and Extrapolated Disturbance Magnitude. In *Smart Grid Conference (SGC)* (pp. 1-5). Academic Press.

Sato, T., Kammen, D. M., Duan, B., Macuha, M., Zhou, Z., Wu, J., & Asfaw, S. A. (2015). *Smart grid standards: specifications, requirements, and technologies*. John Wiley & Sons. doi:10.1002/9781118653722

Seyedfarshi, S., & Pourmostadam, K. (2013). *The challenges in developing a smart grid roadmap for the distribution network of Iran*. Academic Press.

Shabanzadeh, M., & Moghaddam, M. P. (2013, November). What is the smart grid? definitions, perspectives, and ultimate goals. *28th International Power System Conference (PSC).*

Specht, M., Rohjans, S., Trefke, J., Uslar, M., & Vázquez, J. M. G. (2013). International Smart Grid Roadmaps and their Assessment. *EAI Endorsed Trans. Energy Web*, *1*(1), e2.

van der Werff, E., Perlaviciute, G., & Steg, L. (2016). Transition to smart grids: A psychological perspective. In *Smart Grids from a Global Perspective* (pp. 43–62). Cham: Springer. doi:10.1007/978-3-319-28077-6_4

Verbong, G. P. J., Verkade, N., Verhees, B., Huijben, J. C. C. M., & Höffken, J. I. (2016). Smart Business for Smart Users: A Social Agenda for Developing Smart Grids. In *Smart grids from a global perspective* (pp. 27–42). Cham: Springer. doi:10.1007/978-3-319-28077-6_3

Zamani, R., Hamedani-Golshan, M. E., Haes Alhelou, H., Siano, P., & Pota, R, H. (. (2018). Islanding Detection of Synchronous Distributed Generator Based on the Active and Reactive Power Control Loops. *Energies*, *11*(10), 2819. doi:10.3390/en11102819

KEY TERMS AND DEFINITIONS

Implementation: The process of putting a roadmap into action, by carrying out projects and initiatives that address roadmap tasks and priorities, and by monitoring progress using a tracking system.

Roadmap: A specialized type of strategic plan that outlines activities an organisation can undertake over specified time frames to achieve stated goals and outcomes.

Roadmapping: The evolving process by which a roadmap is created, implemented, monitored and updated as necessary.

Setting a Vision: The process of analyzing future scenarios and identifying objectives.

Smart Grid: A developing network of new technologies, equipment, and controls working together to respond immediately to our 21st century demand for electricity. The Smart Grid represents an unprecedented opportunity to move the energy industry into a new era of reliability, availability, and efficiency that will contribute to our Economic and Environmental health.

Stakeholders: Relevant individuals who have an interest in seeing the roadmap developed and implemented, such as representatives from industry, government, academia, and non-governmental organizations (NGOs).

Chapter 9
Electric Vehicles in Smart Grids

Cosmin Darab
Technical University of Cluj-Napoca, Romania

ABSTRACT

Electric vehicles were proposed as a good solution to solving energy crisis and environmental problems caused by the traditional internal combustion engine vehicles. In the last years due to the rapid development of the electric vehicles, the problem of power grid integration was addressed. In order to not put additional pressure onto the power grid several new technologies were developed. This chapter presents the smart grid technology, vehicle-to-grid concept, and electric vehicles grid integration. These technologies made possible the integration of electric vehicles without any major changes in the power grid. Moreover, electric vehicles integration brought new benefits to the power grid like better integration of renewable energy.

INTRODUCTION

Some of the greatest concerns of our era is reducing carbon dioxide emissions and greenhouse gases and resolving the issue of rapid increase of energy demand. Studies present that the most energy demanding sector of the last years is the transport sector. According to the U.S. Energy Information Administration (EIA) this is justified by the increase of population growth and economic sector. In many countries mitigating measures were undertaken in order to impose an emission target. One of those solutions is electrifying the transport sector.

Electric vehicles represent a zero tail pipe emission alternative to the internal combustion engines. Another benefit is that the electric vehicle uses the energy stored in the battery for powering the electric motor which has a lower operation cost and a higher efficiency. Moreover the electric vehicle noise is much lower compared with the classic vehicle. MacKey (2009) presented a research that concluded with the statement that electrifying the whole transport sector will result in cost reductions of 80% from the actual one obtained using internal combustion powered vehicles. Another study made by Short and Denholm (2006) stated that electrifying the transport sector would promote the use of renewable energy sources. The number of electric vehicles will increase rapidly as a result of continuous battery development and charging stations infrastructure development.

DOI: 10.4018/978-1-5225-8030-0.ch009

Nowadays, the fact that the electric vehicle has a high acquisition price and that the charging infrastructure still lacks, are considered drawbacks for potential clients. Also another thing to be considered is the power grid integration. This is a problem due to the fact that usually an electric car's battery is charged at home after work, at peak energy consumption hours according to Alhelou et al (2015). This fact will add extra load to the power grid.

As a solution to the above presented issues, using new technologies was developed. The first and foremost solution is the implementation of the smart grid, a power grid able to use bidirectional communication with all the smart equipment installed with the purpose of optimizing the energy usage.

This chapter focuses on presenting the problems that the electric vehicle power grid integration arise. First part of the chapter presents the background and framework of the electric vehicles, smart grids and charging infrastructure. The second part of this work is dedicated to presenting new technologies and energy management strategies, that were developed to meet this issues. Last part of the chapter is dedicated to presenting some conclusion and future development regarding electric vehicles and their role in smart grids.

BACKGROUND

Integrating a constantly growing number of electric vehicles into the electric power grid is a great challenge. For a successful integration into the power network, observation and than careful assessment of the economical impact is required. A lot of research activity was dedicated into finding the issues, solutions and impact of electric vehicle grid integration. Su (2013) presented a research that proved that most of the vehicles battery is charged at home and the future trend is the development of commercial or work place chargers. This charging scenario is bound to greatly overload the power grid. Another consequence may be the overheating of the power transformers or demanding new investments for the energy distribution system.

Pecas et al. (2011) proved in their work that integrating electric vehicles into the power grid will add value if this action is well planed and technically reorganized to meet the operational standards. In order to confirm the benefits of merging the grid and the vehicle fleets different studies were made. The benefits were divided among the vehicles owners and the utility providers. Bessa and Matos (2012) are the ones to propose a mitigating method using the aggregator. The aggregator is responsible for delivering information and communication between energy service provider, distribution system operator and the transmission system operator. The same idea is presented by Pillai and Bak-Jensen (2011) that propose a virtual power plant concept in which the vehicles fleets are controlled as a distributed energy unit.

All these new concepts developed recently are based on the assumption that the electric vehicles owners will accept to participate in these programs. Basically the vehicle's battery is used as an energy storage or supply source according to the grid demand. There are still some barriers to overcome until these assumptions are feasible. One of those barriers is the economical one. The most expensive part of the vehicle is the battery, which capacity will decrease over time if it is subjected to a large number of charging and discharging cycles. If new battery technologies will be developed, than the production cost will lower and this economical barrier will be easier to overcome. Another approach is the intelligent battery usage. Peterson et al. (2010) made a study in which they compared a battery in a normal duty cycle and one that is used by the grid operator with smart charging applications. Their research revealed that if an intelligent management approach is used the second battery receives minimal capacity loss.

Judging by all the above presented facts it is essential that a new concept of power grid needs to be developed. A grid capable of real time bidirectional communication system between the involved parties. Hence the smart grid concept was introduced, with the advanced communication infrastructure that will easily allow the electric vehicles energy market penetration. The following sections will display a detailed review on electric vehicles to smart grid integration.

ELECTRIC VEHICLES

Electric vehicles were once the best transportation method, but since the nineteenth century (when they first appear), they suffered major changes. Robert Anderson invented the first electric powered carriage. It used non rechargeable batteries and the newly discovered electric motor. A lot of other electric carriages designs followed but were subject to failure due to the lack of battery technology as to the low performance of the electric motor. It was between 1856 and 1881 that the technology evolved and the first DC motor was invented. Also in 1881 the first commercial rechargeable lead acid battery was developed. These two technology development boosted the electric vehicle market. In three years after the first commercial electric vehicle was introduces, the electric vehicles represented almost 30% of the total road vehicles. In 1908 Henry Ford introduced on the market the gasoline powered Ford Model T. It was a great success due to the vehicle low price. Ford managed to industrialized the car manufacturing process, resulting in an affordable vehicle for the mases. In 1912 Charles Kettering developed the starter motor, thus removing the hand crank from the vehicles. These two events had a great impact on vehicle market. Around 25 years later it was stated that there was not a single electric vehicle on the roads.

The interest in electric vehicles increased along with the emission issue of the internal combustion engines. Among the first to take action was the California's Zero Emission Vehicle Mandate that was issued in 1990. This regulatory action stated that 2% of the vehicle fleet should be made out of electric vehicles by the year 1998 and another 8% should be added by 2003. In 1996 General Motors introduces EV1, their first electric vehicle. As oil price kept rising other car manufacturers were turning their attention towards hybrid and electric vehicles. Toyota had a great success in 1997 with their Prius, a hybrid electric vehicle. In the first year they manage to produce and sold over 18000 cars. By 2012 the total electric cars registered were almost 180 000. Nowadays, according to the International Energy Agency's (IEA), Global EV Outlook 2018 (GEVO) report, there are over 3 million electric cars on the road world wide. In 2017 there were sold over 1 million electric cars, with 50% more than 2016. Half of the sales from 2017 were made in China. Figure 1 presents the growth of electric cars in the recent years.

Nowadays there are a handful of countries that have electric vehicles high market share. The leading country is Norway with over 39% of new sales in 2017. Figure 2 presents the electric car market share for 2017.

Electric vehicles registered a huge improvement over the last years. Most of the innovations regarded the power train, battery and charging infrastructure. Electric vehicles, nowadays, may be divided according to their power train topology. Figure 3 presents the most commonly known power train configurations for hybrid electric vehicles.

Figure 1. Number of electric vehicles in circulation

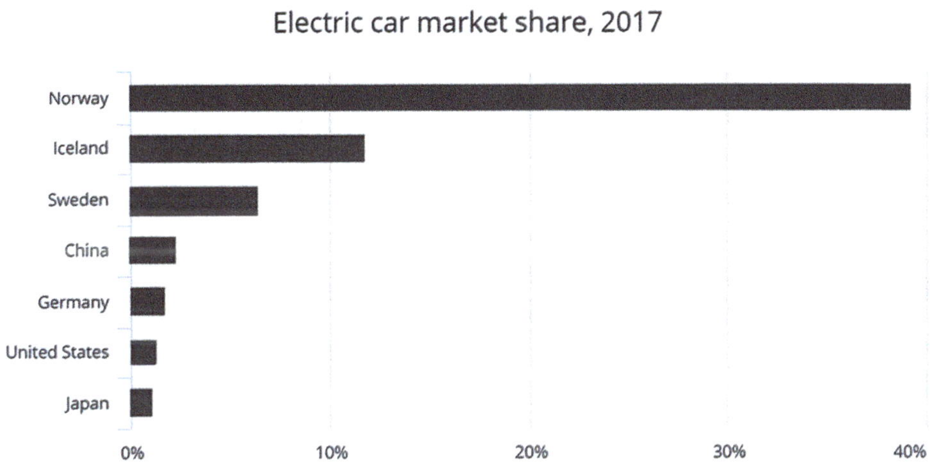

Figure 2. Electric car market share 2017

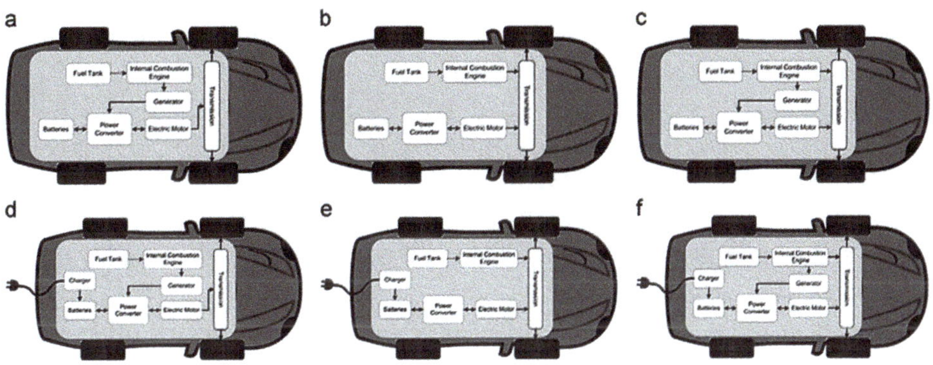

Figure 3. Power train topologies: a – Series Hybrid Electric Vehicle (HEV), b – Parallel Hybrid Electric Vehicle (HEV), c – Series – Parallel Hybrid Electric Vehicle, d – Series Plug in Hybrid Electric Vehicle (PHEV), e – Parallel Plug in Hybrid Electric Vehicle, f – Series – Parallel Hybrid Electric Vehicle

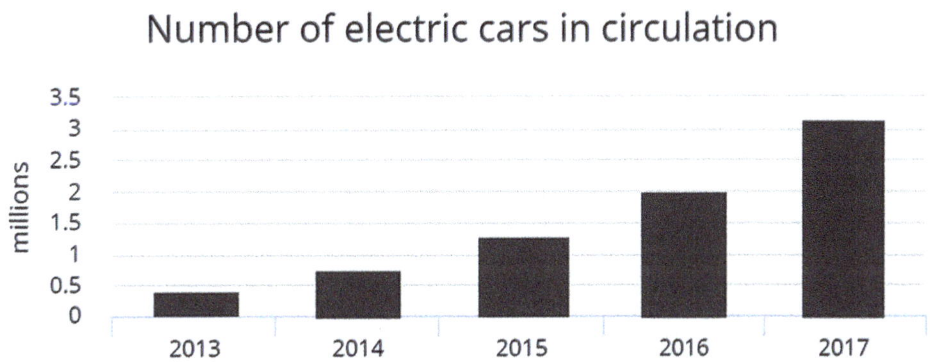

The term Battery Electric Vehicles (BEV), describes the pure electric vehicles that use only the battery stored energy. These vehicles are considered to have 0 tail pipe emissions due to the fact that they do not use fossil fuel. Series hybrid electric vehicles incorporate an electric motor and an internal combustion motor. What distinguishes them from the others is the fact that the internal combustion motor is connected in series with the electric motor and then with the transmission system. Using this topology the conventional internal combustion motor it is not used in the traction system, instead it acts as a generator to produce electric power for the electric motor. The third topology is that of the parallel hybrid electric vehicle. As the name implies both motor types are coupled in parallel. This way the vehicle can operate using either one of the motors, or in some cases using both motors simultaneous. Usually the electric motor is used for low vehicle speeds and the internal combustion motor for speeds greater than a set value. When there is a need for power both motors may be operated for powering the traction system. Series – Parallel hybrid electric vehicle use both of the above topologies combining their advantages. The vehicles energy management system decides when or how each motor will be used. All the above described electric vehicle types use the internal combustion motor as a generator for charging the battery that the electric motor uses. The other three types that follow in the above figure are similar with the already presented ones with the difference that they are also plug-in compatible. Plug-in electric vehicles have the option of charging the battery using a charger. That way, after charging, the vehicle will operate using the stored electric power and switch to fossil fuel after depletion. Nowadays, plug-in electric vehicles are equipped with a greater capacity battery in order to use the internal combustion motor as little as possible, thus reducing emissions.

ELECTRIC VEHICLES GRID INTEGRATION

Electric vehicles and plug-in hybrid electric vehicles are the alternative for reducing the use of fossil fuels and reducing carbon dioxide emissions.

In order to sudden integrate a large number of electric vehicles in the electric grid one needs to consider the aspects of optimal conditions, economic impact and control strategies. Due to the fact that the majority of charging stations are designed for home charging the consequences influence the electric power grid. Also, the fact that the majority of commercial and public charging stations are power charging stations for fast charging is a fact that has to be considered.

Considering these aspects, a lot of studies have been carried out to evaluate the impact of electric vehicles charging on the traditional grid. Pecas J.A. et. al (2011) approaches the issue of electric vehicles to be charged from the medium voltage distribution grid and points out that problems will arise with low voltage and line capacity exceeded. Other studies from the same author (2009) use different approaches like night charging of the vehicles and also show that this is a better solution for grid overcharging (Alshahrestani et al., 2018; Makdisie et al., 2018; Alhelou et al., 2018; Alhelou et al., 2016; Nadweh et al., 2018).

These effects may be countered by new investments and by improving the performance and efficiency of the grid, in other words implementing the concept of smart grid. The general operation method of the smart grid is to manage the electrical energy intelligent using advanced communication systems that provide specific information about the costumers demands. This implies that the system is able to receive information from the conventional energy generating system, consumers, renewable sources, network operators and electric vehicles, and match them in such a way that the network is secure and sustainable. A good example would be a car charging in a peak energy demand and the grid would execute a protocol of stopping the charging just for a few minutes until the demand lowers and then starting the charging process again.

Another strategy to be implemented in the smart grid technology is vehicle to grid concept. This implies that the car's battery could be used as a resource in the grid. Using storage devices to store energy when the energy demand is low and then inject the stored energy when the demand is high is one of the simplest solutions. However, this solution is not yet cost efficient due to the high price and also operating costs of the storage devices. But using the battery of the electric vehicle for this strategy is a cheaper solution.

One of the most important component for the smart grid and also for the electric vehicle is the charging infrastructure. And one of the most challenging factor is implementing a charging infrastructure at the national level. The standards for charging units and cords are described by the International Electrotechnical Commission (IEC) in Europe and in America by Society of Automotive Engineers (SAE). There are three types of charging equipment defined by the standards and are different from each other by the values of power and voltage. Both standards describe that the charging equipment has three functions such as voltage regulation, rectification and a coupling media for charging the vehicle. There is also specified the shock protection in wet and dry conditions. Level 1 and Level 2 chargers convert the AC power into DC power using the on board charging devices. Level 3 charger is used for "fast charging" and it supplies DC power directly from the charging equipment, and is usually referred to as off board charging. The charging infrastructure is divided in two coupling methods. The first one is referred to as "conducted coupling" and it is the traditional way of charging composed of a cable link between the car and the grid. The second method is the "inductive method" and instead of using a wire connection it uses a magnetic connection. As mentioned before, the chargers may be integrated in the vehicle (on board) or may be located outside the vehicle (off board). The on board equipment has limitations regarding the output ratings due to some restrictions (size, weight) but using this equipment the vehicle may be charged anywhere at an electric outlet. The off board charging stations are used for fast charging and they usually use high current values (up to 250 A).

SMART GRID

Smart grid is a term that is used to describe a power grid that integrates communication systems, automation and remote control technologies in order to improve the power supply by making it more reliable, more efficient and more sustainable. Figure 4 presents the proposed Smart Grid framework. Adding all these specs to the electric power network will result in a two-way communication system between the

Figure 4. Proposed smart grid framework

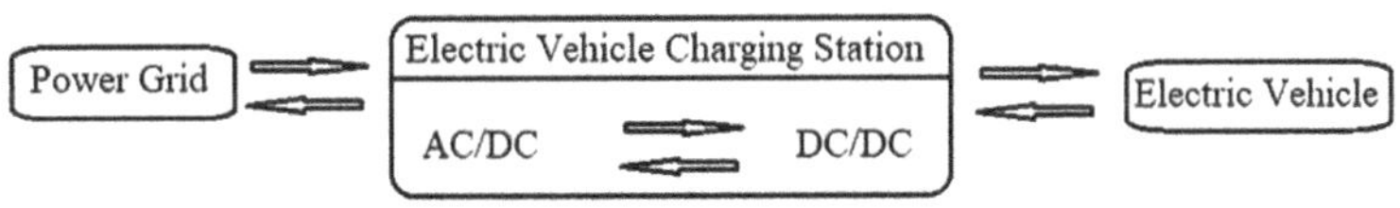

Table 1. Convention grid and smart grid characteristics

Functions	Conventional Power Grid	Smart Grid
Control	Manual	Automated
Communication	One-way	Two-way
Smart equipment	Not mandatory	Mandatory
Recovery	Manual	Self healing
Power generation	Centralized	Distributed
Client role	Passive	Active

supplier and the consumer. Specific equipment is needed in order to obtain a bi-directional link that will interconnect all the grid actors. For this purpose smart meters and various sensors are usually installed.

Real time data acquisition is used hand in hand with smart monitoring and control in order to optimize and control all the interconnected devices. Fini, Yousefi and Alhelou (2016) presented in their work some novel optimization techniques to be integrated in the smart grid control strategies for the load frequency converters. Another specific characteristic of the smart grid is the fact that the consumer is actively involved and is able to interact with the grid. As an example, the user is able to aces grid information like the electricity cost or usage by making use of the smart meter. Based on these information, the client is able to decide on their own energy usage patterns, an action that could be helpful in grid balancing. Smart grids are more stable an reliable due to the self heling capabilities. If a fault occurs, the first action to be taken is to isolate the problem and than to "heal" that part of the grid. This actions are possible due to the fact that a smart grid has intelligent equipment and multiple generation units that are dispersed along the grid. Alhelou et al. (2018) presented some results that were obtained by making use of intelligent control for the grid equipment making the power network more reliable and efficient. Another feature that a smart grid must perform is the fault detection. Some fault detection methods that would be suitable for smart grid equipment were presented by Alhelou et al. (2018) in their work. Table 1 presents all the benefits of having a smart grid and also the drawbacks in order to clearly separate between them.

Due to the benefits of having a smart grid a lot of projects around the world were implemented successfully. Some of these projects have been running from 8-10 years already and have been monitored for assessing the benefits of the smart grid. Each of the implemented project is unique and each developed based on some particular factors like geographical area, type of generation units, clients behavior and so on. According to Global Smart Grid Impact Report a lot of creative and innovative technologies were developed. One of the largest U.S. project is the Pacific Northwest Smart Grid. This grid integrates a lot of wind powered turbines as generation units. In order to improve the response of the grid and to better integrate the renewable energy a new energy management system was developed that integrates wind and load forecasts. Another example is the Compahnia de Electricidade de Macao – CEM Smart Grid implemented in Asia. This grid is special due to the fact that 98% of the generating units are low speed diesel generators and combined cycle gas turbines that has a generation capacity of 472 MW. In order to improve their grid they came up with ingenious technologies like a robust communication system that is fully automated with integrated Supervisory Control and Data Acquisition (SCADA) and Remote Terminal Units (RTU). There are other great examples of successfully implemented Smart Grids all over the world and a general observation of the facts could be that implementing new technologies and using innovative ideas that best suit the specifics of the grid, disregarding the perceived risk, is the right approach. Most of the research in this field, states that projects that avoided innovation due to perceived risk tended to rank lower than those that choose to deploy innovation.

From the beginning of the implementation of smart grid projects, a lot of new technologies were developed. One of those technologies is the advanced metering infrastructure (AMI) which basically refers to a smart equipment that records and measures the information regarding the client energy usage. Also, using these smart metering devices a two-way communication system is possible, allowing the power utility and the end customer to share information. The benefits of smart meters are improving the monitoring and control of the grid resulting in an economical billing system and a economical power network.

Another technology developed within a smart grid is the home automation network (HAN). This smart infrastructure enables the communication of all electric equipment within the house. Also it receives information from the installed smart meter so that it could operate based on the clients wish regarding the energy market cost and also based on the grid real time information. A good example of HAN application logic would be as follows:

- The user sets their preferences for the house appliance, lighting, electric vehicle charging station and renewable sources.
- HAN control equipment receives a energy cost vector for a fixed period of time.
- Based on the client preferences and the provided energy cost all house management is programed so that the house would be more efficient and economical.

Supervisory control and data acquisition (SCADA) is a technology that gathers real time information from all the smart monitoring devices connected within the grid and using the established protocols and strategies operates grid equipment. This is one of the most commonly used technology and is implemented successfully in many power grids. SCADA usually integrates a few basic components like: programable logic controllers, sensors and interface devices, relays and switching equipment, a two way communication system for transferring information and control signals and some data base servers for information storage from all the distributed remote terminal units within the grid.

Demand Response (DR) is one of the latest technology developed for smart grid optimization. This technology supports the end user in involving itself in smart grid optimization operation for earning some incentives as a participation reward. The best example for the DR technology application would be the peak load period, when the customers can limit their energy usage so that the power grid load is reduces and there is no need for additional energy generation. This technology may also be used in case of emergencies for effectively create a low cost alternative and also to sustain the smart grid operation. More details about demand response and also a algorithm logic is presented in a following subchapter.

Another technology used in integrating electric vehicles into smart grids is the vehicle to grid (V2G) technology. This strategy is described in detail in the next subchapter. It defines the energy exchange between the grid and the battery of the electric vehicle using the implemented two way communication system. The proper use of the V2G technology will result in active power regulation, reactive power regulation and also some ancillary service support.

For implementing a robust smart power grid, distributed generation units are scattered across the grid, unlike the classic power grid concept that uses a few main power plants. This technology enables customers that are at distribution level to generate power for the nearby loads. This is a very effective way of reducing power losses since there is no more the requirement for power to be transmitted from the power plant. This concept is well suited for the integration of renewable energy sources like wind turbines and solar panels. The adoption of renewables into the grid can be better integrated using the storage technologies presented like V2G.

VEHICLE TO GRID CONCEPT

Integrating the electrified transport sector into the power grid will bring a lot of issues for the traditional power network. As presented above integrating a large number of electric vehicles will overload the grid during the charging times. As a counter, the strategy of vehicle to grid is thought to bring benefits and make the process of integrating electric vehicles into the power grid a lot easier. Basically, it implies the communication between the power grid and the electric vehicles.

Researchers developed different strategies such as vehicle to vehicle (V2V), vehicle to grid (V2G) and vehicle to home (V2H). The V2V technology was developed as a communication system between EV from a community that would charge and discharge their batteries among them based on the power load characteristics. The idea behind the V2H integration is that the battery would act as a source of power to help the home system in load hours and also as a storage in case of power excess (especially good when integrating renewable sources). V2G usually refers to the interexchange between the local EV community and the power grid using the management and control integrated in the smart grid (Fini et al., 2016; Alhelou et al., 2018; Zamani et al., 2018; Alhelou et al., 2015; Njenda et al., 2018; Haes Alhelou et al., 2018).

All these concepts usually require power sources, power transmission systems, power loads, communication systems, electric vehicles and many others. A schematic representation of the above mention elements and their interactions is presented in Figure 5.

Figure 5. V2G framework representation

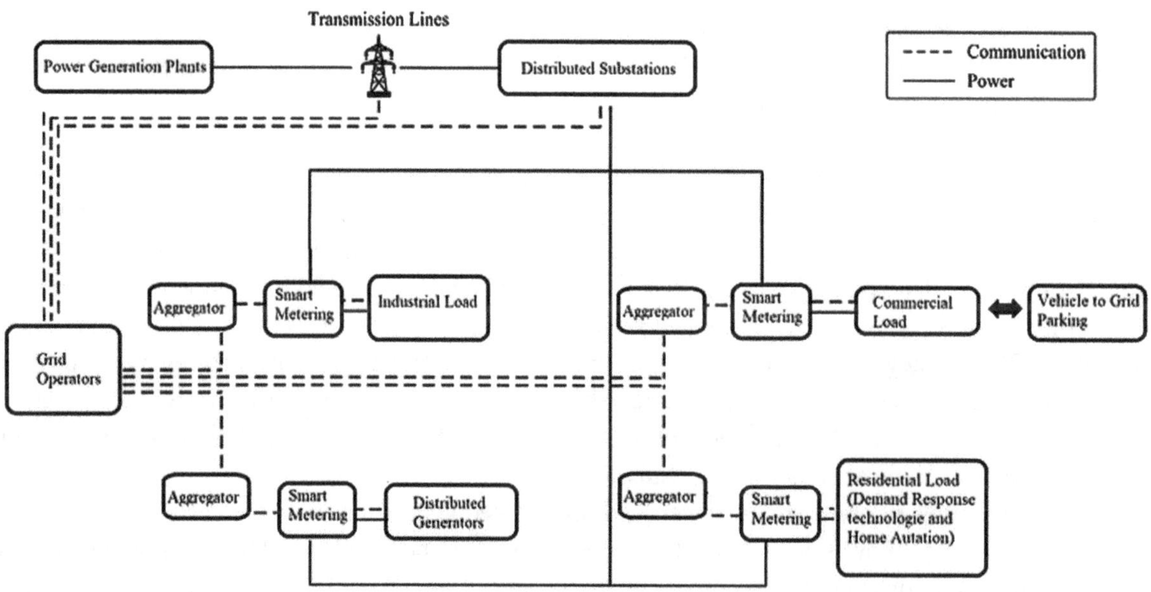

The concept was designed to allow the operator to control the power flow between the EV's battery and the power grid. One method presented by Bhatt J et. al (2014), describes the overall process that will result in a reduced emission gases, an improved grid power quality and in maximizing the profit.

There are two ways of implementing the V2G technology: unidirectional and bidirectional. Unidirectional refers to controlling the charging of the battery in a single power flow between the vehicle and the grid. This way of implementing V2G is the cheapest involving the addition of a controller. The benefits of this implementation will result in power grid regulation. Implementing it will require some trading agreement between the power utility and the EV's owners. These agreements will benefit the owners and the power utility that will be able to avoid overloading. Bidirectional V2G represents the dual power flow between the power utility and the EV. Using bidirectional V2G communication, additional functions can be implemented, such as reactive power regulation, peak load shaving, voltage and frequency regulation. The charging station that is capable of bidirectional flow typically consists of an AC/DC converter that rectifies the AC grid power to DC and also vice-versa and a DC/DC converter that using current control manages the power flow. Figure 6 presents a power flow diagram in a V2G configuration.

Active power support using V2G may be achieved by charging the EV's battery during the off pick hours and also by injecting power into the grid when is high demand, thus resulting in load leveling and pick load shavings. Reactive power regulation may also be achieved using the charger DC link capacitor and switching control. Another great benefit would be the power factor regulation that can reduce the losses in the power grid.

Figure 6. Power Flow Diagram in a V2G configuration

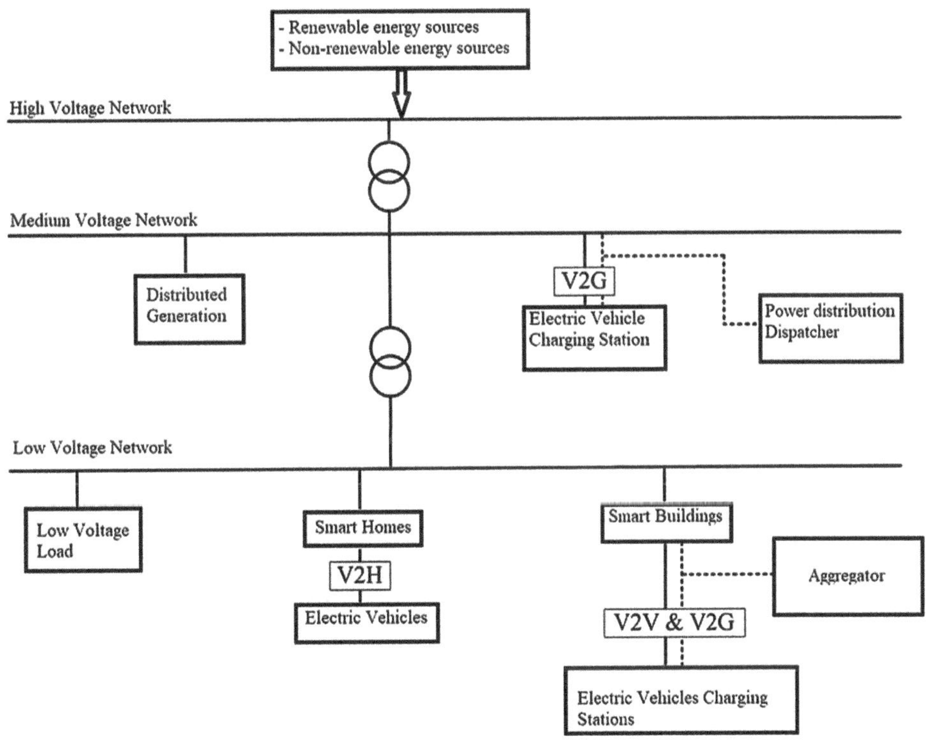

V2G technology works hand in hand with the integration of renewable energies into the grid. The inclusion of renewables into smart grids will develop a greener and sustainable power grid. Usually renewable energy sources adopted are photovoltaics and wind turbines. Renewable energy sources are highly fluctuating, depending on the weather conditions like wind or solar irradiation. This is their main barrier when integrating into the power grid. One study made by Makdisie, Haidar and Alhelou, presented some strategies for improving the renewables conversion system for a better grid integration. There are many research studies on how to prevent and better integrate the renewable sources and almost all of them require a storage system that is able to absorb excess power generated when the weather conditions are favorable and to provide power when the demand increases and the green sources are at low power production. However this storage devices involve a high initial cost. The electric vehicle's battery can perform the role of a buffering system that could work either as a storage or as an energy source depending on the demand, thus solving the unpredictable factor that the renewable energy sources would add to the power grid. EV could be seen as distributed power sources within the smart grid due to the usage of their batteries. Therefore integrating a large number of EV into V2G program will increase the percentage of renewable energy that could be accommodated by the power grid.

As presented, implementing the V2G concept brings a lot of benefits to the power grid, however one should also present the challenges of implementing the technology. First is the social barrier, EV owners would tend to keep the state of charge of their battery high for security reasons or for some unplanned long trips. These differences will keep them from enrolling in the proposed V2G program. For overcoming this issue the V2G network needs to guarantee a minimum level for the state of charge of the battery. Also a good charging network needs to be implemented in order to provide sufficient confidence. Another barrier to be overcome is the investments that need to be directed towards the charging stations. A typical V2G charging station requires some additional hardware and software equipment. Each participant to the V2G program will need a bidirectional charging station capable of switching between charging and discharging the battery. This process will add energy losses due to frequent AC/DC and DC/AC conversions. And another obstacle would be the EV's battery. Implementing the described technology requires high rates of charging and discharging the battery. This process leads to battery degradation faster than a normal use would. Sortomme and El-Sharkawi (2011) show that the process of charging and discharging the battery will increase the irreversible chemical reaction inside the battery which leads to higher internal resistance and reduce the capacity of the battery. In order to reduce these effects studies were made to determine in what conditions V2G will be less detrimental to implement. Dogger et al. (2011) present their results that state that the number of charging and discharging cycles reduces battery life. Also low temperatures and extreme values of state of charge are detrimental for the battery. Another battery parameter to be considered is the depth of discharge which is recommended to be less than 60% for maintaining an acceptable life cycle according to Millner (2011). As a conclusion, for the current technology, batteries should be kept between 30% and 90% state of charge. This is the range in which the V2G operator should consider keeping the batteries for prolonging their life cycle and performance.

The above described obstacles must be overcome for implementing the bidirectional V2G technology. Till now there are very few bidirectional V2G implemented cases, but with the new developments in battery technology the number projects will arise.

On the other hand the number of successful unidirectional V2G that are implemented in many countries is not small. This helps overcoming the social barrier and also encourages electric vehicle into market. Table 2 presents the discussed benefits and drawbacks of each V2G technology.

Table 2. Benefits and drawbacks of each V2G technology

	Unidirectional	Bidirectional
Hardware infrastructure	Communication system	Communication system Bidirectional battery charger
Cost	Low	High
Services	Grid regulation	Active power support Reactive power support Energy storage Filtering harmonics Regulation of frequency Stability Power factor regulation
Benefits	Grid overloading prevention Minimize emissions Profit optimization	Grid overloading prevention Minimize emissions Profit optimization Reduce grid loses Load profile improvement Prevent voltage fluctuation Enhances renewable grid integration
Drawbacks	Limited services	Social barriers Faster battery degradation High costs

DEMAND RESPONSE STRATEGIES

The most used optimization strategies are distributed algorithms, centralized algorithms and dispatch & aggregate algorithms. A general schematic representation of the above described demand response algorithms is presented in the below figure:

First, the distributed algorithms are the most common algorithms used in demand response applications. Their task consist of distributing power over the grid to the enrolled customers. Using this process, when trying to find a suitable solution it looks at all the grid actors using repetitive algorithms and questioning all parties. This algorithm is not mandatory one that can share information between all parties without the use of an operator. Also in most cases the use of a third party is used, usually for communication purposes but sometimes for coordination processes like managing the client's preferences and so on.

Figure 7. Most commonly used DR algorithms

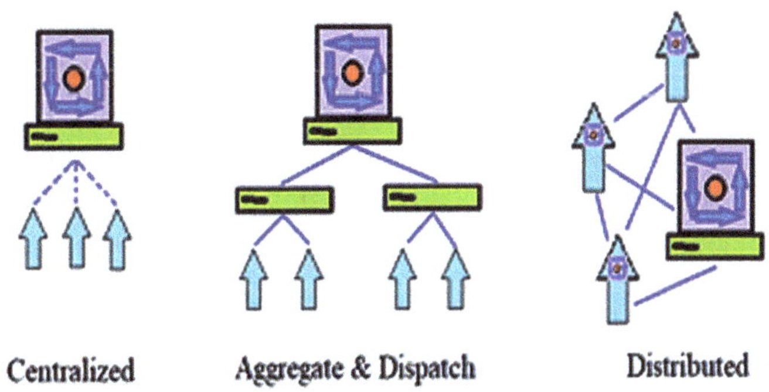

Describing the centralized algorithms one can say that they are the exact opposite of the distributed ones. Within this algorithm all the intelligent devices in the smart grid send information to the coordinator of the process. Usually this information is the one the clients send regarding their preferences. After it adds some information, like forecasts or other constrains, to the ones collected, it starts computing a solution that appease all the conditions. This computation is made just once not like the distributed algorithms that use iterative computing methods. This particularity makes the centralized algorithms the least scalable ones. Another drawback of this algorithm is the fact that the information from and towards a party of the grid may reach a bottleneck and issue errors. But centralized algorithm are often used by researchers who propose solutions in dealing with the described drawbacks.

The solution to the centralized algorithms drawbacks are the use of search techniques coupled with approximation techniques. It is known the fact that these algorithms provide a way of integrating a lot of information in the demand response timetable and then use mathematical computations to solve the optimization process. Making this steps is a foolproof method of taking into account all constrains and information. The solutions proposed using approximation and search techniques are suitable to large scale implementations where the time used for computations is high. Other optimum criteria methods were researched, but the approximation and search criteria is the easiest one to implement among them.

Among the two described algorithms one can introduce the aggregate & dispatch algorithms. These algorithms usually divide the process of optimization and the consequences of their outcome. The meaning of this action is that the information that is transmitted to the central aggregator is aggregated to the level of being of less complexity improving the scalability and the optimization process. Of course that this method also has some constrains and the result may not be the optimal one.

Some demand response algorithms were created by researchers using the standard centralized technique that was divided in order to be distributed to multiple clients in the demand response grid. If a technique as the dual decomposition is used than the work flow of the algorithm will be described as repeatedly trading information between the actors until a consensus is reached. The use of aggregate & dispatch method can be considered the one that is most suitable and with the best results concerning the clients preferences. Regarding the distributed algorithms, their drawbacks are the need to continuously exchange information until the solution is provided. Also the need of an intelligent communication system may be accounted as the infrastructure cost may rise more than in the case of the other two algorithms.

This division between the algorithms is related to the implementation of the algorithm into control strategy. The control architecture for the centralized algorithm will have only one coordinator that will gather all the data provided by the clients and also the grid constrains and this coordinator will be responsible for sending information to the grid devices. In the case of distributed algorithms the problem is simpler due to the fact that the devices may even communicate with each other. The three discussed demand response methods are presented in Figure 8 with their optimal level and expandability level:

The alternative of using the discussed aggregate & dispatch methods offer the possibility of using a model that best describes the grid. This grid may be updated with information from all the parties of the demand response cluster. The timetable and the optimization process are derived from using the model and afterwards the results are transmitted to all devices using preferred methods. Approximating all the grid information into a model and then use some methods of spreading the results also means that some details will clearly be neglected.

This will make the aggregate & dispatch algorithm not perfect with approximate solutions. The fact that makes them the method to choose is the suitability for large demand response clusters. With

Figure 8. Comparison between DR algorithms

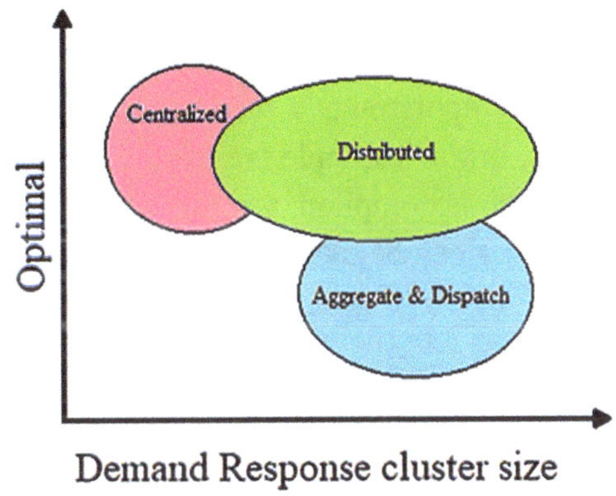

cheaper implementation costs than the other two. This alternative is appropriate for smart grids with a large number of devices where the demand of the algorithm is to satisfy most of the clients and all other constrains but keep the complexity as low as possible.

BALANCING RENEWABLE ENERGY SOURCES

New environmental regulation encourage the transition to green energy sources. The main reason for their likeability is that they have no carbon footprint nor greenhouse impact. National power grid supplier's concern is that these power sources are scattered along the grid. The distributed nature of the green power sources and the fact that their power output cannot be predicted, due to their dependency of the weather conditions, makes them a risk factor for the classical grid. Also, injecting power to grid from distributed points along the lines rises the risk of higher voltage levels than the admissible values. The method that we propose is to use the electric vehicles battery to balance their power generation, or better said to shift this power in time.

In the following paragraphs we will consider the balancing of a renewable energy source using the demand response algorithms and making use of the electric vehicle battery. The proposed solution implies integrating the renewable generating source with electric vehicle charging process. The solution that we present derived from the development of a distributed algorithm. The purpose of the algorithm is to absorb the green generated power, thus increasing the demand, and reducing the environmental harmful emissions. The key actor in this strategy is the use of an third party, named responsible party that will be in charge of matching the power consumption with the power generated during a fixed period of time. This position requires a lot of responsibility, hence if the balance is not kept than the responsible party will pay the generated cost. This algorithm will respect the user's privacy, hence private information will not be revealed. Also the customer's preferences, like departure time, arrival time, consumption limits and so on, need to be accounted.

Firstly we will use just one coordinator for this example; later on this algorithm can be developed and enlarged to the required number of customers and coordinators. In this scenario the customers are defined as the electric vehicle charging places. The coordinator is responsible for the intercommunication of all parties and the responsible party is provided with the weather forecast and the proposed consumption profile. In the presented form there is place also for the inclusion of aggregator that could be placed between the coordinator and the responsible party, but it will not be added in this scheme because we want to keep it as simple as possible and after implementation and validation it will be easy to extend.

The algorithm presented below is categorized as a receding horizon algorithm. For implementation first we need to define a time period. In this example we used a t minutes time period. As such every time period, noted t, a control algorithm is compiled, and this algorithm will consider the next M time periods. The proposed algorithm is described below with presenting the steps in their compilation order following the time periods defined:

- Initialization of the variables:
  - The electric vehicles charging spots are defined.
  - Responsible party receives the wind power source forecast for the upcoming M time periods.
  - The coordinator develops a virtual cost vector c that will be used to initialize demand and also supply.
- Coordinator delivers the computed cost vector c to the customers and also to the responsible party.
- Every customer computes their own power consumption timetable based on the provided c vector and also their own preferences, and delivers that information back to the coordinator for processing.
- Responsible party computes the power production timetable based on the weather forecast and also using the provided c vector. The resulted timetable is send back to the coordinator for processing.
- After receiving the information from both the customers and the responsible party, the coordinator starts the process of comparing them as follows:
  - If the difference between the power generated and the customers demand is lower than a predefined possible level, or the high limit number of iterations is attained, the algorithm finishes and the customers and the responsible party are alerted that the timetables are final.
  - Else, the coordinator modifies the cost vector c and iterates from the start.

The logic presented above can easily be implemented mathematically. If the logic is split into subprograms then it would be easier to implement in multiple parties inclusion. This strategy may be implemented using other types of control mechanisms like electric vehicle hierarchical control proposed by Alhelou and Golshan (2016), making the algorithm more efficient and economical. Some researchers implemented similar logics using dual decomposition. This way the problem becomes easier for multiple sub-grids. This developed distributed algorithm is suited to take on the challenge of successfully integrating the renewable energy sources into the power grid. Of course all the presented technologies are eligible to implement into a smart grid, one that has all the intelligent infrastructure capable of supporting all the actions taken into the solution.

FUTURE RESEARCH DIRECTIONS

For a better management process of electric vehicle grid integration some new ideas were stated. Among those proposals that still need further research is the development of direct current grids or microgrids. A distributed DC grid, theoretically, fits better with electric vehicles and renewable energy sources. Eliminating the need to rectify the current also removes some of the energy losses. This topic still needs research as the idea is bold and not cheap to implement.

Another area that may bring a lot of benefits if properly researched is the implementation of distributed algorithms for a better energy distribution. A well designed algorithm is a good tradeoff among a large number of users and attaining the balancing objective of the power grid. But in order to push these algorithms from the theoretical development to the real world applications there is still the need of practical validation. Also further modeling and simulation tools that incorporate smart grid technologies is required.

CONCLUSION

Electric vehicles use electrical energy to power up their depleted battery. To do that they use specific charging infrastructures or a household electrical outlet. To integrate many electric vehicles into the power grid is difficult due to overloading. One electric vehicle consumes into one year almost the same energy as one household. This sudden energy demand is to much for the power grid. In order to successfully integrate them some innovative technology were developed. As presented above technology like demand response, vehicle to grid, home automaton network, smart metering, developed a new power grid concept: smart grid. Using the new researched technology, a smart grid is capable of integrating the desired number of electric vehicles. The smart grid concept requires some additional equipment like smart meters, smart chargers and some control equipment. Using bidirectional communication it can gather information in real time and also devise a control strategy that best fit the power demand. Additional benefits may be achieved if a vehicle to grid concept is applied. As earlier discussed this technology, if properly used, can achieve load peak flattening, active and reactive power support and also a better renewable energy integration.

This paper reviews the challenges and also the framework of electric vehicle grid integration. First the issue of the power grid overloading is presented. The new technologies that were developed for overcoming this issue are presented next. In the last part of the chapter some new ideas and future research directions are described.

All the above presented technologies aim to accelerate the ability of the communities to meet the emission regulations and optimize the smart grid applications. Another benefit to be accounted is the innovation of infrastructure an services that these changes will bring. Altogether implementing these new technologies and developing the smart grid will greatly increase productivity, generate economical boost ultimately offering tremendous benefits to citizens.

ACKNOWLEDGMENT

This research was supported by the Romanian Ministry of Research and Innovation, CCCDI – UEFISCDI [PN-III-P1-1.2-PCCDI-2017-0404 / 31PCCDI/2018].

REFERENCES

Alhelou, H., Hamedani-Golshan, M. E., Zamani, R., Heydarian-Forushani, E., & Siano, P. (2018). Challenges and Opportunities of Load Frequency Control in Conventional, Modern and Future Smart Power Systems: A Comprehensive Review. *Energies*, *11*(10), 2497. doi:10.3390/en11102497

Alhelou, H. H. (2018). *Fault Detection and Isolation in Power Systems Using Unknown Input Observer. In Advanced Condition Monitoring and Fault Diagnosis of Electric Machines* (p. 38). Hershey, PA: IGI Global.

Alhelou, H. H., Golshan, M., & Fini, M. (2018). Wind Driven Optimization Algorithm Application to Load Frequency Control in Interconnected Power Systems Considering GRC and GDB Nonlinearities. *Electric Power Components and Syst.*

Alhelou, H. H., & Golshan, M. E. H. (2016, May). Hierarchical plug-in EV control based on primary frequency response in interconnected smart grid. In *Electrical Engineering (ICEE), 2016 24th Iranian Conference on* (pp. 561-566). IEEE. 10.1109/IranianCEE.2016.7585585

Alhelou, H. H., Golshan, M. H., & Askari-Marnani, J. (2018). Robust sensor fault detection and isolation scheme for interconnected smart power systems in presence of RER and EVs using unknown input observer. *International Journal of Electrical Power & Energy Systems*, *99*, 682–694. doi:10.1016/j.ijepes.2018.02.013

Alhelou, H. H., Hamedani-Golshan, M. E., Heydarian-Forushani, E., Al-Sumaiti, A. S., & Siano, P. (2018, September). Decentralized Fractional Order Control Scheme for LFC of Deregulated Nonlinear Power Systems in Presence of EVs and RER. In *2018 International Conference on Smart Energy Systems and Technologies (SEST)* (pp. 1-6). IEEE. 10.1109/SEST.2018.8495858

Alhelou, H. S. H., Golshan, M. E. H., & Fini, M. H. (2015, December). Multi agent electric vehicle control based primary frequency support for future smart micro-grid. In *Smart Grid Conference (SGC)* (pp. 22-27). Academic Press.

Bessa, R. J., & Matos, M. A. (2012). Economic and technical management of an aggregation agent for electric vehicles: A literature survey. *European Transactions on Electrical Power*, *22*(3), 334–350. doi:10.1002/etep.565

Bhatt, J., Shah, V., & Jani, O. (2014). An instrumentation engineer's review on smart grid: Critical applications and parameters. *Renewable & Sustainable Energy Reviews*, *40*, 1217–1239. doi:10.1016/j.rser.2014.07.187

Dogger, J. D., Roossien, B., & Nieuwenhout, F. D. J. (2011). Characterization of Li-ion batteries for intelligent management of distributed grid-connected storage. *IEEE Transactions on Energy Conversion*, *26*(1), 256–263. doi:10.1109/TEC.2009.2032579

Fini, M. H., Yousefi, G. R., & Alhelou, H. H. (2016). Comparative study on the performance of many-objective and single-objective optimisation algorithms in tuning load frequency controllers of multi-area power systems. *IET Generation, Transmission & Distribution, 10*(12), 2915–2923. doi:10.1049/iet-gtd.2015.1334

Global Smart Grid Impact Report. (n.d.). Retrieved fromhttps://www.smartgrid.gov/files/global_smart_grid_impact_report_2013.pdf

International Energy Agency's (IEA) Global EV Outlook. (GEVO). (2018). Retrieved from https://www.iea.org/gevo2018/

Krein, P. T. (2007). Battery management for maximum performance in plug-in electric and hybrid vehicles. *Proceedings of the IEEE VPPC 2007, Vehicle Power and Propulsion Conference*, 2–5. 10.1109/VPPC.2007.4544086

Letendre, S., & Watts, R. A. (2009). *Effects of Plug-in Hybrid Electric Vehicles on the Vermont Electric Transmission System.* Transportation Research Board Annual Meeting, Washington, DC.

MacKay, D. J. C. (2009). *Sustainable Energy: Without the Hot Air.* Cambridge, UK: UIT.

Makdisie, C., Haidar, B., & Alhelou, H. H. (2018). An Optimal Photovoltaic Conversion System for Future Smart Grids. In *Handbook of Research on Power and Energy System Optimization* (pp. 601–657). IGI Global. doi:10.4018/978-1-5225-3935-3.ch018

Millner, A. (2010). Modeling Lithium Ion battery degradationin electric vehicles. *Proceedings of the IEEE CITRES 2010 Conference on Innovative Technologies for an efficient and reliable electricity supply*, 349 – 356. 10.1109/CITRES.2010.5619782

Nadweh, S., Hayek, G., Atieh, B., & Haes Alhelou, H. (2018). Using Four – Quadrant Chopper with Variable Speed Drive System Dc-Link to Improve the Quality of Supplied Power for Industrial Facilities. Majlesi Journal of Electrical Engineering.

Njenda, T. C., Golshan, M. E. H., & Alhelou, H. H. (2018, November). WAMS Based Intelligent Under Frequency Load Shedding Considering Online Disturbance Estimation. In *Smart Grid Conference (SGC)* (pp. 1-5). Academic Press.

Njenda, T. C., Golshan, M. E. H., & Alhelou, H. H. (2018, November). WAMS based Under Frequency Load Shedding Considering Minimum Frequency Prediction and Extrapolated Disturbance Magnitude. In *Smart Grid Conference (SGC)* (pp. 1-5). Academic Press.

Pecas, L. J. A., Soares, F. J., & Almeida, P. M. R. (2011). Integration of electric vehicles in the electric power system. *Proceedings of the IEEE, 99*(1), 168–183. doi:10.1109/JPROC.2010.2066250

Peças Lopes, J. A., Soares, F. J., & Rocha Almeida, P. M. (2009). Identifying management procedures to deal with connection of electric vehicles in the grid. *Proceedings of the IEEE PowerTech.* 10.1109/PTC.2009.5282155

Peterson, S. B., Apt, J., & Whitacre, J. F. (2010). Lithium-ion battery cell degradation resulting from realistic vehicle and vehicle-to-grid utilization. *Journal of Power Sources, 195*(8), 2385–2392. doi:10.1016/j.jpowsour.2009.10.010

Pillai, J.R., & Bak-Jensen, B. (2011). Integration of vehicle to grid in the western Danish power system. *IEEE Trans Sustain Energy, 2*(1), 12–9.

Short, W., & Denholm, P. (2006). *A Preliminary Assessment of Plug-in Hybrid Electric Vehicles on Wind Energy Markets*. National Renewable Energy Laboratory, Golden, CO. Tech. Rep. NREL/TP-620-39729.

Sortomme, E., & El-Sharkawi, M. A. (2011). Optimal Charging Strategies for Unidirectional Vehicle-to-Grid. *IEEE Transactions on Smart Grid, 2*(1), 131–138. doi:10.1109/TSG.2010.2090910

U.S.Energy Information Administration (EIA). (2013). International Energy Outlook, Report No.: DOE/EIA-0484(2013). Washington, DC: Office of Energy Analysis, U.S. Department of Energy.

Zamani, R., Hamedani-Golshan, M. E., Haes Alhelou, H., Siano, P., & Pota, H. (2018). Islanding Detection of Synchronous Distributed Generator Based on the Active and Reactive Power Control Loops. *Energies, 11*(10), 2819. doi:10.3390/en11102819

ADDITIONAL READING

Bradley, T. H., & Frank, A. A. (2009). Design, demonstrations and sustainability impact assessments for plug-in hybrid electric vehicles. *Renewable & Sustainable Energy Reviews, 13*(1), 115–128. doi:10.1016/j.rser.2007.05.003

Dallinger, D., & Wietschel, M. *Grid integration of intermittent renewable energy sources using price-responsive plug-in electric vehicles.* Renew Sustain Energy Rev 2012;16(5):3370e82. https://econpapers.repec.org/article/eeerensus/v_3a16_3ay_3a2012_3ai_3a5_3ap_3a3370-3382.htm

Galus, M. D., & Andersson, G. 2012b. Balancing Renewable Energy Sources using Vehicle to Grid Services controlled by MPC in a Metropolitan Area Distribution Network. In: Cigre, Electra (06).

Green, R. C. II, Wang, L., & Alam, M. (2011). The impact of plug-in hybrid electric vehicles on distribution networks: A review and outlook. *Renewable & Sustainable Energy Reviews, 15*(1), 544–553. doi:10.1016/j.rser.2010.08.015

Ortega-Vazquez, M. A., Bouffard, F., & Silva, V. (2013). Electric vehicle aggregator/system operator coordination for charging scheduling and services procurement. *IEEE Transactions on Power Systems, 28*(2), 1806–1815. doi:10.1109/TPWRS.2012.2221750

Richardson DB. *Electric vehicles and the electric grid: a review of modeling approaches, Impacts, and renewable energy integration.* Renew Sustain Energy Rev 2013;19:247e54. .doi:10.1016/j.rser.2012.11.042

Su, W. *Smart grid operations integrated with plug-in electric vehicles and renewable energy resources* [Ph.D. dissertation]. North Carolina: Department of Electrical and Computer Engineering, North Carolina State University; 2013.

Tarroja B, Shaffer B, Samuelsen S. *The importance of grid integration for achievable greenhouse gas emissions reductions from alternative vehicle technologies*. Energy 2015;87:504e19. . doi:10.1016/j.energy.2015.05.012

Weiss J, Hledik R, Lueken R, Lee T, Gorman W. *The electrification accelerator: understanding the implications of autonomous vehicles for electric utilities*. Electr J 30(10) 2017:50e7. . doi:10.1016/j.tej.2017.11.009

Yilmaz M, Krein PT. *Review of the impact of vehicle-to-grid technologies on distribution systems and utility interfaces*. IEEE Trans Power Electron 2013;28(12):5673e89. . doi:10.1109/TPEL.2012.2227500

KEY TERMS AND DEFINITIONS

Aggregator: In this chapter, the term refers to demand response aggregator and is a commercial entity that provides demand response services such as the ones described in this chapter to customers with strategies or technology to reduce their electric consumption and then making use of the electric load reductions in wholesale energy markets.

Bidirectional Charger: The term refers to the electric vehicle smart charging equipment, able to communicate bidirectional with the aggregator, usually found in smart grid applications with vehicle to grid technology.

Plug-in Hybrid Electric Vehicle: Hybrid electric vehicles that has the option to charge its battery from the designated charging stations or directly from an electric outlet.

Power Grid: An interconnected network designed for transporting electric power from the generation plants to the customer.

Renewable Energy: Represents the energy produces from a energy source that is naturally replenished after use. Some examples are: sunlight, wind, rain, waves, tides, geothermal, etc.

Smart Charging: A term used to refer to the intelligent process of electric vehicle charging activity taking into account the power grid information. Electrical vehicle charging process is scheduled based on the clients demands and also on the communication with the responsible aggregator.

Smart Grid: Power grid with smart equipment used for bidirectional communication and control. It refers to the electric power network of the future; the one that will be equipped with intelligent devices and automated control strategies.

Well to Wheels: Represents a through emission analyze that takes into account not only the tail pipe emissions but also the emissions generated to produce the energy used to charge its vehicle battery.

Chapter 10
Issues Associated With Microgrid Integration

Baseem Khan
https://orcid.org/0000-0002-0562-0933
Hawassa University, Ethiopia

Sudeep Tanwar
Nirma University, India

ABSTRACT

Microgrid (MG) is the vital technology that can be utilized to supply electricity to rural areas by fulfilling various aspects of electricity such as sustainability and reliability. Further, MG technology can also be used as localized generation sources and back up supply source. As MG can be worked in interconnected mode, various issues related to interconnection with utility grid are raised. Several issues such as technical, regulatory, and operational are associated with grid integration. Therefore, this chapter deals with the issues that are associated with the grid integration of microgrid.

INTRODUCTION

Traditional electricity grid is converted into the smart structure. The key feature of this smart system is the incorporation of the renewable energy sources at different levels such as distributed level and bulk level. International energy agency predicted that the energy generation from the renewable energy sources is increased up to three times till 2035 (Khan and Singh, 2017; Mulualem and Khan 2017). Further, the total energy production from the renewable energy sources will be increased to 31%, in which hydro, wind and solar will provide 50%, 25% and 7.5%, respectively. The two major issues with renewable energy generation are intermittency and climate dependency of renewable sources. These problems make integration of these sources with conventional grid more difficult and complex. The above discussed problems can be minimized with the help of energy storage devices. These devices incorporated various storage systems such as batteries, heat buffers along with advanced generation techniques such as

DOI: 10.4018/978-1-5225-8030-0.ch010

fuel cell technology, electric vehicle technology etc. Therefore, there was a necessity to develop such a system, which incorporates different renewable energy sources with energy storage options to mitigate the issues related with renewable energy sources. This necessity is fulfilled with the development of Microgrid system (Khan and Singh, 2017; Fanuel et al. 2018). It is the combination of different type of loads (domestic, commercial, industrial) with various renewable energy sources such as solar photo voltaic, wind, micro turbine and small hydro along with energy storage devices such as battery energy storage, heat buffer, flywheel storage, and electric vehicle technology system.. In smart grid structure, micro grid technology provides a holistic approach for the integration of renewable energy sources. It has several benefits over the conventional grid system as it's minimise energy losses, improve reliability and enhance energy management. Further, at distribution level, micro grid technology provided better solution of energy scarcity, generation coordination and control problems due to its better performance with respect to distributed generation technology (Khan and Singh, 2017; Kifle et al. 2018).

Microgrid is not designed to handle the large power being fed by the utility distribution feeders. Further, the characteristics of micro grid components possess big challenges. The issues related to the integration of microgrid raises the challenges to operation and control of main utility grid. Out of various interfacing issues, load frequency control is also one of the important issue. It can be treated as single objective or multi objective load frequency control problem. A comparative analysis of single and multi objective load frequency controllers is presented by Fini et al. (2016). Haes et al. (2015) presented a multi agent primary frequency supporting controller which is based on electric vehicle control. This controller is very useful for future smart micro grid. Alhelou and Golshan (2016) presented a controlling scheme for plug-in electric vehicle to control primary frequency response in interconnected smart grid. Alhelou et al. (2018) presented a comprehensive review on challenges and opportunities for load frequency control in traditional, modern and future smart systems. Other than load frequency control, fault detection and isolation of faulted section is also very important. Alhelou (2018) presented a fault detection and isolation overview in power systems by using unknown input observer. Further, Alhelou et al. (2018) proposed a robust sensor based outage detection and isolation technique by using unknown input observer. This technique is utilized for renewable energy sources and electric vehicle integrated smart power system. Lastly, Makdisie et al. (2018) discussed the photovoltaic conversion system in an optimal way for futuristic smart grid systems. Therefore, this chapter deals with the various micro grid integration issues face by the utilities in the practical power system.

MICROGRID STRUCTURE

It is a distribution network which supply through low and medium voltages distribution lines. Various self sufficient and independent distributed energy sources i.e. PV, wind, Fuel cell, micro hydro etc. and storage devices such as battery storage, flywheel storage etc. along with demands are incorporated and grouped insides micro grid structure. Figure 1 presented a typical overview of micro grid structure. Different distributed energy sources are integrated with in micro grids by its corresponding bus bars equipped with power electronics converter. Point of common coupling (PCC) is the point where micro grid is connected to the upstream network.

There are two modes in which micro grid operate. The first one is the grid connected mode and another one is the stand alone mode or islanded mode. In grid interfaced mode of operation PCC is closed

Figure 1. Micro grid system
(Khan and Singh, 2017)

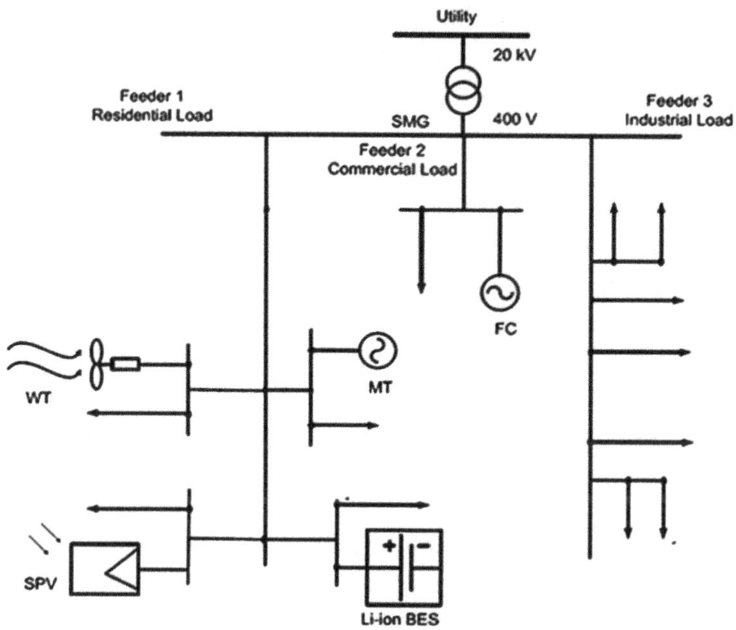

and micro grid is linked with utility grid. Whenever there is any disturbance in utility grid or micro grid, PCC is opened and a micro grid is disconnected to the main grid, than micro grid operated in standalone mode (Singh and Khan, 2017).

The renewable energy production is further classified into dispatchable and non-dispatchable production. Dispatchable production is able to change their power production upon demand and by the request of grid operators. They are micro hydro and mega hydropower; Ocean/marine current power and wave power, geothermal and ocean thermal energy conversion, Biofuel biomass etc. (Khattam and Salama, 2004). Non-dispatchable renewable energy based generators are wind energy and photo voltaic, because wind turbine output depends on the wind speed and solar power available by the radiant light and heat of the sun (Khan and Singh, 2017).

INTEGRATION OF MICROGRID TO THE MAIN GRID

Most of the small scale DG sources in the load side are integrated at medium or low voltage network as low penetration fashion where they are connected as passive systems and they are not involving grid voltage controlling, frequency controlling and stability activities. Still in the case of high penetration, the interfaces can be modified to work as active generators so that DER can participate in the frequency, voltage and system stability control activities of the grid. Power electronic is used to interfaces between the grid and the renewable power source of microgrid so that there are not any negative influences in reliability, stability and power quality of the supply after the interconnection DERs to the grid. Numerous components and constraints are involved in the integration of DER to the utility grid (Huang et. al.

2008). The integration of varying intermittent renewable sources like Solar and Wind Energy Conversion systems to the grid can provide a technical relief in the form of reduced losses, reduced network flows and voltage drops. However, there are also several undesirable impacts due to high penetration of these variable DERs which include voltage swell, voltage fluctuations, reverse power flow, changes in power factor, injection of unwanted harmonics, frequency regulation issues, fault currents and grounding issues and unintentional islanding (Huang et. al. 2008; Alhelou et al., 2018; Njenda et al., 2018; Zamani et al., 2018). Advanced protection system should be included in the DG units to disconnect the units in case of fault or unfavorable grid conditions.

Grid integration of distributed renewable sources are classified depending on the resource availability, load demand, and existing electrical power system, into three categories namely low penetration with existing grid, high penetration with existing grid and high penetration with future smart-micro-grid configuration.

Low Penetration With Existing Grid

In low penetrated networks, the distributed generators units are not involving in frequency control activities and voltage control activities of the PCC point. Grid operator are responsible for managing the overall system stability and DG operators can send the maximum available power to main grid and local loads without major consideration of grid constraints.

The DG operators have to deliver the power based by grid synchronization via PLL systems with correct phase sequence. Whenever grid frequency is exceeded the allowable limit the inverters are required to disconnect from the grid. And it operates in power factor (PF) correction mode, where PF keep closer to unity. Most of PV units and Wind generators can inject the maximum available active power into the grid; most existing VSC are operating in power factor correction mode (zero reactive power).

The network operators face real problem when DG sources are connected to low voltage lines since microgrids have dispersed generation units; sizes of the DGs are very small and low inertia characteristic, especially frequency deviations. The amount of DG units connecting to particular distribution network is limited by the voltage control margins of that distribution network; to overcome these challenges static synchronous compensator (STATCOM), voltage source converter (VSC), Automatic tap control transformers and special control mechanism are used by operators to control the network voltage.

High Penetration With Existing Grid

When growing the renewable energy source penetration cause complication in the system constraints due to the intermittency of RES; that the percentage of the renewable power injected into the existing grid is relatively high as compared to the power assigned to the conventional power plant. Therefore, in such type of situation intermittent power sources cannot work as passive generators, but they have to actively participate in grid frequency and voltage control activities. In addition to grid synchronization with phase sequence matching and protection system, controls and inverters should be more intelligent.

The grid operator cannot transfer the energy to or from main grid in the case of islanded power systems with a significant penetration of RES power, so the isolated system has to deal with intermittency issue. Since the amount power delivered is considerably effect to the grid stability, Phase balanced operations and proper VSC inverters connection strategies have to implement in the system. Voltage control loop can be included in VSC inverters to provide the required reactive power to the grid, in this way VSC will intelligently response to the grid conditions. On the other hand; inverter have to operate within defined power factor range not unit power factor so that VSC will have the capability to control the grid voltage at the PCC point.

High Penetration With Smart Grid Concepts

The combination of different renewable energy generation resources (such as micro hydropower, photovoltaic arrays, geothermal, wind-turbine generators) in a microgrid can be integrating to the grid and increase the penetration of renewable energies to change the whole system into a smart grid with advanced technologies. Upcoming smart grid networks will provide a real-time, multi-directional flow of energy and information. Smart intelligent equipment's with modern digital controls are used in entire electricity grid from central control office to end customer levels (Huang et. al. 2008; Alshahrestani et al., 2018; Makdisie et al., 2018; Alhelou et al., 2018; Alhelou et al., 2016; Nadweh et al., 2018; Njenda et al., 2018).

However, maintaining the stability and reliability of the network become a problem when the contribution from DGs is maximizing, then solution may be using smart grid concepts such as micro grids, large scale energy storage with advanced energy management system systems, smart homes with demand response management etc. This will help to better communication and coordination between all the participants in the electricity business such as power plant operators, network operators, the end-consumers and government.

MICRO GRID INTEGRATION ISSUES

Regulatory and Legal Issues

Mainly, two major legal issues impacted the micro grids. These are as follows:

1. Whether these micro grids are deemed to be power distribution companies so that oversight by regulatory authorities.
2. Whether Micro grids are feasible to work under legal governing frameworks of sale and procure energy and generate and distribute electricity as the state regulations about utilities are not applicable on Micro grids,

For making micro grid projects bankable, a clear legal framework with regulatory certainty is must be required. In the absence of this, the investment cost will be very high and gains will also uncertain and this will not justify the time and capital require for such projects (CEMP, 2010). There are various contexts such as legal, regulatory, electricity generation and distribution on which the Micro grid projects can be evaluated. Studies of these contexts are good initial point for the development of future distribution system integrated with micro grid (Kema, 2014; Burr et al., 2013).

Policies of Interconnection

The main issue which has a legal uncertainty is the rules and regulations related to the interconnection of micro grid with the utility grid. In 1990s with the application of deregulation, independent power producers (IPPs) come into the picture. But, there was no requirement to connect these IPPs equipments with the main utility grid (Khan et al., 2014, 2016). For fulfilling the requirements of different utilities after the deregulation, manufacturers and developers developed their plan by supplying considerable amount of capital and time for the advancement of micro grid technology. The publication of the IEEE 1547 standards in the year 2003 is the important step in the direction of integration of distributed generator up to the capacity of 10MVA. These standards provide the set of rules for interfacing of distributed generators to the utility grid in a secure manner (Basso, 2014). The major spotlight of all the interconnection policies together with IEEE 1547 is on the isolation of distributed generation sources from the grid in the case of main grid failure to secure safety of operators and workers. In the year 2011, IEEE approved standard 1547.4, which standardized the protocols for secure intended islanding and re-closure of micro grid technology. Further, it incorporated regulations for modelling, operation, planning and interfacing of renewable energy sources with the main utility grid. The following guidelines are also included by the IEEE 1547.4 standard.

1. Micro grids functionalities in islanded as well as grid connected mode
2. Change over to intended islanded operation mode
3. islanded mode of operation
4. Re-closure to the utility by providing accurate frequency, voltage and phase angle.

Moreover, it also provides regulations for monitoring, communications, power quality, control, protection and safety.

There is one more standard namely California's rule 21, which provides various standards for interfacing of renewable energy generation to the utility grid. It addressed various interfacing requirements for removing hurdles that is placed by utility suppliers. It established a clear process of review, certification and testing practices, fees, standardized technology and efficient process of application for eradicating the above discussed hurdles. Interconnection of micro grid is of great importance for achieving the advantages of grid services and revenues, otherwise micro grid will work in islanded mode and loss the benefits.

Regulations for Utility

Micro grid can be considered as electric utility if the following functions are performed by it.

1. Intended to supply multiple retail customers
2. Cross a public path with electricity supply lines
3. Obtained a franchise from state authority

The above summary depends on the following detailed discussion. If the state authority finds that the facilities offered by micro grid comes under the services of utilities, than state authority can be standardized the rates for selling electricity and make a decision on approval of facility construction. It has the great implications for micro grid developers. Micro grid utility may presume a commitment to supply potential consumer on the basis of written or oral demand, if micro grid is deemed to be a distribution utility. Permission from local municipal authority is required by micro grid utility to supply electricity to customers through public paths (CEMP, 2010). The type of this permission can be franchise or other lesser consent. The ability of micro grid utility to get this permission largely depends on the pre existing utility, whether it has provided exclusive franchise or efficiently blocking the opponents. Because of limited scope of facilities and small size structure, most of the time micro grid utility wouldn't need franchise and thus micro grid utility wouldn't be governed by the existing utility regulatory agency. Though this condition is varying according to the project and is being decided by courts. Further, different consumers protection regulations also comply by the micro grid utility for selling electricity to retail consumers. At last, if micro grids are generating electricity through combustion of fuel such as diesel generator or micro turbines than the central and state laws of emissions regulate the generation of micro grid. Further in some cases permit is also required. The selection of business model also affects the degree by which utility franchise come in to operation.

Oppositions From Utilities

Through utility connected micro grid customers will still connected to the grid for more reliable supply. Therefore, self utilization of the energy by the micro grid could minimize the revenue of micro grid utility. Further, a huge unwillingness is made by the various utility for the large integration of distributed energy sources due to issues related to safety, protection and management. Therefore extra charges are applied on distributed generators owners by the utilities for halting the program of net metering. Electricity market deregulation is required for changing the situation in which micro grids are considered as threats to the valued energy sources those are fairly compensated (Khan et al. 2014 and 2013, Khan and Agnihotri, 2012). All the three sectors i.e. generation, transmission and distribution are fully unbundled into different utility services and further, IPP are allowed to supply and compete in whole sale and retail electricity market. For managing distributed energy sources and facilitating grid services such as frequency regulation, congestion management and black start, micro grid utility requires fair economic signals, which can be possible with the application of real time or time of use pricing. With the help of this micro grid utilities can enhance their revenues.

MICRO GRID CONTROL

Generally, the control system must place at different level of the system and a consistent communication between several control unit is required since there is a continuous change of power production in the DGs and the load demand in fluctuation with time. MG Central Controller (MGCC) installed at the Medium/Low Voltage substation, which has a supervisory task of centrally control and manage the MG and integrates with the main grid. The MGCC includes several key functions, such as economically managing functions and control functionalities and is the head of the hierarchical control systems; communicates between network operators.

The MG is intended to operate in the following two different operating conditions: the normal interconnected mode with a distribution network and the emergency mode in islanding operation via a central switch, which must also implement the synchronization between both power systems.

Figure 2. The microgrid control architecture
(Kifle et al. 2018)

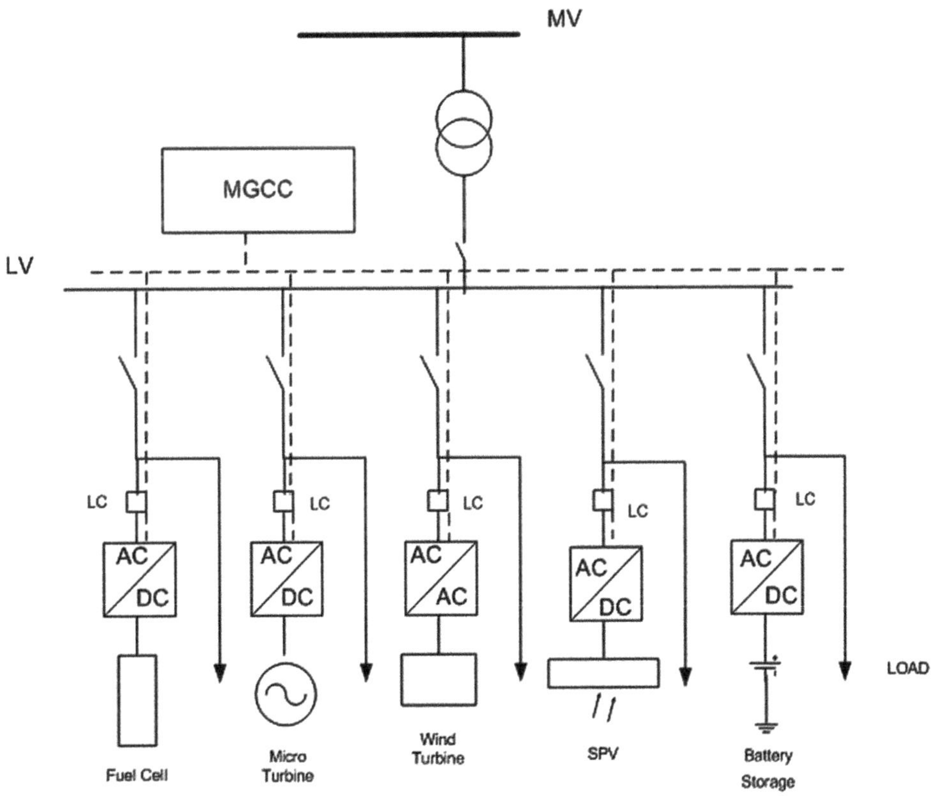

The typical single line structure of a microgrid control system as described in Figure 2. It is clear that a direct connection of the microgrid LV line to distributed renewable energy sources (PV, Wind generator, micro turbine) and to the electrical grid network is not possible so power electronic interfaces (DC/AC or AC/DC/AC) are required due to the characteristics of the energy produced. Inverter control and circuit protection is thus an important concern in MG operation.

In the microgrid control system there are main parts including: Micro Source Controllers (MC) on the counsumer production side and Load Controllers (LC) on the consumer demand side; Microgrid System Central Controller (MGCC) on the middle of the main grid and microgrid structures and Distribution Management System (DMS) in the grid network side.

The different DGs sources and energy storage devices are connected to the low feeder lines through the micro source controllers (MCs). MC has a function of controlling the power flow and bus voltage profile of the microsources according to the load changes or any other disturbances. These feeders are also supplied with several sectionalizing circuit breakers (SCBs) which help in isolating a part of the microgrid as needed in case trouble. Power electronic interfaces and inverters (AC/DC,AC/AC DC/AC) are important mean for controlling and monitoring the loads using load controllers (LC).

The overall operation and management in both the modes (isolated and grid-tied) is controlled and coordinated with the help of microsources controllers (MCs) at the local level and Microgrid System central controller (MGCCs) at the global level; there is a point of common coupling (PCC) through the circuit breakers (CB) between the microgrid and the medium voltage level utility grid. The MGCC has list of responsible for the overall control of microgrid operation and protection; like maintaining specified bus voltages and frequency of the entire microgrid; energy optimization for the microgrid. On the utility side there is a Distribution Management Systems (DMS) having several feeders including several Microgrids; function for distribution areas management and control.

So there is two parts Control tasks: first one is Microgrid-Side Controller (MC & LC) to take the maximum power from the input source and the protection of input side converter must be in considered. Second part Grid-Side Controller (MGCC & DMS) having the following main tasks: a) Input active power control derived for network; b) Control of the reactive power transferred between network and micro-grid; c) DC link voltage control; d) Synchronization of network; e) Assurance of power quality injected to the network (Passey et al. 2011).

CONCLUSION

The main hurdle for the expansion of MG technology is the different issues which are associated with the integration of this technology. The integration of microgrids with RES in the current utility grids is the first step towards the transition from the conventional power system to smart grid system. Most of the existing power system overall cost is also becoming expensive in the near future; RES technological improvement; advancement in energy storage systems can help the new microgrid system based on DG to become economically viable to consumer. More penetration of RESs is expected in microgrid systems as they are almost pollution-free and thus environment friendly.

ACKNOWLEDGMENT

The authors would like to thank all of colleagues for their support and cooperation. Authors would also like to thank Department of Electrical and Computer Engineering, Hawassa University for providing the environment and support to carry out this work.

REFERENCES

Alhelou, H., Hamedani-Golshan, M. E., Zamani, R., Heydarian-Forushani, E., & Siano, P. (2018). Challenges and Opportunities of Load Frequency Control in Conventional, Modern and Future Smart Power Systems: A Comprehensive Review. *Energies*, *11*(10), 2497. doi:10.3390/en11102497

Alhelou, H. H. (2018). *Fault Detection and Isolation in Power Systems Using Unknown Input Observer. In Advanced Condition Monitoring and Fault Diagnosis of Electric Machines* (p. 38). Hershey, PA: IGI Global.

Alhelou, H. H., Golshan, M., & Fini, M. (2018). Wind Driven Optimization Algorithm Application to Load Frequency Control in Interconnected Power Systems Considering GRC and GDB Nonlinearities. *Electric Power Components and Syst*.

Alhelou, H. H., & Golshan, M. E. H. (2016, May). Hierarchical plug-in EV control based on primary frequency response in interconnected smart grid. In *Electrical Engineering (ICEE), 2016 24th Iranian Conference on* (pp. 561-566). IEEE. 10.1109/IranianCEE.2016.7585585

Alhelou, H. H., Golshan, M. H., & Askari-Marnani, J. (2018). Robust sensor fault detection and isolation scheme for interconnected smart power systems in presence of RER and EVs using unknown input observer. *International Journal of Electrical Power & Energy Systems*, *99*, 682–694. doi:10.1016/j.ijepes.2018.02.013

Alhelou, H. H., Hamedani-Golshan, M. E., Heydarian-Forushani, E., Al-Sumaiti, A. S., & Siano, P. (2018, September). Decentralized Fractional Order Control Scheme for LFC of Deregulated Nonlinear Power Systems in Presence of EVs and RER. In *2018 International Conference on Smart Energy Systems and Technologies (SEST)* (pp. 1-6). IEEE. 10.1109/SEST.2018.8495858

Alhelou, H. S. H., Golshan, M. E. H., & Fini, M. H. (2015, December). Multi agent electric vehicle control based primary frequency support for future smart micro-grid. In *Smart Grid Conference (SGC)* (pp. 22-27). Academic Press.

Alshahrestani, A., Golshan, M. E. H., & Alhelou, H. H. (2018, November). WAMS Based Online Estimation of Total Inertia Constant and Damping Coefficient for Future Smart Grid Systems. In *Smart Grid Conference (SGC)* (pp. 1-5). Academic Press.

Burr MT, Zimmer MJ, Meloy B, Bertrand J, Levesque W, Warner G. et al. (2013) Minnesota microgrids; 2013.

Center for Energy, Marine Transportation and Public Policy (CEMP) at Columbia University. (2010). Microgrids: an assessment of the value, opportunities and barriers to deployment in New York State. New York State Energy Research and Development Authority.

Edward, J. N., Ramadan, A., & Shatshat, E. (2010). Multi-Microgrid Control Systems (MMCS). *IEEE Power and Energy Society General Meeting, 25*, 1-6.

Fanuel, M., Khan, B., Singh, N., & Singh, P. (2018). Energy Production in Smart Cities by Utilization of Kinetic Energy of Vehicles over Speed Breaker. *International Journal of Civic Engagement and Social Change*, *5*(2), 1–35. doi:10.4018/IJCESC.2018040101

Fini, M. H., Yousefi, G. R., & Alhelou, H. H. (2016). Comparative study on the performance of many-objective and single-objective optimisation algorithms in tuning load frequency controllers of multi-area power systems. *Generation, Transmission & Distribution, 10*(12), 2915-2923.

Haes, H. S. A., Golshan, M. E. H., & Fini, M. H. (2015). Multi agent electric vehicle control based primary frequency support for future smart micro-grid. *2015 Smart Grid Conference (SGC)*, 22-27. 10.1109/SGC.2015.7857385

Huang, J. Y., Jiang, C. W., & Xu, R. (2008). A review on distributed energy resources and Microgrid. *Renewable & Sustainable Energy Reviews*, 2472–2483.

Jariso, M., Khan, B., Tesfaye, D., & Singh, J. (2017). Modelling and Designing of Stand-Alone Photovoltaic System (Case Study: Addis Boder Health Center south west Ethiopia). *IEEE International conference on Electronics, communication, and aerospace Technology, ICECA 2017*, 168-173.

Kema, D.N.V. (2014). *Microgrids – benefits, models, barriers and suggested policy initiatives for the commonwealth of Massachusetts*. Academic Press.

Khan, B., & Agnihotri, G. (2012). A Novel Transmission Loss Allocation Method based on Transmission Usage. *IEEE Fifth Power India Conference*. 10.1109/PowerI.2012.6479479

Khan, B., Agnihotri, G., & Gupta G. (2013). A multipurpose matrices methodology for transmission usage, loss and reliability margin allocation in restructured environment. *Electrical & Computer Engineering: An International Journal, 2*(3).

Khan, B., Agnihotri, G., Gupta, G., & Rathore, P. (2014). A Power Flow Tracing based Method for Transmission Usage, Loss & Reliability Margin Allocation. *AASRI Procedia, 7*, 94–100. doi:10.1016/j.aasri.2014.05.035

Khan, B., Agnihotri, G., & Mishra, A. S. (2016). An Approach for Transmission Loss and Cost Allocation by Loss Allocation Index and Cooperative Game Theory. *J. Inst. Eng. India Ser. B, 97*(1), 41–46. doi:10.100740031-014-0165-1

Khan, B., Agnihotri, G., Rathore, P., Mishra, A., & Naidu, G. (2014). A Cooperative Game Theory Approach for Usage and Reliability Margin Cost Allocation under Contingent Restructured Market. *International Review of Electrical Engineering, 9*(4), 854–862.

Khan, B., & Singh, P. (2017c). The Current and Future States of Ethiopia's Energy Sector and Potential for Green Energy: A Comprehensive Study. *International Journal of Engineering Research in Africa, 33*, 115–139. doi:10.4028/www.scientific.net/JERA.33.115

Khan, B., & Singh, P. (2017a). Optimal Power Flow Techniques under Characterization of Conventional and Renewable Energy Sources: A Comprehensive Analysis. *Journal of Engineering*.

Khan, B., & Singh, P. (2017b). Selecting a Meta-Heuristic Technique for Smart Micro-Grid Optimization Problem: A Comprehensive Analysis. IEEE Access, 5, 13951-13977.

Khattam, W. E., & Salama, M. M. A. (2004). Distributed generation technologies, definitions and benefits. *Electric Power Systems Research, 71*(2), 119–128. doi:10.1016/j.epsr.2004.01.006

Kifle, Y., Khan, B., & Singh, J. (2018). Designing and Modelling Grid Connected Photovoltaic System: (Case Study: EEU Building at Hawassa city). *Int. J. of Convergence Computing, 3*(1), 20–34. doi:10.1504/IJCONVC.2018.091113

Kifle, Y., Khan, B., & Singh, P. (2018). Assessment and Enhancement of Distribution System Reliability by Renewable Energy Sources and Energy Storage. *Journal of Green Engineering, 8*(3), 219–262. doi:10.13052/jge1904-4720.832

Makdisie, C., Haidar, B., & Alhelou, H. H. (2018). An Optimal Photovoltaic Conversion System for Future Smart Grids. In *Handbook of Research on Power and Energy System Optimization* (pp. 601–657). IGI Global. doi:10.4018/978-1-5225-3935-3.ch018

Mulualem, T., & Khan, B. (2017). Design of an off-grid hybrid PV/wind power system for remote mobile base station: A case study. *AIMS Energy*, *5*(1), 96–112. doi:10.3934/energy.2017.1.96

Nadweh, S., Hayek, G., Atieh, B., & Haes Alhelou, H. (2018). Using Four – Quadrant Chopper with Variable Speed Drive System Dc-Link to Improve the Quality of Supplied Power for Industrial Facilities. Majlesi Journal of Electrical Engineering.

Njenda, T. C., Golshan, M. E. H., & Alhelou, H. H. (2018, November). WAMS Based Intelligent Under Frequency Load Shedding Considering Online Disturbance Estimation. In *Smart Grid Conference (SGC)* (pp. 1-5). Academic Press.

Njenda, T. C., Golshan, M. E. H., & Alhelou, H. H. (2018, November). WAMS based Under Frequency Load Shedding Considering Minimum Frequency Prediction and Extrapolated Disturbance Magnitude. In *Smart Grid Conference (SGC)* (pp. 1-5). Academic Press.

Passey, R., Spooner, T., MacGill, I., Watt, M., & Syngellakis, K. (2011). The potential impacts of grid-connected distributed generation and how to address them: A review of technical and non-technical factors. *Energy Policy*, *39*(10), 6280–6290. doi:10.1016/j.enpol.2011.07.027

Saleh, M. S., Althaibani, A., Esa, Y., Mhandi, Y., & Mohamed, A. A. (2015). Impact of clustering micro grids on their stability and resilience during blackouts. *2015 International Conference on Smart Grid and Clean Energy Technologies (ICSGCE)*, 195–200. 10.1109/ICSGCE.2015.7454295

Setiawan, A. A., Zhao, Y., Lee, R. S., & Nayar, C. V. (2009). Design, Economic Analysis and Environmental Considerations of Mini-Grid Hybrid Power System with Reverse Osmosis desalination Plant for Remote Areas. *Renewable Energy-Elsevier.*, *34*(2), 374–383. doi:10.1016/j.renene.2008.05.014

Singh, P., & Khan, B. (2017). Smart Microgrid Energy Management Using a Novel Artificial Shark Optimization. Complexity. doi:10.1155/2017/2158926

Zamani, R., Hamedani-Golshan, M. E., Haes Alhelou, H., Siano, P., & Pota, H. (2018). Islanding Detection of Synchronous Distributed Generator Based on the Active and Reactive Power Control Loops. *Energies*, *11*(10), 2819. doi:10.3390/en11102819

KEY TERMS AND DEFINITIONS

Electric Grid: An electrical grid is an interconnected network for delivering electricity from producers to consumers. It consists of generating stations that produce electrical power high voltage transmission lines that carry power from distant sources to demand centers distribution lines that connect individual customers.

Micro-Grid PPC: The point where the microgrid is connected with the main grid through a breaker mechanism.

Microgrids: A microgrid is a localized group of electricity sources and loads that normally operates connected to and synchronous with the traditional wide area synchronous grid, but can also disconnect to "island mode"—and function autonomously as physical or economic conditions dictate.

Chapter 11
Power Quality of Electrical Power Systems

Feras Youssef Mahfoud
Universitatea Politehnica Bucureşti, Romania

Basarab Dan Guzun
Universitatea Politehnica Bucureşti, Romania

George Cristian Lazaroiu
University Politehnica of Bucharest, Romania

H. H. Alhelou
Tishreen University, Syria

ABSTRACT

Power quality problems can cause processes and equipment to malfunction or shut down. And the consequences can range from excessive energy costs to complete work stoppage. Obviously, power quality is critical. There are many ways in which a power feed can be poor quality, and so no single figure can completely quantify the quality of a power feed. In this chapter, the authors present all definitions, classifications, and problems related to power quality. Finally, they do a comparison between the practical measurements and standards related to power quality.

INTRODUCTION

Electrical energy is one of the most important of raw materials for the time being used in all areas of life Industrial, commercial, agricultural and domestic. Electrical energy is a product, but a product unusual, this is due to two reasons:

1. Cannot be stored in quantities.
2. Cannot be subject to quality safety tests.

DOI: 10.4018/978-1-5225-8030-0.ch011

Therefore, the electrical energy is transferred from generation place to consumption directly without any tests or examinations, so that electricity must be able to fulfill their function properly without any problems and this is what is called the power quality.

An electrical power system is expected to deliver undistorted sinusoidal rated voltage and current continuously at rated frequency to the end users. Electric power quality has captured increasing attention in power engineering in recent years.

Background of Power Quality

Power Quality is defined as "any power problem manifested in voltage, current, and/or frequency deviations that results in the failure and/or mal-operation of end user's equipment. PQ is simply the interaction of electric power with electrical equipment.

Electromagnetic Compatibility ("EMC") and is defined as: "the ability of an equipment or system to function satisfactorily in its electromagnetic environment without introducing excessive electromagnetic disturbances to anything in that environment." (Ferracci, 2001).

Sometimes, Power quality is a term used to discuss events on electric power grids that can damage or disrupt sensitive electronic devices. There are many ways in which a power feed can be poor quality, and so no single figure can completely quantify the quality of a power feed.

Power quality problems can cause processes and equipment to malfunction or shut down. And the consequences can range from excessive energy costs to complete work stoppage. Obviously, power quality is critical.

Poor power quality can result in lost productivity, lost and corrupt data, damaged equipment and poor power efficiency. When added up, U.S. companies waste an estimated $26 billion on electrical power-related issues each year.

Power quality problems make their effects felt in three general areas: downtime, equipment problems, and energy costs.

The Electric Power Research Institute (EPRI) conducted a five-year (1990-1995) monitoring program for distribution power quality (DPQ-I) among 24 utilities throughout the United States of America. Another program DPQ-II was conducted in 2001-2002. These study results that voltage sags (dips) and swells, transient over-voltages (due to capacitor switching), harmonics and grounding Power Quality related problems are the most common PQ complaints among the American customer as presented in Figure 1 (Eberhard, 2011 & Alhelou et al 2018).

The best measure of power quality is the ability of electrical equipment to operate in a satisfactory manner, given proper care and maintenance and without adversely affecting the operation of other electrical equipment connected to the system (Ferracci et al 2001 & Makdisi et al 2018).

Terms and Definitions of Power Quality

- **Variation:** is defined as small deviation from nominal or desired value. e.g.harmonic distortion, voltage fluctuations.
- **Voltage Fluctuations:** are mainly due to rapidly varying industrial loads such as welding machines, arc furnaces or rolling mills (Ferracci et al 2001 & Eberhard 2011).

Figure 1. PQ problems experienced by the American customers

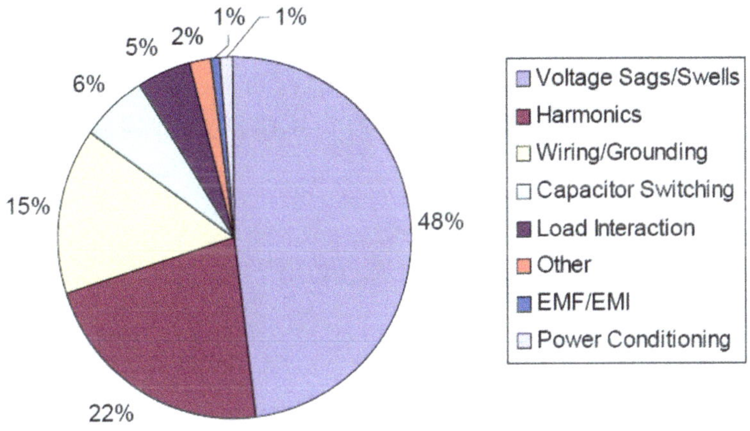

- **Distortion:** Qualitative term indicating the deviation of a periodic wave from its ideal waveform characteristics (See figure 2).
- **Events:** is defined as sudden(large) deviations from the normal voltage or current i.e. Interruptions, voltage sags and inrush currents.

Power quality disturbances that are common in a power system include: voltage sags, voltage swells, short-term interruptions, transients, voltage unbalance, harmonics, and voltage fluctuations.

- **Interruption:** Complete loss of voltage or current for a time period.
- **Inrush:** Large current that a load draws when initially turned on.
- **Swell:** RMS increase in AC voltage at power frequency from half of a cycle to a few seconds' duration (SANKARAN, 2002).

Figure 2. Waveform with distortion

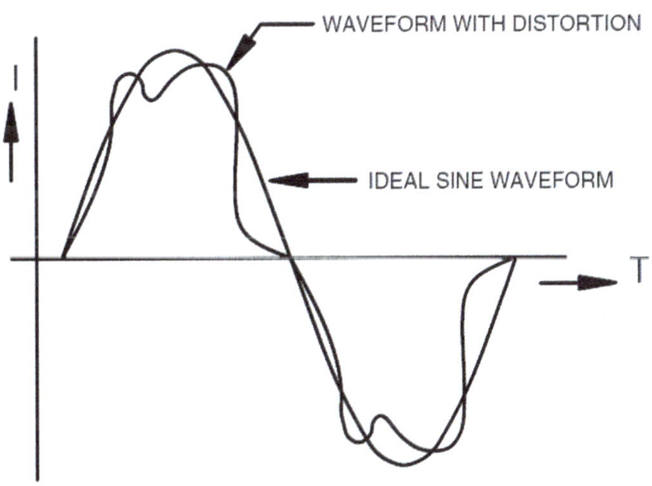

- **Overvoltage:** When the peak value of the applied voltage to a device exceeds the limits defined in a standard or specification (Ferracci et al, 2001).

Voltage unbalance: if terms value of the phase voltages or the phase angles between consecutive phases are not equal. Unbalanced loading of the network normally causes voltage unbalance. All standards, calculate voltage unbalance as following:

$$\frac{Maximum\ Deviation\ From\ Average\ Voltage}{Average\ Voltage} \times 100\left(for\ percent\right).$$

Example:

Measured Voltages are: 243 Volts RMS, 241 Volts RMS and 233 Volts RMS.

Average Volts= (243 + 241 + 233) / 3 = 239 v

Voltage Unbalance = (239-233/239) × 100 = 2.5%

- **A Voltage Dip (Sag):** is a sudden reduction of the voltage at a point in an electrical power system followed by voltage recovery after a short period of time from a few cycles to a few seconds (*Standard EN 50160).*

A voltage dip is regarded as occurring on a 3-phase system if at least one phase is affected by the disturbance.

There is a dip to x % if the rms value falls below the dip threshold x % of the reference value Uref. The threshold x is typically set below 90 (*Standard EN 50160 &* IEEE Std. 1159-1995). The reference voltage Uref is generally the nominal voltage for LV power systems and the declared voltage for MV and HV power systems (Ferracci 2001& Fini, 2016).

- **Voltage Interruptions:** are a special type of voltage dip to a few percentages of Uref (typically within the range 1-10%). often result from tripping and automatic enclosure of a circuit breaker designed to avoid long interruptions which have longer duration (Ferracci, 2001 & Alhelou et al 2016 & *Standard EN 50160 &* IEEE Std. 1159-1995).
- **Harmonic:** Harmonics are periodic distortions of voltage, current, or power sine waves. Each waveform can be considered as a combination of various sine waves with different frequencies and magnitudes.

Sinusoidal component of a periodic wave having a frequency that is an integral multiple of the fundamental frequency. If the fundamental frequency is 60 Hz, then the second harmonic is a sinusoidal wave of 120 Hz, the fifth harmonic is a sinusoidal wave of 300 Hz, and so on (SANKARAN, 2002).

Because of the above property, the Fourier series concept is universally applied in analyzing harmonic problems. When both the positive and negative half cycles of a waveform have identical shapes, the Fourier series contains only odd harmonics.

In fact, the presence of even harmonics is often a clue that there is something wrong—either with the load equipment or with the transducer used to make the measurement. Usually, the higher-order harmonics (above the range of the 25th to 50th, depending on the system) are negligible for power system analysis (Roger C. Dugan & Mark F. (2004).

Total Harmonic Distortion THD, it is a term used to describe the net deviation of a nonlinear waveform from ideal sine waveform characteristics. It is defined as the ratio of the root-mean square of the harmonic content to the root-mean square value of the fundamental quantity. Frequently the THD is expressed in percent.

- The THD is zero for a perfectly sinusoidal wave. It increases indefinitely as the waveform distortion increases.
- A THD of 5% is commonly cited as the border line between high and low distortion for distribution circuits.

Figure 3. Fourier series representation of a distorted waveform

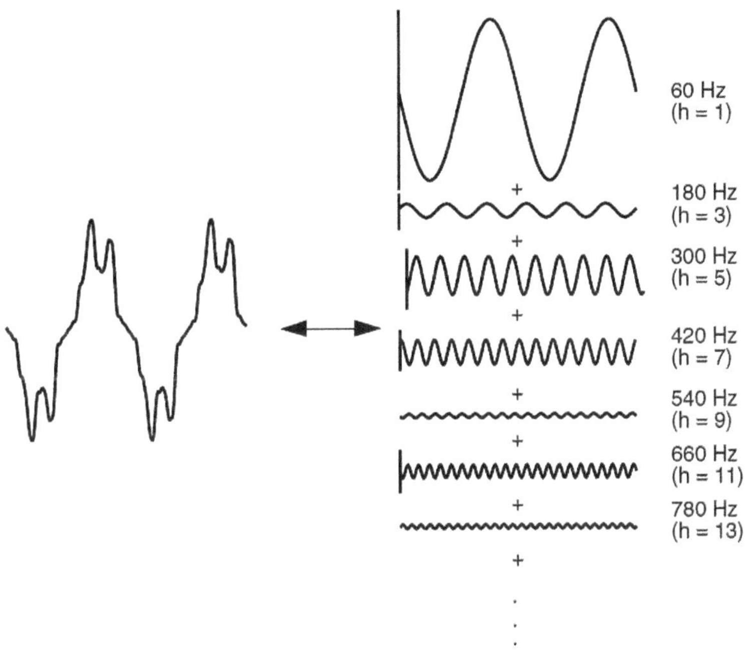

Table 1. Voltage harmonic distortion limits (IEEE Standard 519-1992)

Bus Voltage at Point of Common Coupling	Total Harmonic Distortion THD (%)
69 kV and below	5.0
69.001Kv through 161 kV	2.5
161.001 kV and above	1.5

- **Notch:** Disturbance of the normal power voltage waveform lasting less than a half cycle; the disturbance is initially of opposite polarity than the waveform and, thus, subtracts from the waveform (SANKARAN, 2002), which can be seen in figure 4.

crest factor A value reported by many power quality monitoring instruments representing the ratio of the crest value of the measured waveform to the root mean square of the fundamental. For example, the crest factor of a sinusoidal wave is 1.414 (Roger et al, 2004 & Alshahrestani et al, 2018).

Interruption sustained (power quality) A type of long-duration variation. The complete loss of voltage (<0.1 pu) on one or more phase conductors for a Time greater than 1 min (Roger et al, 2004). Oscillatory transient A sudden, nonpower frequency change in the SteadyState

Condition of voltage or current that includes either positive- or negative polarity value (Roger et al, 2004).

- **Plt:** The long-term flicker severity level as defined by IEC 61000-4-15, based on an observation period of 2 h (Roger et al, 2004).
- **Pst:** The short-term flicker severity level as defined by IEC 61000-4-15, based on an observation period of 10 min. A Pst value greater than 1.0 corresponds to the level of irritability for 50 percent of the persons subjected to the measured Flicker (Roger et al, 2004).

Classification of Various Power Quality Disturbances

Power quality problems have been classified in different forms. Its analysis and classification are very important. The classification of Power Quality disturbances in electric power systems has become an important task for proper developing and designing the preventive and corrective procedures.

Table 2. Summary definition of power quality disturbances

	Disturbance	Short Definition
1	Interruption	voltage magnitude is zero
2	Under voltage	voltage magnitude is below its nominal value
3	Over voltage	voltage magnitude is above its nominal value
4	Voltage sag	A reduction in RMS voltage over a range of 0.1–0.9 pu for a duration greater than 10 ms but less than 1 s
5	Voltage swell	An increase in RMS voltage over a range of 1.1–1.8 pu for a duration greater than 10 ms but less than 1 s
6	Flicker	A visual effect of frequency variation of voltage in a system
7	Voltage/Current unbalance	Deviation in magnitude of voltage/current of any one or two of the three phases
8	Ringing waves	A transient condition which decays gradually
9	Outage	Power interruption for not exceeding 60 s duration due to fault or mal tripping of switchgear/system
10	Transients	Sudden rise of signal
11	Harmonics	Non-sinusoidal wave forms
12		

Figure 4 shows the profile of a transients, together with the associated definitions.

Figure 5 shows the profile of a voltage variation, together with the associated definitions also.

Figure 6 shows the profile of harmonically distorted waveform.

Table 3 illustrates a classification of the disturbances in different power Systems, main causes of them and corresponding impact on equipment (Ferracci, 2001 &EEE Standard 519-1992 & *Standard EN 50160* & IEEE Standard 1250-1995 & IEEE Standard 446-1995 & IEEE Standard 142-1991).

The typical phenomena related to wave amplitude, and its main characteristics are defined in table I, the table 4 lists the limits given in standards IEEE std 1159 – 1995 (IEEE Std. 1159-1995) according to the classification of various power quality events (Eberhard, 2011 & Philip, 2010).

Figure 4. Transients profile

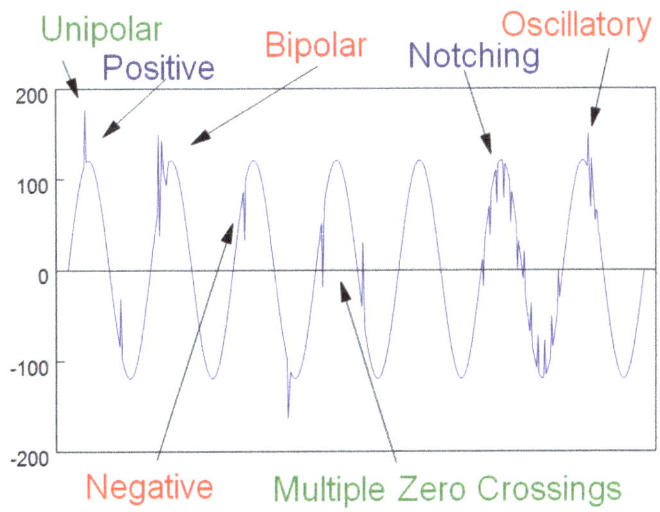

Analyzing Power Quality Data by PC Applications Software (Power Acceptability Curve)

Power Acceptability Curves regulate the minimum PQ level that utilities must provide to consumers and the immunity level that equipment should have to operate properly when the power supplied is within the standards. The power acceptability curves may be used as references curves to identify the severity of Short-term voltage variations.

The most commonly used curves are CBEMA curve and ITIC Curve, these set of curves are the most frequently employed displays of data to represent the power.

On the CBEMA and ITIC, the height on the vertical axis shows the severity of the dip or swell relative to the nominal voltage, the horizontal position shows the duration of the dip or swell, see figure12 And figure 13 (*Network Protection & Automation Guide*. 2011), (Golovanov, 2008).

The CBEMA curve (Fig 12), created by the computer and Business Equipment Manufacturer's Association. This standard specifies the minimum withstanding capability of computer equipment to voltage sags, micro interruptions and over voltages (De Almeida, 2011).

Figure 5. Voltage variations profile

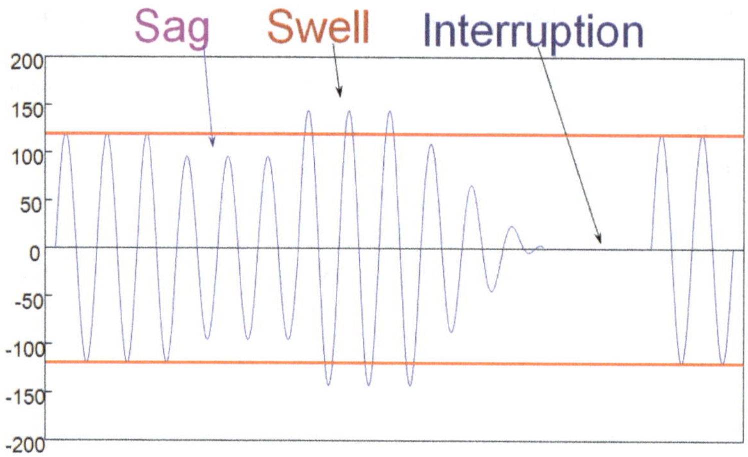

Figure 6. Harmonically distorted waveform profile

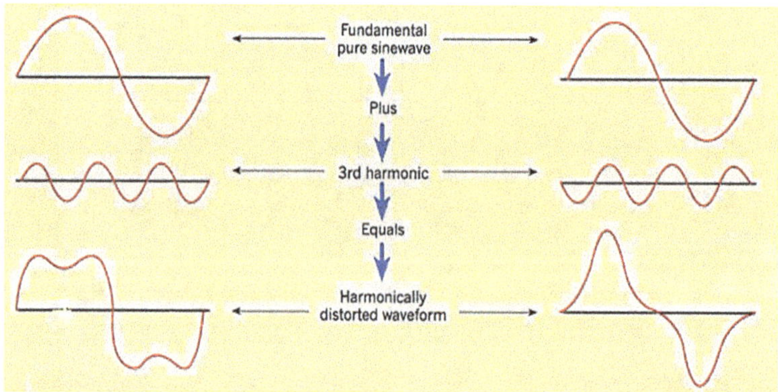

Figure 7. Imbalance waveform profile

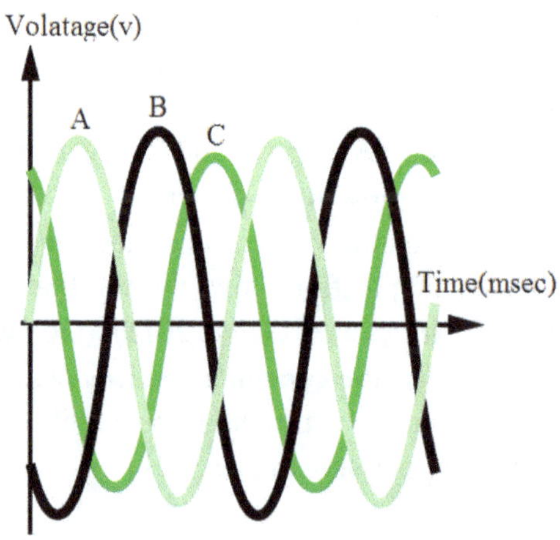

Figure 8. Fluctuation waveform profile

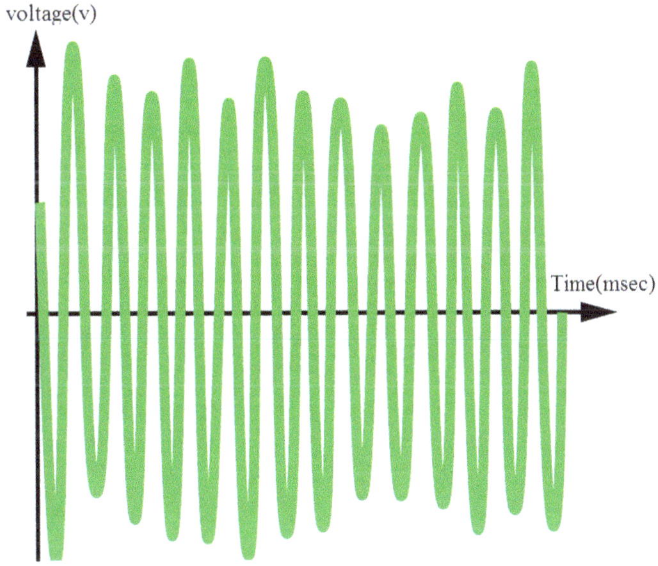

Figure 9. Examples of voltage signals and disturbances: (a) nominal voltage signal, (b) oscillatory transient, (c) harmonics, (d) notching, (e) sag, (f) swell, (g) spikes and (h) outage

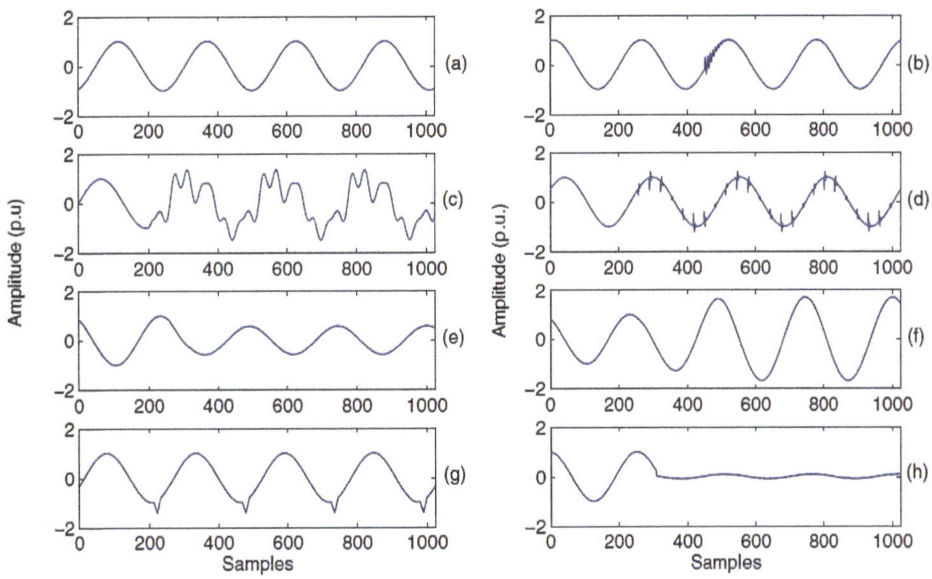

The capability profile (fig 12) implies that computers can tolerates low variations from -13% to + 5.8%, and greater amplitude disturbances can be tolerated as their durations become shorter.

The axes represent magnitude and duration of the event. Points below the envelope are presumed to cause the load to drop out due to lack of energy. Points above the envelope are presumed to cause other malfunctions such as insulation failure, overvoltage trip, and overexcitation.

Figure 10. Examples of multiple disturbances: (a) sag with harmonics, (b) sag with oscillatory transient and harmonics, (c) oscillatory transient with harmonics, (d) sag with notching and harmonics, (e) swell with spikes, (f) swell with notching and harmonics, (g) sag with notching, and (h) swell with oscillatory transient and harmonics

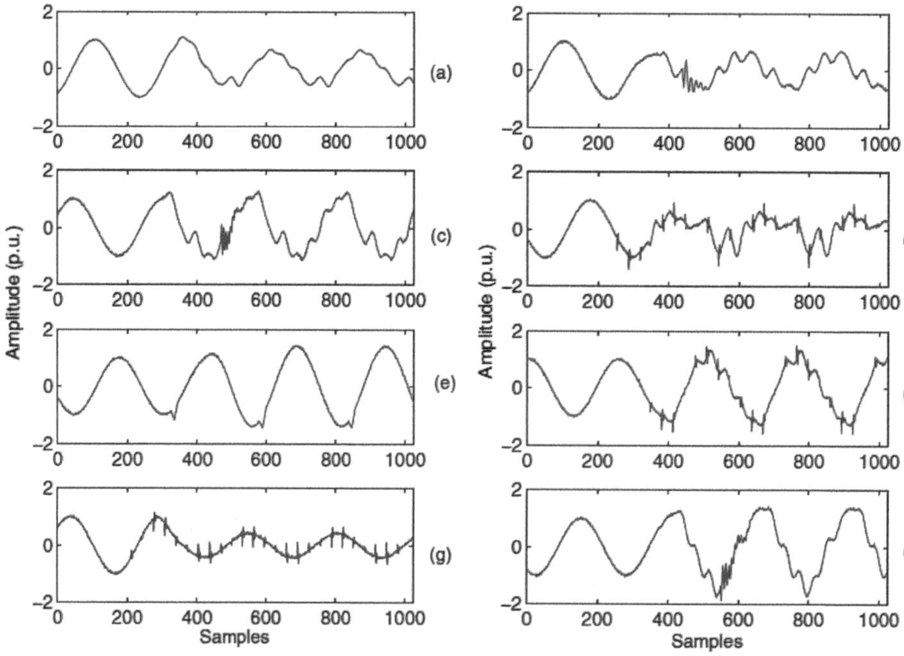

The upper curve is actually defined down to 0.001 cycle where it has a value of about 375 percent voltage (Roger et al, 2004).

Other standard is created by Information Technology Industry Council called (ITIC), illustrated in Figure 13. This curve is divided into two regions (*Network Protection & Automation Guide*.2011), (Surajit et al, 2011 & Heydt, 2000):

- withstand disturbance: any voltage disturbance lie in that area should not cause malfunction at all.
- affected by disturbance: any voltage disturbance lie in that area will cause malfunction.

It may be necessary to check carefully which standards are applicable when considering equipment susceptibility (*Network Protection & Automation Guide*.2011), (Golovanov, 2008 & Njenda et al, 2018).

An Algorithm for Common Power Quality Problem

Customer's nonlinear devices participate in Many PQ problems in the network, other PQ problems such as voltage dips, transients are mainly originated in the network side.

Therefore, to improve PQ performance in the network, we proposed an algorithm to determine the responsibility of the parties involved in this problem in the network with general solutions as shown in Figure 14.

Table 3. Classification of the disturbances in different power systems

Category	Causes	Impact
Voltage dips	Local and remote faults Inductive loading Switch on of large loads	Tripping of sensitive equipment Resetting of control systems Motor stalling/tripping
Voltage surges	Capacitor switching Switch off of large loads Phase faults	Tripping of sensitive equipment Damage to insulation and windings Damage to power supplies for electronic equipment
Overvoltage	Load switching Capacitor switching System voltage regulation	Problems with equipment that requires constant steady-state voltage
Harmonics	Industrial furnaces Non-linear loads Transformers/generators Rectifier equipment	Mal-operation of sensitive equipment and relays Capacitor fuse or capacitor failures Telephone interference
Power frequency variation	Loss of generation Extreme loading conditions	Negligible most of time Motors run slower De-tuning of harmonic filters
voltage fluctuation	AC motor drives Inter-harmonic current components Welding and arc furnaces	Flicker in: Fluorescent lamps Incandescent lamps
Rapid voltage change	Motor starting Transformer tap changing	Light flicker Tripping of equipment
Voltage imbalance	Unbalanced loads Unbalanced impedances	Overheating in motors/generators Interruption of 3-phase operation
Short and long voltage interruptions	Power system faults Equipment failures Control malfunctions CB tripping	Loss of supply to customer equipment Computer shutdowns Motor tripping
Undervoltage	Heavy network loading Loss of generation Poor power factor Lack of var support	All equipment without backup supply facilities
Transients	Lightning Capacitive switching Non –linear switching loads System voltage regulation	Control system resetting Damage to sensitive electronic components Damage to insulation

Monitoring of Power Quality

Power quality monitoring is the process of gathering, analyzing, and interpreting raw measurement data into useful information. The process of gathering data is usually carried out by continuous measurement of voltage and current over an extended period. The process of analysis and interpretation has been traditionally performed manually.

Figure 11. Equipment affected by PQ problems in different sectors

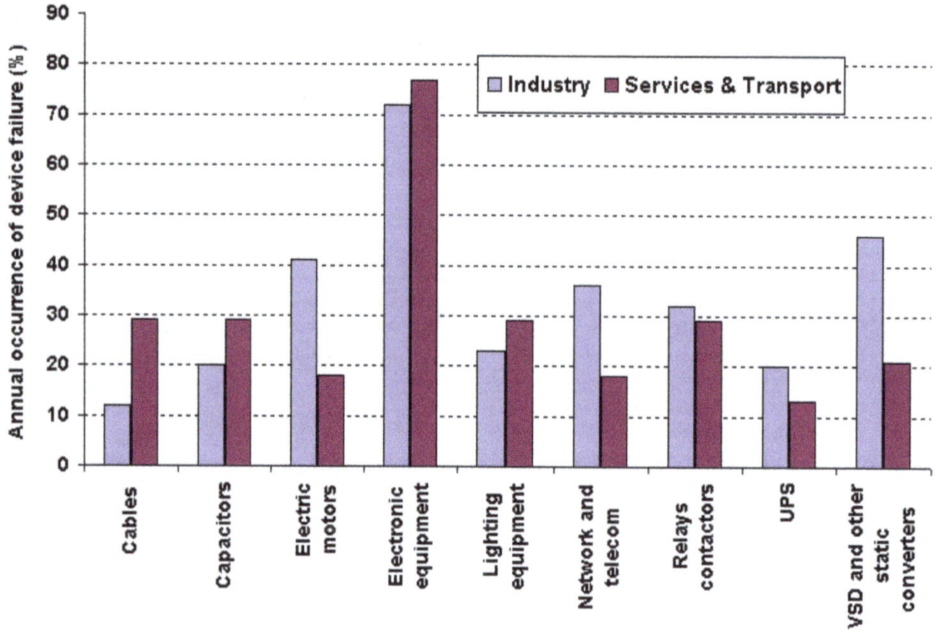

Measurements done at BARAGAN Photovoltaic Farm

we execute some measurements at BARAGAN Photovoltaic farm by using FLUKE SET for two times, each once for one week and after one week we download the data by using Power Log software version 4.3.1. The capacity of that farm is 4MW/h to 32MW/h. Figure 15 shows the interface for this program.

Case 1: Number of Events According to Values of Measured Voltages

Figure 16 shows the number of events according to the average of Vrms Ph-n, these measurements were recorded during the date from 12/12/2014 2:54:04 PM to 19/12/2014 9:44:04 AM.

Case 2: Power Frequency Variations

Figure 17 shows frequency deviations, the maximum value of frequency is 50.039Hz, the minimum value is 49.952 Hz, these values are within acceptable limits listed in the table 4. Acceptable limits listed in table 5.

Case 3: Voltage Fluctuation

Figure 18 shows voltage fluctuations, the maximum value of voltage for three phases is 0.330%, the minimum value is 0.125%, and these values are within acceptable limits listed in the table 4.

Acceptable limits listed in table 4.

Table 4. Power system disturbances according to IEEE Std 1159-1995 (IEEE Std. 1159-1995)

Disturbance Subtype	Categories	Typical Duration	Typical Voltage Magnitude
1. Transients			
1.1. Impulsive	Nanosecond	<50 nseconds	
	Microsecond	50 nseconds -1 msecond	
	Millisecond	>1msecond	
1.2. Oscillatory	Low Frequency	0.3 -50 msec	0-4pu
	Medium Frequency	20 μsec	0-8pu
	High Frequency	5 μsec	0-4pu
2. Short Duration Variations			
2.1. Instantaneous	Interruption	0.5 -30 Cycles	<0.1 Pu
	Sag	0.5 -30 Cycles	0.1 -0.9 Pu
	Swell	0.5 -30 Cycles	1.1 -1.8 Pu
2.2. Momentary	Interruption	30 Cycles -3 Seconds	<0.1 Pu
	Sag (Dip)	30 Cycles -3 Seconds	0.1 -0.9 Pu
	Swell	30 Cycles -3 Seconds	1.1 -1.4 Pu
2.3. Temporary	Interruption	3sec -1min	
	Sag (Dip)	3sec -1min	
	Swell	3sec -1min	
3. Long Duration Variations			
3.1. Interruption Sustained		>1min	0.0 Pu
3.2. Under-Voltages		>1min	0.8 -0.9 Pu
3.3. Over-Voltages		>1min	1.1 -1.2 Pu
4.Voltage Unbalance		Steady state	0.5 -2%
5. Wave Distortion			
5.1.Dc Offset		Steady State	0-0.1%
5.2. Harmonics		Steady State	0 -20%
5.3. Inter-Harmonics		Steady State	0-2%
5.4. Notching		Steady State	0.2%
5.5. Noise		Steady State	0.1%
6. Voltage Fluctuations		Intermittent	0.1-7%
7. Power Frequency Variations			
7.1. Slight Deviation		<10 S	49.5 Hz-50.5 Hz
7.2. Severe Deviation		<10 S	47.0 Hz- 52.0 Hz

Figure 12. CBEMA curve for equipment susceptibility

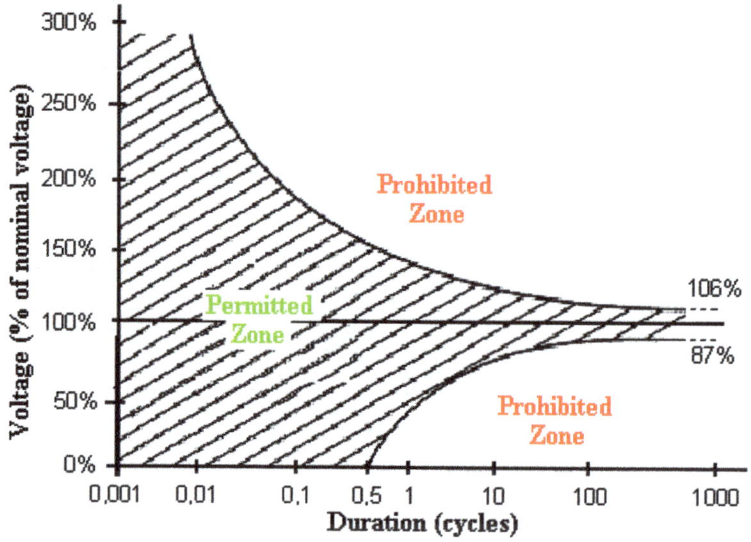

Figure 13. ITI curve for equipment susceptibility

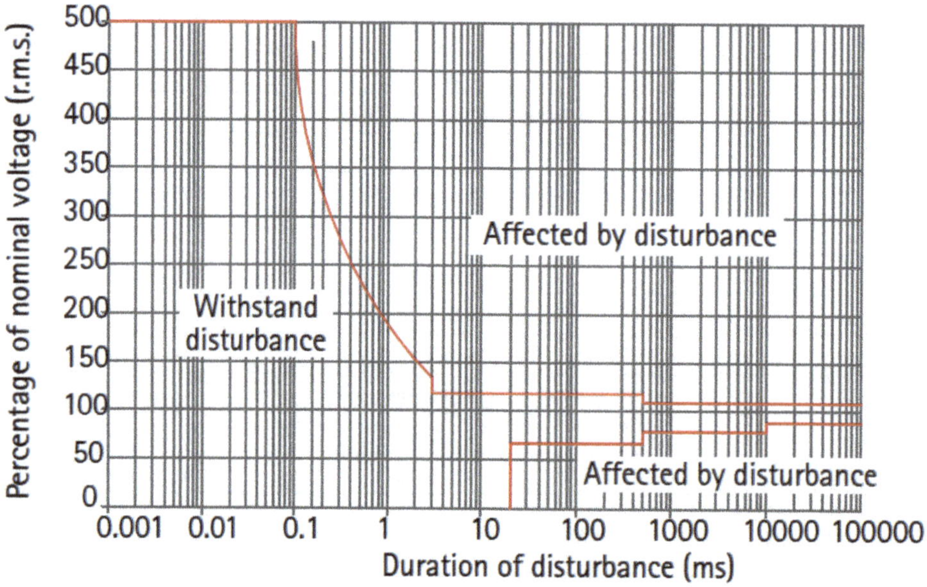

Case 4: Over Voltage

Figure 19 shows recorded voltage waveform of duration 200ms, we note that the maximum value of voltage is 20950V and this value is within limits mentioned in table 4.

Figure 14. Flowchart on PQ solutions

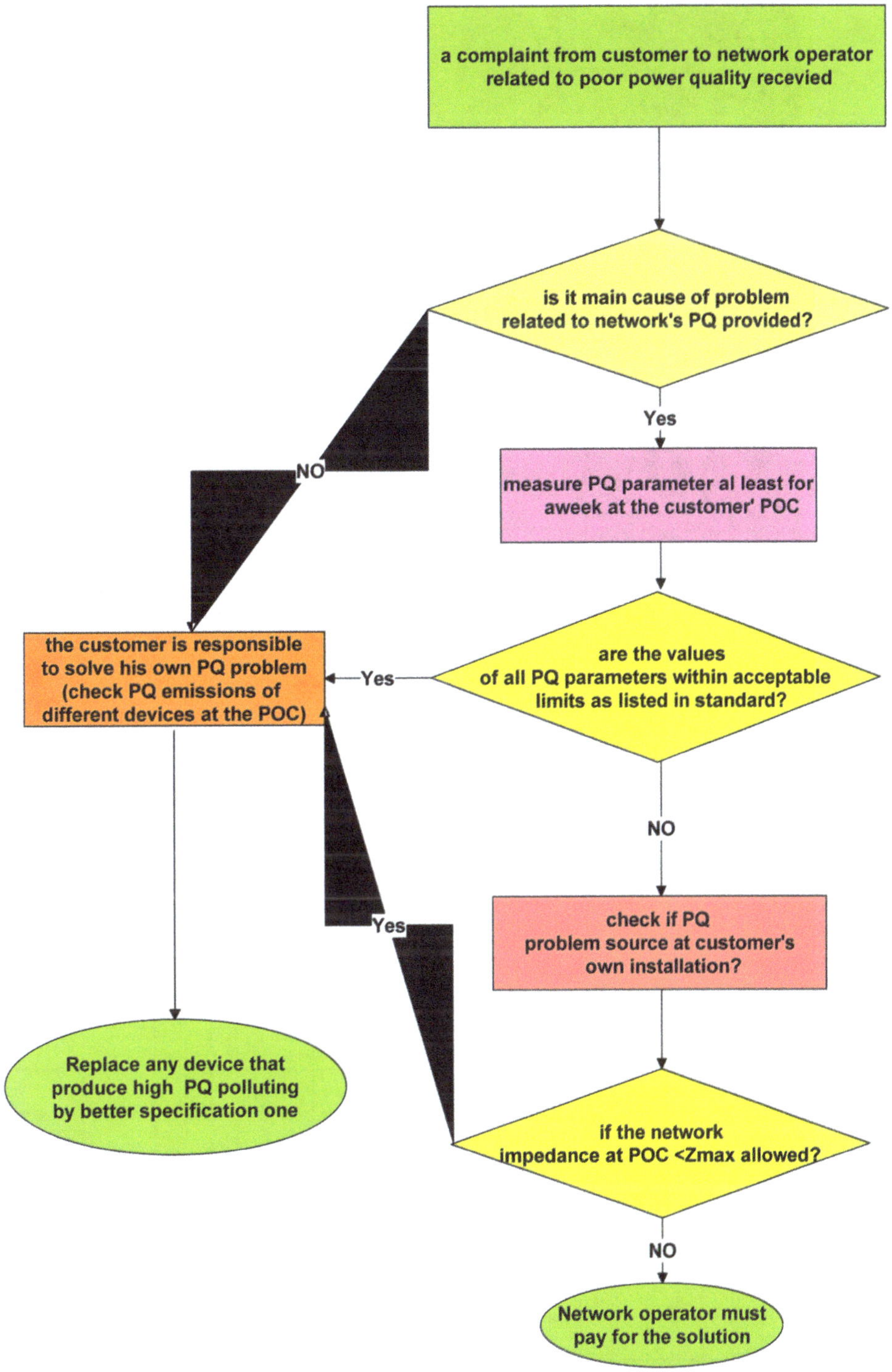

Figure 15. Interface of power log software

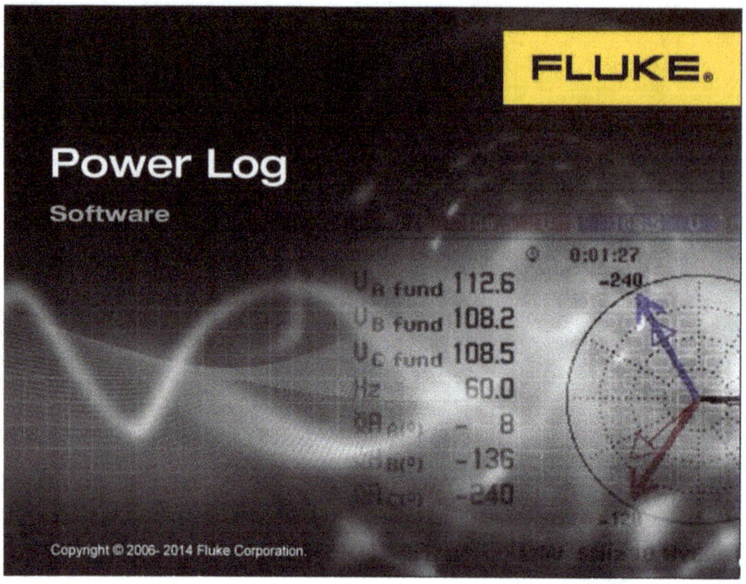

Figure 16. Number of events

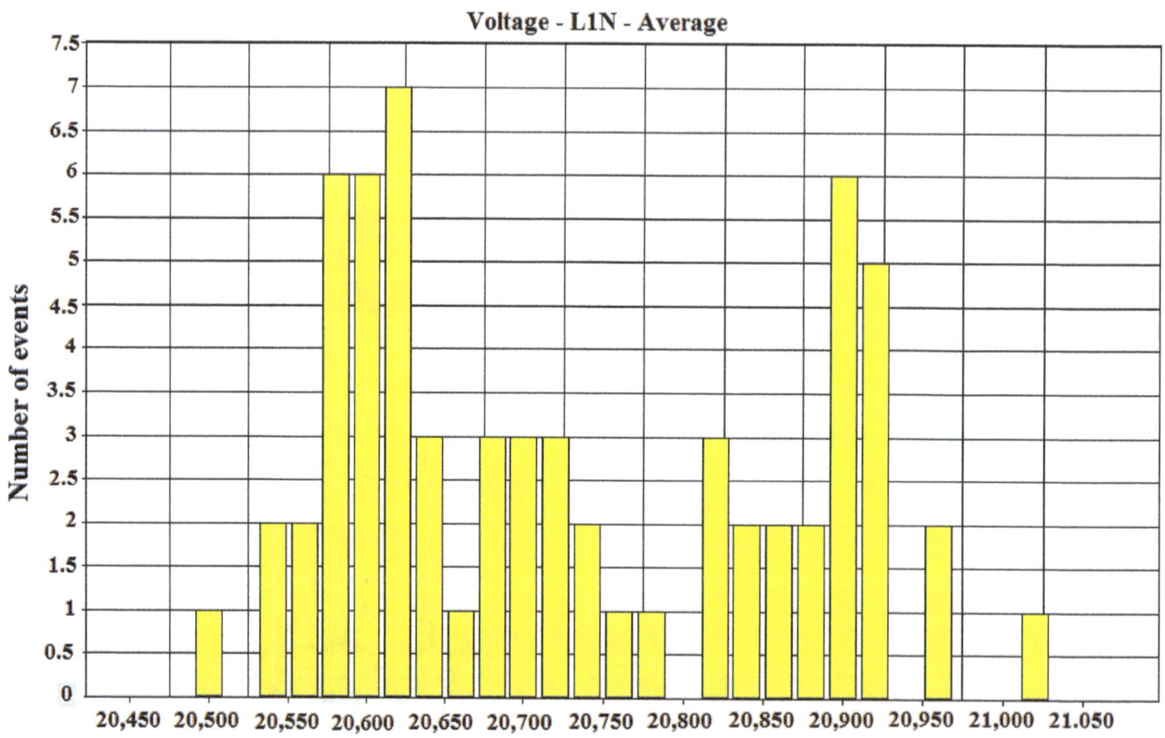

Table 5.

Frequency Deviation	Duration	Range
Slight Deviation	<10 S	49.5 Hz-50.5 Hz
Severe Deviation	<10 S	47.0 Hz- 52.0 Hz

Figure 17. Frequency deviations

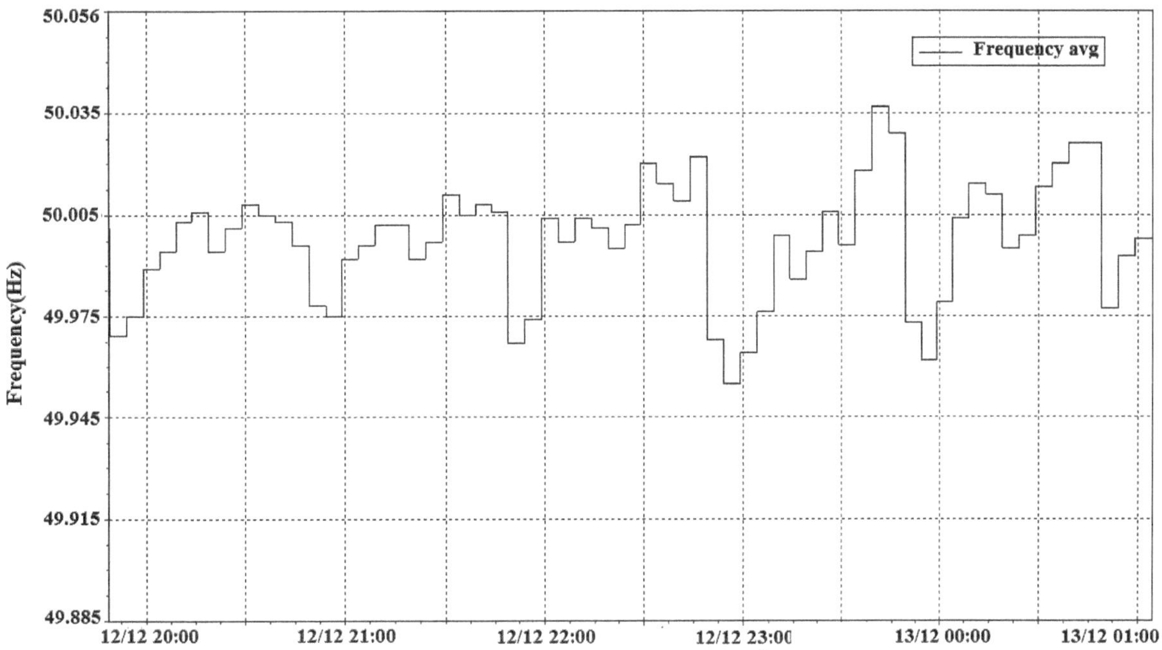

Table 6.

Disturbance Type	Duration	Range
Voltage Fluctuations	Intermittent	0.1-7%

Case 5: Harmonics

Figure 20 shows total harmonic distortion in percent (THD %). we note that the maximum value of total harmonic distortion is 3.7% and this value is within limits mentioned in table 1 and table 4.

Case 6: Dips and Swells

Figure 21 shows dips points, we note that there are two dips points and that points located into permitted zone according to table 4, figure 12 and figure 13.

Figure 18. Voltage fluctuations (flickers)

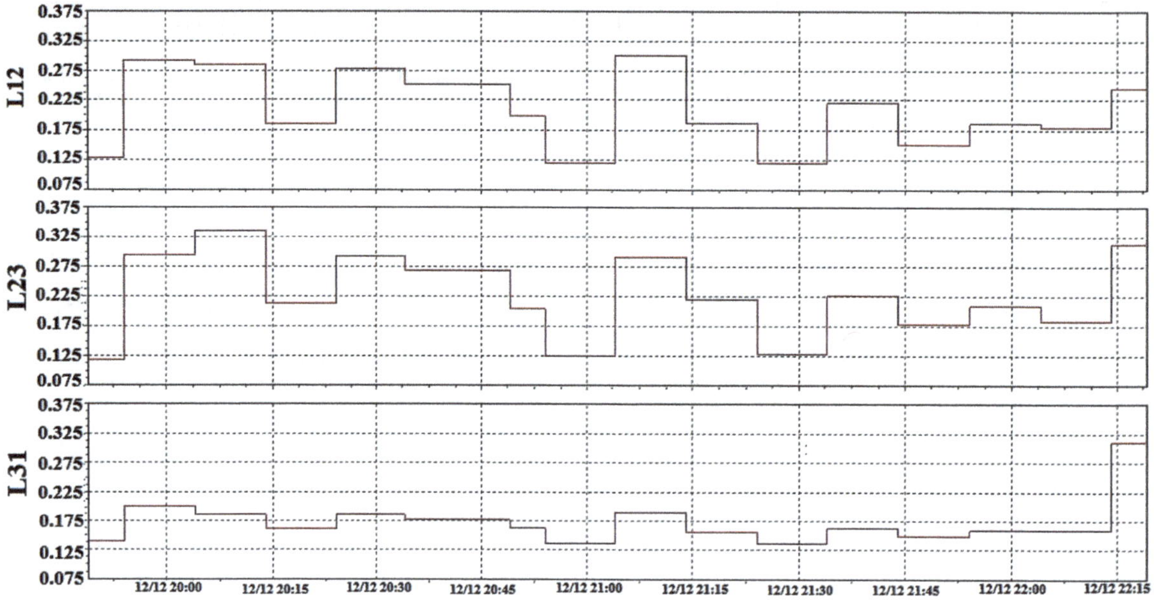

Figure 19. Over voltage waveform

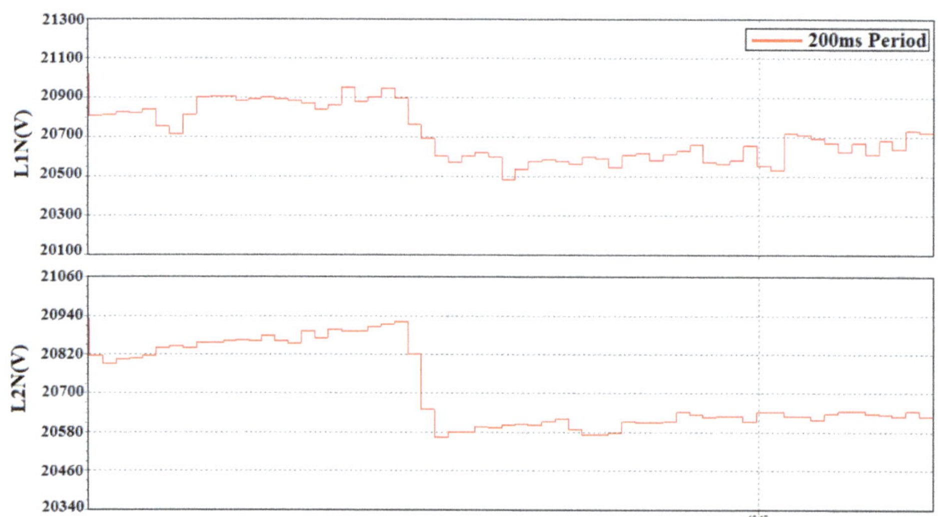

Table 7.

Bus Voltage at Point of Common Coupling	Total Harmonic Distortion THD (%)
69 Kv and below	5.0

Figure 20. Total harmonic distortion

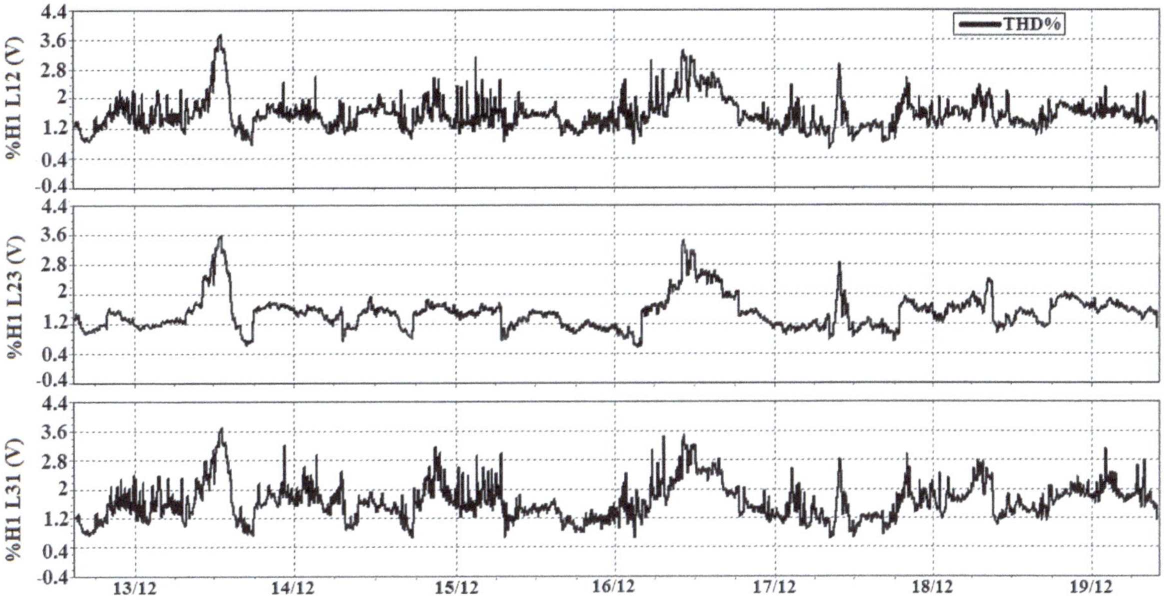

Figure 21. Shows dips points

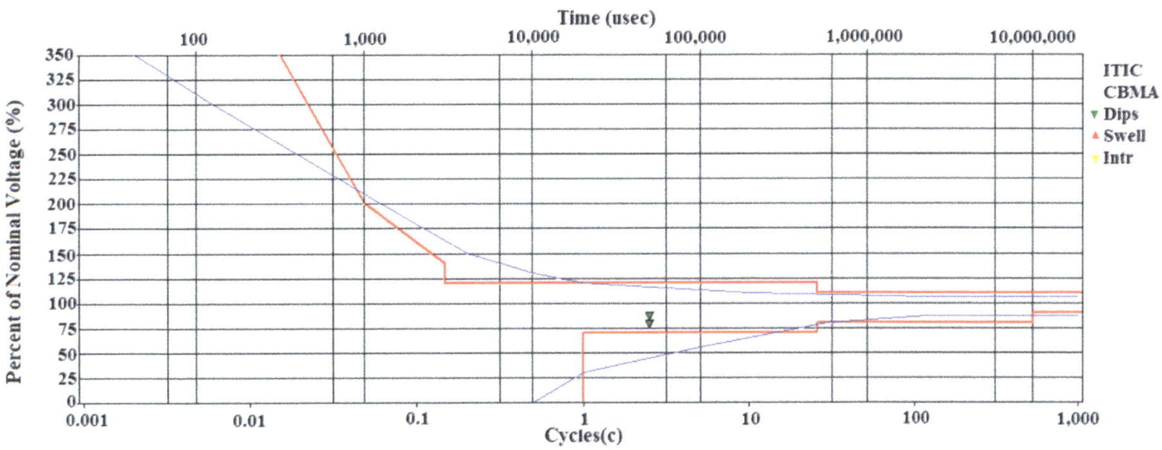

POWER QUALITY STANDARDS

Standards and guidelines have been given by different technical bodies like IEEE, ANSI, IEC, etc. Those guidelines are very helpful in EPQ study and practice.

The designers of computers and micro-process controllers are not versed in power system and power quality issues. By the same token, power system designers and operators have limited knowledge of the operation of sensitive electronics.

This environment has led to a need for power quality standards and guidelines. Nowadays, several engineering organizations all over the world are spending a large amount of resources to generate power quality standards (Surajit, 2011& Alhelou et al 2018 & Alhelou et al 2016 & Nadweh 2018 & Njenda et al, 2018).

In Europe, the most relevant standards in PQ are the EN 50160 (by CENELEC) and IEC 61000.

Following is a list of power quality and related standards from two such organizations; some of the standards listed are in existence at this time, while others are still in process:

The IEEE Standard (Institute of Electrical and Electronic Engineers (IEEE); Piscataway, NJ; http://www.ieee.org;): IEEE documents are a widespread alternative to the IEC series. It identifies The major sources of power quality in power systems. Following main features are referred in this series (Surajit 2011 & Alhelou 2018):

- IEEE 100 Standard dictionary of electrical and electronic engineering.
- IEEE 120 Master Test guide for electrical measurements in power circuits.
- IEEE 141 Recommended practice for electric power distribution for industrial
- Plants with effect of voltage disturbances on equipment within an industrial area.
- IEEE 142 Recommended Practice for Grounding of Industrial and Commercial Power Systems.
- IEEE 142 Recommended Practice for Grounding of Industrial and Commercial Power Systems.
- IEEE 241 Recommended Practice for Electric Power Systems in Commercial Buildings.
- IEEE 430 Standard procedures for the measurement of radio noise from overhead power lines and substations.
- IEEE 472 Event recorders.
- IEEE 644 Standard Procedure for Measurement of Power Frequency Electric and Magnetic Fields from AC Power Lines.
- IEEE C63.12 Recommended Practice for Electromagnetic Compatibility Limits.
- IEEE 518 Guide for the Installation of Electrical Equipment to Minimize Electrical Noise Inputs to Controllers from External Sources.
- IEEE 519 -1992 Recommended Practices and Requirements for Harmonic Control in Electrical Power Systems, document is a widespread alternative to the IEC series. It identifies the major sources of harmonic in power systems.
- IEEE 859 Standard terms for reporting and analyzing outage occurrences and outage states of electrical transmission facilities.
- IEEE 1100 Recommended Practice for Powering and Grounding Sensitive Electronic Equipment.
- IEEE 1159 Recommended Practice for Monitoring Electric Power Quality.
- IEEE 1250 Guides for service to equipment sensitive to momentary voltage disturbances.
- IEEE 1346 Recommended practice for evaluating electric power system compatibility with electronics process equipment.
- IEEE C57.110 Recommended Practice for Establishing Transformer Capability when Supplying Nonsinusoidal Load.
- IEEE P1433 Power Quality Definitions.
- IEEE P1453 Voltage Flicker.
- IEEE P1564 Voltage Sag Indices.

The IEC Standard (International Electrotechnical Commission (IEC); Geneva, Switzerland; http://www.iec.ch):

- IEC/TR3 61000-2-1 Electromagnetic Compatibility — Environment: Power system equipment, industrial loads and residual loads as considered here as three major sources of harmonics in this series.
- IEC/TR3 61000-3-6 Electromagnetic Compatibility — Limits: Limits for the harmonic currents produced by equipment connected to low voltage systems with input currents equal to and below 75A per phase are discussed in this section.
- IEC 61000-4-7 Electromagnetic Compatibility — Testing and Measurement Techniques — General Guides on Harmonics and Inter-harmonics Measurements and Instrumentation: This section deals with testing and measurement techniques
- IEC 61642 Industrial A.C. Networks Affected by Harmonics — Application of Filters and Shunt Capacitors.
- IEC SC77A Low Frequency EMC Phenomena.
- IEC TC77/WG1 Terminology.
- IEC SC77A/WG1 Harmonics and Other Low Frequency Disturbances.
- IEC SC77A/WG6 Low Frequency Immunity Tests.
- IEC SC77A/WG2 Voltage Fluctuations and Other Low Frequency Disturbances.
- IEC SC77A/WG8 Electromagnetic Interference Related to the Network Frequency.
- IEC SC77A/WG9 Power Quality Measurement Methods.
- IEC 38 Standard voltages.
- IEC 816 Guides on methods of measurement of short-duration transients on low-voltage power and signal lines. Equipment susceptible to transients.
- IEC 868 Flicker meter. Functional and design specifications.
- IEC 868-0 Flicker meter. Evaluation of flicker severity. Evaluates the severity of voltage fluctuation on the light flicker.
- IEC 1000-3-2 Electromagnetic compatibility Part 3: Limits Section 2: Limits for harmonic current emissions (equipment absorbed current <16 A per phase).
- IEC 1000-3-6 Electromagnetic compatibility Part 3: Limits Section 6: Emission limits evaluation for perturbing loads connected to MV and HV networks.
- IEC 1000-4 Electromagnetic compatibility Part 4: Sampling and metering techniques.
- EN 50160 Voltage characteristics of electricity supplied by public distribution systems.
- EC/EN 60868 Flicker meter implementation.
- IEC 61000 Electromagnetic compatibility (EMC).

CONCLUSION

Ongoing standards development should help make all parties more aware of power quality concerns and provide better tools and techniques for developing the optimum solutions to problems. Equipment manufacturers must be able to provide information describing the sensitivity of their equipment to these variations. Ongoing monitoring efforts and case studies will provide the information to characterize system performance and to understand the susceptibility of different types of customer systems. Analytical tools will also benefit from the increased level of monitoring and characterization. Models should be improved and the tools themselves should become easier to use.

REFERENCES

Alhelou, H., Hamedani-Golshan, M. E., Zamani, R., Heydarian-Forushani, E., & Siano, P. (2018). Challenges and Opportunities of Load Frequency Control in Conventional, Modern and Future Smart Power Systems: A Comprehensive Review. *Energies*, *11*(10), 2497. doi:10.3390/en11102497

Alhelou, H. H. (2018). *Fault Detection and Isolation in Power Systems Using Unknown Input Observer. In Advanced Condition Monitoring and Fault Diagnosis of Electric Machines* (p. 38). Hershey, PA: IGI Global.

Alhelou, H. H., Golshan, M., & Fini, M. (2018). Wind Driven Optimization Algorithm Application to Load Frequency Control in Interconnected Power Systems Considering GRC and GDB Nonlinearities. *Electric Power Components and Syst.*

Alhelou, H. H., & Golshan, M. E. H. (2016, May). Hierarchical plug-in EV control based on primary frequency response in interconnected smart grid. In *Electrical Engineering (ICEE), 2016 24th Iranian Conference on* (pp. 561-566). IEEE. 10.1109/IranianCEE.2016.7585585

Alhelou, H. H., Golshan, M. H., & Askari-Marnani, J. (2018). Robust sensor fault detection and isolation scheme for interconnected smart power systems in presence of RER and EVs using unknown input observer. *International Journal of Electrical Power & Energy Systems*, *99*, 682–694. doi:10.1016/j.ijepes.2018.02.013

Alhelou, H. H., Hamedani-Golshan, M. E., Heydarian-Forushani, E., Al-Sumaiti, A. S., & Siano, P. (2018, September). Decentralized Fractional Order Control Scheme for LFC of Deregulated Nonlinear Power Systems in Presence of EVs and RER. In *2018 International Conference on Smart Energy Systems and Technologies (SEST)* (pp. 1-6). IEEE. 10.1109/SEST.2018.8495858

Alhelou, H. S. H., Golshan, M. E. H., & Fini, M. H. (2015, December). Multi agent electric vehicle control based primary frequency support for future smart micro-grid. In *Smart Grid Conference (SGC)* (pp. 22-27). Academic Press.

Alshahrestani, A., Golshan, M. E. H., & Alhelou, H. H. (2018, November). WAMS Based Online Estimation of Total Inertia Constant and Damping Coefficient for Future Smart Grid Systems. In *Smart Grid Conference (SGC)* (pp. 1-5). Academic Press.

Barker & Unruh. (2010). *A Standard Glossary of Power Quality Terminology*. IEEE.

Chattopadhyay, S., Mitra, M., & Sengupta, S. (2011). *Electric Power Quality*. Department of Applied Physics. University of Calcutta. doi:10.1007/978-94-007-0635-4

De Almeida, A., Moreira, L., & Delgado, J. (2011). *Power Quality Problems and New Solutions*. University of Coimbra.

Dugan. (2004). Electrical Power Systems Quality (2nd ed.). McGraw-Hill.

Ferracci, P. (2001). *Cahiers Techniques Schneider Electric no 199, Power Quality. Andreas Eberhard.*

Fini, M. H., Yousefi, G. R., & Alhelou, H. H. (2016). Comparative study on the performance of many-objective and single-objective optimisation algorithms in tuning load frequency controllers of multi-area power systems. *IET Generation, Transmission & Distribution*, *10*(12), 2915–2923. doi:10.1049/iet-gtd.2015.1334

Golovanov, N., Postolache, P., Iordănescu, I., Toader, C., Popescu, S., Porumb, R., & Lipan, L. (2008). Instalaţii Electroenergetice ŞI Elemente De Audit Industrial. Bucharest, Romania: Academic Press.

Heydt, G. T. (2000). *Contemporary Topics in Electric Power Quality*. Arizona State University.

IEEE Standard 1250-1995, *IEEE Guide for Service to Equipment Sensitive to Momentary Voltage Disturbances.*

IEEE Standard 142-1991, *IEEE Recommended Practice for Grounding of Industrial and Commercial Power Systems.*

IEEE Standard 446-1995, *IEEE Recommended Practice for Emergency and Standby Power Systems for Industrial and Commercial Applications.*

IEEE Standard 519-1992, *IEEE Recommended Practices and Requirements for Harmonic Control in Electrical Power Systems*

IEEE Standards Board, IEEE Std. 1159-1995, "*IEEE Recommended Practice for Monitoring Electric Power Quality*". New York: IEEE, Inc. June 1995.

Makdisie, C., Haidar, B., & Alhelou, H. H. (2018). An Optimal Photovoltaic Conversion System for Future Smart Grids. In *Handbook of Research on Power and Energy System Optimization* (pp. 601–657). IGI Global. doi:10.4018/978-1-5225-3935-3.ch018

Nadweh, S., Hayek, G., Atieh, B., & Haes Alhelou, H. (2018). Using Four – Quadrant Chopper with Variable Speed Drive System Dc-Link to Improve the Quality of Supplied Power for Industrial Facilities. Majlesi Journal of Electrical Engineering.

Njenda, T. C., Golshan, M. E. H., & Alhelou, H. H. (2018, November). WAMS Based Intelligent Under Frequency Load Shedding Considering Online Disturbance Estimation. In *Smart Grid Conference (SGC)* (pp. 1-5). Academic Press.

Njenda, T. C., Golshan, M. E. H., & Alhelou, H. H. (2018, November). WAMS based Under Frequency Load Shedding Considering Minimum Frequency Prediction and Extrapolated Disturbance Magnitude. In *Smart Grid Conference (SGC)* (pp. 1-5). Academic Press.

Sankaran, C. (2002). *Power quality*. Academic Press.

Standard EN 50160 Voltage characteristics of electricity supplied by public distribution systems. issued by CENELEC, November 1994.

Zamani, R., Hamedani-Golshan, M. E., Haes Alhelou, H., Siano, P., & Pota, H. (2018). Islanding Detection of Synchronous Distributed Generator Based on the Active and Reactive Power Control Loops. *Energies*, *11*(10), 2819. doi:10.3390/en11102819

KEY TERMS AND DEFINITIONS

Frequency.: In statistics the frequency (or absolute frequency) of an event is the number of times the event occurred in an experiment or study. These frequencies are often graphically represented in histograms.

Good Power Quality: Good power quality can be defined as a steady supply voltage that stays within the prescribed range, steady AC frequency close to the rated value, and smooth voltage curve waveform (resembles a sine wave).

Power Quality: Electric power quality, or simply power quality, involves voltage, frequency, and waveform.

Power Quality Compression Algorithm: A power quality compression algorithm is an algorithm used in the analysis of power quality.

Voltage Quality: Voltage quality is used to refer to all disturbances in the supply of electricity, excluding interruptions that are covered.

Chapter 12
An Overview of Wide Area Measurement System and Its Application in Modern Power Systems

H. H. Alhelou
Tishreen University, Syria

ABSTRACT

In this chapter, wide area measurement systems (WAMS), which are one of the cornerstones in modern power systems, are overviewed. The WAMS has great applications in power system monitoring, operation, control, and protection systems. In the modern power systems, WAMS is adopted as a base for the modern monitoring and control techniques. Therefore, an introduction of WAMS is firstly provided. Then, phasor measurement unit (PMU), which is the base of WAMS, is described. Afterward, the most recent developments in power system estimation, stability, and security techniques, which are based on WAMS, are introduced. Later, general system setup for WAMS-based under-frequency load shedding (UFLS) is provided. Finally, the required communications infrastructures are comprehensively discussed.

INTRODUCTION

Wide area measurement systems (WAMS) are measurement systems that are based on the transmission of analogue and/or digital information using telecommunication systems and allowing synchronization (time stamping) of the measurements using a common time reference. The devices that are used for measuring by WAMS have their own clocks synchronized with the common time reference using synchronizing devices. This concept has been used for radio signals as well. Supervisory Control and Data Acquisition (SCADA) also depend on the accuracy time reference of transmitters. An example of a transmitter is the DCF77, which produces signals that are used for synchronization (Machowski et al., 1988). There are many communication channels for use with WAMS such as GPS, optic fiber, telephone lines, power lines and microwave links. The most important factor to consider when using WAMS is the

DOI: 10.4018/978-1-5225-8030-0.ch012

delay associated with the communication channel. The delay gives the time that is taken before a control or protection action is taken. These delays must be included in power system design and simulations in order to stabilize the grid in practical applications. WAMS assists power systems operators by providing real time measurements that are used in maintaining/fixing stability of a power system/network. The WAMS using Phasor Measurement Unit gather data from system. Data is communicated to the control centre, processed for control protection, and monitoring.

The following are some of the factors that have driven the need to use WAMS:

- Optimum utilisation of the grid to meet economical and technical demands.
- Increase in interconnections of grids within regions and continents.
- Improvements in technology for measurement, control and protection.
- Mixed sources of power i.e. wind, hydro and solar and independent power providers now involved.
- Sufficient Research on the power system phenomena causing widespread blackouts,

The old technologies like SCADA posed some challenges hence the need was found for improved technology. The challenges like delays in data acquisition using SCADA makes it difficult to have real time picture of the power system. The data sources used by SCADA, which are traditional relay, IEDs and meters, depend on local clocks, which make it difficult to compare data from two different source due inconsistent local clocks. The SCADA data involves use of estimations of quantities like voltage or line flow to obtain the phase angle. Estimated data can only give close to accurate phase angle if system is stable. The SCADA setback is summed up to be:

Insufficient time synchronised data to help plan or change system according to dynamic changes, not time synchronised hence phase is not obtained directly when power systems is frequently changing, topography of voltages and currents also affect SCADA performance.. After considering the developments that have led to the adoption of WAMS next is the physical architecture and it can vary with networks. The utilities have different topologies and technology hence the need is to design and deploy unique WAMS that meet the unique needs of the given utility. The architectures can be different but the components of a WAMS include.

- Phasor Measurement Units
- Communication means/link
- System/network Monitoring Centre

The system protection terminal SPT is part of WAMS made of PMUs, communications and system protection centre. The below diagram helps to illustrate the typical architecture of an SPT (Fini et al., 2016; Alhelou et al., 2018; Zamani et al., 2018; Alhelou et al., 2015; Njenda et al., 2018; Haes Alhelou et al., 2018).. Monitoring control and protection of a grid or power network covering wider geographical area requires WAMS, which combines measuring, control and communications (Morison et al., 2004). The WAMS setup allows acquiring all the data at the same time in real time for use by WAMS applications shown in Fig. 1.

Figure 1. WAMS supervision set-up

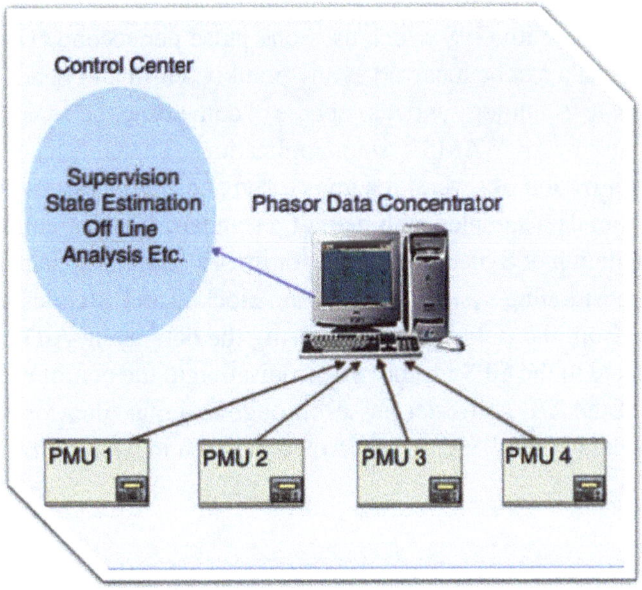

PHASOR MEASUREMENT UNIT

A Phasor Measurement Unit (PMU) according to IEEE is "a device that produces Synchronized Phasor, Frequency, and Rate of Change of Frequency (ROCOF) estimates from voltage and/or current signals and a time synchronizing signal". In simple terms, a measurement system that allows the measurement of the phasors of voltages and currents in a power system is known as the phasor measurement unit (PMU). Synchronized phasor measurement units (PMUs) from time of inception have become the ultimate data acquisition technology, which is used in wide area measurement systems with many applications that are currently under development around the globe. Phasor Measurement Units (PMUs) are currently the most accurate and advanced synchronized measurement technology on the market. The parameters measured are voltage and current phasors and frequency information synchronized with high precision to a common time reference. The measurement data from the PMUs is conveyed via Wide-Area Network (WAN).

Operating Principles and Structure of a PMU

The PMUs require a network protocol and serial communication RS-232 has been the most commonly used. Considering the challenges with serial communications, changes have been made to use internet protocols. Limits in data sent has been the major drive for utilities to migrate from serial communication to new methods based on Ethernet. The protocols include TCP only, UDP only and a combination of both. TCP uses handshake method and missing pockets of data are resend thus introducing a delay whilst for UDP no resending of data eliminating the delay introduced in TCP protocol. The combination of using TCP and UDP is a hybrid system for WAMS communication link to avoid delay but still have data with integrity. The Important factor is to have WAMS collect the real-time data of operation parameters of wide-area network with high precision, synchronized and using high-speed communication network, thus the security situation on the grid can be maintained effectively. Fig. 2 shows a typical PMU structure. The ability to choose a common reference is an in important factor on the application of PMUs in power systems using WAMS. With GPS which uses one pulse per second (1PPS) for synchronization of measurements, GPS signals can be received at any point on earth and hence any measurement. The PMUs is based on solid-state technology and it's capable of converting the analogue data into digital data for communications and use by the WAMS system/applications. The GPS receivers provide the mechanism to allow the synchronization of several readings taken at different distant locations. Synchronized phasora are calculated from data samples with help of a standard time signal as the reference point for measurements. Implementation of Synchronized phasor involves PMUs synchronizing from a reference time source of a global positioning system (GPS) radio clock. The PMU has analogue filters to filter the analogue signal/data from the field before processing the data using A/D converter. The analogue to digital converter is linked to the GPS for the synchronisation to the common time. The signal or data after processing by use of the A/D converter passes through a digital filter for further processing in the PMU. The main components of the PMU are described in depth in what follows.

Anti-Aliasing Filter

As per Nyquist-Shannon sampling theorem, to reconstruct a signal after sampling, sampling frequency must greater than twice the maximum frequency of the signal to be sampled. If lower sampling rates are used then the original signal's information may not be recoverable from the sampled signal and they may appear as aliases. So to avoid this effect, Anti-aliasing filter (AAF) is used which restricts the bandwidth of the signal to approximately satisfy the sampling theorem, for the fixed sampling frequency of the system.

Phase Locked Oscillator

Usually in PMU, the pulse signals from the satellite are phase locked with the sampling clock. Phase lock oscillator accomplishes this job. The Phase locked oscillator system is analyzed by dividing the one pulse per second signal from GPS into the required number of pulses per second for sampling. At present in most systems, this is 12 per cycle of fundamental frequency. The sampling instant could be identified, as the GPS time tag identifies the pulse number within one-second interval.

Figure 2. PMU structure

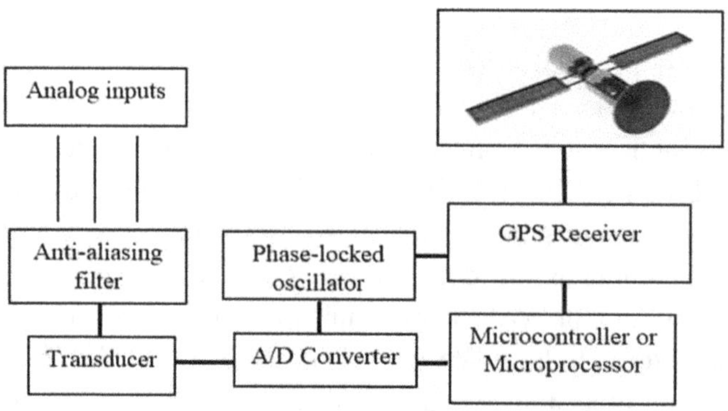

A/D Converter

The Analog to digital converter digitizes the analog signal, from the AAF,at sampling instants defined by the sampling time signals from PLO. These digitized samples are then fed to the phasor microprocessor.

Phasor Microprocessor

It is programmed to calculate the positive sequence components from the digitized sampled data by using a recursive algorithm which is usually Discrete Fourier Transform (DFT). This calculated phasor is time-tagged. All the measured data are transmitted to the remote location through a proper communication channel using modems.

GPS

Global positioning system is a satellite based navigation system, which gives information about location and time irrespective of weather conditions. It consists of a network of 24 satellites orbiting in 6 geo-synchronous orbits such that at any given instant 4 satellites are visible from any point on the earth surface. GPS provides accurate information about location and time (could be local time or Universal Time Coordinate) of the GPS receiver's location. Along with its satellite is capable of transmitting common access 'one pulse per second' with an identifier, which is accurate about 1μs at any location on earth. This pulse combined with time-tag is crucial for the application considered here (Alshahrestani et al., 2018; Makdisie et al., 2018; Alhelou et al., 2018; Alhelou et al., 2016; Nadweh et al., 2018; Njenda et al., 2018).

Sampling Process of a PMU

The diagram below shows the point we start from to attain the required data. The analogue data is derived from sensors in the system for sampling. The sinusoidal waveform with the use of GPS synchronising and Analogue to digital converter in the PMUs to produce synchrophasor component. The instruments or sensors that provide the data can include current transformers, voltage transformers, and Capacitor voltage transformers. The status of equipment like circuit breakers and relays are also signals PMUs access for control and protection. The sampling techniques depends on standards established by IEEE.

The various methods for phasor estimation as found in the literature are Zero Crossing, Discrete Fourier Transform (DFT), Sliding Discrete Fourier Transform (SDFT), Least Error Squares, Kalman Filter, Demodulation and Phasor measurement angle changing. These can be categorised/classified into DFT based, wavelet based, Optimization based and Taylor series based.Phasor estimation techniques widely implemented are Zero crossing, DFT and Sliding DFT. Variety of techniques are available in theory but few are characterized by an affordable computational complexity needed for real-time applications. The major issue arise when $f(t) \neq f0$ and during dynamic conditions of the power network considering that "phasor" concept is by definition static. A majority of the algorithms are window-based, that is for instance the synchrophasor is based on a previously acquired set estimations (data) representing a portion of the acquired waveform (current or voltage). The algorithm typical parameters include window length (half-cycle, full-cycle, multi-cycle), window type (rectangular, flat-top, raised cosine, etc.) and Sampling process (coherent, incoherent, high, slow rate)

The Categorised Techniques Used by PMUs Include:

1. **DFT Based:** The advantages are Low computational complexity, harmonic rejection. The disadvantages Spectral leakage, Harmonic interference, Off-nominal frequency Interpolated DFT
2. **Wavelet Based:** Which use Recursive wavelet. Its advantage is Harmonic rejection and its drawback Computational complexity.
3. **Optimization Based:** Usually provide accurate estimates in combination with other methods non deterministic but major drawback is that it's driven Kalman Filter by optimality criteria.
4. **Taylor Series Based:** It intrinsically reflects the dynamic behaviours of power systems but the setback of the algorithm is computational complexity.

The Fourier analysis DFT because of low computational complexity is favourable for implementation. The possible corrections on the technique includes adequate aliasing filters, increased sampling frequency, and iterative compensation of self-interaction.

$$V = V_{\mathrm{Re}} + jV_{lm} = \frac{\sqrt{2}}{N}\sum_{k=1}^{N} u_k \left[\cos\left(k\frac{2\pi}{N}\right) - j\sin\left(k\frac{2\pi}{N}\right)\right] \quad (1)$$

With DFT the calculated orthogonal components of the phasor are those of the first (fundamental) harmonic other higher harmonics and the DC component are washed out. The three-phase phasors of voltages and currents are replaced by their positive sequence components. Equation (1) show basic calculation of voltage components when using DFT.

Figure 3. PMU data acquisition

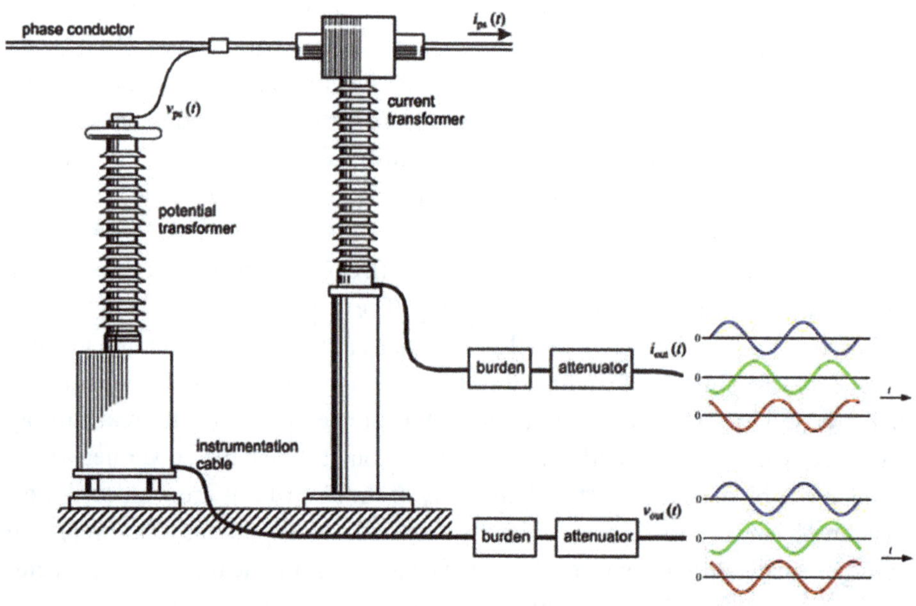

Application of PMU in Power Systems

PMU has made it possible for a number of functions and roles to be possible today in the power system field. Explained below are some of the major applications of the PMU.

State Estimation

PMUs have become particularly useful and vital in power system state estimation (SE). Real power, reactive power injections and flows are utilized in conventional state estimation algorithms for estimation of system states (voltage amplitude and phase angle). State estimation is very important in real-time monitoring and control of the power system because it processes redundant measurements and provides steady-state operating state for advance Energy Management System (EMS) application programs. It is known that traditional state estimation uses measured voltage, current, real power and reactive power to determine the operating condition of the electric network. Traditional state estimation has some limitations associated with it and these limitations are: (i) It is technically challenging and computationally more expensive to estimate the likely state of the power system based on measured parameters such as voltages, current, real power and reactive power. (ii) Traditional state estimation is usually solved in one minute interval, and this means that the result provided by the approach may be old. To overcome these technical and challenging difficulties, improved computational ability together with synchronized phasor measurement units (PMUs) are deployed to provide globally time synchronized phasor measurements with accuracy of one microseconds for bus voltages and line currents.

PMUs have the ability to improve SE accuracy by measuring voltage angles directly, which are the state variables to be estimated. The addition of voltage phase angle measurements to a traditional SE is capable of greatly increasing the accuracy of SE.

This is the most important part of monitoring a power system and PMUs can transmit information to data acquisition centers continuously. Based on the PMU data a state vector is developed to trace the dynamics of the power system. One phasor measurement every 2-5 cycles (33.3 - 83.33msec) can be sustained though with advancement in technology this has significantly improved.

Figure 4. PMU sampling process

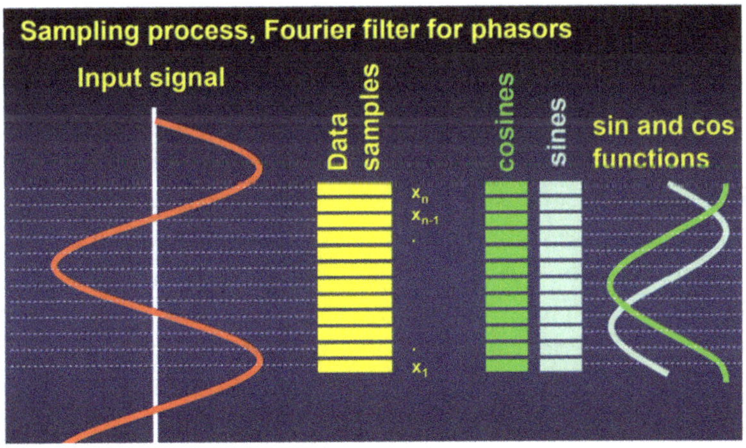

Oscillation Detection and Control

The increase in the amount of electric power that is transmitted through transmission lines is capable of causing transmission bottlenecks and oscillations of power transmission systems. Basically, system oscillations originate from interconnected generators in power system. The deviation of one generator from the synchronous speed will prompt other generators in the system to provide power in order to reduce the speed deviation. However, the effect of inertia of the generators could creep in causing the entire system or part of the system to swing. In most cases, this disturbance might not be significant and would decay slowly. However, in other times, it might be very significant and does not decay which causes the system to lose synchronism and a final collapse of the system. PMUs could be used as the eye of the power system to detect oscillations early enough before they lead to critical consequences. One good example is the detection of a 0.8 Hz oscillatory mode in Eastern Denmark believed to have been of inter-area nature involving generators in Sweden and Eastern Denmark. This discovery was accomplished by the DTU PMU produced at the Technical University of Denmark.

Voltage Stability Monitoring and Control

Severe changes in system conditions is enough to change the operating system of a power system dynamically. Such changes affect the voltage phasor, current phasor and system frequency. Voltage stability has been known to be related to the loadability of a transmission network. Different research has shown that voltage magnitude and phase angles are the best indicators of voltage stability margin. The use of synchronized phasor measurements units (PMUs) to improve voltage stability monitoring and control has become very efficient. This is possible through the following applications: Before WAMS stability analysis of power systems was done off-line. Nowadays synchronized phasor measurements allows real-time stability analysis of the power system. Based on state vector behavior instability prediction can easily be done.

- Voltage Instability Load Shedding (VILS)
- Wide Area Voltage Stability Monitoring and Control

Under the VILS approach, PMUs are used to monitor the electric network to track how close the transmission system is to its loadability limit. If the system is very close to that limit, and every other approach has been exhausted to avoid voltage collapse without significant improvement, the VILS scheme is then deployed as the last safety option to prevent voltage collapse in the system. The shedding of load is done in pre-defined blocks that are usually triggered in stages due to scheduled maintenance, and unexpected disturbances. PMUs provide measurement-based on-line voltage stability monitoring and control which has the ability to improve the power transfer limits and increase security of the system operation. Under the Wide Area Voltage Stability Monitoring and Control scheme, computer simulation tools are used to help operators monitor and control system voltage stability. One good tool used is a Voltage Stability Assessment (VSA) program which relies on state estimators to provide steady-state solution.

Controllers such as variable series capacitors (VSC), unified power flow controllers (UPFCs) and power system stabilizers (PSS) regulate the grid based on a local feedback. The challenge with this scheme is model parameter might be different from the actual system conditions at any time. Using PMU the power system model can be realized in real time.

Protection of Power Systems

Examples of protection problems include: Protection of series compensated lines, protection of multi-terminal lines, and the inability to satisfactorily set out-of-step relays. The reliable measurement of voltage or current is possible to offer substantial improvement in protection functions. Furthermore, the communication of such measurements from one end of a protected line to another is necessary to improve protection functions. These measurements offered by PMUs are effective in improving protection functions, which have relatively slow response time. The latency of remote measurements for such protection functions is not really a significant issue. The differential protection of buses, transformers and generators is a well-established protection principle that emphasizes the importance of synchronized phasor measurements. True differential protection was not possible before the introduction of synchronized phasor measurements. Differential protection is important for series compensated lines and tapped lines. Differential protection involves communication wires or communication band channels. The easy availability of synchronized measurements using GPS technology and the improvement in communication technology makes it possible to consider true differential protection of transmission lines and cables (Fini et al., 2016; Alhelou et al., 2018; Zamani et al., 2018; Alhelou et al., 2015; Njenda et al., 2018; Haes Alhelou et al., 2018).

Adaptive Relaying

Power swings on the grid are observed by a distance relay, which checks the line impedance changes (Centeno et al., 1997). Before WAMS, the relays were set based on possible system disturbances. This poses a threat if the disturbance is different from the anticipated ones. With PMUs, relays can be set to be adaptive to any system condition.

PMU Link to Power System

To understand how the PMU is connected to the power system we use Fig. 5. From the diagram it can be seen that the current transformer (CT) is connected in series and the voltage transformer (VT) is connected in parallel across any two phases. The CT and VT measures the current and voltage of the power system respectively. The measured signals are passed through the burden attenuation and transformed to a format readable by the PMU. The burden attenuator also makes sure that the PM is not damaged by signal from the VT and CT.

WAMS IN POWER SYSTEMS

The ever-increasing demand for power has led to power systems being operated near steady state stability limit. Consequently, this calls for almost real-time monitoring and control of power systems to maintain or improve their stability. Employing wide area measurement systems (WAMS) will go a long way in achieving this monitoring and control objectives. Conventional SCADA (supervisory control and data acquisition) / EMS (energy management systems) do not provide for dynamic coverage of a wide area power network as they are mainly designed for local operation. A further comparison between SCADA and PMU use is given in Table 1. With PMUs, more samples are obtained per second, there is measure-

Figure 5. Basic representation of PMU link to power network

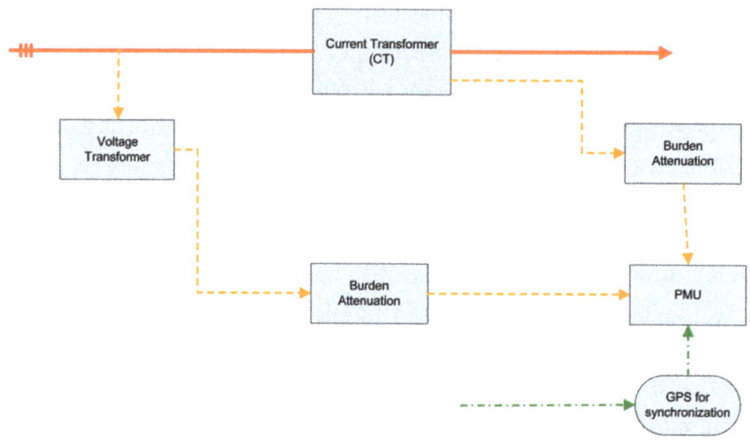

ment of phasors and are used in wide area monitoring and control which indicates their superiority of SCADA (Alshahrestani et al., 2018; Makdisie et al., 2018; Alhelou et al., 2018; Alhelou et al., 2016; Nadweh et al., 2018; Njenda et al., 2018).

With WAMS based on PMU measurements, this total coverage can be realized. In the event of a serious contingency that threatens the stability of the power system or that may lead to a severe frequency decline, PMU measurements can be used to set up a real time WAMS-based UFLS. A good example on the need to use PMUs is the WECC system break up which occurs in 1996. Fig. 6 shows the estimated and what happed in reality.

With WAMS, the advantage of measuring voltage and current phasors with such a high accuracy has brought more advantages in a power system control. Some of the general applications include

1. Wide Area Monitoring (WAM) of power systems, which cover vast geographical areas. Here voltage angles, magnitudes, and frequency are easily monitored in real time.
2. Wide Area Protection (WAP) of the power system. Based on phasor measurements special protection systems can be developed.
3. Wide Area Control (WAC) of power systems. Control actions can be given from a control location to control or regulate any part of the system.

Figure 6. Comparisons between SCADA and PMU

ATTRIBUTE	SCADA	PMU
Resolution	1 sample every 2-4 Seconds (steady State Observability)	10-60 samples per second (Dynamic Observablity)
Measured Quantities	Magnitude only	Magnitude and phase angle
Time Synchronization	No	Yes
Total Input/output Channels	100+ Analog & Digital	~10 Phasors, 16+ digital, 16+ analog
Focus	Local monitoring and control	Wide area monitoring and control

WAMS combine WAM, WAP and WAC and is also known as Wide Area Measurement, Protection and Control (WAMPAC). In this thesis we will simply use the term WAMS. WAMS rely on phasor measurements for improved stability and system reliability. With PMUs used in WAMS, the following can be easily achieved.

In summary the application of WAMS are

1. Analysis of faults, the capacity to store data continuously helps to carry out investigations after every fault or disturbance. The valuable information assist in making improvements and plan for expansion and reinforcement to the grid/system.
2. Management and monitoring of transmission grid.
3. Analysis of dynamic system behaviour.
4. Distribution and costumer side automation.

General System Setup for WAMS Based UFLS

Using WAMS, PMUs are located at specific positions throughout a wide network coverage area. These PMUs form part of local devices commonly referred to as system protection terminals (SPT). SPTs are the backbone of control algorithms, which facilitates communication with other SPTs, substation equipment and system protection centers (SPC). PMUs use the IEEE 1344 data format for communication and they measure voltage, current, and frequency phasors using the discrete Fourier transform (DFT). Using PMUs, power system transients or surges are detected quickly. This advantage to detect fast transients is the main reason behind the use of WAMS in UFLS. Fig. 7 shows the general setup of WAMS (Terzija et al., 2011). At a real time monitoring unit center specific control commands are issued.

COMMUNICATION IN WAMS

As a vital part of WAMS, communication plays a critical role. The communication link makes it possible to achieve WAMS. The delay in communication affects effectiveness of WAMS system. A delay in data transfer affects or results as a delay in the response of the system. The below pictorial explanation helps to highlight the impact of communication links.

Figure 7. Comparison of actual system dynamics and estimated result from 1996 WECC event

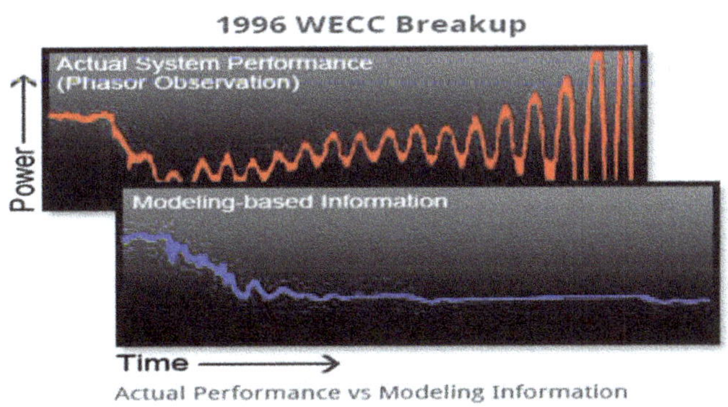

Figure 8. WAMS set-up in power systems

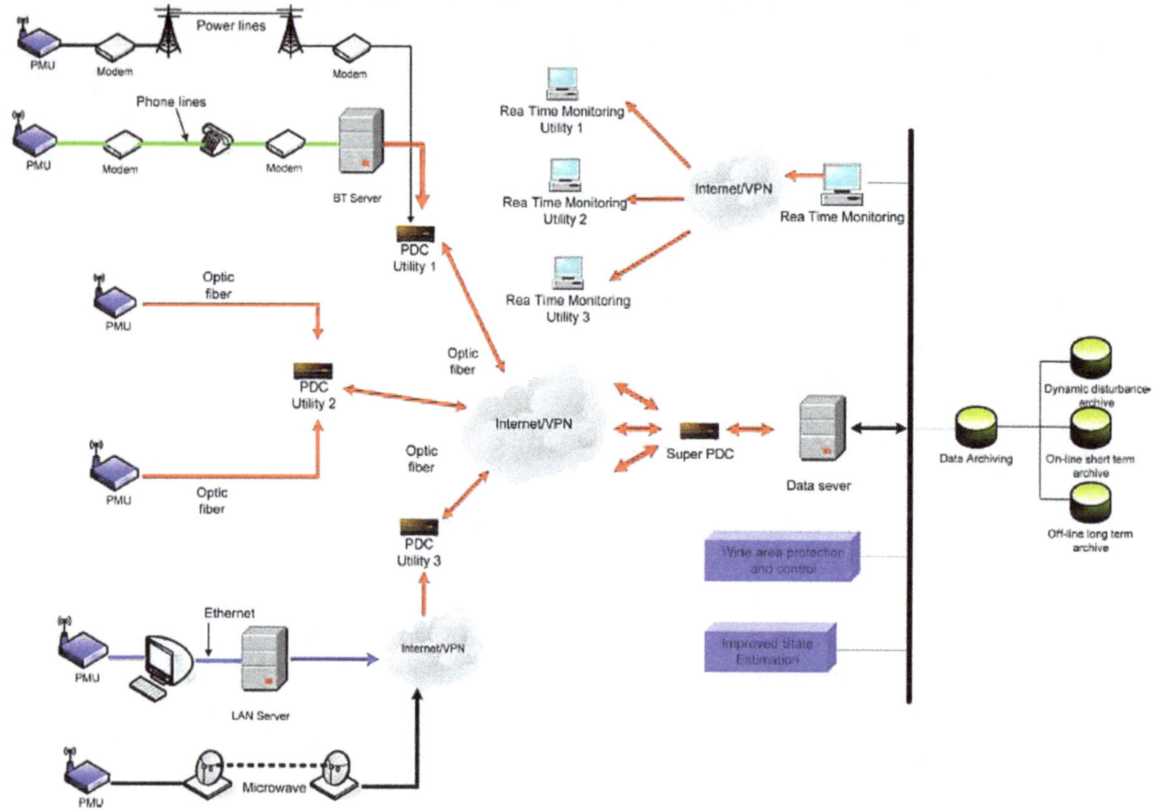

The above illustration shows that a break in the communication link has serious impact on WAMS. Special attention should be paid to communication equipment/infrastructures that is the medium, security, cost and propagation delay. Reliability and capacity of the communication link affects the overall reliability and availability of the grid when WAMS is involved (Fini et al., 2016; Alhelou et al., 2018; Zamani et al., 2018; Alhelou et al., 2015; Njenda et al., 2018; Haes Alhelou et al., 2018).

Figure 9. Communication link in WAMS

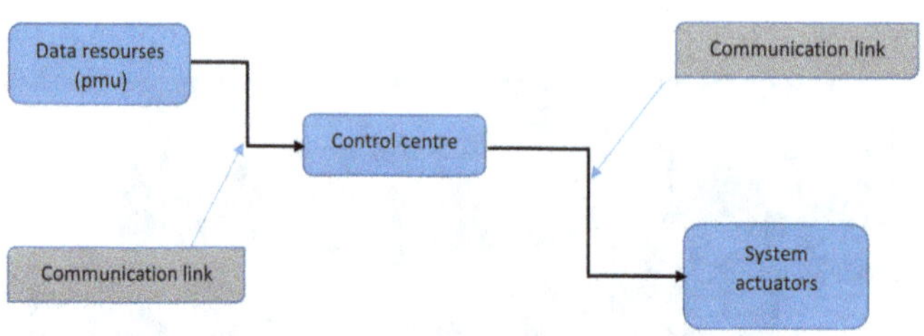

Available Communication Options

WAMS Based on GPS Signal

Here Low-earth orbiting (LEO) satellites and GPS referencing are used to exchange data between the PMUs and the SPCs. The high accuracy of the GPS reference time is about 1µs. This accuracy is very good to measure AC phasors, which have a 50 or 60 Hz frequency. If we consider a 50 Hz system, the period time corresponding to a full 360orotation is 20 ms = 20 000 µs. By propotion a time error of 1 µs corresponds to the angle error of 360o/20 000 which is equal to 0.018o. If we say 360/0.018o, we get an error of 0.005% which is small enough for phasor measurement. Such an error is small enough from the point of view of phasor measurement. With WAMS, which rely on satellite, GPS (Global Positioning System) a very high accuracy of at least 1µs is obtained which makes WAMS the best choice to use when implementing quick response commands. The main disadvantages with using this channel are its high cost and narrow bandwidth. A good example where this has been used is the remote substation SCADA. The Southern California Edison (SCE) is recognized as the first, to use satellite system for substation SCADA (Moore, 1999).

Telephone Lines

Telephone lines form the basis of most utility communications because they are fairly cheap and easy to set up. They offer up to 56 kbps analog data. Dial-up or leased lines can also be used though they are expensive.

Figure 10. WAMS based on GPS

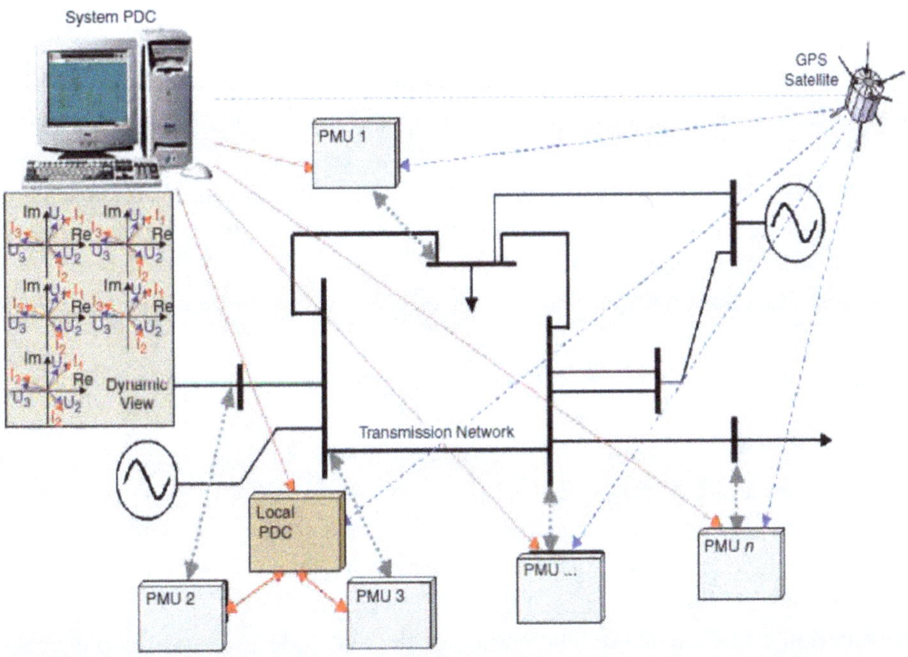

Fiber-Optic Cables

This is the most preferred option by power system operators. Optic fiber offer data rates ranging from 50 million to 1 billion bits per second. Only a small part of its large bandwidth is required for use in WAMS. The other portion of bandwidth can be used for other services such as internet access and videoconferencing. This is an added advantage as power utilities can generate more revenue in that way (Lamarre, 1997). The main advantages of optic fiber are its immunity to radio frequency atmospheric interference, and the large amounts of bandwidth. The high initial investment and the lack of expertise within many utilities are the disadvantages of this technology.

Power Lines

This technology utilizes the medium and low voltage electric supply grid for transmission of data and voice. It is a considerable option for WAMS as it offer data rate up to of 4Mbps via the electricity supply grid (Radford, 1996). Spectrum techniques such as the direct sequence spread spectrum (DSSS) and frequency hopping spread spectrum (FHSS) or orthogonal frequency division multiplexing (OFDIVI) can be used to improve the bit error rate (BER) (Dostert, 1990).

Microwave Links

These are also used by utility companies and are preferred to leased lines. Digital microwave links with frequency bands less than 10 GHz have propagation distances of up to 30miles while those with greater than 10 GHz only cover less than 5miles. The main disadvantages with this technique is signal fading thereby requiring boosters and multipath propagation.

Causes of Communication Delay

Communication delays are always inevitable and can affect the performance of WAMS (Yook et al., 2000). Delays, which arise with PMUs and the communication link, are as follows.

- **Transducer Delays:** Voltage transducers (VT) and current transducers (CT) which are used to convert high voltages and currents to smaller values introduce some delays for transferring quantities from primary and secondary.
- **DFT Window Size:** This refers to the number of samples required to compute the phasors using DFT.
- **Processing Time:** The time required in converting the transducer data, into phasor information using DFT.
- **Data Size of the PMU Output:** The size of the information bits contained in the data frame, header frame and the configuration frame.
- **Multiplexing and Transitions:** Transitions between the communication link and the data processing equipment leads to delays that are caused at the instances when data is retrieved or emitted by the communication link.
- **Communication Link Involved:** The choice or the available link and the physical distance involved in transmitting the PMU output to the central processing unit can add to the delay.

- **Data Concentrators:** These are data collecting centers that are 1ocated at the central processing unit. They are responsible for collecting all the PMU data transmitted over the communication link.

Delay Calculations

Delay is a critical factor with WAMS as it indicate the usability of a particular communication channel. Large amounts of communication delay result in slower control and protective actions. Communication delay can be calculated as follows (Naduvathuparambil et al., 2002):

$$\tau = \tau_f + \tau_p + \frac{L}{R} + \theta \qquad (2)$$

where τ is the total link delay, τ_f is the fixed delay associated with transducers used, DFT processing, data concentration and multiplexing, τ_p is the link propagation delay, L is the amount of data transmitted, R is the data rate of the link, and θ is the associated random delay jitter.

For standardization purposes, delays can be calculated by assuming that there are 10 - 12 phasor measurements, each 4 bytes in length, 10 input status channels, each 2 bytes in length. The combined delay caused by processing, concentrators, multiplexing and transducers is approximately 75ms, which is a fixed delay regardless of the communication medium used. Nevertheless, in today's systems this delay has significantly reduced. As an example table 2 shows some delays associated with various communication channels when using PMUs in WAMS. The communication links are options that can be used for real-time stabilization of power system in the event of a contingency that can lead to large area cascaded outages.

For example if we consider a WAMS network based on optic fiber

$$\tau_f = 75ms$$

$$\tau_p = 25ms$$

$$\frac{L}{R} + \theta \approx 0,$$

this assumption holds true because optic cables do not contribute to the delay associated with the length of the PMU packet because of their high data rates, and hence the associated delay can be considered negligible. Therefore,

$$\tau_{optic-fiber} = 75 + 25 + 0 = 100ms \qquad (3)$$

Table 1. Communication link delays

Communication Link	Associated Delay on Way (Milliseconds)
Fiber-optic cables	≈100-150
Digital microwave links	≈100-150
Power line (PLC)	≈150-350
Telephone lines	≈200-300
Satellite link	≈500-700

RECOMMENDED IEEE DATA FORMAT

Based on the IEEE 1344 standard, issues to do with synchronization of data sampling, data to phasor conversions, and formats for timing input and phasor data output from the PMU are addressed (Martin et al., 1998). The standard helps in setting a basis for phasor measurements and to simplify exchange of data between power system operators that are involved in monitoring and control. The PMU transmits messages (according to the IEEE 1344 format) in three types of frames which are:

- **Data Frame:** This provides the information related to phasor data and the state of the digital inputs on each channel. Frequency, angle, over-current, under-voltage and rate of frequency changes are define here.
- **Header Frame:** This is an ASCII file, it contains identification of information about the PMU data source, algorithms, and the analog filters used.
- **Configuration Frame:** The PMU data is converted to machine-readable binary file. Each PMU has an identification code, number of phasors, and number of digital channels, channel and phasor names, nominal line frequency, and the transmission period of the phasors. The PMU also receives messages that indicate the instance of start or stop the transmission of data from the PMU. The data format for this primarily consists of the time-tag and the command code, which informs the PMU of the action that needs to be taken.

CONCLUSION

With WAMS real time control of power systems can be easily done. In the event of a fault which causes frequency to below the acceptable thresholds based on the center of inertia frequency the disturbance magnitude is easily calculated. Adaptive tuning of relays is also done and there is no over or under shedding like in conventional schemes.

REFERENCES

Alhelou, H., Hamedani-Golshan, M. E., Zamani, R., Heydarian-Forushani, E., & Siano, P. (2018). Challenges and Opportunities of Load Frequency Control in Conventional, Modern and Future Smart Power Systems: A Comprehensive Review. *Energies*, *11*(10), 2497. doi:10.3390/en11102497

Alhelou, H. H. (2018). *Fault Detection and Isolation in Power Systems Using Unknown Input Observer. In Advanced Condition Monitoring and Fault Diagnosis of Electric Machines* (p. 38). Hershey, PA: IGI Global.

Alhelou, H. H., Golshan, M., & Fini, M. (2018). Wind Driven Optimization Algorithm Application to Load Frequency Control in Interconnected Power Systems Considering GRC and GDB Nonlinearities. *Electric Power Components and Syst.*

Alhelou, H. H., & Golshan, M. E. H. (2016, May). Hierarchical plug-in EV control based on primary frequency response in interconnected smart grid. In *Electrical Engineering (ICEE), 2016 24th Iranian Conference on* (pp. 561-566). IEEE. 10.1109/IranianCEE.2016.7585585

Alhelou, H. H., Golshan, M. H., & Askari-Marnani, J. (2018). Robust sensor fault detection and isolation scheme for interconnected smart power systems in presence of RER and EVs using unknown input observer. *International Journal of Electrical Power & Energy Systems*, *99*, 682–694. doi:10.1016/j.ijepes.2018.02.013

Alhelou, H. H., Hamedani-Golshan, M. E., Heydarian-Forushani, E., Al-Sumaiti, A. S., & Siano, P. (2018, September). Decentralized Fractional Order Control Scheme for LFC of Deregulated Nonlinear Power Systems in Presence of EVs and RER. In *2018 International Conference on Smart Energy Systems and Technologies (SEST)* (pp. 1-6). IEEE. 10.1109/SEST.2018.8495858

Alhelou, H. S. H., Golshan, M. E. H., & Fini, M. H. (2015, December). Multi agent electric vehicle control based primary frequency support for future smart micro-grid. In *Smart Grid Conference (SGC)* (pp. 22-27). Academic Press.

Alshahrestani, A., Golshan, M. E. H., & Alhelou, H. H. (2018, November). WAMS Based Online Estimation of Total Inertia Constant and Damping Coefficient for Future Smart Grid Systems. In *Smart Grid Conference (SGC)* (pp. 1-5). Academic Press.

Centeno, V., Phadke, A. G., Edris, A., Benton, J., Gaudi, M., & Michel, G. (1997). An adaptive out-of-step relay for power system protection. *IEEE Transactions on Power Delivery*, *12*(1), 61–71. doi:10.1109/61.568226

Dostert, K. M. (1990). Frequency-hopping spread-spectrum modulation for digital communications over electrical power lines. *IEEE Journal on Selected Areas in Communications*, *8*(4), 700–710. doi:10.1109/49.54466

Fini, M. H., Yousefi, G. R., & Alhelou, H. H. (2016). Comparative study on the performance of many-objective and single-objective optimisation algorithms in tuning load frequency controllers of multi-area power systems. *IET Generation, Transmission & Distribution*, *10*(12), 2915–2923. doi:10.1049/iet-gtd.2015.1334

Lamarre, L. (1997). At home with telecommunications. *EPRI Journal*, *22*(1), 6–16.

Machowski, J., Janus, W. B., & Bumby, J. R. (1988). *Power System Dynamics*. Academic Press.

Makdisie, C., Haidar, B., & Alhelou, H. H. (2018). An Optimal Photovoltaic Conversion System for Future Smart Grids. In *Handbook of Research on Power and Energy System Optimization* (pp. 601–657). IGI Global. doi:10.4018/978-1-5225-3935-3.ch018

Martin, K. E., Benmouyal, G., Adamiak, M. G., Begovic, M., Burnett, R. O., Carr, K. R., ... Michel, G. L. (1998). IEEE standard for synchrophasors for power systems. *IEEE Transactions on Power Delivery*, *13*(1), 73–77. doi:10.1109/61.660853

Moore, T. (1999). Utility communications go into orbit. *EPRI Journal*, *24*(3), 18–27.

Morison, K., Wang, L., & Kundur, P. (2004). Power system security assessment. *IEEE Power & Energy Magazine*, *2*(5), 30–39. doi:10.1109/MPAE.2004.1338120

Naduvathuparambil, B., Valenti, M. C., & Feliachi, A. (2002). Communication delays in wide area measurement systems. In *System Theory, 2002. Proceedings of the Thirty-Fourth Southeastern Symposium on* (pp. 118-122). IEEE. 10.1109/SSST.2002.1027017

Nadweh, S., Hayek, G., Atieh, B., & Haes Alhelou, H. (2019). Using Four – Quadrant Chopper with Variable Speed Drive System Dc-Link to Improve the Quality of Supplied Power for Industrial Facilities. Majlesi Journal of Electrical Engineering.

Njenda, T. C., Golshan, M. E. H., & Alhelou, H. H. (2018, November). WAMS Based Intelligent Under Frequency Load Shedding Considering Online Disturbance Estimation. In *Smart Grid Conference (SGC)* (pp. 1-5). Academic Press.

Njenda, T. C., Golshan, M. E. H., & Alhelou, H. H. (2018, November). WAMS based Under Frequency Load Shedding Considering Minimum Frequency Prediction and Extrapolated Disturbance Magnitude. In *Smart Grid Conference (SGC)* (pp. 1-5). Academic Press.

Radford, D. (1996). Spread spectrum data leap through AC power wiring. *IEEE Spectrum*, *33*(11), 48–53. doi:10.1109/6.542275

Terzija, V. V., Valverde, G., Cai, D., Regulski, P., Madani, V., Fitch, J., ... Phadke, A. G. (2011). Wide-area monitoring, protection, and control of future electric power networks. *Proceedings of the IEEE*, *99*(1), 80–93. doi:10.1109/JPROC.2010.2060450

Yook, J. K., Tilbury, D. M., & Soparkar, N. R. (2000). A design methodology for distributed control systems to optimize performance in the presence of time delays. In *American Control Conference, 2000. Proceedings of the 2000* (Vol. 3, pp. 1959-1964). IEEE. 10.1109/ACC.2000.879544

Zamani, R., Hamedani-Golshan, M. E., Haes Alhelou, H., Siano, P., & Pota, H. (2018). Islanding Detection of Synchronous Distributed Generator Based on the Active and Reactive Power Control Loops. *Energies*, *11*(10), 2819. doi:10.3390/en11102819

KEY TERMS AND DEFINITIONS

Energy Management: Energy management includes planning and operation of energy production and energy consumption units. Objectives are resource conservation, climate protection, and cost savings, while the users have permanent access to the energy they need.

Phasor Measurement Unit: A phasor measurement unit (PMU) is a device used to estimate the magnitude and phase angle of an electrical Phasor quantity like voltage or current in the electricity grid using a common time source for synchronization.

Power System: An electric power system is a network of electrical components deployed to supply, transfer, and use electric power. An example of an electric power system is the grid that provides power to an extended area.

SCADA: Supervisory control and data acquisition (SCADA) is a system of software and hardware elements that allows industrial organizations to control industrial processes locally or at remote locations and monitor, gather, and process real-time data.

Selectivity: The property by which only the faulty element of the system is isolated, and the remaining healthy sections are left intact.

Wide Area Measurement System: Wide area measurement system (WAMS) is technology to improve situational awareness and visibility within power system of today's and future grids. It uses real time synchro phasor data to measure the state of grid that enables improvement in stability and reliability of power grid.

Chapter 13
EMC Installation for Variable Speed Drive Systems (VSDs):
Fields, Emissions, Coupling, and Shielding

Safwan Nadweh
Tishreen University, Syria

Zeina Barakat
Tishreen University, Syria

Ghassan Hayek
Tishreen University, Syria

ABSTRACT

This chapter introduces EMC installation for variable speed drive systems. As an introduction, EMC standards have been mentioned in order to define the requirements characteristics, besides the fundamentals of static, electric, magnetic, and electromagnetic fields. Both inductive and capacitive coupling have been discussed to deal with shielding. Finally, VSDS emission and electromagnetic interferences were studied with installation requirement in VSDS (supply cables, cable between converter and motor, control cables, earthing requirements, and grounding).

INTRODUCTION

Today, alternating driving has become the system used in most modern industries, so the development of these systems has led to the emergence of driving systems operating at variable speeds to ensure the requirements of torque and speed (Armstrong, 2006; Armstrong, 2007; Alhelou et al., 2016).

The variable speed drive systems are used in industrial facilities for many purposes, the most important of which are energy saving, compatibility between driving speed and desired speed requirements, improving the life of electronic equipment as a result of smoothing start and increasing the life of mechanical equipment. Figure: 1 show the Basic elements of VSD power circuit.

DOI: 10.4018/978-1-5225-8030-0.ch013

Figure 1. Basic elements of VSD power circuit

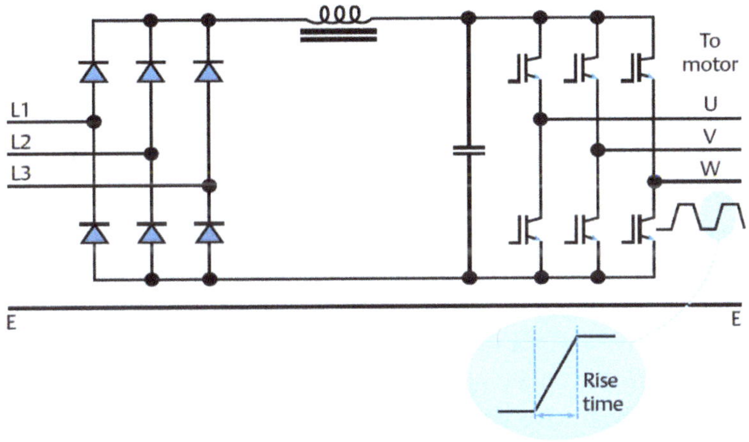

The variable-speed drive systems, like other equipment, are subject to EMC, then enters the phase of the smoothing, and finally returns to the alternating state by means of the different modulation techniques (PWM, SPWM, SVPWM, ... etc.).

The inverter consists of IGBT transistors or other electronic components with a on and off time between 50-200 ns at high values for the desired voltage and currents (IEC TS 61000-1-2, 2008).

Usually the switching frequency is between 2-20 kHz. The inverter uses different electronic elements and sometimes they are controlled by different electron elements and microcontroller with up to 40 MHZ of on/off frequency (Steenstra et al., 2008; Williams et al., 2012).

Variable speed drive systems use large cooling fans for electronic equipment, which is considered to be a very sensitive VSD infrastructure which usually forms the basic structure of drive and is convenient low inductance base for EMC of cable screen, etc.

Harmonics generated by an individual drive are unlikely to cause interference, but they are cumulative so that an installation containing a high proportion of drive loads may cause difficulties (Fini et al., 2016; Alhelou et al., 2018; Zamani et al., 2018; Alhelou et al., 2015; Njenda et al., 2018; Haes Alhelou et al., 2018).

Apart from supply harmonics, emission also occurs as a result of the switching of the power output stage over a wide range of frequencies which are harmonics of the basic switching frequency – that is, size times the supply frequency for a 6-pulse DC drive, and the PWM carrier frequency for a PWM drive. This covers a range extending from 300Hz, for DC drives, up to many MHz for AC drives. Unwanted electromagnetic coupling is relatively unusual at frequencies below about 100kHz. Few standards set limits in that range, and interference problems are unusual.

The power stage of a variable speed drive is a potentially powerful source of electromagnetic emission (noise), because of the high voltage and current which is subject to rapid switching. The power output connections of a drive carry the highest level of high frequency voltage. They can be a powerful source of electromagnetic emission.

EMC VARIABLE SPEED DRIVE STANDARDS

EN/IEC 61800-3

EN/IEC 61800-3 includes requirements for VSDS input and/or output voltages (line-to-line) of up to 35 kV AC rms, rated from a few hundred watts to hundreds of MW, connected in residential, commercial or industrial locations. It claims to achieve EMC for such applications, but it has warned that it cannot cover risky cases that could occur. It also warns that it does not cover the EMC performance in the case of faults happening. Specifies emissions limits, immunity levels, and test methods for a VSDS according to its intended application and the competency of its installer. It does this by defining two different electromagnetic "environments" or locations, and four different "Categories" of installer competency.

A full definition of the different environments, product categories and performance requirements can be found in the standard, but below is a summary.

TESTING TO EN/IEC 61800-3

This Guide is only concerned with the use of good EMC engineering installation practices to help VSDS comply with EN/IEC 61800-3, it doesn't cover the real testing techniques that are used. REO have seventeen Guides that detail how to actually perform the various tests required including the situation where the test cannot be carried out at a "controlled EMC site", such as inside a test laboratory anechoic chamber.

The guidance on what to do when testing has to be done on-site, or when the mains current required by a VSDS exceeds 100A. If further guidance is required on on-site testing, may prove useful.

COMPLYING WITH THE EMC DIRECTIVE

It is not the aim of this Guide to discuss how a manufacturer should go about legally complying with the EMC Directive, as implemented in the various national laws of the European Union's (EU's) Member States, for the United Kingdom (UK).

Table 1. Classification of categories C1 to C4 as per AS 61000.3.2005

Classification by Category	C1	C2	C3	C4
Environment	1st Environment	1st & 2nd Environment	2nd Environment	2nd Environment
Voltage/Current	>1000v	>1000v	>1000v	<1000v In400 < A Connection To It Network.
EMC Expertise	No Requirement	Installation And Commissioning By An EMC Expert		EMC Plan Required
Limits According To En55011	Class B	Class A1 (Plus Warning Notice)	Class A2 (Plus Warning Notice)	Values Exceed Class A2

STATIC ELECTRIC AND MAGNETIC FIELDS

Static Electric

Electrical Charges Protons, neutrons and electrons are very different from each other. They have their own properties, or characteristics. One of these properties is called an electrical charge. Protons have what we call a "positive" (+) charge. Electrons have a "negative" (-) charge. Neutrons have no charge, they are neutral (Hosseini et al., 2011; Alhelou et al., 2018).

The charge of one proton is equal in strength to the charge of one electron. When the number of protons in an atom equals the number of electrons, the atom itself has no overall charge, it is neutral. Static electricity is the imbalance of positive and negative charges as shown in Fig. 2.

Static electricity is an imbalance of electric charges within or on the surface of a material. The charge remains until it is able to move away by means of an electric current or electrical discharge. Static electricity is named in contrast with current electricity, which flows through wires or other conductors and transmits energy.

A static electric charge can be created whenever two surfaces contact and separate, and at least one of the surfaces has a high resistance to electric current (and is therefore an electrical insulator). The effects of static electricity are familiar to most people because people can feel, hear, and even see the spark as the excess charge is neutralized when brought close to a large electrical conductor (for example, a path to ground), or a region with an excess charge of the opposite polarity (positive or negative). The familiar phenomenon of a static shock – more specifically, an electrostatic discharge – is caused by the neutralization of charge.

ELECTRIC FIELDS

Fundamentals of Electric Fields

An electric field is a vector field nearby an electric charge that applies force on other charges, attracting or preventing them. Mathematically the electric field is a vector field that companions to each point in space the force, which called Coulomb force, that would be practiced per unit of charge, by an infinitesimal test charge at that point as shown in figure: 3.

Figure 2. Positive and negative charges

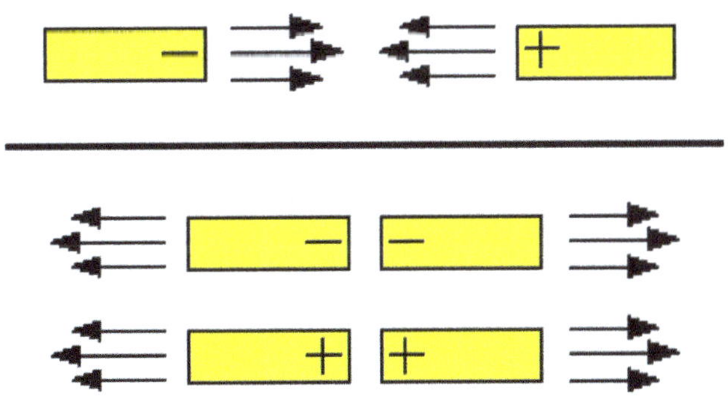

Figure 3. Electric field emanating from a point positive electric charge suspended over an infinite sheet of conducting material

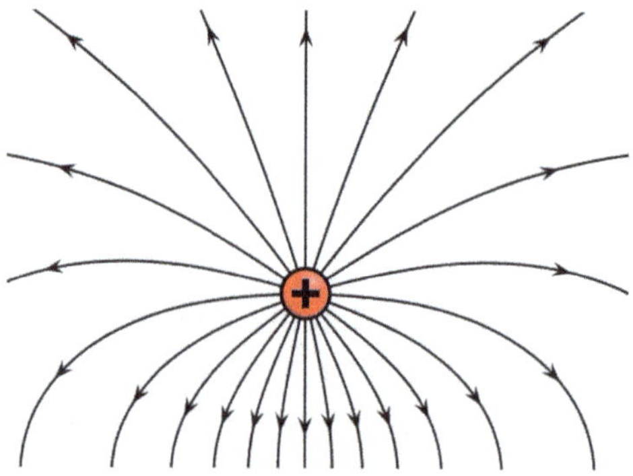

Electric fields are produced by electric charges, and by time-varying magnetic fields. Electric fields are important in many areas of physics, and are dominated practically in electrical technology. On nuclear scale, the electric field is responsible for the attractive force between the atomic nucleus and electrons that holds atoms together, and the forces between atoms that cause chemical bonding. The electric field and the magnetic field together form the electromagnetic force, one of the four fundamental forces of nature.

ELECTRIC FIELDS CAUSES AND DESCRIPTION

Electric fields are produced by electric charges, described by Gauss's law, or varying magnetic fields, defined by Faraday's law of induction. Together, these laws are sufficient to define the performance of the electric field as a function of charge repartition and magnetic field. However, since the magnetic field is described as a function of electric field, the equations of both fields are coupled and together form Maxwell's equations that describe both fields as a function of charges and currents (Fini et al., 2016; Alhelou et al., 2018; Zamani et al., 2018; Alhelou et al., 2015; Njenda et al., 2018; Haes Alhelou et al., 2018).

MAGNETIC FIELDS

Magnetic Fields Definition

A region around a magnetic material or a moving electric charge within which the force of magnetism acts.

A magnetic field is a vector field which defines the magnetic effect of electrical currents and magnetized materials. In everyday life, the effects of magnetic fields are most readily come across with nearby permanent magnets, which pull on magnetic materials and attract or repel other magnets. Magnetic fields surround and are produced by magnetized material and by moving electric charges such as those used in electromagnets. Magnetic fields apply forces on nearby moving electrical charges and torques on nearby magnets. In addition, a magnetic field that varies with location exerts a force on magnetic materials.

Figure 4. Electric field (lines with arrows) of a charge (+) induces surface charges (red and blue areas) on metal objects due to electrostatic induction

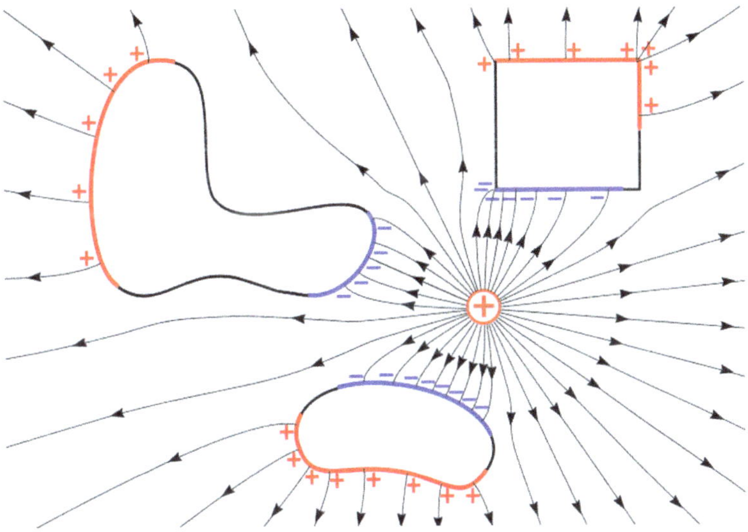

Both the strength and direction of a magnetic field varies with location. As such, it is an example of a vector field.

The term 'magnetic field' is used for two different but closely related fields denoted by the symbols B and H. In the International System of Units, H is measured in units of amperes per meter and B is measured in teslas or newton's per meter per ampere. H and B differ in how they account for magnetization. In a space, B and H are the same separately from units; but in a magnetized material, and H change by the magnetization M of the material at that point in the material. Figure: 5 introduce a comparison of B, H and M inside and outside a cylindrical bar magnet (Alshahrestani et al., 2018; Makdisie et al., 2018; Alhelou et al., 2018; Alhelou et al., 2016; Nadweh et al., 2018; Njenda et al., 2018).

Magnetic Fields Producing

Magnetic fields are produced by moving electric charges and the intrinsic magnetic moments of elementary particles associated with a fundamental quantum property, their spin. Magnetic fields and electric fields are interrelated, and are both components of the electromagnetic force, one of the four fundamental forces of nature.

Magnetic Fields Uses

Magnetic fields are widely used throughout modern technology, particularly in electrical engineering and electro mechanics. Rotating magnetic fields are used in both electric motors and generators. The collaboration of magnetic fields in electric devices such as transformers is studied in the discipline of magnetic circuits. Magnetic forces give information about the charge carriers in a material through the Hall Effect.

Figure 5. Comparison of B, H and M inside and outside a cylindrical bar magnet

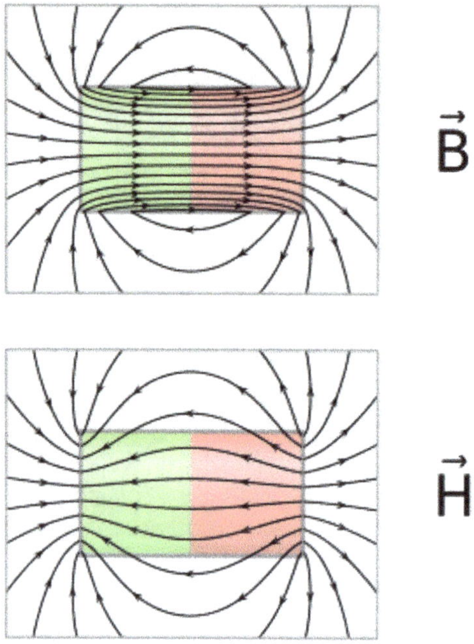

Electromagnetic Fields

An electromagnetic field (EMF or EM field) is a field formed by electrically charged objects. It affects the behaviour of charged objects in the locality of the field. The electromagnetic field spreads indefinitely throughout space and defines the electromagnetic interaction. It is one of the four fundamental forces of nature (the others are attraction, weak and strong interaction).

The field can be viewed as the mixture of an electric and magnetic field. The electric field is created by stationary charges, and the magnetic field by moving charges (currents); these two are often described as the sources of the field. The way in which charges and currents cooperate with the electromagnetic field is described by Maxwell's equations and the Lorentz force law. The force created by the electric field is much stronger than the force created by the magnetic field.

From a classical perspective in the history of electromagnetism, the electromagnetic field can be considered as a smooth, continuous field, propagated in a wave like manner, whereas from the perspective of quantum field theory, the field is seen as quantized, being composed of individual particles.

The flow of electricity through a conductor produces both an electric and magnetic field around the conductor. Collectively, these two fields are referred to as an electromagnetic field or EMF. The strength of the electric field is measured in volts per meter and varies with the amount of the source voltage. The higher the source voltage, the higher the strength of the field. Electric field strength decreases rapidly with distance from the source.

Electric fields are produced both naturally and by any conductor carrying electricity. The strength of the earth's natural electric field varies, but on average is about one-thousandth of a volt per meter. Electric field strength typically varies from 10 to 150 volts per meter under electric distribution lines and 5 to 100 volts per meter inside homes and workplaces. The strength of a magnetic field is typi-

Figure 6. Electromagnetic field

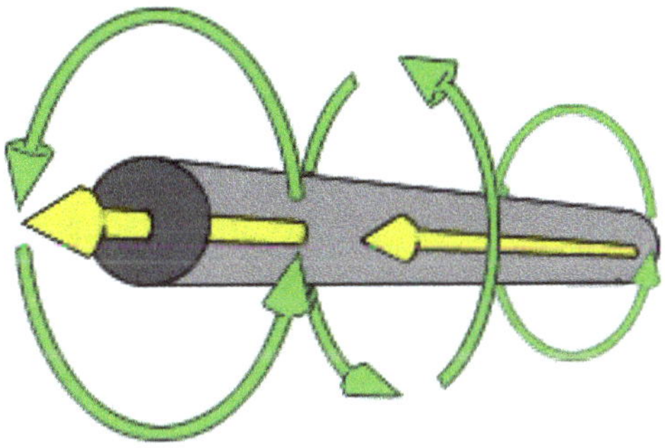

cally measured in units of gauss or milligauss and varies with the amount of current moving through a conductor. Lines or devices requiring high levels of current flow produce stronger magnetic fields than those with low current flow. For example, the measure of a magnetic field directly under a high voltage transmission line is somewhere between 20 to 650 milligauss. The magnetic field measured underneath a lower power distribution line is .5 to 30 milligauss (Joffe, 2010).

Magnetic fields produced by electrical circuits drop off rapidly with distance from the source. The magnetic field produced by a microwave at 1 foot is 70 to 100 milligauss while at five feet away, the magnetic field strength drops to five milligauss.

Electric fields are blocked by shielding such as walls, houses, trees, other vegetation, soil, and other large dense objects. Magnetic fields, on the other hand, pass easily through most objects and are only blocked by structures containing large amounts of iron or iron alloy metals as shown in figure 7.

Figure 7. Shielding blocked electric fields

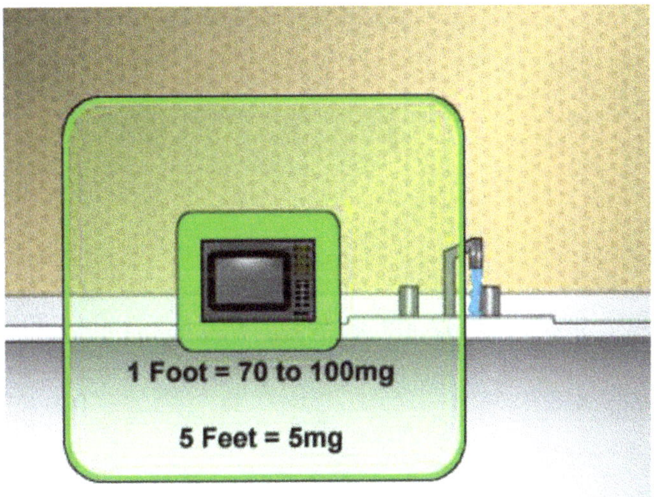

COUPLING (INDUCTIVE COUPLING)

Inductive Coupling

Magnetic field coupling (also called inductive coupling) occurs when energy is coupled from one circuit to another through a magnetic field. Since currents are the sources of magnetic fields, this is most likely to happen when the impedance of the source circuit is low (IEC/TR 61000-2-5; IEC 61000-4-6).

Consider the two circuits sharing a common return plane, coupling between the circuits can occur when the magnetic field lines from one of the circuits pass through the loop formed by the other circuit. Schematically, this can be represented by a mutual inductance between the two signal wires as shown in Figure: 8.

In most cases, a convenient closed-form equation for calculating the mutual inductance will not be available. However, we can often estimate the mutual inductance by estimating the percentage of the total magnetic flux generated by the first loop that couples the second loop. For example, suppose the two wires in the example above are 20 mm above the plane and separated by 5 mm. We could visualize the magnetic flux lines that wrap the current in line 1 as shown in Figure: 10.

Capacitive Coupling

The transfer of energy within an electrical network or between distant networks by means of displacement current between circuit(s) nodes, induced by the electric field. This coupling can have an intentional or accidental effect.

Figure 8. Two circuits above a signal return plane

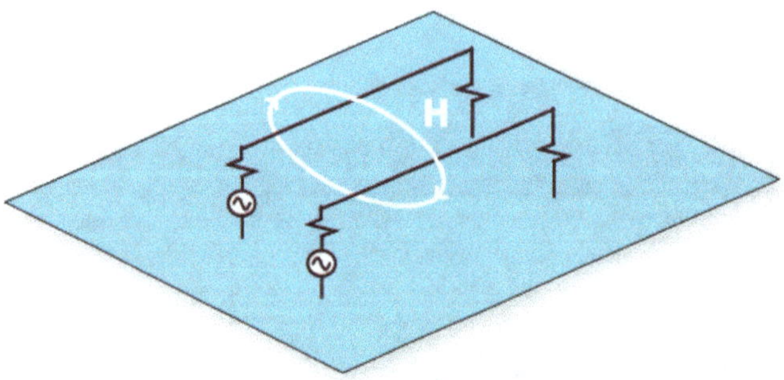

Figure 9. Schematic representations of the circuits in Fig. 1 including inductive coupling

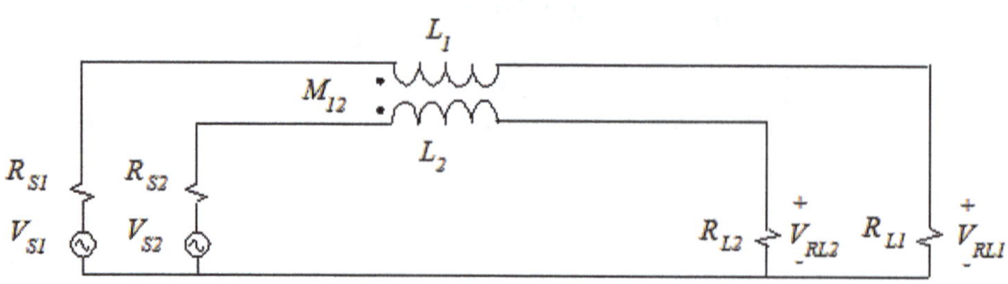

Figure 10. More intuitive schematic representations of the circuits in figure

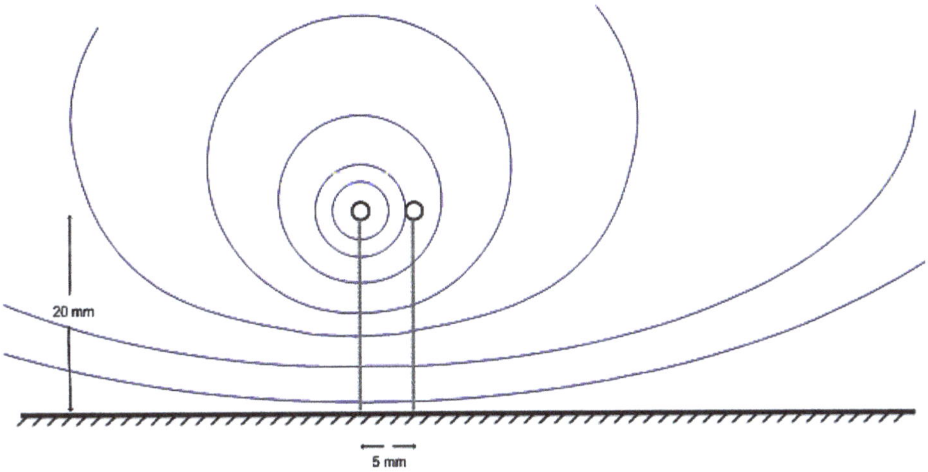

Capacitive coupling with high-voltage power lines can light a lamp continuously at low intensity. In its simplest implementation, capacitive coupling is achieved by placing a capacitor between two nodes. Where analysis of many points in a circuit is carried out, the capacitance at each point and between points can be described in a matrix form.

Using Capacitive Coupling in Analog Circuits

In Analog circuits, a coupling capacitor is used to connect two circuits such that only the AC signal from the first circuit can pass through to the next while DC is blocked. This technique helps to isolate the DC bias settings of the two coupled circuits. Capacitive coupling is also known as AC coupling and the capacitor used for the purpose is also known as a DC-blocking capacitor. Figure: 11 shows using capacitive coupling in Analog circuits

A coupling capacitor's ability to prevent a DC load from interfering with an AC source is particularly useful in Class A amplifier circuits by preventing a 0 volt input being passed to a transistor with additional resistor biasing; creating continuous amplification (Alshahrestani et al., 2018; Makdisie et al., 2018; Alhelou et al., 2018; Alhelou et al., 2016; Nadweh et al., 2018; Njenda et al., 2018).

Using Capacitive Coupling in Digital Circuits

AC coupling is also widely used in digital circuits to transmit digital signals with a zero DC component, known as DC-balanced signals. DC-balanced waveforms are useful in communications systems, since they can be used over AC-coupled electrical connections to avoid voltage imbalance problems and charge accumulation between connected systems or components.

For this reason, most modern line codes are designed to produce DC-balanced waveforms. The most common classes of DC-balanced line codes are constant-weight codes and paired-disparity codes.

Figure 11. Using capacitive coupling in analog circuits

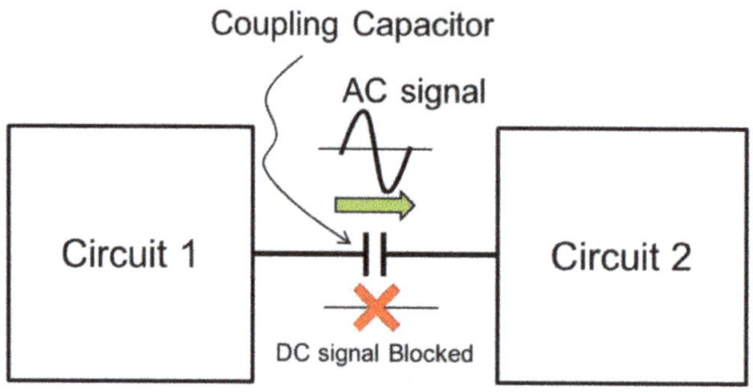

Shielding to Decouple and Coupling to Shield

Defining Electromagnetic Coupling to Cable Shield

The role, effects, and function of electromagnetic coupling are among the most important topics in cable shielding. They are also among the most puzzling and stimulating issues. The existing approaches span two extremes:

1. Reducing all the shielding mechanisms to different forms of coupling.
2. Disregarding completely the role of coupling in shielding. Although, as usual, the "truth" resides between the extremes, it is worthwhile to briefly investigate what they amount to.

THE OBJECTIVE OF SHIELDING

The objective of shielding is to avert the unwanted coupling between the circuits and the EMI environment, including other circuits. On the other hand, being a system circuit element itself, the cable shield couples into both the source and the victim of EMI. Even the shield transfer function may be thought of as representing signal coupling between the shield's two opposite surfaces. But, of course, to make sure of their "acceptability," these concepts must be implemented in combination with the appropriate exploration. Because this is not always done, many misunderstood and/or irrelevant effects are too often "blamed" on coupling phenomena.

The opposite extreme is to completely neglect the contribution of coupling to shielding process. One often-used such "technique" is to reduce the scope of shielding effectiveness to the shield transfer function. However useful and convenient are the transfer parameters, they represent only one, dimension of the cable shielding performance. But at that time, we did not have enough background to trace down the causes of the divergences. A common denominator to these and many other misconceptions is heavy dependence on "common sense," combined with disrespect to system EMC analysis. Previously, we proved that the critical worth of a cable shield is determined by its system shielding usefulness, which is a function not just of the shield transfer function but also of the ambient environment and electronic system particulars. Thus, the shielding effectiveness may be affected by factors that are either indirectly

Figure 12. Equivalent circuit of shield conductor

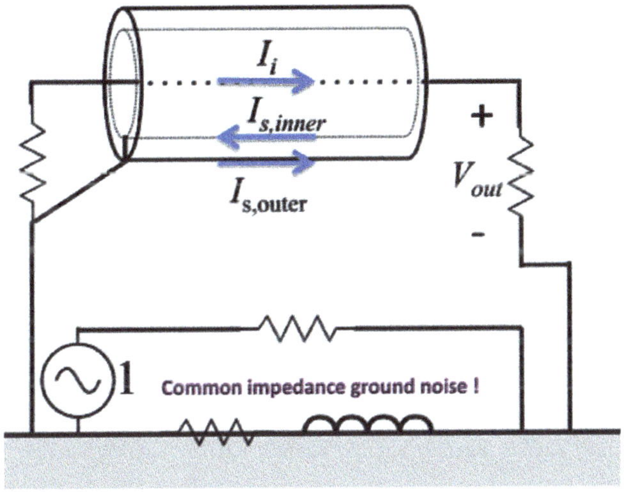

related to, or even independent of, the particular shield design and transfer characteristics. To account for the most important of such factors, we have identified five basic ingredients / steps involved in cable shielding and used them to build a system model of cable shielding performance. And even if the shield transfer function is abstractly and explicitly placed at the center of our model, these other factors also are crucial to shielded cable EMC performance. Electromagnetic coupling to the shield is one of the most decisive such factors. The processes of EMI coupling to the shield represent the link between the electromagnetic environment outside and inside the shielding enclosure and the shield.

EMISSION FROM VSDS

There are several sources of emission from VSDS the most important of these are:

Flicker

VSDS don't produce flicker by themselves, but it is possible for user to design a system with modulated loads, so the drive provider must be warn of this possibility.

Figure 13. Magnetic field induction in shielded cable

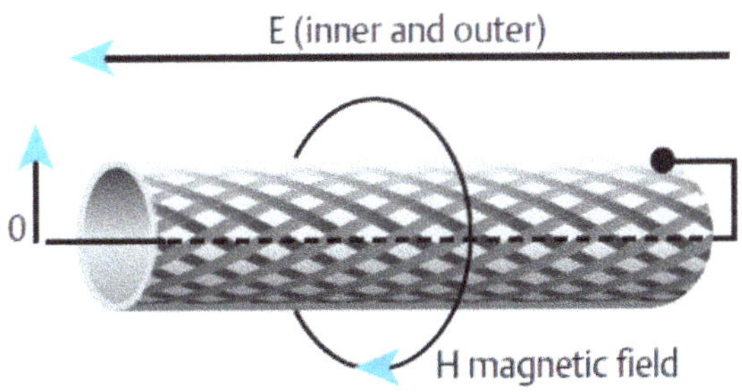

Conducted Radio Frequency

Inverter output power frequencies range between 2 kHz and 50 MHz, which covers the range where RF emission is calculated by the conducted method. It must be connected through the user's cables to the motor, which nearby with distance up to 200m. Since an output filter sufficient to meet conducted emission standards would be excessively costly, it is usual to rely on the use of a screened power cable thereby the installation is subjected to conducted emission limits.

Drive input power carries a large RF because of involving the coming back path for common-mode RF current which flows from the output cable cores through the capacitance of the cable and motor to earth as shown in Fig14.

An input RFI filter affords a low-impedance coming back path to decrease emission into the supply. figure indicates how the filter and screened power cable complete the loop for rf current and avert it from circulating outside the drive system.

Line Harmonics

Line harmonics are generated by input rectifier, which keeping under control by using large inductors in DC-LINK and AC line chokes. Most drives are yonder the space of EN 61000-3-2 so installations are considered on a case-by case basis under the electricity providers' recommendations.

ELECTROMAGNETIC INTERFERENCE

The Nature of Electromagnetic Interference

Electromagnetic interference (EMI), also called radio-frequency interference (RFI) when in the radio frequency spectrum, is a disturbance generated by an external source that affects an electrical circuit by electromagnetic induction, electrostatic coupling, or conduction. EMI is caused by an electromagnetic disturbance affecting the performance of a device, transmission channel, or system.

Figure 14. Main RF emission paths

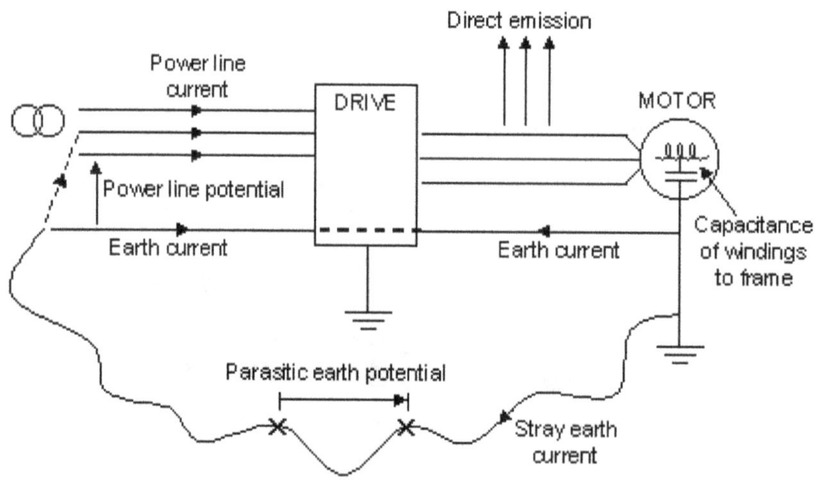

EMI is actually any frequency of electrical noise, whereas RFI is a specific subset of electrical noise on the EMI spectrum. There are two types of EMI. Conducted EMI is unwanted high frequencies that ride on the AC wave form. Radiated EMI is similar to an unwanted radio broadcast being emitted from the power lines.

Solution: Electromagnetic shielding is the practice of reducing the electromagnetic field in the space of blocking the field with fences made of conductive or magnetic materials.

A conductive insertion used to block electrostatic fields is also known as a Faraday cage.

Interference Nature in VSDS

VSDS control the speed of AC motors by changing the frequency and voltage of the electrical source to motor. The most common VSDs use Pulse Width Modulation PWM for the generation of a 3 phase voltage with agreeing frequency. PWM produce a variable output frequency and voltage by switching transistors on and off at a fast rate (classically at 3,000 times / second and up to 16,000 times / seconds) and with very short rise times of each pulse (~0.2microsec). Figure 16 shows an inverter sector with six IGBT transistors.

Installation Requirements for VSDS

The use of a VSD_S certainly involves the installer in providing some special EMC types, in particular the correct increasing of the drive module in relation to its input filter and screened motor cable. The manufacturer brings accountability for confirming that clear instructions are given to the installer (Mahat et al., 2006).

The typical industrial control panel comprised a painted steel back-plate mounted into a steel cubicle which had no earthing provisions beyond the requirements for electrical safety bonding. Although the screening offered by such a cubicle is generally adequate, the inconsistent and inadequate RF earth bonding is not acceptable because the circulating RF currents are high. Installers were reduced to scraping off paint in the key contact areas.

Figure 15. Complete drive, motor, filter system showing RF current paths

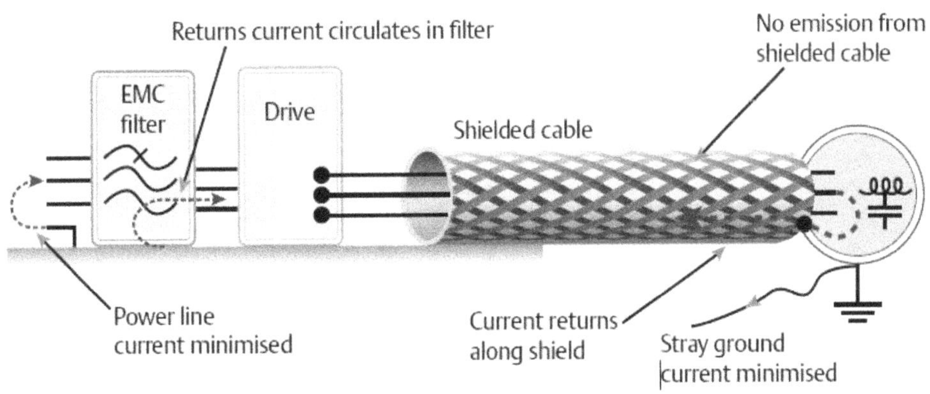

In accordance with accepted practice and the local, power is normally provided to a VSD from a distribution board (DB) or a motor control center (MCC). Adequate arrangements should be made to provide safety isolation switches and short-circuit protection in the connection point to the power supply (Paul et al., 2006; Purcell et al., 2012).

The short-circuit protection is required to protect the power cable to the AC converter and the input rectifier bridge at the converter. The converter provides down-stream protection for the motor cable and the motor itself.

Adequate safety earthing should also be provided in accordance with the local wiring rules and codes of practice. The metal frames of the AC Converter and the AC motor should be earthed as shown in Figure 8.3 to keep touch potentials within safe limits. The chassis of the AC converter is equipped with one or more protective earth (PE) terminals, which should be connected back to common safety earth bar.

VSDS Power Supply Cables

The variable speed drive should be connected to the power supply by means of a cable that is adequate for the current rating of the VSDS. Reference can be made to Australian standard AS 3008 when selecting cables. The AC converter requires a 3-phase supply cable (red/white/blue) and a protective earth conductor (green/yellow), which means a 4- core cable with copper or aluminium conductors. A neutral conductor is not necessary and is usually not brought to the frequency converter.

The AC converter is a source of harmonic currents that flow back into the low impedance of the power supply system. This conducted harmonic current is carried into other electrical equipment, where it causes additional heat losses and interference.

Sensitive electronic instrumentation, such as magnetic flow-meters, thermocouples and other microprocessor based equipment, ideally should not be connected to the same power source, unless via a filtered power supply. Also, interference can be radiated from the power supply cable and coupled into other circuits, so these cables should be routed well away from sensitive control circuits.

The power supply cable should preferably be laid in a metal duct or cable ladder and shielded in some way to reduce the radiation of EM fields due to the harmonic currents. Steel wire armoured (SWA) cables, are particularly suitable for this purpose. If the power cable is unshielded, control and communications cables should not be located within about 300 mm of the power cable. The conductor sizes should be selected in accordance with normal economic cable selection criteria, which take into account

Figure 16. Inverter section with 6 IGBTs

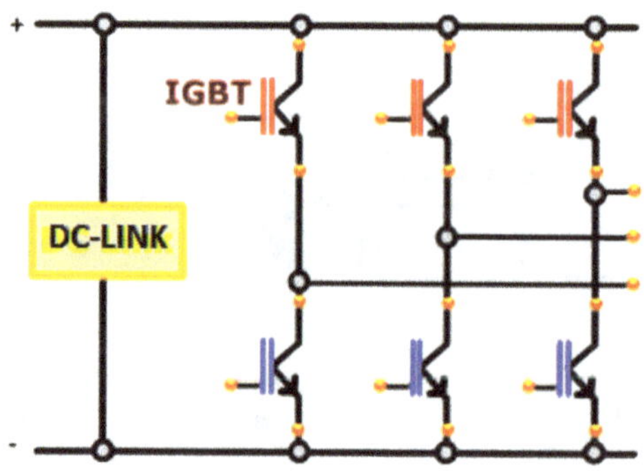

Figure 17. Power supply, motor, and earthing connections

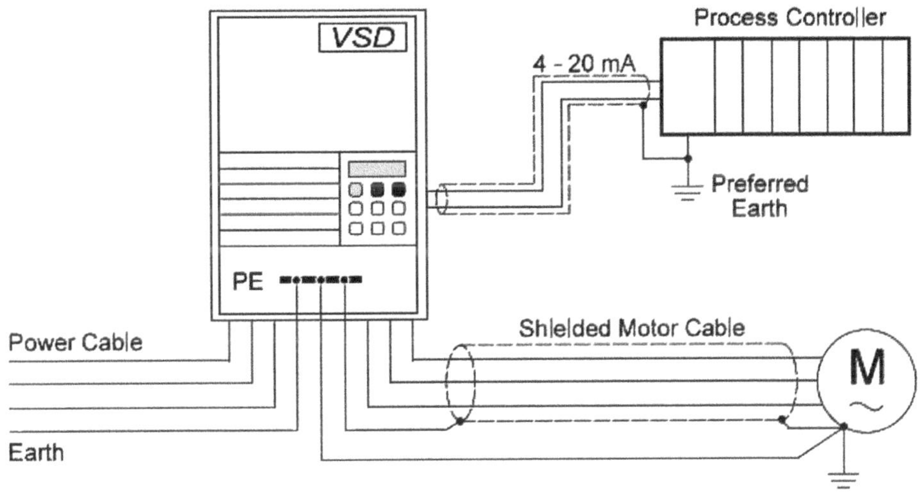

the maximum continuous current rating of the VSD, the short-circuit rating, the length of the cable and the voltage of the power supply system. The relevant local safety regulations should be strictly observed.

However, when selecting the cable cross-sectional area for the power supply cables and upstream transformers, a de-rating factor of at least 10% should be included to accommodate the additional heating due to the conducted harmonic currents. If a supply side harmonic filter is fitted at the converter, this may not be necessary. Three phase systems composed of three single-conductor cables should be avoided if possible. Power cables with a trefoil configuration produce a lower radiated EM field.

Cables Between Converter and Motor

The cable from the AC converter to the motor carries a switched PWM voltage, which is modulated at high frequency by the inverter. This results in a higher level of harmonics than the power supply cable. Harmonic frequencies are in the frequency spectrum of 100 kHz to 1 MHz The motor cable should preferably be screened or located inside a metal duct. Control and communications cables should not be located close to this cable.

The level of radiated EM fields is higher for cables with 3 separate single cores, laid horizontally on a cable ladder, than a trefoil cable with a concentric shield. The recommended size for the cable between the AC converter and the motor should preferably be the same as the power supply cable.

The reasons are:

It will be easier to add a bypass device in parallel with the frequency converter at a later date, using the same cable, cable lugs and connections.

The load-carrying capacity of the motor cable is also reduced by harmonic currents and additionally by the capacitive leakage currents.

It should be borne in mind that the AC converter VSD provides short-circuit and overload protection for the cable and motor.

A separate earth conductor between the converter and motor is recommended for both safety and noise attenuation. The earth conductor from the motor must be connected back to the PE terminal of the converter and should not be connected back to the distribution board. This will avoid any circulating high frequency currents in the earth system. When armored or shielded cables are used between the converter and motor, it may be necessary to fit a barrier termination gland at the motor end when the cable is longer than about 50 m.

The reason is that the high frequency leakage currents flow from the cable through the shunt capacitance and into the shield. If these currents return via the motor and other parts of the earthing system, the interference is spread over a larger area. It is preferable for the leakage currents to return to the source via the shortest route, which is via the shield itself. The shield or steel wire armour (SWA) should be earthed at both the converter end and to the frame of the motor.

Control Cables

The control cables should be provided in accordance with normal local practice. These should have a cross-sectional area of at least 0.5 mm2 for reasonable volt drop performance. The control and communications cables connected to the converter should be shielded to provide protection from EMI. The shields should be earthed at one end only, at a point remote from the converter.

Earthing Requirements

However, when earthing a drive system, two aspects need to be considered – personal protection and EMC. The protective earth exists to provide a low impedance path to ground for any 50 Hz current. Similarly, the functional earth for EMC compliance needs to provide a low impedance path to earth for any high frequency current. With a bit of planning, both objectives can be achieved with the same earthing system.

While the earth lead provided in the four-core cable is adequate for personal protection, the four-core cable is unsuitable for ensuring EMC, as it is not symmetrical.

In a three-core cable, the cores are symmetrically arranged and so are the magnetic fields that surround each conductor. With a four-core cable, the lack of symmetry can cause current to be induced in the earth conductor and propagate through the system.

An earthing system to provide both personal and functional earthing should include a three-core supply cable combined with earthing by stranded or braided copper conductors and an external equipotential bonding cable.

Another earthing practice likely to cause disturbance is when the earth cable is terminated on the earth bar of the drive. While this may seem simple and intuitive, it does not address the need for functional earthing of the enclosure. The earth conductor needs to be terminated at the PE terminal on the VSD metalwork, as well on the PE terminal of the motor. Similarly, the cable braid, or armour, should be terminated at the VSD metalwork. This provides functional earthing and by cross bonding from the VSD metalwork to the enclosure earth bar, safety earthing requirements are met, while ensuring good EMC performance.

As mentioned earlier, both the AC converter and the motor must be provided with a safety earth according to the requirements of local standards. The main purpose of this earthing is to avoid dangerous voltages on exposed metal parts under fault conditions.

When designing and installing these earth connections, the requirements for the reduction of EMI should also be achieved with these same earth connections. The main earthing connections of an AC converter are usually arranged as shown below. The PE terminal on the converter should be connected back to the system earth bar, usually located in the distribution board. This connection should provide a low impedance path back to earth.

Grounding

Electrical systems are grounded for:

1. Safety.
2. Interference Reduction.

No significant current flows through a grounding conductor under normal conditions.

The signal grounding conductor is NOT the same as the signal routing conductors. Don't confuse signal routing with signal grounding. For MHz noise frequencies, grounding is not as important as field containment.

There are Two Grounding Techniques

1. **Ground Grid:** Uses a low resistance to "short out" the ground voltage difference.
2. **Isolated Grounding (Single Point Grounding):** Avoids connecting to the ground voltage difference.

The ground grid is best for safety under high current faulting or lightning conditions. Isolated (single point) grounding is best for reducing kHz noise.

Common Cabling Errors

The following are some of the common cabling errors made when installing VSDs:

The earth conductor from the AC converter is run in the same duct or cable ladder as other cables, such as control cables and power cables for other equipment. Harmonic currents can be coupled into sensitive circuits. Ideally, instrument cables should be run in separate metal ducts or steel conduit.

Running unshielded motor cable next to the supply cable to the AC converter or the power cables for other equipment. High frequency harmonic currents can be coupled into the power cable, which can then be conducted to other sensitive electronic equipment. Other cables should be separated from the motor cable or converter power cable by a minimum of 300 mm.

Running excessively long cables between the AC converter and the motor, these should be no longer than 100 m. If longer cables are necessary, motor filters are necessary to reduce the leakage current. Alternatively, the switching frequency may be reduced.

FUTURE RESEARCH DIRECTIONS

Modeling variable speed drive system with EMC consideration by using MATLAB/SIMULINK.

Figure 18. Using single ground

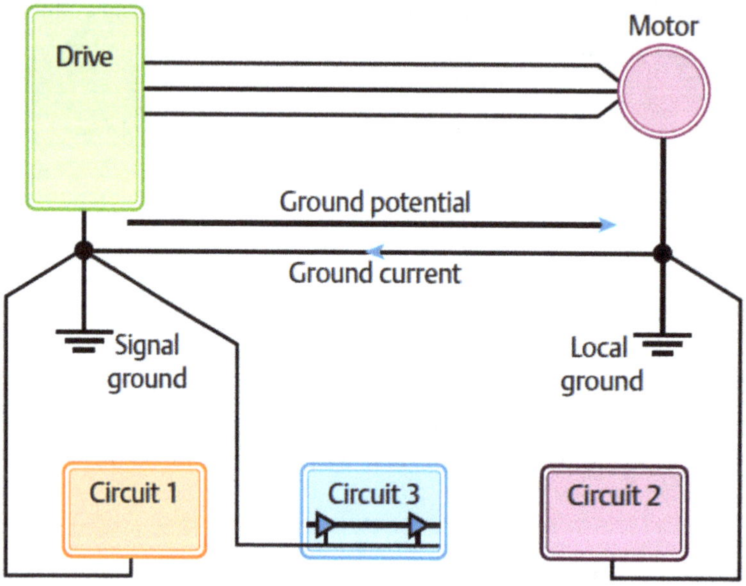

CONCLUSION

Best EMC installation was discussed for variable speed drive systems. EMC standards have been in different sides of variable speed drive systems, in addition to mentioned the fundamentals of static, electric, magnetic, and electromagnetic fields.

VSD_S emission, and electromagnetic interferences were studied with installation requirement in VSDS (supply cables, cable between converter and motor, control cables, earthing requirements, and grounding), in order to achieve better performance for variable speed drive systems.

REFERENCES

Alhelou, H., Hamedani-Golshan, M. E., Zamani, R., Heydarian-Forushani, E., & Siano, P. (2018). Challenges and Opportunities of Load Frequency Control in Conventional, Modern and Future Smart Power Systems: A Comprehensive Review. *Energies*, *11*(10), 2497. doi:10.3390/en11102497

Alhelou, H. H. (2018). *Fault Detection and Isolation in Power Systems Using Unknown Input Observer. In Advanced Condition Monitoring and Fault Diagnosis of Electric Machines* (p. 38). Hershey, PA: IGI Global.

Alhelou, H. H., Golshan, M., & Fini, M. (2018). Wind Driven Optimization Algorithm Application to Load Frequency Control in Interconnected Power Systems Considering GRC and GDB Nonlinearities. *Electric Power Components and Syst*.

Alhelou, H. H., & Golshan, M. E. H. (2016, May). Hierarchical plug-in EV control based on primary frequency response in interconnected smart grid. In *Electrical Engineering (ICEE), 2016 24th Iranian Conference on* (pp. 561-566). IEEE. 10.1109/IranianCEE.2016.7585585

Alhelou, H. H., Golshan, M. H., & Askari-Marnani, J. (2018). Robust sensor fault detection and isolation scheme for interconnected smart power systems in presence of RER and EVs using unknown input observer. *International Journal of Electrical Power & Energy Systems*, *99*, 682–694. doi:10.1016/j.ijepes.2018.02.013

Alhelou, H. H., Hamedani-Golshan, M. E., Heydarian-Forushani, E., Al-Sumaiti, A. S., & Siano, P. (2018, September). Decentralized Fractional Order Control Scheme for LFC of Deregulated Nonlinear Power Systems in Presence of EVs and RER. In *2018 International Conference on Smart Energy Systems and Technologies (SEST)* (pp. 1-6). IEEE. 10.1109/SEST.2018.8495858

Alhelou, H. S. H., Golshan, M. E. H., & Fini, M. H. (2015, December). Multi agent electric vehicle control based primary frequency support for future smart micro-grid. In *Smart Grid Conference (SGC)* (pp. 22-27). Academic Press.

Alshahrestani, A., Golshan, M. E. H., & Alhelou, H. H. (2018, November). WAMS Based Online Estimation of Total Inertia Constant and Damping Coefficient for Future Smart Grid Systems. In *Smart Grid Conference (SGC)* (pp. 1-5). Academic Press.

Armstrong, K. (2006). Design Techniques for EMC, in six parts. *The EMC Journal, 62*. Retrieved from www.theemcjournal.com

Armstrong, K, (2007). *EMC for Printed Circuit Boards – Basic and Advanced*. Academic Press.

EN/IEC 61800-5-2, *Adjustable speed electrical power drive systems* – Part 5-2: Safety requirements – Functional.

Fini, M. H., Yousefi, G. R., & Alhelou, H. H. (2016). Comparative study on the performance of many-objective and single-objective optimisation algorithms in tuning load frequency controllers of multi-area power systems. *IET Generation, Transmission & Distribution*, *10*(12), 2915–2923. doi:10.1049/iet-gtd.2015.1334

Hosseini, S. A., Karami, M. S., & Madahi, S. K. (2011). Optimal capacity, location and number of distributed generation at 20 kv substations. *Australian Journal of Basic and Applied Sciences.*

IEC 61000-6-4, *Electromagnetic compatibility (EMC) – Part 6-4: Generic standards – Emission standard for industrial environments.*

IEC TS 61000-1-2,(2008), *basic safety publication, draft second edition, Electromagnetic Compatibility (EMC) – Part 1-2: Design and Layout Techniques*, Nutwood/ Armstrong.

IEC/TR 61000-2-5, Electromagnetic compatibility *(EMC) – Part 2-5: Environment – Description and classification* of electromagnetic environments

Joffe, E. (2010). Grounds for Grounding. A Circuit to System Handbook. Wiley-IEEE.

Mahat, N. P., & Mithulananthan, N. (2006). An analytical approach for DG allocation in primary distribution network. *Electrical Power and Energy Systems*, *28*(10), 669–678. doi:10.1016/j.ijepes.2006.02.013

Makdisie, C., Haidar, B., & Alhelou, H. H. (2018). An Optimal Photovoltaic Conversion System for Future Smart Grids. In *Handbook of Research on Power and Energy System Optimization* (pp. 601–657). IGI Global. doi:10.4018/978-1-5225-3935-3.ch018

Nadweh, S., Hayek, G., Atieh, B., & Haes Alhelou, H. (2018). Using Four – Quadrant Chopper with Variable Speed Drive System Dc-Link to Improve the Quality of Supplied Power for Industrial Facilities. Majlesi Journal of Electrical Engineering.

Njenda, T. C., Golshan, M. E. H., & Alhelou, H. H. (2018, November). WAMS Based Intelligent Under Frequency Load Shedding Considering Online Disturbance Estimation. In *Smart Grid Conference (SGC)* (pp. 1-5). Academic Press.

Njenda, T. C., Golshan, M. E. H., & Alhelou, H. H. (2018, November). WAMS based Under Frequency Load Shedding Considering Minimum Frequency Prediction and Extrapolated Disturbance Magnitude. In *Smart Grid Conference (SGC)* (pp. 1-5). Academic Press.

Paul, C. R. (2006). *Introduction to Electromagnetic Compatibility* (2nd ed.). Wiley Series in Microwave and Optical Engineering.

Purcell, E. M. (2012). *Electricity and magnetism* (3rd ed.). Cambridge, UK: Cambridge Univ. Press.

REO Guide on Good EMC Engineering Practices in the Design and Construction of Industrial Cabinets. (n.d.). Available from www.reo.co.uk/

Steenstra, H. T., & van Deursen, A. P. J. (2008). Reduction of conducted interference by steel armor in buried cables: Measurements and modelling. *IEEE Transactions on Electromagnetic Compatibility*, *50*(3), 678–686. doi:10.1109/TEMC.2008.926902

Williams, K., & Armstrong, M. (2012). Newnes EMC for Systems and Installations. Academic Press.

Zamani, R., Hamedani-Golshan, M. E., Haes Alhelou, H., Siano, P., & Pota, R, H. (. (2018). Islanding Detection of Synchronous Distributed Generator Based on the Active and Reactive Power Control Loops. *Energies*, *11*(10), 2819. doi:10.3390/en11102819

KEY TERMS AND DEFINITIONS

Capacitive Coupling: Is the transfer of energy within an electrical network or between distant networks by means of displacement current between circuit(s) nodes, induced by the electric field. This coupling can have an intentional or accidental effect.

Earthing: Is used to protection from an electric shock.

Grounding: In electrical engineering, ground or earth is the reference point in an electrical circuit from which voltages are measured, a common return path for electric current, or a direct physical connection to the earth. Electrical circuits may be connected to ground (earth) for several reasons.

Shielding: Electromagnetic shielding is the practice of reducing the electromagnetic field in a space by blocking the field with barriers made of conductive or magnetic materials.

Variable Speed Drives (VSDs): Also known as adjustable speed drives, are large industrial electric motors whose speed can be adjusted by means of an external controller. They are used in process control and help saving energy in plants that use many powerful electric motors.

Chapter 14
Economic Operation of Smart Micro-Grid:
A Meta-Heuristic Approach

Baseem Khan
https://orcid.org/0000-0002-0562-0933
Hawassa University, Ethiopia

Pawan Singh
Amity University, India

ABSTRACT

Presently, economic operation of micro-grid is a major concern in smart grid environment. This is a very complex problem, which can be solved with the help of various meta-heuristic techniques. Therefore, this chapter provides a comparative analysis of four different renowned meta-heuristic techniques with reference to the problem of optimal operation of micro-grid (MG). Genetic algorithm (GA), particle swarm optimization (PSO), differential evaluation (DE), and firefly (FF) algorithm are utilized for this purpose.

1. INTRODUCTION

Since 1990s, the deregulation of the electricity supply industry started in different parts of the world. The deregulated electricity market structure can be retail or wholesale, in which generation, transmission and distribution sectors functioned as separated entities (Khan et al., 2014, 2016). Electricity market deregulation is required for changing the situation in which micro-grids are considered as threats to the valued energy sources those are fairly compensated (Khan et al. 2014 and 2013, Khan and Agnihotri, 2012). All these three sectors are fully unbundled into different utility services and are allowed to supply and compete in wholesale and retail electricity market. After deregulation, traditional electricity grid is converted into the smart structure since 2010. The key feature of this smart system is the incorporation of the renewable energy sources at different levels such as distributed level and bulk level. International energy agency predicted that the energy generation from the renewable energy sources is increased up

DOI: 10.4018/978-1-5225-8030-0.ch014

to three times till 2035 (Khan and Singh, 2017; Mulualem and Khan 2017). Further, the total energy production from the renewable energy sources will be increased to 31%, in which hydro, wind and solar will provide 50%, 25% and 7.5%, respectively. The two major issues with renewable energy generation are intermittency and climate dependency of renewable sources. These problems make integration of these sources with conventional grid more difficult and complex. The above discussed problems can be minimized with the help of energy storage devices. These devices incorporated various conventional storage systems such as batteries and heat buffers along with advanced storage techniques such as fuel cell technology, electric vehicle technology etc. Therefore, there was a necessity to develop such a system which incorporates different renewable energy sources with energy storage options to mitigate the issues related with renewable energy sources. This necessity is fulfilled with the development of micro-grid system (Khan and Singh, 2017; Fanuel et al. 2018). It is the combination of different type of loads (domestic, commercial, industrial) with various renewable energy sources such as solar photo voltaic, wind, micro turbine and small hydro along with energy storage devices such as battery energy storage, heat buffer, flywheel storage, and electric vehicle technology. Moreover it also incorporates advanced generation techniques for example fuel cell technology and combined heat and power (CHP) technology. In smart grid structure, micro-grid technology provides a holistic approach for the integration of renewable energy sources. It has several benefits over the conventional grid system as it's minimise energy losses, improve reliability and enhance energy management. Further, at distribution level, micro-grid technology provided better solution of energy scarcity, generation coordination and control problems due to its better performance with respect to distributed generation technology (Khan and Singh, 2017; Kifle et al. 2018). There are two different modes of operation available in micro-grid technology. These are as follows:

1. Grid connected operation mode
2. Islanded operation mode

As per the requirement of operation, any mode can be selected. Therefore the problems of coordination, control and operation are very important from micro-grid point of view. Moreover costs are also associated with micro-grid's generation, operation and maintenance in a similar way as conventional grid. Therefore, economic operation of micro-grid refers the optimization of the micro-grid costs associated with generation, maintenance and operation etc (Khan and Singh, 2017). It is a cost minimization problem under various systems constrains.

Different Meta-heuristics methods are proposed by various researchers for solving the cost optimization problem of micro-grid. Meta-heuristic optimization techniques, which are depended on population, is mainly categorised into two categories:

1. Swarm intelligence
2. Evolutionary algorithms

Some renowned evolutionary methods are further classified as follows:

1. Differential evaluation (DE)
2. Evolution Programming (EP)
3. Evaluation strategy (ES)

4. Genetic Algorithm (GA) etc.

Further, some swarm intelligence based methods are as follows:

1. Fire fly (FF)
2. Particle swarm optimization (PSO)
3. Ant colony optimization (ACO)
4. Artificial bee colony (ABC) etc.

Other then these above discussed two classifications; some natural phenomena based meta-heuristic techniques are also available. These are as follows:

1. Gravitational Search (GSA)
2. Harmony Search (HS) algorithm
3. Flower Pollination (FPA)
4. Biogeography-Based Optimization (BBO) etc.

There are numerous advantages of Meta-heuristic techniques, out of which some of them are as follows:

1. No gradient information is required by Meta-heuristic techniques.
2. These techniques have capability to recover from local optima because of their inherent stochasticity.
3. These techniques can better handle the uncertainties in objectives.
4. By making few changes in algorithms, these techniques can handle multiple objectives.

Generally, these techniques are probabilistic in nature and can be controlled by some parameters such as population size, population, number of generations etc. These parameters are general parameters, which are required in every technique. Other than these parameters, some methodology specific parameters are also required. These parameters are selected according to the particular methodology. Some of technique specific parameters are as follows:

1. **GA:** mutation probability, cross over, selection of operator
2. **PSO:** social factors, cognitive factors, weight of inertia
3. **ABC:** count of various types of bees such as employed, onlooker, scout, and their limits

Moreover, tuning of specific parameters is also required by the Meta-heuristic techniques. Tuning of parameters is extremely critical issue in these techniques as it directly related to the performance of these methodologies. Increased computation time and local optima are the some disadvantages of improper tuning of the parameters in Meta-heuristic techniques. Table 1 presents the time line for the evolution of various Meta-heuristic techniques.

This chapter discussed the micro-grid operational cost minimization problem with the help of different Meta-heuristic algorithms such as PSO, GA, FF and DE. These selected techniques belong to two different areas of Meta-heuristic approach such as evolutionary and swarm intelligence. These techniques are discussed and compared on the basis of micro-grid operation cost minimization problem.

Table 1. Timeline of meta-heuristic algorithms

Year	Algorithm
1975	Holland introduced the Genetic Algorithm (GA)
1977	Glover proposed Scatter Search (SS).
1980	Smith elucidated genetic programming.
1983	Kirkpatrick et al. proposed Simulated Annealing (SA)
1986	Glover and McMillan offered Tabu Search (TS).
1986	Farmer et al. suggested the Artificial immune system (AIS).
1988	Koza registered his first patent on genetic programming.
1989	Evolver provided the first optimization software using the GA.
1989	Moscato presented Memetic Algorithm.
1992	Dorigo proposed the Ant Colony Algorithm (ACO).
1993	Fonseca and Fleming provided Multi-Objective GA (MOGA).
1994	Battiti and Tecchiolli introduced Reactive Search Optimization (RSO)
1995	Kennedy and Eberhart proposed Particle Swarm Optimization (PSO).
1997	Storn and Price suggested Differential Evolution (DE).
1997	Rubinstein presented the Cross Entropy Method (CEM).
1999	Taillard and Voss proposed POPMUSIC.
2001	Geem et al. provided Harmony Search (HS).
2001	Hanseth and Aanestad offered Bootstrap Algorithm (BA).
2004	Nakrani and Tovey presented Bees Optimization (BO).
2005	Krishnanand and Ghose introduced Glowworm Swarm Optimization (GSO).
2005	Karaboga proposed Artificial Bee Colony Algorithm (ABC).
2006	Haddad et al. suggested Honey-bee Mating Optimization (HMO).
2007	Hamed Shah-Hosseini offered Intelligent Water Drops (IWD).
2007	Atashpaz-Gargari and Lucas introduced Imperialist Competitive Algorithm (ICA).
2008	Yang presented Firefly Algorithm (FA).
2008	Mucherino and Seref suggested Monkey Search (MS).
2009	Husseinzadeh- Kashan provided League Championship Algorithm (LCA).
2009	Rashedi et al. introduced Gravitational Search Algorithm (GSA)
2009	Yang and Deb offered Cuckoo Search (CS).
2010	Yang developed Bat Algorithm (BA).
2011	Shah-Hosseini introduced the Galaxy-based Search Algorithm (GbSA).
2011	Tamura and Yasuda designed Spiral Optimization (SO).
2011	Rao et al. presented Teaching Learning-Based Optimization (TLBO) algorithm.
2012	Gandomi and Alavi proposed the Krill Herd (KH) Algorithm.
2012	Çivicioglu introduced Differential Search Algorithm (DSA).
2013	Seyedali Mirjalili introduced Grey Wolf Optimizer
2014	Zheng introduced water wave optimization
2015	Seyedali Mirjalili introduced Ant Lion Optimization algorithm
2016	Seyedali Mirjalili introduced Whale optimization algorithm

(Khan and Singh, 2017)

2. MICRO-GRID STRUCTURE

Micro grid consists of various energy generation sources like photo voltaic, micro turbine, wind turbine, fuel cell and diesel generator. Further, along with generation sources, MG incorporates energy storage devices and systems such as battery energy storage and electric vehicle technology (Khan and Singh, 2017). Further, in micro-grid cost minimization problem both charging and discharging modes of batteries can be incorporated. The structure of this chapter is as follows. The next section discussed the various technologies, incorporated into MG system.. Fourth section provides the different Meta-heuristic techniques utilized to solve the problem of economic operation of MG. Fifth section presents the comparison between various Meta-heuristic techniques followed by the conclusion.

3. MICRO-GRID TECHNOLOGIES

It is a complex and practical problem. As it already discussed that MG is the combination of renewable energy sources, storage devices and various types of load, mathematical models like integer and liner programming are cumbersome and need more time to solve the problem. On the other hand Meta-heuristic approaches are easy to implement and provide optimal solution of complex problem within permissible time limit.

Different micro-grid technologies are available for renewable energy production. Some of them are as follows:

Micro-Grid Generation Technologies

Renewable Energy Sources and Battery Storage Systems Are Incorporated in Micro-Grid

Presently a lot of work is performed in the area of micro-grid with energy storage system. A scheme for energy management of MG that incorporated loads, WT, diesel generator and battery storage is presented by Shen et al. (2016). A practical experience of battery energy storage system is presented by Xu et al. (2016). It presented the modelling and implementation of conversion systems for energy. Model of optimal power rating and energy capacity is developed by Alharbi and Bhattacharya, 2014 to coordinate the operation of MG. For installing various types of battery based storage systems, the economical and technical assessment is performed by Graditi et al., 2016. This will lower the cost of energy for a application on customer side. A generalized framework of multi objective optimal sizing problem of distributed energy sources in MG is presented by Silvestre et al., 2014. This framework is based on the swarm based indicator. An ant colony based optimization is applied by Favuzza et al. 2006 for energy management of distribution system. A grey wolf based optimization approach is applied by Sharma et al. 2016 for finding the capacity of battery storage system in MG. Thompson et al. 2016, discussed the techniques for the optimal investment in the storage capacity of data centres. Technical literature related to optimization methods is reviews by the Gamarra and Guerrero, 2015 for the planning of MG. Singh and Khan, 2017 presented the Micro grid optimization problem with the help of novel artificial shark optimization approach. Jariso et al. 2017 presented the modeling of standalone PV system for rural area of Ethiopia. Kifle et al. presented the reliability analysis of practical distribution system with the incorporation of MG system (Alshahrestani et al., 2018; Makdisie et al., 2018; Alhelou et al., 2018; Alhelou et al., 2016; Nadweh et al., 2018; Njenda et al., 2018).

Electric Vehicle Technologies

A great impact is made by advance electric vehicle technology on the operation of micro-grid. Electric vehicle technology can be utilized as both demand and back up. This section provides a detailed literature review of the research carried out in the field of electric vehicle technology. An energy management system for residential use in smart grid is developed by Melhem et al., 2016 to consider vehicle to grid (V2G) integration and renewable energy generation. An urban context in smart grid is utilized for this purpose. Impact of EV mobility on demand response management is investigated by Yu et al., 2016 in the smart grid environment. A mobile energy system of V2G technology is developed that is dynamic in nature and can be moved in various cites. Therefore a mobile generating station can be presented in the form of EV technology. For integrating electric vehicles in smart grid, optimization techniques are discussed by Laureri et al. 2016. The objective of these methods is to minimize the smart grid total cost of operation by integrating the electric vehicles. Optimal operation of smart households in terms of the minimization of the total energy procurement cost is presented by Paterakis et al. 2016. Bi directional power flow is utilized at household as well as neighbourhood level. Optimal scheduling of EV which is based on cost is performed by Li et al. 2017. It is an online technique, which will also supply the power to the MG. A stochastic optimization is developed for minimizing cost of energy from utility and micro-grid, cost associated with batteries and emission and renewable energy harvesting charges. A real time charging scheme is developed by Yao et al. 2017 for managing the charging of EV. It also integrated the approaches of demand response for parking station.

Plug In Type Electric Vehicle

A power management scheme for the system of super capacitor, fuel cell, battery and hybrid vehicle system is optimized both in real time and offline mode of operation by Odeim et al. 2016. A multi objective GA is utilized for the optimization in real time mode, while dynamic programming and pontryagin's minimum thermo is utilized in offline mode. An optimal charging control method is developed by Xu, 2015 for PEV. It is utilized to apportion existing energy of charging. Multi agent system is utilized to implement the develop scheme. A universal method of optimization is developed by Vinot et al., 2016 to model the hybrid energy vehicle with power split along with electric variable transmission. Moreover, network's parameters are optimized with the help of GA. In smart grid system, for user adaption and demand responses management, a framework is proposed by Fan, 2012. This framework is associated with the new method of PHEV charging.

Electric Vehicles Based on Fuel Cell Technology

Energy management strategy for fuel cell electric vehicle is discussed by Ettihir et al. 2016. Efficiency and maximum power points are unique for FCEV, but they are varying with operating conditions. Further, an extremum seeking process is developed to track the above discussed points. A bi directional full bridge voltage multiplier is incorporated by Chakraborty et al. 2017 to develop a FCEV. Further, this multiplier has zero current commutation. A power management approach and optimum component sizing is developed by Hu et al. 2015 for fuel cell based hybrid bus. Various limiting conditions are analysed by Morales et al., 2015 for enforcing the uncertainty's existence in the modelling of FCEV energy magement system.

Fuel Cell

Fuel cell is the most recent and popular technology that is incorporated in many MG systems now a days. The feasibility of cost and workability of grid integrated MG is analysed by Patterson et al. 2015. It is associated with PV, FC and batteries. The study of super capacitor and aluminium air fuel cell in hybrid power system is performed by Zhang et al. 2016. Various similar parameters are computed and analysed by the authors. Moreover, energy allocation scheme is also developed for hybrid power system. For islanded MG that contained FC-electrolyser, BES and PV, an energy management system is developed by Sun et al. 2016. This system is based on modified droop control characteristics. A hybrid power system is realized by Ramírez-Murillo et al. 2015 that is series-parallel in nature. Further, it contained FC and auxiliary storage system. A reconfigurable MG system's architecture is developed by Thale et al. 2015 that contained FC, WT, PV and micro hydro.

Diesel Generator Integrated MG

Diesel generator is the important component of MG that is utilized as backup energy source in the absence of intermittent renewable energy sources (Fini et al., 2016; Alhelou et al., 2018; Zamani et al., 2018; Alhelou et al., 2015; Njenda et al., 2018; Haes Alhelou et al., 2018). A MG control strategy is presented by Vidyanandan and Senroy, 2016 to regulate frequency in diesel generator and WT based MG. An energy allocation scheme for diesel generator and batteries is developed by Tang et al. 2014. Furthermore, NSGA-II is utilized to solve the optimal operation problem of MG. An optimization algorithm is applied by Hajar et al. 2015 on MG system, which incorporated three energy areas namely batteries, diesel generator and renewable energy. An optimum operation and cost and emission minimization scheme is developed by Mohamed and Koivo 2007 for MG system. An optimal design of islanded MG system integrated with diesel generator is proposed by Afshar et al., 2013.

4. MICRO-GRID OPTIMIZATION: META HEURISTIC TECHNIQUES

These techniques are flexible and robust in their nature to solve the complex MG optimization problem. This section provides the detailed literature about the various Meta-heuristic techniques utilized for solving micro-grid optimization problem.

GA

It is introduced by Holland for understanding the naturals systems' adaptive processes (Holland, 1975). Crossover and mutation are the operators associated with GA for encouraging diversity. It also utilized the probabilistic techniques of selection. Offspring replaced their parents methodically in GA. It is the most applied technique in various power system problems (Fini et al., 2016; Alhelou et al., 2018; Zamani et al., 2018; Alhelou et al., 2015; Njenda et al., 2018; Haes Alhelou et al., 2018).

A commercial MG interfaced with dynamically priced main grid is developed by Jayadev and Shanti Swarup, 2013. It consists of PV, battery energy storage and diesel generator. Optimization problem is formed by discontinuous functions, which is solved by GA. Vergara et al. 2015 proposed a real time energy magement system for a MG. NSGA-II is utilized to solve the multi objective problem of proposed system. A coordinated operational system is developed by Siqi et al. 2014, which consists of diesel generator and hybrid energy storage. Running characteristics of hybrid storage and diesel generator is also incorporated by this system. Moreover for the optimization of power outputs, adaptive genetic algorithm (AGA) is used. Santis et al., 2013 proposed a computational intelligence based technique for the optimization of power flow and demands side. For time of use cost management system, hybrid fuzzy is utilized with GA. Quantum GA is utilised by Liao et al. 2010 for economic dispatch problem of smart MG system. A novel MG energy trading model is developed by Changsong et al. 2009 to realize an optimal schedule for different MG subsystems over a time horizon. It utilized GA to find the optimal MG schedule. For enhancing the load sharing between various distributed energy sources in MG, a control method is developed by Zolfaghari et al. 2016. Proportional integral controller tuned with GA is utilized for this problem. NSGA-II is employed by Shi et al. 2014 for developing multi objective optimization problem of various energy sources and loads consisted by MG. A MG system, which consist of WT, CHP, FC, PV and MT is studied and modelled by Deng et al. 2011. For making optimal modelling and controlling, GA is utilized. Optimal design scheme is proposed by Nasser and Reji, 2016 for reliable and cost effective micro-grid system. For optimal results, genetic particle swarm optimization is utilized. MG optimization with the help of GA is presented by Eldessouky and Gabbar, 2015.

PSO

It consists of the particles' population, which is termed as a swarm. Every member in swarm related to the position and velocity vector. These vectors' sizes are equivalent to the dimensions of search space. At any iteration t, the term velocity represents the directional distance, which a particle covered in (t+1) th iteration. On the basis of particle's individual best i.e. "pbest" and swarm's global best 'gbest', the directional velocity is computed (Kennedy and Eberhart, 1995). Various power system problems are solved by PSO technique. The optimization problem of smart MG with controllable loads by PSO is presented in Saber et al. (2013). Enhanced PSO is utilized by Cao et al. 2016 to solve the problem of MG economic load dispatch. An upgraded PSO with adaptive acceleration and weight coefficients are utilized by Chen et al. 2015 to solve the environmental, health and economic dispatch of MG. Economic operation of MG under uncertain conditions is analyzed with the help of PSO technique in Liang, (2014). For smart home energy management system in smart grid environment, an enhanced PSO is proposed and utilized by Yang et al. 2015. The optimal operation of distinctive grid integrated MG is studied by Hao et al. 2014 with the help of PSO. A multi-agent structure of smart micro-grid is developed by Elamine et al. 2014. It uses neural network that is taught by hybrid PSO and back propagation method. A comprehensive review of smart home energy management system based on various Meta-heuristic techniques is presented by Molla et al. 2018.

DE

A robust and easily parallelizable method, differential evaluation (DE) is proposed by Storn and Price, 1997. This method is utilized to solve the global optimization problem. It is a population dependent technique that imitates with randomly initialled vectors of solutions. To improve existing results in the population, DE used the vector of difference of two randomly selected members. Various power system optimization problems are solve with the help of DE. A differential evolution based energy management system is developed by Tiwari and Srivastava, 2016 for optimizing the working of MG. Operation cost of distinctive MG system is optimized with the help of DE in Shuai, 2013. The planned scheduling of CHP based MG for economic power sharing is performed by DE in Basu et al. 2012.

FF

It is the nature inspired Meta-heuristic optimization technique. It is motivated from the behaviors of fire files (Yang, 2009). There are three basic rules on which this optimization technique is depends. These rules are as follows:

1. With disregard to the gender, all fireflies are attracted to each other.
2. Brightness is co-related to the attractiveness.
3. Objective function is proportional to the brightness.

Optimization of the power management scheme of super capacitor, fuel cell and battery based hybrid vehicular system is done by Odeim et al. 2016.

5. COMPARISON BETWEEN DIFFERENT META HEURISTIC TECHNIQUES

Table 2 presented the comprehensive comparison between selected Meta-heuristic techniques on the problem of economic control of MG.

6. CONCLUSION

This chapter presents a comprehensive comparison between renowned Meta-heuristic techniques namely genetic algorithm (GA), Particle Swarm Optimization (PSO), Differential Evaluation (DE), and firefly (FF) algorithm. Economic operation of MG is selected as the reference for presenting the feasibility of all these methodologies. These methodologies are selected from two different categories i.e. evaluation based approach and swarm based approach. A detailed comparison is also presented in this chapter. This chapter helps the researchers to select the appropriate Meta-heuristic technique for MG optimization problem.

Table 2. Comparison between different meta-heuristic techniques

Measures	FF	GA	PSO	DE
Input Parameters	Variation of attractiveness Distance between two files. step size Light absorption coefficient	Crossover rate Mutation rate Population size	Population size Velocity of each particle Learning factors Inertia weight maximum change of a particle velocity position of best particle the current position of particle Best position of particle in previous cycle	upper bound of scaling factor lower bound of scaling factor cross over probability
Convergence Rate	Fast	Fast	Rapid but less than GA	Average
Intensification and Diversification Component	Attraction Mutation Mutation coefficient damping	Crossover Mutation Natural selection	Local search fitness	Scaling Cross over Mutation Natural selection
Advantages	Less complex and easy to implement automatical subdivision able to deal with multimodality tuned to control the randomness Deal with non-linear, problems Good initial solution is not required	The ability of Solving different kinds of optimization problems Finding a global best solution in many problems Easy to combining with other algorithms Design for real and binary search space	Simple structure Easy to implement Fast performance Having few parameters to adjust Efficient global search approach Less dependent on initial points	More stable Easily applicable on wide variety of real valued problems despite noisy, multi-modal, multi-dimensional space
Disadvantages	Improper tuning of algorithm-specific parameters leads to increases the computational effort. High dependency of algorithm-specific parameters leads to local optimal solution Complexity for combinatorial optimization Higher execution time	Using complex operators for selection and cross over Unpredicted results Premature convergence rate Trapping into local optima Taking long run-time Weak local search Difficult encoding scheme Slow convergence rate Finding sub-optimal solution Dependency of the crossover and mutation rates on the stability and convergence Weak local search	Slow convergence rate Parameter selection problem Easily get trapped in a local optimum poor exploration when solving complex multimodal problems If a particle falls into a local optimum, sometimes it cannot get rid of itself from the position Weak local search	Tuning of the algorithm-specific parameters is crucial Improper tuning of algorithm-specific parameters leads to increases the computational effort. High dependency of algorithm-specific parameters leads to local optimal solution a greater degree of computational complexity for combinatorial problems
Applications	Digital Image Compression and Image Processing Feature selection and fault detection Trail neural network Semantic Web Composition Classification and Clustering Rigid Image Registration Problems Parameter Optimization of SVM	Scheduling Chemistry, Chemical Engineering Medicine Data Mining and Data Analysis Geometry and Physics Economics and Finance Networking and Communication Electrical Engineering and Circuit Design Image Processing Combinatorial Optimization	Machine Learning Function Optimization, Geometry and Physics Operations Research Chemistry, Chemical Engineering, Electrical Engineering and Circuit Design	Engineering, Structural Optimization, and Design Chemistry, Chemical Engineering Scheduling, Function Optimization, Electrical Engineering and Circuit Design

(Khan and Singh, 2017)

ACKNOWLEDGMENT

The authors would like to thank all of colleagues for their support and cooperation. Authors would also like to thank Department of Electrical and Computer Engineering, Hawassa University for providing the environment and support to carry out this work.

REFERENCES

Afshar, H., Moravej, Z., & Niasati, M. (2013). Modeling and optimization of micro-grid considering emissions. *Proc. Smart Grid Conf. (SGC)*, 225-229. 10.1109/SGC.2013.6733812

Alharbi, H., & Bhattacharya, K. (2014). Optimal sizing of battery energy storage systems for micro-grids. *Proc. IEEE Electr. Power Energy Conf.*, 275-280. 10.1109/EPEC.2014.44

Alhelou, H., Hamedani-Golshan, M. E., Zamani, R., Heydarian-Forushani, E., & Siano, P. (2018). Challenges and Opportunities of Load Frequency Control in Conventional, Modern and Future Smart Power Systems: A Comprehensive Review. *Energies*, *11*(10), 2497. doi:10.3390/en11102497

Alhelou, H. H. (2018). *Fault Detection and Isolation in Power Systems Using Unknown Input Observer. In Advanced Condition Monitoring and Fault Diagnosis of Electric Machines* (p. 38). Hershey, PA: IGI Global.

Alhelou, H. H., Golshan, M., & Fini, M. (2018). Wind Driven Optimization Algorithm Application to Load Frequency Control in Interconnected Power Systems Considering GRC and GDB Nonlinearities. *Electric Power Components and Syst.*

Alhelou, H. H., & Golshan, M. E. H. (2016, May). Hierarchical plug-in EV control based on primary frequency response in interconnected smart grid. In *Electrical Engineering (ICEE), 2016 24th Iranian Conference on* (pp. 561-566). IEEE. 10.1109/IranianCEE.2016.7585585

Alhelou, H. H., Golshan, M. H., & Askari-Marnani, J. (2018). Robust sensor fault detection and isolation scheme for interconnected smart power systems in presence of RER and EVs using unknown input observer. *International Journal of Electrical Power & Energy Systems*, *99*, 682–694. doi:10.1016/j.ijepes.2018.02.013

Alhelou, H. H., Hamedani-Golshan, M. E., Heydarian-Forushani, E., Al-Sumaiti, A. S., & Siano, P. (2018, September). Decentralized Fractional Order Control Scheme for LFC of Deregulated Nonlinear Power Systems in Presence of EVs and RER. In *2018 International Conference on Smart Energy Systems and Technologies (SEST)* (pp. 1-6). IEEE. 10.1109/SEST.2018.8495858

Alhelou, H. S. H., Golshan, M. E. H., & Fini, M. H. (2015, December). Multi agent electric vehicle control based primary frequency support for future smart micro-grid. In *Smart Grid Conference (SGC)* (pp. 22-27). Academic Press.

Alshahrestani, A., Golshan, M. E. H., & Alhelou, H. H. (2018, November). WAMS Based Online Estimation of Total Inertia Constant and Damping Coefficient for Future Smart Grid Systems. In *Smart Grid Conference (SGC)* (pp. 1-5). Academic Press.

Basu, Bhattacharya, Chowdhury, & Chowdhury. (2012). Planned scheduling for economic power sharing in a CHP-based micro-grid. *IEEE Trans. Power Syst., 27*(1), 30-38.

Cao, H. (2016). Economic dispatch of micro-grid based on improved particle-swarm optimization algorithm. *Proc. North Amer. Power Symp. (NAPS)*, 1-6. 10.1109/NAPS.2016.7747875

Chakraborty, Breaz, Rathore, & Gao. (2017). Parasitics-assisted soft-switching and secondary modulated snubberless clamping currentfed bidirectional voltage doubler for fuel cell vehicles. *IEEE Trans. Veh. Technol., 66*(2), 1053-1062.

Changsong, C., Shanxu, D., Tao, C., & Bangyin, L. (2009). Energy trading model for optimal micro-grid scheduling based on genetic algorithm. *Proc. IEEE 6th Int. Power Electron. Motion Control Conf.*, 2136-2139.

Chen, J., Wang, J., Chen, Q., & Wu, D. (2015). Optimal dispatch of medium voltage micro-grid using an adaptive PSO algorithm. *Proc. 7th Int. Conf. Intell. Human-Mach. Syst. Cybern.*, 324-329. 10.1109/IHMSC.2015.48

Deng, Q., Gao, X., & Zhou, H. (2011). System modeling and optimization of micro-grid using genetic algorithm. *Proc. 2nd Int. Conf. Intell. Control Inf. Process.*, 540-544.

Elamine, D. O., Nfaoui, E. H., & Jaouad, B. (2014). Multi-agent architecture for smart micro-grid optimal control using a hybrid BP-PSO algorithm for wind power prediction. *Proc. 2nd World Conf. Complex Syst. (WCCS)*, 554-560. 10.1109/ICoCS.2014.7060951

Eldessouky, A. S., & Gabbar, H. A. (2015). Micro grid renewables dynamic and static performance optimization using genetic algorithm. *Proc. IEEE Int. Conf. Smart Energy Grid Eng. (SEGE)*, 1-6. 10.1109/SEGE.2015.7324596

Ettihir, Boulon, & Agbossou. (2016). Energy management strategy for a fuel cell hybrid vehicle based on maximum efficiency and maximum power identification. *IET Electr. Syst. Transp., 6*(4), 261-268.

Fan, Z. (2012). A distributed demand response algorithm and its application to PHEV charging in smart grids. *IEEE Trans. Smart Grid, 3*(3), 1280-1290.

Fanuel, M., Khan, B., Singh, N., & Singh, P. (2018, July). Energy Production in Smart Cities by Utilization of Kinetic Energy of Vehicles over Speed Breaker. *International Journal of Civic Engagement and Social Change*, *5*(2), 1–35. doi:10.4018/IJCESC.2018040101

Favuzza, Graditi, & Sanseverino. (2006). Adaptive and dynamic ant colony search algorithm for optimal distribution systems reinforcement strategy. *Appl. Intell., 24*(1), 31-42.

Fini, M. H., Yousefi, G. R., & Alhelou, H. H. (2016). Comparative study on the performance of many-objective and single-objective optimisation algorithms in tuning load frequency controllers of multi-area power systems. *IET Generation, Transmission & Distribution*, *10*(12), 2915–2923. doi:10.1049/iet-gtd.2015.1334

Gamarra & Guerrero. (2015). Computational optimization techniques applied to micro-grids planning: A review. *Renew. Sustain. Energy Rev., 48*, 413-424.

Graditi, Ippolito, Telaretti, & Zizzo. (2016). Technical and economical assessment of distributed electrochemical storages for load shifting applications: An Italian case study. *Renew. Sustain. Energy Rev., 57*, 515-523.

Hajar, K., Hably, A., Elrafhi, A., Obeid, Z., & Bacha, S. (2015). Optimization of a micro-grid with renewable energy and distributed generation: A case study. *Proc. 19th Int. Conf. Syst. Theory, Control Comput. (ICSTCC)*, 662-665. 10.1109/ICSTCC.2015.7321369

Hao, G., Cong, R., & Zhou, H. (2014). PSO applied to optimal operation of a micro-grid with wind power. *Proc. 6th Int. Symp. Parallel Architectures, Algorithms Programm.*, 46-51. 10.1109/PAAP.2014.61

Holland, J. H. (1975). *Adaptation in Natural and Arti_cial Systems*. Ann Arbor, MI: Univ. Michigan Press.

Hu, Jiang, Egardt, & Cao. (2015). Advanced power-source integration in hybrid electric vehicles: Multicriteria optimization approach. *IEEE Trans. Ind. Electron., 62*(12), 7847-7858.

Jariso, M., Khan, B., Tesfaye, D., & Singh, J. (2017). Modelling and Designing of Stand-Alone Photovoltaic System (Case Study: Addis Boder Health Center south west Ethiopia). *IEEE International conference on Electronics, communication, and aerospace Technology, ICECA 2017*, 168-173.

Jayadev, V., & Swarup, K. S. (2013). Optimization of micro-grid with demand side management using Genetic Algorithm. *Proc. IET Conf. Power Unity, Whole Syst. Approach*, 1-6. 10.1049/ic.2013.0124

Kennedy, J., & Eberhart, R. C. (1995). Particle swarm optimization. *Proc. IEEE Int. Conf. Neural Netw.*, 1942-1948. 10.1109/ICNN.1995.488968

Khan, B., & Agnihotri, G. (2012). A Novel Transmission Loss Allocation Method based on Transmission Usage. *IEEE Fifth Power India Conference*, Haryana, India. 10.1109/PowerI.2012.6479479

Khan, B., Agnihotri, G., & Gupta, G. (2013). A multipurpose matrices methodology for transmission usage, loss and reliability margin allocation in restructured environment. *Electrical & Computer Engineering: An International Journal, 2*(3).

Khan, B., Agnihotri, G., Gupta, G., & Rathore, P. (2014). A Power Flow Tracing based Method for Transmission Usage, Loss & Reliability Margin Allocation. *AASRI Procedia, 7*, 94–100. doi:10.1016/j.aasri.2014.05.035

Khan, B., Agnihotri, G., & Mishra, A. S. (2016). An Approach for Transmission Loss and Cost Allocation by Loss Allocation Index and Cooperative Game Theory. *J. Inst. Eng. India Ser. B, 97*(1), 41–46. doi:10.100740031-014-0165-1

Khan, B., Agnihotri, G., Rathore, P., Mishra, A., & Naidu, G. (2014). A Cooperative Game Theory Approach for Usage and Reliability Margin Cost Allocation under Contingent Restructured Market. *International Review of Electrical Engineering, 9*(4), 854–862.

Khan, B., & Singh, P. (2017). The Current and Future States of Ethiopia's Energy Sector and Potential for Green Energy: A Comprehensive Study. *International Journal of Engineering Research in Africa, 33*, 115–139. doi:10.4028/www.scientific.net/JERA.33.115

Khan, B., & Singh, P. (2017). Selecting a Meta-Heuristic Technique for Smart Micro-Grid Optimization Problem: A Comprehensive Analysis. IEEE Access, 5, 13951-13977.

Khan, B., & Singh, P. (2017). Optimal Power Flow Techniques under Characterization of Conventional and Renewable Energy Sources: A Comprehensive Analysis. *Journal of Engineering*.

Kifle, Y., Khan, B., & Singh, J. (2018). Designing and Modeling Grid Connected Photovoltaic System: (Case Study: EEU Building at Hawassa city). *Int. J. of Convergence Computing, 3*(1), 20–34. doi:10.1504/IJCONVC.2018.091113

Kifle, Y., Khan, B., & Singh, P. (2018, July). Assessment and Enhancement of Distribution System Reliability by Renewable Energy Sources and Energy Storage. *Journal of Green Engineering, 8*(3), 219–262. doi:10.13052/jge1904-4720.832

Laureri, F., Puliga, L., & Robba, M. (2016). An optimization model for the integration of electric vehicles and smart grids: Problem definition and experimental validation. *Proc. IEEE Int. Smart Cities Conf. (ISC)*, 1-6.

Li, Wu, Hu, Li, Hossain, & Ghoneim. (2017). HELOS: Heterogeneous load scheduling for electric vehicle-integrated micro-grids. *IEEE Trans. Veh. Technol., 66*(7), 5785-5796.

Liang, G., Liyuan, P., Ruihuan, L., Fen, Z., Jinhui, L., & Xin, W. (2014). Study on economic operation for micro-grid based on scenario and PSO. *Proc. Int. Conf. Power Syst. Technol.*, 3152-3156. 10.1109/POWERCON.2014.6993642

Liao, G. C. (2010). Using chaotic quantum genetic algorithm solving environmental economic dispatch of smart micro-grid containing distributed generation system problems. *Proc. Int. Conf. Power Syst. Technol.*, 1-7. 10.1109/POWERCON.2010.5666468

Makdisie, C., Haidar, B., & Alhelou, H. H. (2018). An Optimal Photovoltaic Conversion System for Future Smart Grids. In *Handbook of Research on Power and Energy System Optimization* (pp. 601–657). IGI Global. doi:10.4018/978-1-5225-3935-3.ch018

Melhem, F. Y., Moubayed, N., & Grunder, O. (2016). Residential energy management in smart grid considering renewable energy sources and vehicleto- grid integration. *Proc. IEEE Electr. Power Energy Conf. (EPEC)*, 1-6.

Mohamed, F. A., & Koivo, H. N. (2007). System modelling and online optimal management of MicroGrid using multiobjective optimization. *Proc. Int. Conf. Clean Electr. Power*, 148-153. 10.1109/ICCEP.2007.384202

Molla, T., Khan, B., & Singh, P. (2018). A Comprehensive Analysis of Smart Home Energy Management System Optimization Techniques. *Journal of Autonomous Intelligence, 1*(1).

Morales-Morales, Cervantes, & Cano-Castillo. (2015). On the design of robust energy management strategies for FCHEV. *IEEE Trans. Veh. Technol., 64*(5), 1716-1728.

Mulualem, T. Y., & Khan, B. (2017). Design of an off-grid hybrid PV/wind power system for remote mobile base station: A case study. *AIMS Energy*, *5*(1), 96–112. doi:10.3934/energy.2017.1.96

Nadweh, S., Hayek, G., Atieh, B., & Haes Alhelou, H. (2018). Using Four – Quadrant Chopper with Variable Speed Drive System Dc-Link to Improve the Quality of Supplied Power for Industrial Facilities. Majlesi Journal of Electrical Engineering.

Nasser, A., & Reji, P. (2016). Optimal planning approach for a cost effective and reliable micro-grid. *Proc. Int. Conf. Cogeneration, Small Power Plants District Energy (ICUE)*, 1-6. 10.1109/COGEN.2016.7728966

Njenda, T. C., Golshan, M. E. H., & Alhelou, H. H. (2018, November). WAMS Based Intelligent Under Frequency Load Shedding Considering Online Disturbance Estimation. In *Smart Grid Conference (SGC)* (pp. 1-5). Academic Press.

Njenda, T. C., Golshan, M. E. H., & Alhelou, H. H. (2018, November). WAMS based Under Frequency Load Shedding Considering Minimum Frequency Prediction and Extrapolated Disturbance Magnitude. In *Smart Grid Conference (SGC)* (pp. 1-5). Academic Press.

Odeim, Roes, & Heinzel. (2016). Power management optimization of a fuel cell/battery/supercapacitor hybrid system for transit bus applications. *IEEE Trans. Veh. Technol., 65*(7), 5783-5788.

Paterakis, Erdinç, Pappi, Bakirtzis, & Catalão. (2016). Coordinated operation of a neighborhood of smart households comprising electric vehicles, energy storage and distributed generation. *IEEE Trans. Smart Grid, 7*(6), 2736-2747.

Patterson, Macia, & Kannan. (2015). Hybrid micro-grid model based on solar photovoltaic battery fuel cell system for intermittent load applications. *IEEE Trans. Energy Convers., 30*(1), 359-366.

Pawan & Baseem. (2017). Smart Microgrid Energy Management Using a Novel Artificial Shark Optimization. Complexity.

Ramírez-Murillo, Restrepo, Calvente, Romero, & Giral. (2015). Energy management of a fuel-cell serial_parallel hybrid system. *IEEE Trans. Ind. Electron., 62*(8), 5227-5235.

Saber, A. Y., & Venayagamoorthy, G. K. (2013). Smart micro-grid optimization with controllable loads using particle swarm optimization. *Proc. IEEE Power Energy Soc. General Meet.*, 1-5. 10.1109/PESMG.2013.6672780

Santis, E. D., Rizzi, A., Sadeghiany, A., & Mascioli, F. M. F. (2013). Genetic optimization of a fuzzy control system for energy flow management in micro-grids. *Proc. Joint IFSA World Congr. NAFIPS Annu. Meet. (IFSA/NAFIPS),* 418-423. 10.1109/IFSA-NAFIPS.2013.6608437

Sharma, Bhattacharjee, & Bhattacharya. (2016). Grey wolf optimisation for optimal sizing of battery energy storage device to minimise operation cost of micro-grid. *IET Generat., Transmiss. Distrib., 10*(3), 625-637.

Shen, J., Jiang, C., Liu, Y., & Wang, X. (2016). A Microgrid Energy Management System and Risk Management Under an Electricity Market Environment. IEEE Access, 4, 2349-2356. doi:10.1109/ACCESS.2016.2555926

Shi, Z., Peng, Y., & Wei, W. (2014). Optimal sizing of DGs and storage for micro-grid with interruptible load using improved NSGA-II. *Proc. IEEE Congr. Evol. Comput. (CEC),* 2108-2115. 10.1109/CEC.2014.6900414

Shuai, D., Nian, L., & Yingmiao, C. (2013). Optimal operation of the island micro-grid with renewable energy and desalination. *Proc. Int. Conf. Mech. Sci., Electric Eng. Comput. (MEC),* 3718-3722.

Silvestre, Di, Graditi, & Sanseverino. (2014). A generalized framework for optimal sizing of distributed energy resources in micro-grids using an indicator-based swarm approach. *IEEE Trans Ind. Informat., 10*(1), 152-162.

Siqi, C. (2014). Optimal coordinated operation for micro-grid with hybrid energy storage and diesel generator. *Proc. Int. Conf. Power Syst. Technol.*, 3207-3212.

Storn, R., & Price, K. (1997). Differential evolution: A simple and efficient heuristic for global optimization over continuous spaces. *J. Global Optim., 11*(4), 341-359.

Sun, Liu, Cai, Zhang, Zhu, & Wang. (2016). Frequency based power management for photovoltaic/battery/fuel cell-electrolyser stand-alone micro-grid. *IET Power Electron., 9*(13), 2602-2610.

Tang, Q., Liu, N., & Zhang, J. (2014). Optimal operation method for micro-grid with wind/PV/diesel generator/battery and desalination. *Journal of Applied Mathematics, 2014*(Jun), 857541.

Thale, Wandhare, & Agarwal. (2015). A novel reconfigurable micro-grid architecture with renewable energy sources and storage. *IEEE Trans. Ind. Appl., 51*(2), 1805-1816.

Thompson, Oikonomou, Etemadi, & Sorger. (2016). Optimization of data center battery storage investments for micro-grid cost savings, emissions reduction, and reliability enhancement. *IEEE Trans. Ind. Appl., 52*(3), 2053-2060.

Tiwari, N., & Srivastava, L. (2016). Generation scheduling and micro-grid energy management using differential evolution algorithm. *Proc. Int.Conf. Circuit, Power Comput. Technol. (ICCPCT),* 1-7. 10.1109/ICCPCT.2016.7530218

Vergara, P. P., Torquato, R., & da Silva, L. C. P. (2015). Towards a real-time energy management system for a micro-grid using a multi-objective genetic algorithm. *Proc. IEEE Power Energy Soc. General Meet.*, 1-5. 10.1109/PESGM.2015.7285956

Vidyanandan, K. V., & Senroy, N. (2016). Frequency regulation in a wind diesel powered micro-grid using flywheels and fuel cells. *IET Generat., Transmiss. Distrib., 10*(3), 780-788.

Vinot, Reinbold, & Trigui. (2016). Global optimized design of an electric variable transmission for HEVs. *IEEE Trans. Veh. Technol., 65*(8), 6794-6798.

Xu, Bishop, Oikarinen, & Hao. (2016). Application and modeling of battery energy storage in power systems. *CSEE Journal of Power and Energy Systems, 2*(3), 82-90.

Xu, Y. (2015). Optimal distributed charging rate control of plug-in electric vehicles for demand management. *IEEE Trans. Power Syst., 30*(3), 1536-1545.

Yang, H.-T., Yang, C.-T., Tsai, C.-C., Chen, G.-J., & Chen, S.-Y. (2015). Improved PSO based home energy management systems integrated with demand response in a smart grid. *Proc. IEEE Congr. Evol. Comput. (CEC),* 275-282. 10.1109/CEC.2015.7256902

Yang, X.-S. (2009). Firefly algorithms for multimodal optimization. In Stochastic Algorithms: Foundations and Applications (SAGA). Berlin, Germany: Springer.

Yao, Lim, & Tsai. (2017). A real-time charging scheme for demand response in electric vehicle parking station. *IEEE Trans. Smart Grid, 8*(1), 52-62.

Yu, Zhong, Xie, Yuen, Gjessing, & Zhang. (2016). Balancing power demand through EV mobility in vehicle-to-grid mobile energy networks. *IEEE Trans. Ind. Informat., 12*(1), 79-90.

Zamani, R., Hamedani-Golshan, M. E., Haes Alhelou, H., Siano, P., & Pota, H. (2018). Islanding Detection of Synchronous Distributed Generator Based on the Active and Reactive Power Control Loops. *Energies, 11*(10), 2819. doi:10.3390/en11102819

Zhang, Mou, & Yang. (2016). An energy management study on hybrid power of electric vehicle based on aluminum air fuel cell. *IEEE Trans. Appl. Supercond., 26*(7), 1-6.

Zolfaghari, M., Habil, H. B., Abyaneh, H. A., & Abedi, M. (2016). Load sharing improvement between parallel-connected inverter based DGs using a GA based optimization control strategy in micro-grids. *Proc. IEEE PES Asia-Pacific Power Energy Eng. Conf. (APPEEC),* 320-323. 10.1109/APPEEC.2016.7779520

KEY TERMS AND DEFINITIONS

Electric Grid: An electrical grid is an interconnected network for delivering electricity from producers to consumers. It consists of generating stations that produce electrical power high voltage transmission lines that carry power from distant sources to demand centers distribution lines that connect individual customers.

Micro-Grid PPC: The point where the micro-grid is connected with the main grid through a breaker mechanism.

Microgrids: A micro-grid is a localized group of electricity sources and loads that normally operates connected to and synchronous with the traditional wide area synchronous grid, but can also disconnect to "island mode"—and function autonomously as physical or economic conditions dictate.

Chapter 15
Introduction to Smart Grid and Micro-Grid Systems:
Related Environmental Issues to Global Changes Are the Major Concerns to the Globe Interest

Safwan Nadweh
Tishreen University, Syria

Zeina Barakat
Tishreen University, Syria

ABSTRACT

This chapter describes the upcoming technology for electrical power systems that gives the appropriate solution for the integration of the distributed energy resources. In this chapter, different categories of smart grids have been classified, and the advantages, weakness, and opportunities of each one, are given in addition to determining its own operating conditions. Micro-grids are the most common kind of smart grid. It has been classified under different criteria, such as architecture with different topology (connected mode, island mode, etc.) and demand criteria (simple micro grids, multi-DG, utility) and by capacity into simple micro-grid, corporate micro-grid, and independent micro-grid, and by AC/DC type to DC micro-grids, AC micro-grids, Hybrid micro-grids. Finally, most familiar Micro-grid components have been discussed such as an energy management system along with several types of control and communication systems in addition to the economic study of a micro-grids.

INTRODUCTION

Related environmental issues to Global changes are the major concerns to the globe interest. This lead to increase the growing efforts to reduce the carbon emissions from the energy sector by using renewables, new technologies and generating substantial commitment to develop and deploy these technologies.

DOI: 10.4018/978-1-5225-8030-0.ch015

Figure 1. Renewable energy participation in electricity production for 2020

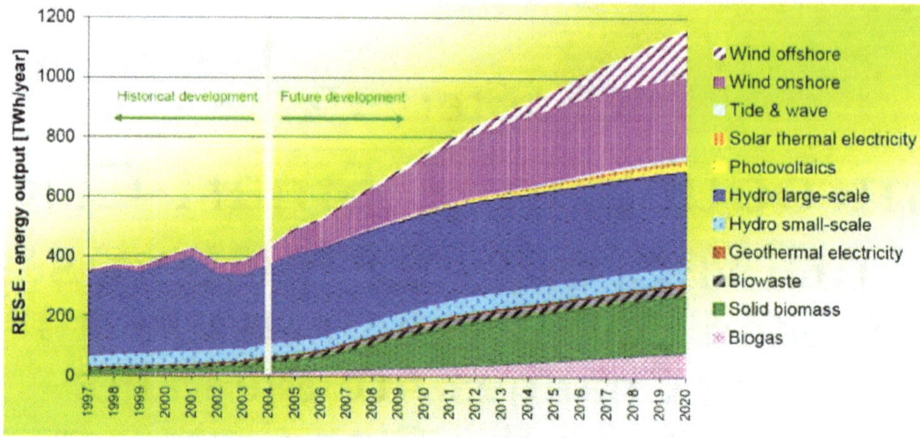

According to the statics (Raffael, 2010), in 2020 the proportion of using renewable energy (RS) in electrical generation will grow up to triple time. Subsequently, a new method and technologies should be developed.

Distributed Resources (DRs) is the new aspect for generation technology. This new technology along with controllable loads presents a serious challenge for traditional power systems planning and their operation rules. Typically, RSs were used to installed to the power system by ''Fit and forget'' approach. Based on this approach the DRs are less visible to the power system.

This rises a main case '' Increasing the RSs and theirs implement in providing electricity, what is the best scheme to be integrated with the present power systems??''.

Smart grids is the most recent solution to improve the integration. These kind of networks give the solution for the fluctuated output characteristics of DRs and allow them to access to the electricity domain and its market (Alshahrestani et al., 2018; Makdisie et al., 2018; Alhelou et al., 2018; Alhelou et al., 2016; Nadweh et al., 2018; Njenda et al., 2018).

DISTRIBUTED ENERGY RESOURCES

During the last decades, Implementation of DRs has been gradually increasing in electricity generation. Typically DRs refer to small scales and storage technologies with capability less than 250 Kw. such as Renewable energy (solar power, wind power, etc), electrical vehicles and Combined heat power units. Fig (1) shows the contribution of renewable energy in electricity production for 2020 (CEN-CENELEC-ETSI, 2012).

Although the huge benefits introduced by DERs, it still show a many challenges that can Restrict its deployment.

In (Colson et al., 2011), reverse power flow as the generation capacity for one area with number of DRs exceeds local demand is one of the biggest challenges that affect the protection schemes and could result in network congestion, in addition the technical concerns.

Installed DRs with power systems has many advantages that can improve the performance of power systems. Benefits related to using DRs can be classified as:

1. **For Costumers:** DRs increase the potential of participation of end users in both electricity market transactions and power system operation with bigger space of freedom.
2. **For Power Systems:**
   a. The overall losses through transmission system is reduced as the results of placing generation units closer to the end users.
   b. High power quality and independent power.

SMART GRID

(Olivares et al., 2011), Up to date there is no a common definition for Smart grid, but it typically is understood as a concept involve technologies, and operating Practices which intended to bring electricity to our century . Broadly Speaking, this involves two main goals:

1. Using technologies to overhead the electricity system.
2. Giving the consumer new applications, using and conserving electricity.

Smart grid can be classified for three different categories:

1. **Micro Grids (MG):** It can be defined as a group of DRs and loads connected to the grid or even work in island mode.
2. **Virtual Power Plant (VPP):** This term has released at 1997 as a virtual utility" VPP is an aggregate of Different kinds of DRs''.

Energy Hubs (EH): (Geidl et al., 2007) described as multiple energy units such as electricity, gas, heat….etc. These different forms of energy can be converted, conditioned, and even be stored.

Each one of the mentioned technologies has its own advantages and disadvantages. They can be summarized as flow.

MICRO GRIDS

Advantages:

1. Improved reliability of power systems.
2. Find solutions for electrical situation of local areas.
3. It smooth the integration of energy production and distribution systems.

Figure 2. Example of a microgrid with a load classification into three reliability types.
Source: http://certs.lbl.gov/certs-der.html

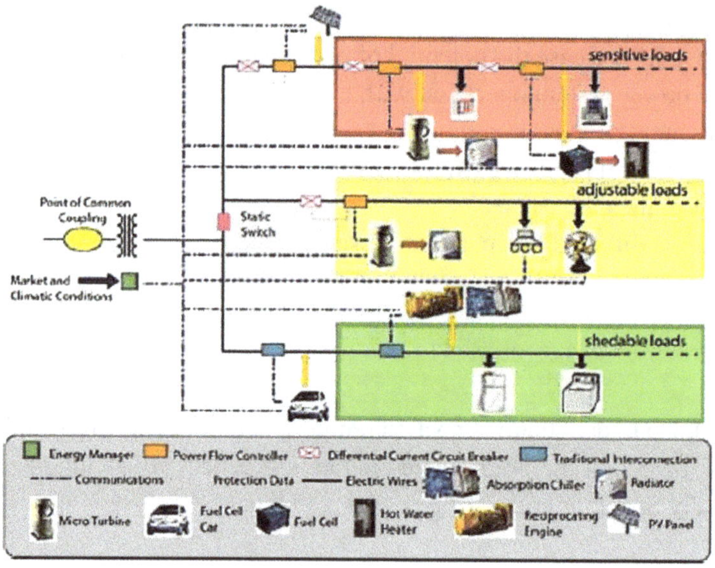

Figure 3. Example of virtual power plant

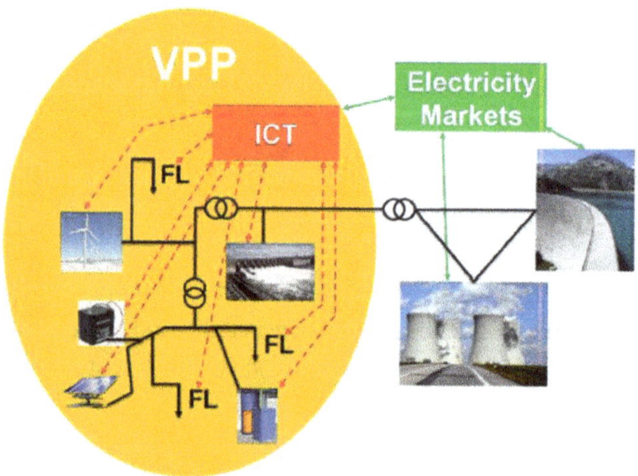

Figure 4. The hospital of Baden modelled as energy hub

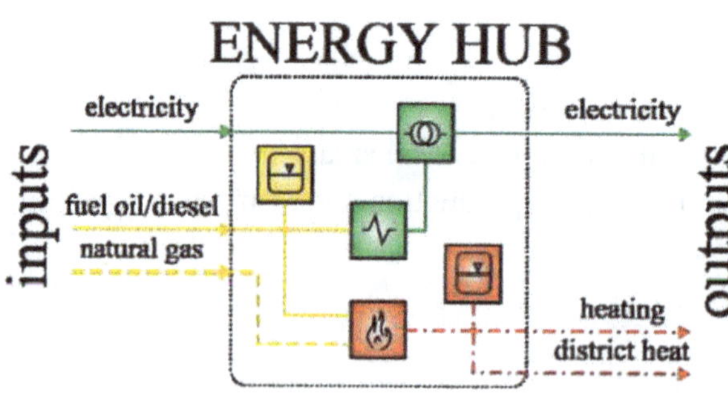

Weakness:

1. Realistically, with increasing the size and number of DRs units, will lead to more complexity with microgrid and the proportion of enteral faults increase.

Opportunities:

1. Improve the electricity supply situation in landed and remote regions and in less development countries.
2. A decent solution for the emergency integrated DERs to face the increased load domain instead of fossil fuel units.

VIRTUAL POWER PLANT (VPP)

Advantages:

1. Small units accessing to electricity market.
2. DRs able to participate in VPP despite their geographical distance.

Weakness:

1. Cost rule, where the cost of controlling VPP should always be less than energy markets.
2. Basics equipment for communication and informatics technology.

Opportunities:

1. Improving the economical efficient for small units by increasing market incentives.
2. Balance the peak load.
3. Industrial field participation in control energy markets.

ENERGY HUBS

Advantages:

1. Huge potentials for modeling every new technology due to its flexible and scalable features.
2. Inclusive view for energy field.

Opportunities:

1. View on multi-energy carriers could bring new scenarios and synergy benefits.

Weakness:

1. It has more theoretical concepts. It is a modeling and planning tool which requires more efforts or realization.
2. The required input data can't be available with the frame of time (day to day).

Variety features for each form of smart grid allow to choose the suitable form depending on the conditions and the purpose of work.

In last few Years Microgrid has been growing gradually as a result of its Potential benefits that provide reliable, secure, efficient, environmentally friendly, and sustainable electricity from renewable energy sources (Fini et al., 2016; Alhelou et al., 2018; Zamani et al., 2018; Alhelou et al., 2015; Njenda et al., 2018; Haes Alhelou et al., 2018).

MICROGRIDS (MG)

There are different definitions for the concept Micro Grid (MG), such as:

The MICROGRID, as defined by the U.S. Department of Energy, is *"a group of interconnected loads and distributed energy resources (DERs) with clearly defined electrical boundaries that acts as a single controllable entity with respect to the grid and can connect and disconnect from the grid to enable it to operate in both grid-connected or island modes* "(Parhizi et al., 2015).

Despite the number of definitions, all of them share a common points that characterize MicroGrid such as: the combining renewable energy and other types of generation technologies (Combined heat power plant, electrical vehicles, storage.... etc) along with demand in small power system that can be work independently or connected to the traditional electrical grids (Dissertation in University of Vaasa, 2011).

Another common characteristics are the voltage rates and geographical Location, some definitions don't come to determine it, on the other hand, some definitions classify the MG as a low voltage electrical network, beside microgrid in not necessary geographically small (SMB Smart Grid Strategic Group, 2010).

Normally, Most of the generation units connected to the micro grid via electronic interface, consisting of either a DC/AC inverter or an AC/DC rectifier and a DC/AC inverter. These converters can provide an additional layer of control and protection to the unit. The third components of Microgrid is an advance control and monitoring system that fulfilled the supervision goals (National Institute of Standards and Technology, 2010; Lasseter, 2002). http://www.smartgrids.eu/sites/default/files/private/2011_03_01_mandate_m490_en.

MICROGRID ARCHITECTURE

The physical connection between the components of the microgrid (generation units, Costumer loads) is done according to three different schemes. These schemes known as radial, ring, and mesh configurations. As shown in fig (5). The structure topology may differ from the operation topology (CEN-CENELEC-ETSI Smart Grid Coordination Group, 2012).

1. Radial scheme

Figure 5. Different microgrid architecture

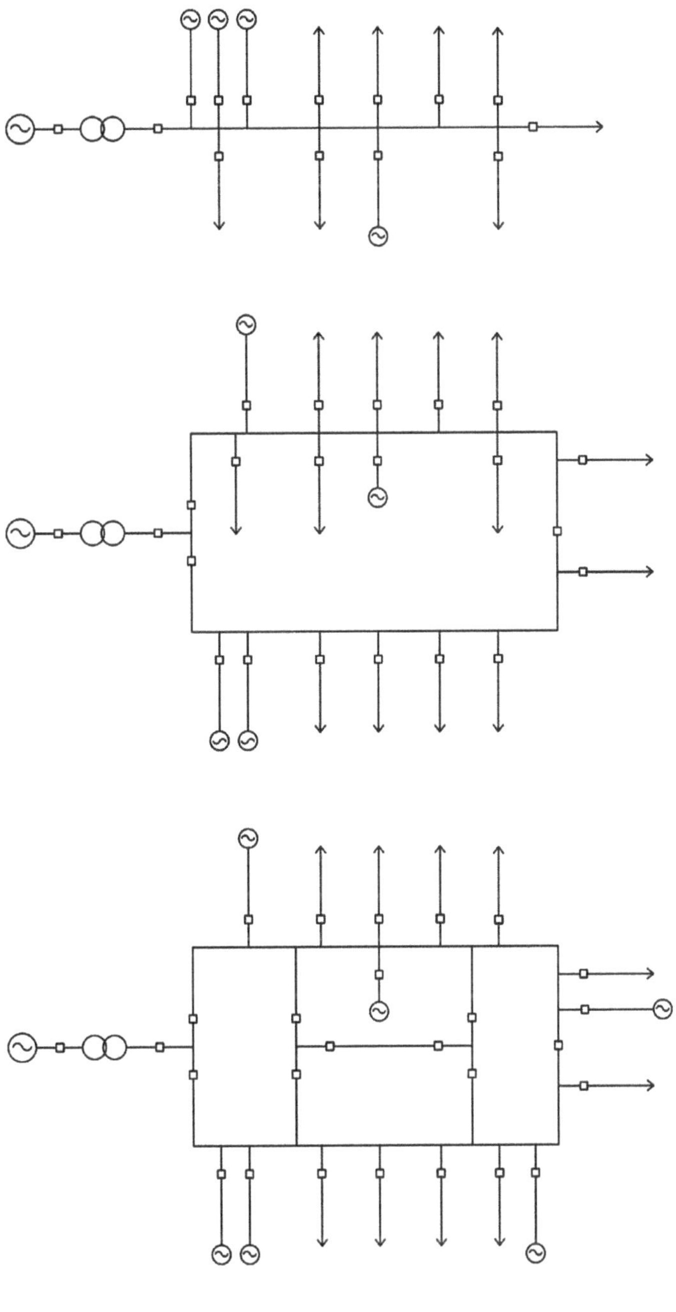

2. Ring scheme
3. Mesh scheme

Radial topology is the simplest structure and the most common topology in the rural areas. It consists of one main line (or many parallel main lines) which DERs and loads connected to this line as branches. The main characteristic of this topology is making the technical implementation of control and protection system simpler. Besides, DERs can be connected closer to the transformer and the flow of power maintains in one direction, which allows keeping the control and protection units at the substations. This topology applicable to low voltage network where fuse is currently the primary protection equipment (Turunen, 2016; Farhangi, 2017).

Ring topology forms from set of lines which creates a loop or a ring shape. This kind of topologies are very common in residential areas. The loop shape allows the power to flow in two directions in any point of the grid, as sequence, the stability feature and power losses in the network improve, but it requires more complex protection system. Because of the complicated required protection system the ring topology often turn into radial one with two main parallel lines by keeping the loop open. This will improve the selectivity of the network by allowing the isolation of a faulty network segment and *restoration* of supply to all customers behind the faulty area, while maintaining system simplicity.

Mesh topology offers multi alternative connections to all nodes of network which increases the redundancy and rises the flexibility of the network. This kind of topology more expensive due to the additional equipment such as additional power lines and breaker. This also will introduce a challenge for the protection system same as ring topology.

Although microgrid may differ in structure but, Micro Grids typically have two operation modes. Connected mode and island one.

Connected Mode

Connected mode allows the micro grid to install to the main electrical grid. Connection point between micro grid and electrical system often referred as a point of common coupling PCC.

In this mode MG has the ability to inject or receive the power from the grid. The MG supplies the local loads. When the generated power exceed the load demand, the MG will deliver the spare power to the main grid. On the other hand, if a disturbance occurs, the microgrid is disconnected from the distribution network as soon as possible in order to avoid any further damage. In that case, the microgrid will operate in an island mode.

Island Mode

In this mode the production characteristics should meet the local load demand. This operating mode a challenge due to the low inertia inherent to small DRs and lack thereof for inverter-based energy sources (such as renewables). This mode fit more properly with islands.

Islanding May happened normally due to a decision of operator or suddenly due to a fault in the utility of grid. The main difference between the reasons is the power flow between the microgrid and the grid, where in the first case the active and reactive power flow is controlled to zero before the transition in order to achieve the power balance inside the micrgrid (Fini et al., 2016; Alhelou et al., 2018; Zamani et al., 2018; Alhelou et al., 2015; Njenda et al., 2018; Haes Alhelou et al., 2018).

In the second case, micro grid has three different operation possibilities depending on the Distribution System Operator (DSO):

1. **Disconnection Required:** Microgrid seen as a one generation unit according to the utility grid, therefor it should be disconnected to prevent energized islands in the main gird. This disconnected operation is done in the frame of protection system principles for generation units.

2. **Disconnection Possible:** The operator in this situation has the freedom to decide the action should been taken. For examples allowing the protection system to operate before disconnecting from the grid.
3. **Disconnection Prohibited:** Microgrid should support the utility grid as a normal generation unit during the fault. This would require the characteristics of power system for micro grid.

Distributed command and control strategy used to apply because it makes the requirements of performance more simple where the upper control level only need to be informed of occurrence of the events.

Along with the intelligence control system, microgrid should has an optimum communication system to deal with the immense data which is in real-time for each node in the system. In other word, the communication system should provide a certain requirements such as band width, security level and latency requirements.

MICROGRID CLASSIFICATION

There are many different categories that used to classify microgrid. According to that we will mention is this chapter most common classification:

By Demand Category

Classifying microgrids by function demand clearly defines the ownership of a microgrid during operation. According to this criteria, we defined a simple microgrid and multi –DG micro grid and utility one.

- **Simple Microgrid:** This kind of grids is contain one type of DERs, specifically combined heat power plant CHPP, or the kind of grid that continually supply a local load.
- **Multi –DG Microgrid:** This kind can be defined as set of simple microgrid, or a cluster of different DERs. Such grids are more complex and loads need to classify as sheddable ones for emergency situations in order to maintain the power balances.
- **Utility Microgrid:** Different kinds of DERS and simple micro grids which achieve a particular technical conditions can work as utility microgrid. Loads in this type classified according to the priority.

By Capacity

Microgrids are classified into simple microgrid, corporate microgrid, feeder area microgrid, substation area microgrid, and independent microgrid by capacity, as shown in table (1)

AC/DC Type

There are three classification DC microgrid, AC microgrid, and Hybrid microgrid (Fusheng et al., 2016).

- **DC Microgrids:** Components of this kind of Microgrid (DER, Storage systems, DC loads) is connected to DC bus via converter. And the AC loads connected to the DC bus via inverter DC/AC as shown in fig (6)

Coordinated operations in such grids is easier since the DERs control depending on DC voltage. The control system for this grid is simpler because it doesn't need synchronizing operation between the DERs. But the AC required inverter to supply AC loads can be a weak point in this grid.

- **AC Microgrid:** Fig (7) shows the structure of AC Microgrid which connects to the distribution system via AC bus that control the operation of connecting / disconnecting to the distribution system. This is done by circuit breaker in PCC.

DERs and storage system connected to the AC bus using DC/AC inverter. This kind of microgrid has the ability to connect to the grid and doesn't need inverter to supply AC loads. On the other hand this microgrid need more complex control and management systems.

- **Hybrid Microgrid:** This kind of microgrid combined both mention microgrid (DC, AC). It contains a DC Bus and AC bus as shown in fig (8). This will allows direct supplying for AC loads and DC loads

Table 1. Microgrids according to capacity criteria

MicroGrids According to Capacity Criteria		
Type	**Capacity**	**Grid to Be Connected**
Simple micro grid	< 2 MW	Local grid
Corporate microgrid	2-5 MW	Local grid
Feeder area microgrid	5-20 MW	Local grid
Substation area microgrid	>20 MW	Local grid
Independent micro grid	Depending on the loads of the area such as village or island	Diesel – fueled grids

Figure 6. DC microgrid

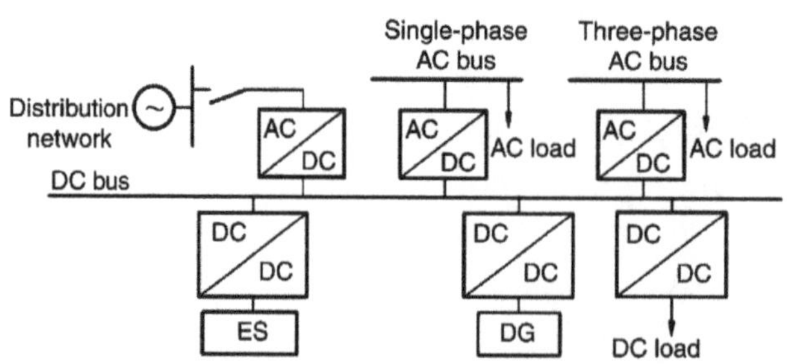

Figure 7. AC microgrid

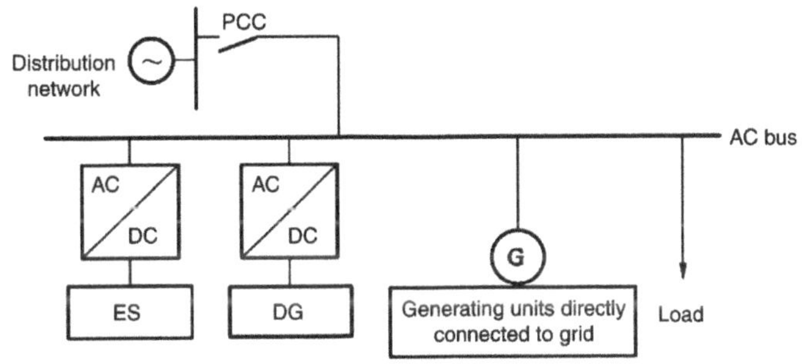

Figure 8. Hybrid microgrid

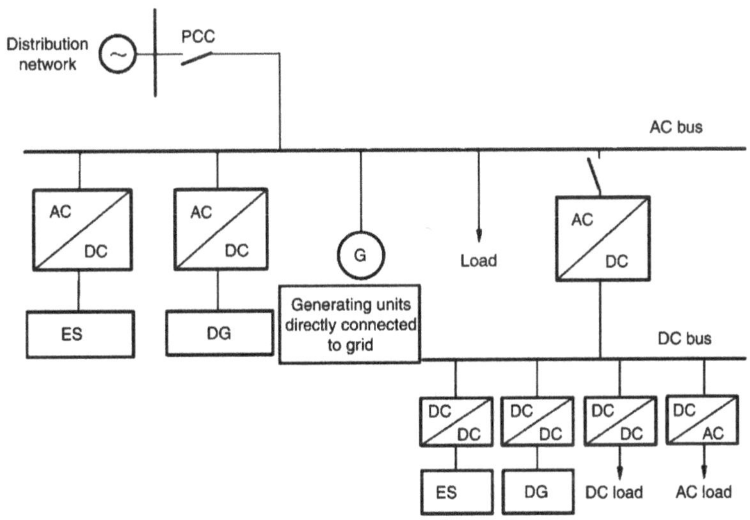

ENERGY MANAGEMENT SYSTEM OF MICROGRID

Management system is an essential component to the microgrid in order to impose the power allocation among DER. Fig (9) shows a diagram representing the management system. This system can be seen as transition of distribution management system.

Each DERs of micro grid is equipped with electronic metering beside to a micro source controller, an intelligent electronic device, that scheduling the DERs Data and controlling it. While the load demand is controlled using load controllers at the connection point of costumers with the grid (CHalise, 2016)

This monitoring tasks allows the EMS to have an overview of the power flow in micro grid. As a result, EMS will be able to optimize the performance of the microgrid i.e EMS is capable to do the following tasks:

1. Achieving integration of renewable energy plants with the traditional ones.
2. Enabling integration of storage systems such as batteries with the power system.
3. Optimizing the operation of different kind of DERs in order to lower the energy costs.

Figure 9. Energy management system (EMS)

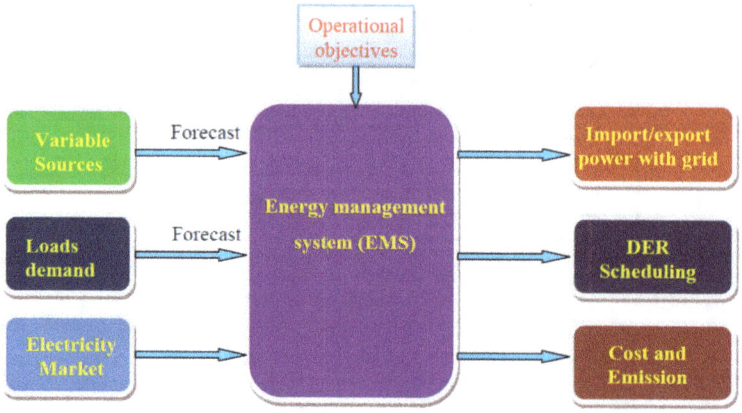

4. Monitoring the electrical power system in real time to enhance its performance.
5. Managing different load types (e.g., fixed, controllable, interruptible and reserve load).
6. Ability of controlling for operator via user interface.

EMS involves number of optimization strategies that regulate the energy plans when system operates in different modes.

In grid connected mode, EMS determines the operation constrains depending on the power flow between the micro grid and the network. When the local load of microgrid lacks the power, the power will be derived from the grid. The maximum power imported from the grid determine the negative limit. When the generating power is exceeds the local demand of the micro grid the micro grid can share its power with network. The minimum amounted of power that microgrid exported to the grid determines the positive limit. Having those limits will allow EMS planning the energy process and reducing the load peak demand.

In island mode, EMS is responsible for managing the loads and other control tasks. EMS enables system operators to schedule intentional islanding periods through an operator portal. When a decision made to move to the island mode, EMS evaluates the availability of entering in such mode through examining the available energy and preparing the storage system. At the requested time, the EMS opens or closes the appropriate contactors and circuit breakers to disconnect Energy from the grid. This strategy also done when the mode automatically turn to island one due to sudden fault in the grid.

At the end of the mode, EMS detects that the system is not able to supply the local load so it will reconnect the microgrid with the grid. Control capabilities of EMS will help switch between operating in grid-connected and islanded modes easily.

CONTROL AND COMMUNICATION SYSTEM

Designing control and communication system is very important part of smart grids. This technologies will help to add an intelligence layer to microgrid. Adding such layer to the power system is what makes microgrid smart or intelligence. This will allow real-time monitoring and control across all the nodes of the system. regarding the location of the node it has to enable the real- time monitoring, in other word monitoring preferably has to be end to end or in each nodes. Data sent by real time monitoring will be used in control application such EMS (Peas Lopes et al., 2010).

The basic control level is set with the individual components. Each components (storage system, converter) has its own control tool which linked to the upper level of control layer such as EMS.

Due to the difference in resources availability, location, and load requirement, there are three different kinds of control techniqes.

CENTRALIZED CONTROL

Central control unit is utilized to communicate between micro grids, this central unit contain the information relevant to all microgrid. This information includes: forecasted values of the nondispatchable sources, load, operational limits (maximum, minimum and most efficient region) of dispatchable sources, state of charge of the battery, and the state of the components (ON or OFF). Using reliable communication channel is a very important requirement in order to achieve an optimum operation for Microgrid such as telephone lines, power line carriers, or a wireless medium. Control strategies to obtain optimal operation can be accomplished by supervisory control and data acquisition (SCADA) systems this type of control is most suitable for situations where all components are located in one central station. An advantage of central control is the ability for central monitoring and the availability of a large amount of system data that can be used to optimize microgrid operation. Since, the entire microgrid system depends upon a single controller, the failure of that controller will cause system failure. Other disadvantages include the inability to support plug and play flexibility and the high computational power and memory requirements necessary for manipulating a large number of data points.

A typical architecture for the central control method is shown in Fig (10) each component accepts the command and performs the operation accordingly. Example commands could include active and reactive power dispatching values and load shedding. Studies, presents various test cases using this type of control technique.

DISTRIBUTED CONTROL

In a distributed control, the local controllers (LC) independently manage the microgrids components. Distributed control can be divided into two types depending on whether or not the LCs communicates with each other. First is droop-based communication-less control. In this type of control, local measurement of voltage and frequency, which does not require a communication link as shown in Fig (11) is used for load sharing among the generators .Frequency droop is typically used to control the active power and voltage droop to control the reactive power. When there is change in load or generation, frequency changes and master unit (battery or generator) adjust the power accordingly.

In (Arboleya et al., 2011), no regular update on droop setting is provided. This method is useful when the resources are dispersed across the microgrid. Droop control also enables plug and play flexibility to expand the system with additional DGs.

Figure 10. Centralized control system

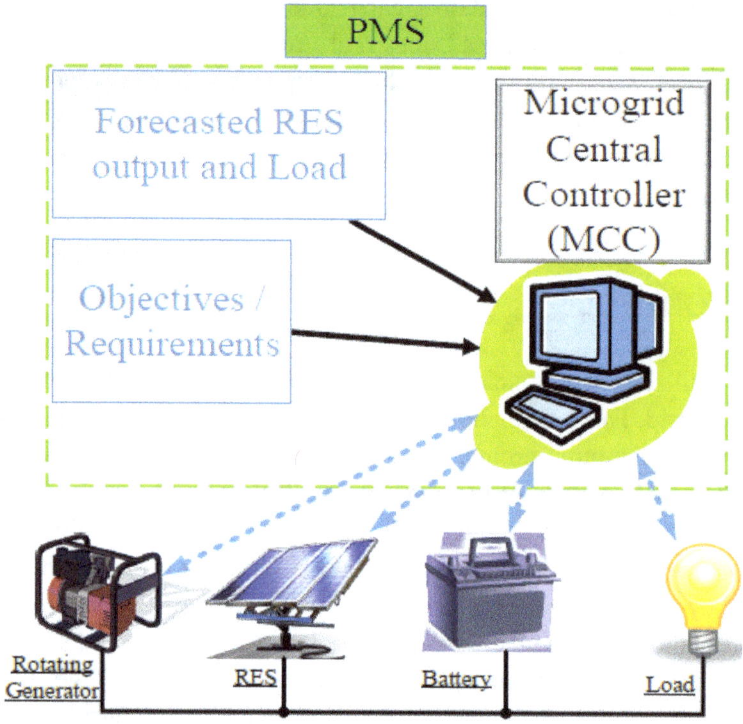

Figure 11. Decentralized control

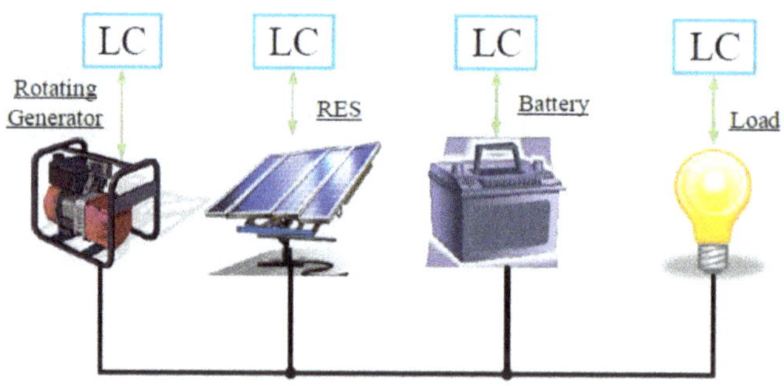

Second type of distributed control is multi-agent system (MAS). A limitation of communication-less system is the inability to optimize the utilization of microgrid resources. The addition of a communication link between the LCs enables the optimal dispatching of DGs to better utilize RES and reduce fuel consumption. A typical architecture for MAS is shown in Fig (12) where each component is assigned to the respective agent and all agents communicate with each other (Arboleya et al., 2012).

Figure 12. Multi-agent system MAS

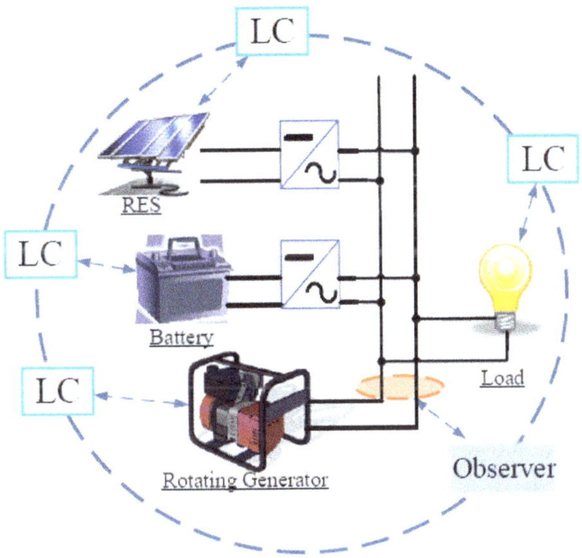

HYBRID CONTROL

Hybrid control is a method where a central controller is used to modify droop parameter of LCs via low cost, slow communication link. A typical architecture for the hybrid control method is shown in Fig (13). The central controller sets steady state parameters while the LC provides transient response without relying on communication. An example of a hybrid system is the one used in the Consortium for Electric Reliability Technology Solution (CERTS) microgrid. In another example, a method utilizing frequency partition instead of droop control was presented (Colson et al., 2011).

Figure 13. Hybrid control

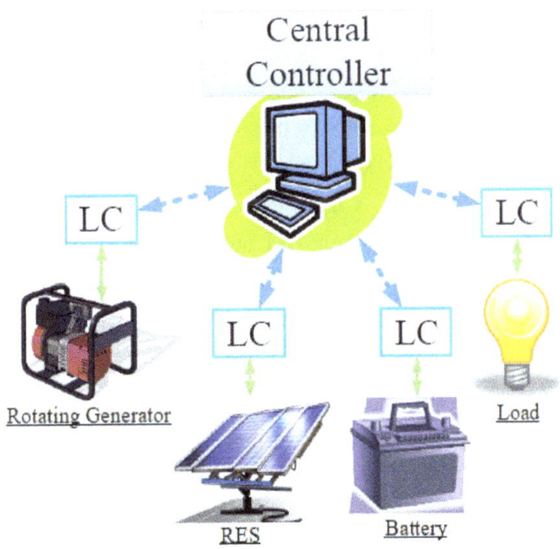

ECONOMIC STUDIES FOR MICROGRID

Improving the quality and the reliability of the entire grid by this all-things-to- all-people approach that target the middle ground where the infrastructure and the investment strategies take a place. One of the offered benefits of microgrids is that the sources of the power located closer to the costumers, subsequently reducing the losses related to distribution and transmission systems (Hatziargyriou et al., 2007).

Microgrids provide the capability to tailor local requirements, facilitate multiple decision-making, and improve overall reliability. In order to fully utilize these benefits, however, they first need to be quantified and integrated into applicable regulatory frameworks.

The economical features for micro grids can be summarized:

1. **At Local Level:** The benefits of microgrids can contribute to the deferral of potential infrastructure investments via the reduction or the shifting of peak loads thereby reducing peak demand charges and electrical consumption costs. Energy cost reductions can come from reduced energy use, for example, from increased efficiency and demand response or from reductions in peak charges and consumption costs.
2. Falling cost of renewable energy technologies, then the cost benefits of improved resiliency, security, and disaster recovery from the integration of microgrids make conventional economic sense and become fairly easy to justify. From a socioeconomic perspective, where people have experienced a crisis firsthand, the prospect of selfreliance for power offered by microgrids becomes an easy sell.
3. **Techno Economic Perspective:** A number of studies have been done around how business cases and cost-benefits analysis for microgrids can work, for example, in Canada the Natural Sciences and Engineering Research Council Smart Microgrid-Net project, 10 in the United States, the Lawrence Berkeley National Laboratory studies, 7 and in Europe, the More Microgrids11 project. Recent large-scale blackouts have made it clear that the top-down architecture and unidirectional power flow of the existing North American electricity grid is becoming less and less sustainable.

ACHIEVEMENTS AND EXPECTATIONS

In (Olivares et al., 2011), EMS component helps achieving the goals of Microgrids, where in small microgrid the output of renewable energy differ each sec and deviate from the forecasting schedule besides that, large load swings. Such conditions commonly occur in micro grid yet the EMS adaptive the power flow so the stability of the system is preserved.

The optimization operation for small microgrid done by EMS needs for detailed energy models of power sources, sinks, and losses. EMS planning and power flow control in a small microgrid should account for losses such as AC/DC and DC/AC conversion losses, battery climate control, battery self-discharge, and standby losses of power subsystems and microgrid (De Brabandere et al., 2007).

As a side benefit of a detailed energy and energy losses model in the EMS, the EMS energy cost optimization logic could seek to minimize the standby power losses (phantom load) by techniques such as following:

- Switching off load feeders upstream of the transformers if possible.
- Turning smart power systems (such as bidirectional inverters) down to their lowest standby stopped operating mode during planned intervals during the day when the power conversion is not needed.

FUTURE RESEARCH DIRECTIONS

- Minimize the standby power losses of microgrids.
- Optimization the operation for small microgrid.

CONCLUSION

The new upcoming technology for electrical power systems that gives appropriate solution for the integration of the distributed energy resources has been describes in this chapter. In addition a different categories of smart grids have been classified, and giving the advantages, weakness, and opportunities of each one, besides determining its own operation conditions. Most familiar Microgrid's components have been discussed such as energy management system, and several types of control and communication systems in addition to economic study of micro grid.

REFERENCES

Alhelou, H., Hamedani-Golshan, M. E., Zamani, R., Heydarian-Forushani, E., & Siano, P. (2018). Challenges and Opportunities of Load Frequency Control in Conventional, Modern and Future Smart Power Systems: A Comprehensive Review. *Energies, 11*(10), 2497. doi:10.3390/en11102497

Alhelou, H. H. (2018). *Fault Detection and Isolation in Power Systems Using Unknown Input Observer. In Advanced Condition Monitoring and Fault Diagnosis of Electric Machines* (p. 38). Hershey, PA: IGI Global.

Alhelou, H. H., Golshan, M., & Fini, M. (2018). Wind Driven Optimization Algorithm Application to Load Frequency Control in Interconnected Power Systems Considering GRC and GDB Nonlinearities. *Electric Power Components and Syst.*

Alhelou, H. H., & Golshan, M. E. H. (2016, May). Hierarchical plug-in EV control based on primary frequency response in interconnected smart grid. In *Electrical Engineering (ICEE), 2016 24th Iranian Conference on* (pp. 561-566). IEEE. 10.1109/IranianCEE.2016.7585585

Alhelou, H. H., Golshan, M. H., & Askari-Marnani, J. (2018). Robust sensor fault detection and isolation scheme for interconnected smart power systems in presence of RER and EVs using unknown input observer. *International Journal of Electrical Power & Energy Systems, 99*, 682–694. doi:10.1016/j.ijepes.2018.02.013

Alhelou, H. H., Hamedani-Golshan, M. E., Heydarian-Forushani, E., Al-Sumaiti, A. S., & Siano, P. (2018, September). Decentralized Fractional Order Control Scheme for LFC of Deregulated Nonlinear Power Systems in Presence of EVs and RER. In *2018 International Conference on Smart Energy Systems and Technologies (SEST)* (pp. 1-6). IEEE. 10.1109/SEST.2018.8495858

Alhelou, H. S. H., Golshan, M. E. H., & Fini, M. H. (2015, December). Multi agent electric vehicle control based primary frequency support for future smart micro-grid. In *Smart Grid Conference (SGC)* (pp. 22-27). Academic Press.

Alshahrestani, A., Golshan, M. E. H., & Alhelou, H. H. (2018, November). WAMS Based Online Estimation of Total Inertia Constant and Damping Coefficient for Future Smart Grid Systems. In *Smart Grid Conference (SGC)* (pp. 1-5). Academic Press.

Arboleya, P., Gonzalez-Moran, C., & Coto, M. (2012, September). A hybrid central-distributed control applied to microgrids with droop characteristic based generators. In *Power Electronics and Motion Control Conference (EPE/PEMC), 2012 15th International* (pp. LS7a-5). IEEE. 10.1109/EPEPEMC.2012.6397503

Bühler, R. (2010). *Integration of renewable energy sources using microgrids, virtual power plants and the energy hub approach* (MSc Master thesis). Swiss Federal Institute of Technology (ETH), Zurich, Switzerland.

Chalise, S. (2016). *Power management of remote microgrids considering battery lifetime*. Academic Press.

Chalise, S., Sternhagen, J., Hansen, T. M., & Tonkoski, R. (2016). Energy management of remote microgrids considering battery lifetime. *The Electricity Journal*, *29*(6), 1–10. doi:10.1016/j.tej.2016.07.003

Colson, C. M., Nehrir, M. H., & Gunderson, R. W. (2011, August). Multi-agent microgrid power management. In *18th IFAC world congress* (pp. 3678-3683). Academic Press. 10.3182/20110828-6-IT-1002.01188

De Brabandere, K., Bolsens, B., Van den Keybus, J., Woyte, A., Driesen, J., & Belmans, R. (2007). A voltage and frequency droop control method for parallel inverters. *IEEE Transactions on Power Electronics*, *22*(4), 1107–1115. doi:10.1109/TPEL.2007.900456

European Commission. (2011). *Standardization Mandate to European Standardisation Organisations (ESOs) to support European Smart Grid deployment*. Author.

Fini, M. H., Yousefi, G. R., & Alhelou, H. H. (2016). Comparative study on the performance of many-objective and single-objective optimisation algorithms in tuning load frequency controllers of multi-area power systems. *IET Generation, Transmission & Distribution*, *10*(12), 2915–2923. doi:10.1049/iet-gtd.2015.1334

Geidl, M., Koeppel, G., Favre-Perrod, P., Klockl, B., Andersson, G., & Frohlich, K. (2007). Energy hubs for the future. *IEEE Power & Energy Magazine*, *5*(1), 24–30. doi:10.1109/MPAE.2007.264850

Hatziargyriou, N., Asano, H., Iravani, R., & Marnay, C. (2007). Microgrids. *IEEE Power & Energy Magazine*, *5*(4), 78–94. doi:10.1109/MPAE.2007.376583

Laaksonen, H. (2011). Technical solutions for low-voltage microgrid concept. *Acta Wasaensia*, *241*, 251.

Lasseter, R. H. (2002). Microgrids. In *Power Engineering Society Winter Meeting, 2002. IEEE* (Vol. 1, pp. 305-308). IEEE.

Lopes, J. P., Moreira, C. L., & Madureira, A. G. (2006). Defining control strategies for microgrids islanded operation. *IEEE Transactions on Power Systems, 21*(2), 916–924. doi:10.1109/TPWRS.2006.873018

Makdisie, C., Haidar, B., & Alhelou, H. H. (2018). An Optimal Photovoltaic Conversion System for Future Smart Grids. In *Handbook of Research on Power and Energy System Optimization* (pp. 601–657). IGI Global. doi:10.4018/978-1-5225-3935-3.ch018

Nadweh, S., Hayek, G., Atieh, B., & Haes Alhelou, H. (2018). Using Four – Quadrant Chopper with Variable Speed Drive System Dc-Link to Improve the Quality of Supplied Power for Industrial Facilities. Majlesi Journal of Electrical Engineering.

National Institute of Standards and Technology. (2010). *NIST Framework and Roadmap for Smart Grid Interoperability Standards, Release 1.0*. NIST Special Publication 1108.

Nikos, H. (2007). Microgrids: an overview of ongoing research, development, and demonstration projects. *IEEE Power & Energy, 5*(4).

Njenda, T. C., Golshan, M. E. H., & Alhelou, H. H. (2018, November). WAMS Based Intelligent Under Frequency Load Shedding Considering Online Disturbance Estimation. In *Smart Grid Conference (SGC)* (pp. 1-5). Academic Press.

Njenda, T. C., Golshan, M. E. H., & Alhelou, H. H. (2018, November). WAMS based Under Frequency Load Shedding Considering Minimum Frequency Prediction and Extrapolated Disturbance Magnitude. In *Smart Grid Conference (SGC)* (pp. 1-5). Academic Press.

Olivares, D. E., Cañizares, C. A., & Kazerani, M. (2011, July). *A centralized optimal energy management system for microgrids*. IEEE Power and Energy Society General Meeting.

Parhizi, S., Lotfi, H., Khodaei, A., & Bahramirad, S. (2015). State of the Art in Research on Microgrids: A Review. *IEEE Access: Practical Innovations, Open Solutions, 3*(1), 890–925. doi:10.1109/ACCESS.2015.2443119

Siemens and RWE Energy. (2008). *First virtual power plant operated by Siemens and RWE Energy*. Retrieved from http://w1.siemens.com/press/en/pressrelease

SMB Smart Grid Strategic Group (SG3). (2010). *IEC Smart Grid Standardization Roadmap*. International Electrotechnical Commission.

Turunen, S. I. (2016). *Protection of Microgrids and Distributed Energy Resources based on IEC 61850*. Academic Press.

You, S., Træholt, C., & Poulsen, B. (2010). *Developing virtual power plant for optimized distributed energy resources operation and integration*. Academic Press.

Zamani, R., Hamedani-Golshan, M. E., Haes Alhelou, H., Siano, P., & Pota, H. (2018). Islanding Detection of Synchronous Distributed Generator Based on the Active and Reactive Power Control Loops. *Energies, 11*(10), 2819. doi:10.3390/en11102819

KEY TERMS AND DEFINITIONS

Energy Management System: An energy management system (EMS) is a system of computer-aided tools used by operators of electric utility grids to monitor, control, and optimize the performance of the generation and/or transmission system. The computer technology is also referred to as SCADA/EMS or EMS/SCADA.

Hybrid Micro-Grid: Northern Power Systems has a proven record in integrating our distributed wind turbines with solar, diesel gensets or other power sources in microgrids.

Smart Grid: An electricity supply network that uses digital communications technology to detect and react to local changes in usage.

Chapter 16
Management of Electrical Maintenance of University Buildings Using Deterioration Models

Bassel Mohamed Alhassan
Tishreen University, Syria

Jamal Younes Omran
Tishreen University, Syria

Fayez Ali Jrad
Tishreen University, Syria

ABSTRACT

Buildings maintenance has received increasing international attention in various fields of scientific research. As a result, there has been a change in the maintenance of buildings from the preventive to the predictive approach. This is done through an evaluation model to support and assist the management of the facility in selecting alternatives and making appropriate decisions in maintenance according to building status and maintenance budget. This chapter investigated the reasons for the electrical maintenance of the university buildings and the degree of importance of each element of electrical maintenance through the design of a questionnaire in which the electrical components were divided into elements and then each element was linked to all maintenance items that related to it. At the end of the research, mathematical models were developed; these models help to forecasting the electrical maintenance items and distribution of the maintenance budget, and to verify the validity of these models, they have been applied to study the case of dorm buildings in Tishreen University.

DOI: 10.4018/978-1-5225-8030-0.ch016

INTRODUCTION

Maintenance is defined as a continuous process and a permanent activity to preserve the buildings and the keeping of the equipment in the best case for normal use. The maintenance includes the entire building with its various parts (structural, health, mechanical, electrical and ventilation). (Bin Akasah, 2007).

Maintenance can also be defined as the protection of the buildings from damage in order to perform its function fully as long as possible with the repair of defects in order to preserve the building and to reduce the costs resulting from negligence or deferral by sharing all technical and administrative means. These include preparation, planning and implementation of the maintenance process as well as follow-up and control during and after implementation (Al-Hassan; jrad, 2015).

Maintenance is the set of technical and executive works that aim at keeping a building items in the state in which it can perform its required functions Maintenance of any component or material must not be less than the necessary minimum limit.

There have been many views that the maintenance process is complex and requires high costs (Bin Akasah, 2007), but it leads to a longer life of the facility and the keeping of cost if implemented according to the correct schedule (Arditi; Nawakorawit, 1999).

To shifting the maintenance work from the approach that depends on the repair of the failure when it occurs, which leads to the accumulation of faults to the predictive approach, which helps management to anticipate the failures that can occur in the construction and forecast the budget required for maintenance. To achieve that modeling method was proposed. The mathematical model is a simplified version of a complex thing and is used to solve problems and make predictions and as a basis for related ideas (Olagunju, 2012). The mathematical model can be described as a process of simplification of complex life issues, which is expressed as mathematical equation. The predictive model can be described as the construction of an evaluation model, helping to evaluate different alternatives.

There has been a lot of research on deterioration models. One of these researches (Abdelaty, 2012) The aim of the research was to set up deterioration models for all important components of the metro to determine the time between tow periodic maintenance and maintenance budget. The researcher set priorities for the maintenance of the subway components through the impact of the component on the service level in the metro.

In another search in the Niger state of Nigeria (Olagunju, 2012), deterioration models have been developed for residential buildings, these models based on the basic components of the building. To determine the most significant components of the building, the researcher put (25) variables expressing the different components of the building and conducting a questionnaire. The researcher found that (8) variables expressed by (94) % of the status or condition of the building.

In another study (Frangopol, 2003; Fini et al., 2016; Alhelou et al., 2018; Zamani et al., 2018; Alhelou et al., 2015; Njenda et al., 2018; Haes Alhelou et al., 2018), The aim of this study was to develop a reliable deterioration model to predict the maintenance work and optimize maintenance cost. The researcher worked on a mathematical model for optimize the cost of maintenance of infrastructure under uncertainty. Where the researcher studied several facilities that need maintenance work with a focus on bridges.

THE AIM AND IMPORTANCE OF RESEARCH

In order to preserve our real estate wealth and due to the fact that it is impossible to build a building that does not require maintenance and in the light of the lack of maintenance manual in our country Syria, The aim of this research is to develop deterioration models of electrical maintenance of university buildings, which helps the administration to quickly evaluate the construction situation in terms of the electrical component, as well as identify the main reasons for the electrical maintenance and the expected electrical maintenance work for the next year, arrange these works and implement the most urgent items according to the budget set for the maintenance work.

RESEARCH METHODS AND MATERIALS

In this study, we used a descriptive approach. A questionnaire on maintenance was designed. The data obtained were then subjected to several methods of analysis using business intelligence by power bi application and using fuzzy logic, Before the questionnaire was designed, the building was divided as it came:

FUNCTIONAL DIVISION OF THE BUILDING

Due to the large number of electrical elements in the construction, and for the ease of control of these elements, we will divide the construction, faults and electrical faults that can be happened according to two methods:

1. Division of electrical components into components.
2. Spatial division of construction.

1. The Division of The Building Into Elements

The electrical component is functionally divided into a set of basic elements. Each of these elements will be linked to all faults that can occur in this element. This division is shown in Figure (1):

Thus, we have 3 electrical elements associated with 20 defects that can occur in these elements. We have taken advantage of the French maintenance manual for residential buildings to get these defects (Bornand, 1991) after making the necessary adjustments to these items to fit the reality and nature of university buildings in our country through the distribution of an Experimental questionnaire on engineers with experience in the field of electrical maintenance of buildings until We have reached the final version of the questionnaire, and this division will help us to develop deterioration models for electrical maintenance of university buildings.

2. Spatial Division of The Building

The structure within the questionnaire was divided spatially into 5 main sections showing in the figure (2), In the questionnaire, the defects were distributed according to their location.

Figure 1. Elements and maintenance items in the electrical component

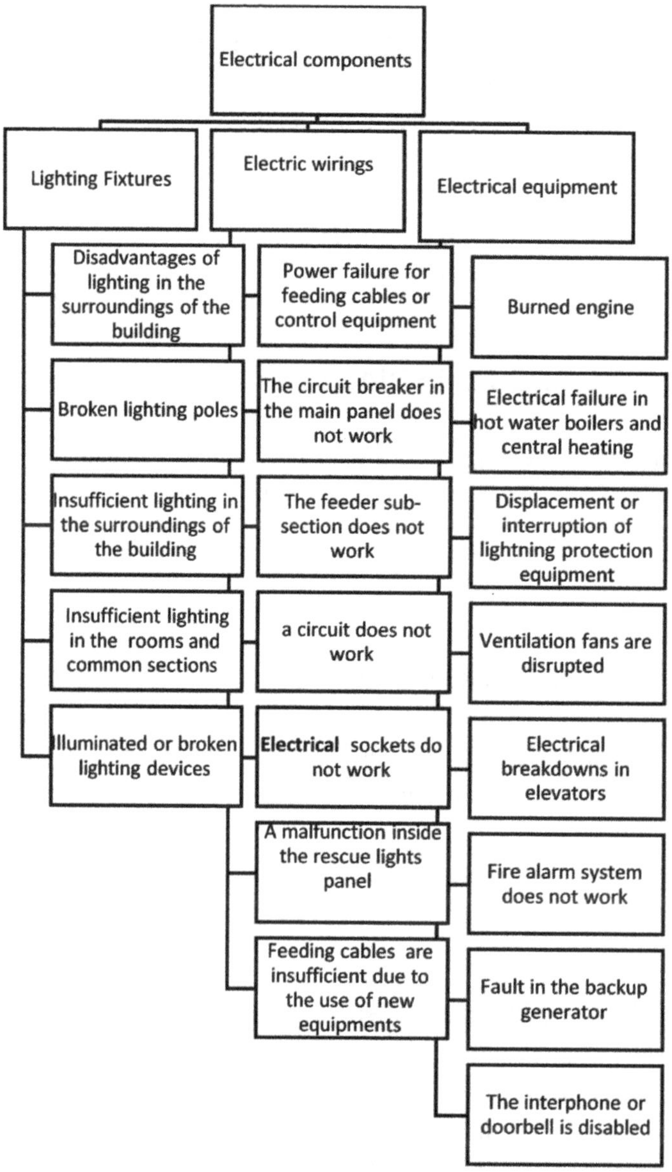

THE QUESTIONNAIRE

The deterioration models for the electrical components of the university buildings were based on mathematical equations, each equation consists of two parts. The first is a constant coefficient of weight associated with each item of maintenance and it remains constant in all buildings. The second is variable and relates to the status and condition of the item or component (component condition). It depends on the evaluation of the engineer supervising the maintenance works.

To obtain the first part of the deterioration model, which is the constant or the weighting coefficient, a questionnaire was drawn up which includes all the electrical maintenance items described in Figure 1, which we called (defect) and we linked each item of maintenance with a set of criteria. criterias that reflects the importance of the item for maintenance work, whether preventive maintenance or repair in case of failure Which have been observed in the previous studies are: Probability of occurrence, risk to the building, time of maintenance, cost of maintenance, Its impact on the level of service in construction, its impact on the environment, its impact on the security and safety of the enterprise.

After returning to the experience, we found that the first 5 criteria (probability of occurrence, risk to the building, time of maintenance, cost of maintenance, service level) are most appropriate for the nature of the facilities under study (university buildings).

The first phase was the development of electrical maintenance items, by distributing a set of experimental questionnaires to engineers with experience in the field of electrical maintenance to obtain maintenance items (defects) that are more appropriate to the reality and nature of university buildings in Syria. After making the necessary adjustments, we obtained the final version of the questionnaire. After that, 112 questionnaires were distributed to the engineers working in the maintenance field in various government departments in Lattakia Governorate, we get (95) completed questionnaire.

After obtaining all the questionnaires, these questionnaires were processed using business intelligence through the application of power bi, the statistical analysis of the results of the questionnaire is given below (Alshahrestani et al., 2018; Makdisie et al., 2018; Alhelou et al., 2018; Alhelou et al., 2016; Nadweh et al., 2018; Njenda et al., 2018).

Figure 2. Spatial division of construction

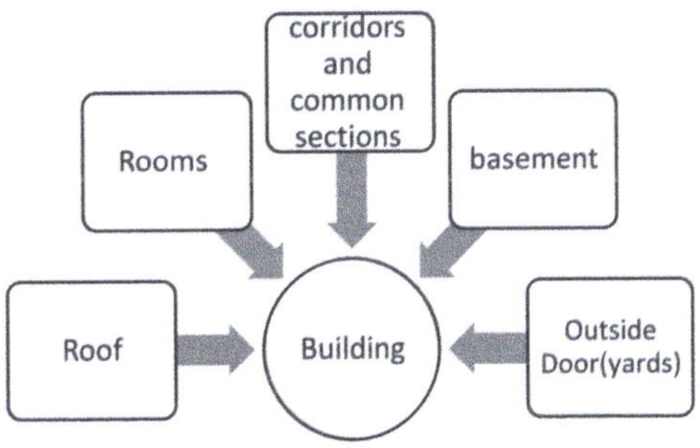

Figure 3. Deterioration model equation

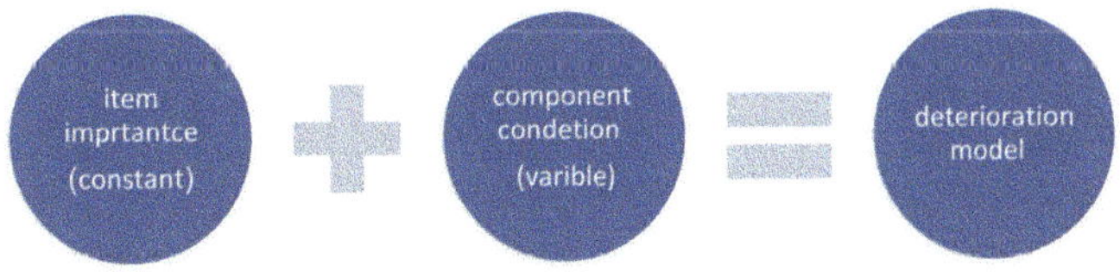

SURVEY ANALYSIS

Importance Degree

After processing the questionnaires, we obtained about maintenance and calculating averages, each watch or item of maintenance has been linked to a set of percentages that express:

1- Probability of occurrence.
2- Risk to the building.
3- Time of maintenance.
4- Cost of maintenance.
5- Its impact on the level of service.

To obtain the constant or coefficient of weighting that falls into the deterioration model for each maintenance item, we must combine these values with each other. For values in this integration, we use the fuzzy logic that allows the use of ambiguous and uncertain concepts.

By using fuzzy tech 6,06 application, fuzzy logic, we get degree of importance for each item in electrical component as shown in the next table:

Element-Level

The electrical component was divided into 3 elements and we will analyze the data related to these elements obtained from the survey results. It should be noted that the lower the percentage of the item, the greater indication of this item:

Figure 4. Input and output in the fuzzy logic

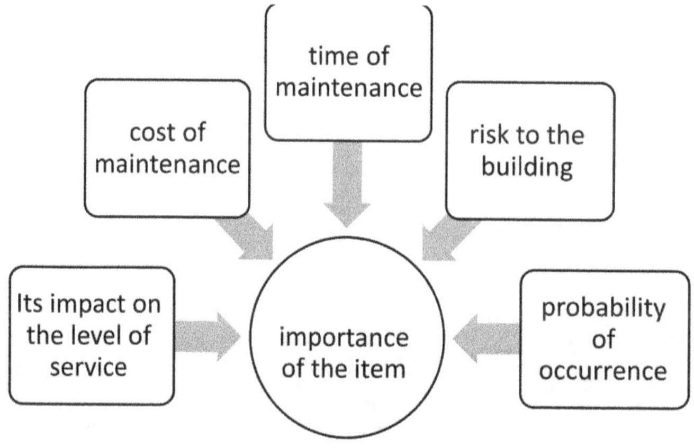

Table 1. Importance degree for electrical component items

Maintenance Item Name	Importance Degree
Disadvantages of lighting in the surroundings of the building	77.5
Broken lighting poles	81
Insufficient lighting in the surroundings of the building	81
Insufficient lighting in the rooms and common sections	83
Illuminated or broken lighting devices	83
Power failure for feeding cables or control equipment	55.5
The circuit breaker in the main panel does not work	62
The subsection circuit breaker does not work	67
a circuit does not work	81
Power sockets do not know	83
A malfunction inside the rescue lights panel	83
Feeding cables are insufficient due to the use of new equipment's	64.5
Burned engine	52
Electrical failure in hot water boilers and central heating	62.5
Displacement or interruption of lightning protection equipment	83
Ventilation fans are disrupted	62
Electrical breakdowns in elevators	50
Fire alarm system does not work	68.5
Fault in the backup generator	45.5
The interphone or doorbell is disabled	83

Probability of Occurrence:

Figures 5-14 correlate to the probability of the following items.

1- Risk to the Building:
2- Time of Maintenance:
3- Cost of Maintenance:
4- Impact on the Level of Service:

Item Level

1- Probability of Occurrence:
2- Risk to the Building:
3- Time of Maintenance:
4- **Cost of Maintenance**:
5- Impact on the Level of Service:

Figure 5. Electrical component elements distribution according to the probability of occurrence

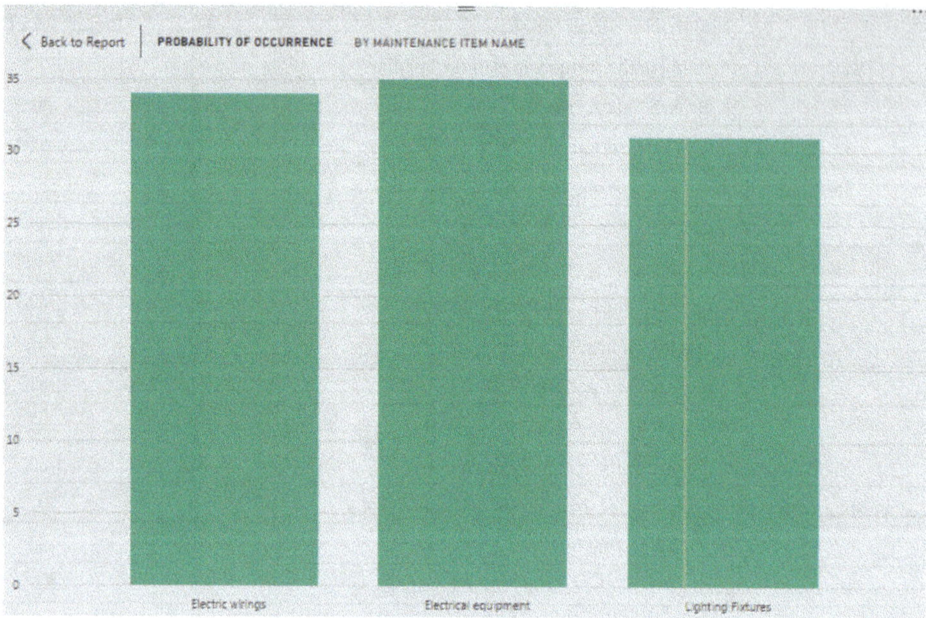

Figure 6. Electrical component elements distribution according to the risk to the building

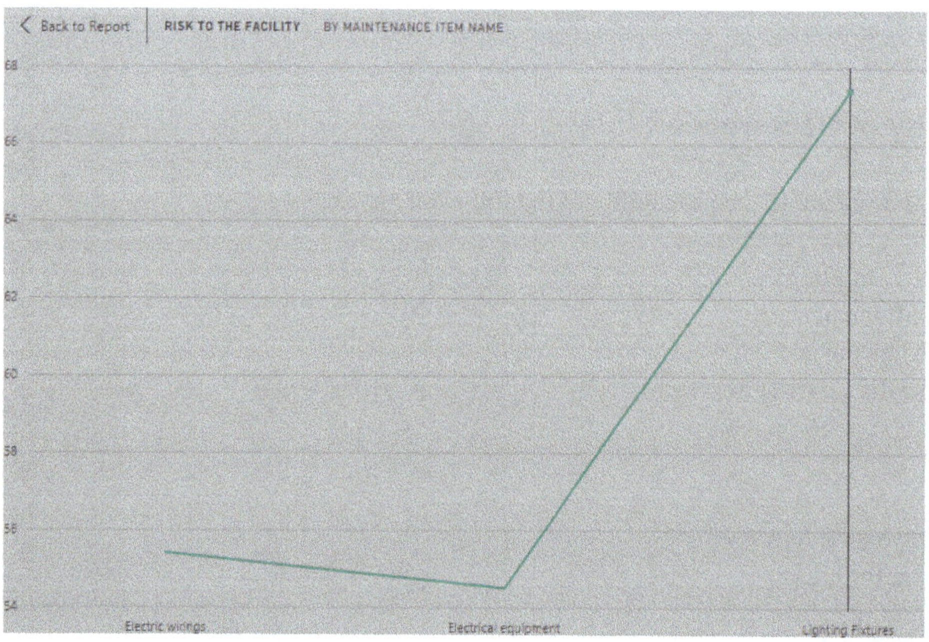

Figure 7. Electrical component elements distribution according to the time of maintenance

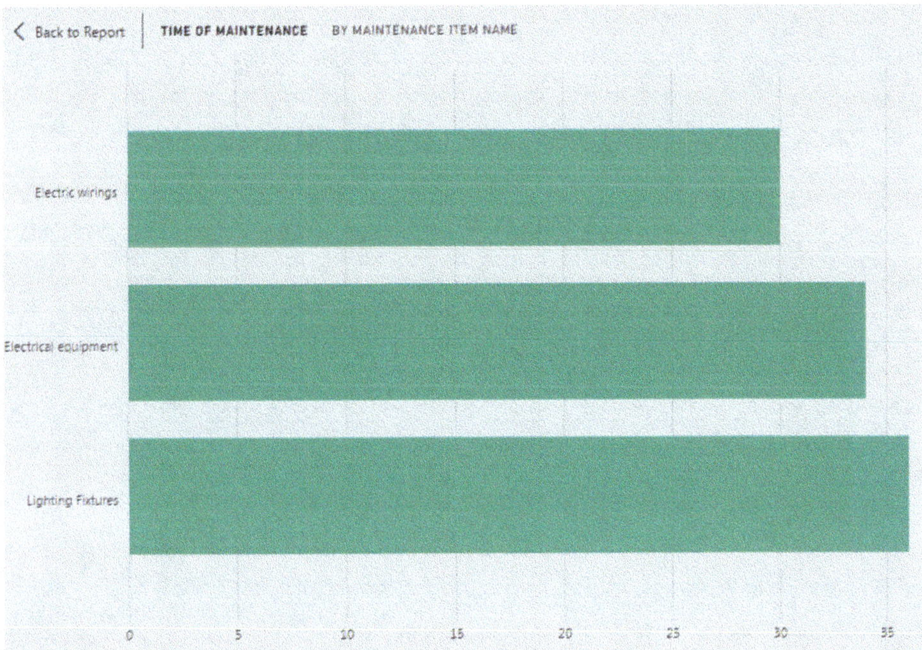

Figure 8. Electrical component elements distribution according to the cost of maintenance

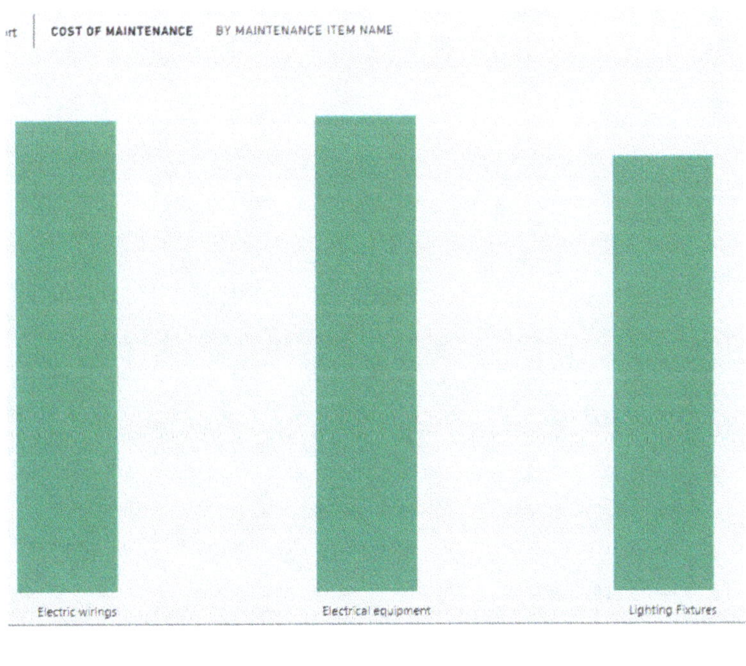

Figure 9. Electrical component elements distribution according to the impact on the level of service

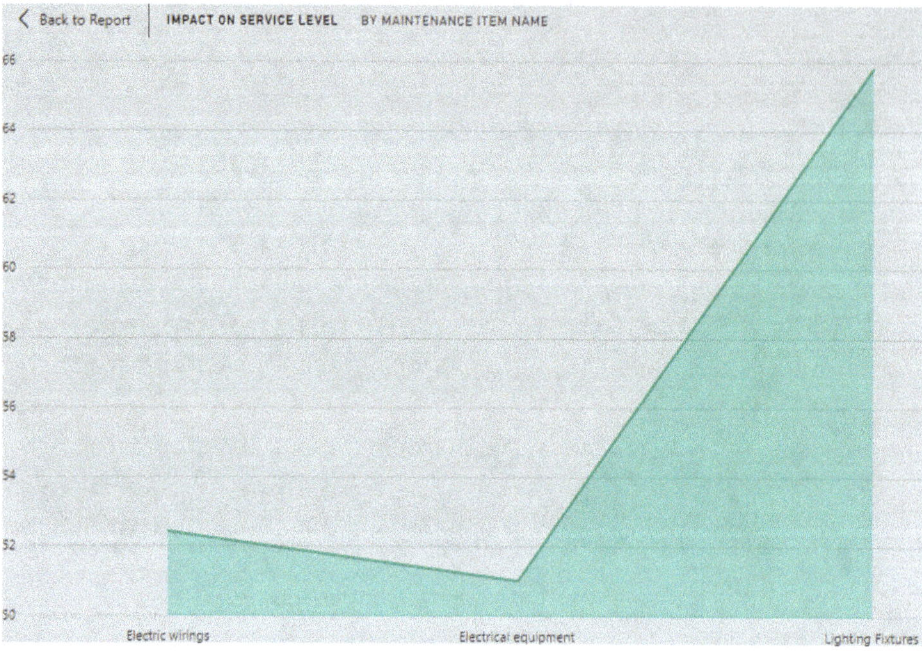

Figure 10. Electrical component items distribution according to the probability of occurrence

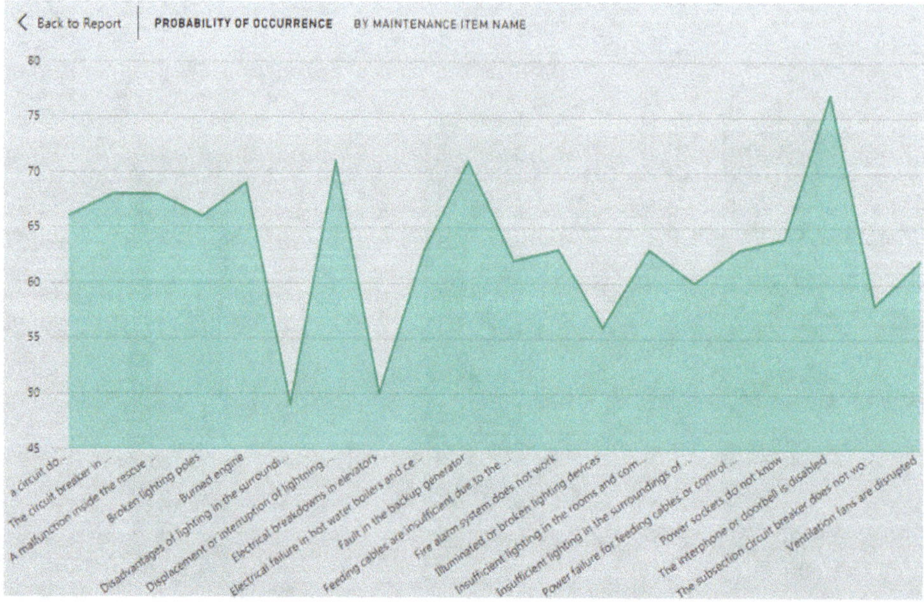

Figure 11. Electrical component items distribution according to the risk to the building

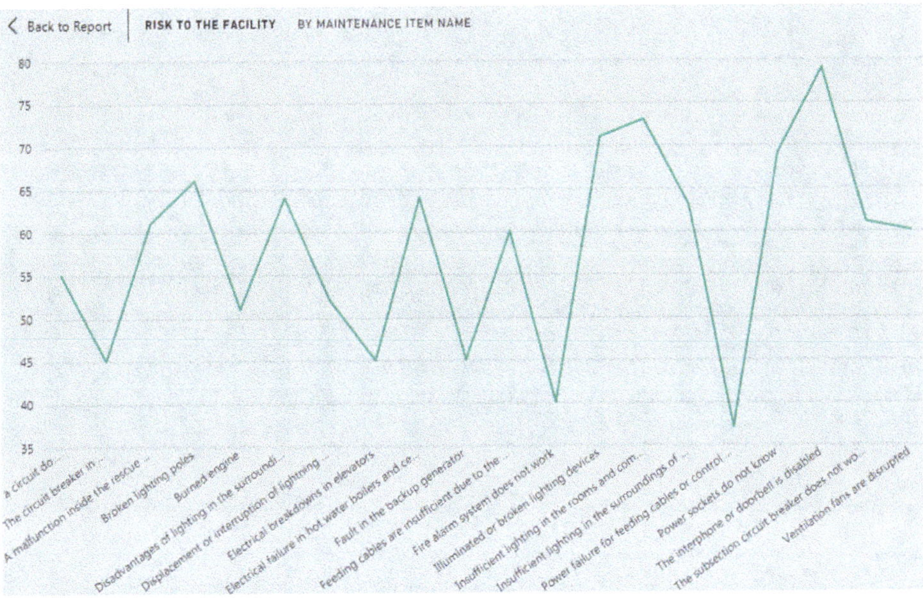

Figure 12. Electrical component items distribution according to the time of maintenance

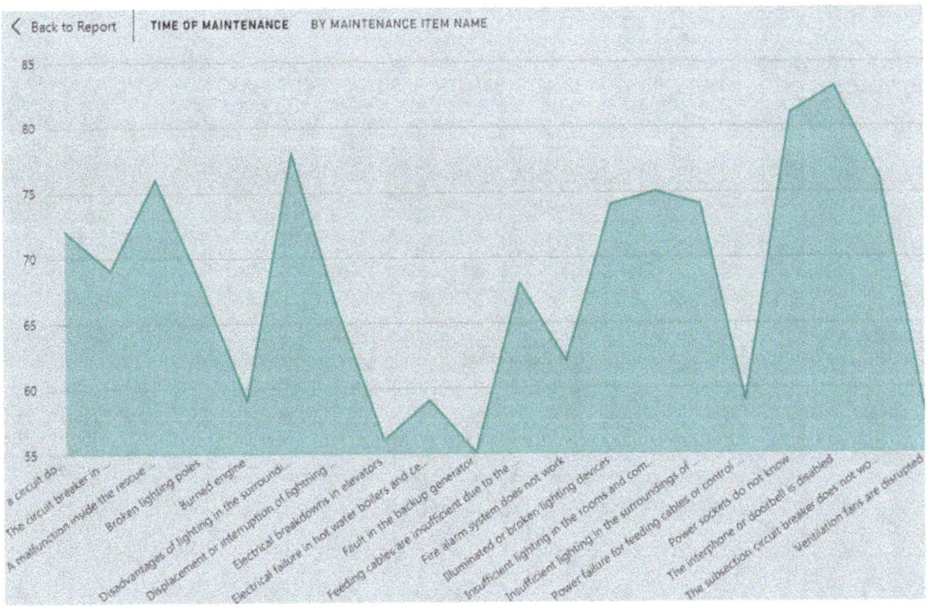

Figure 13. Electrical component items distribution according to the cost of maintenance

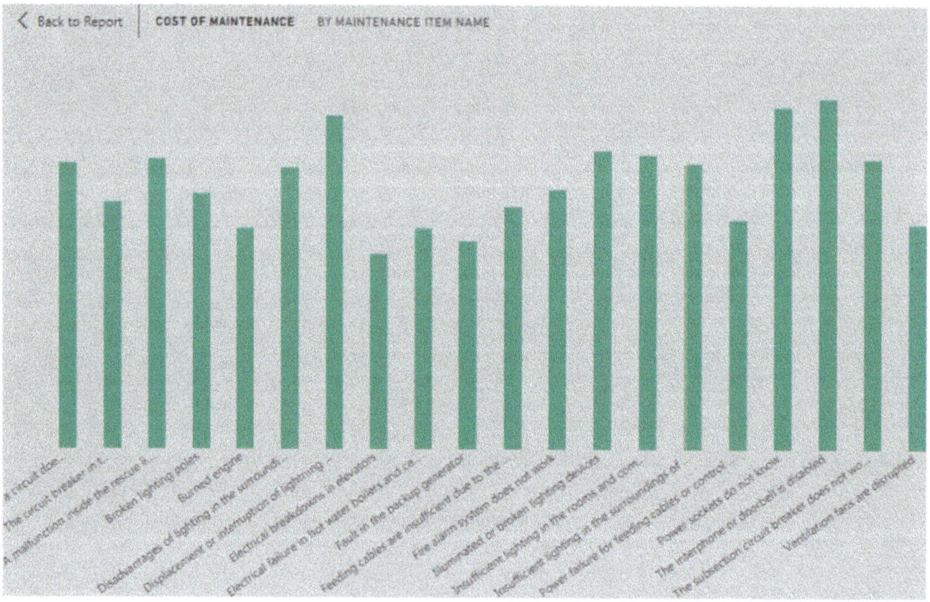

Figure 14. Electrical component items distribution according to the impact on the level of service

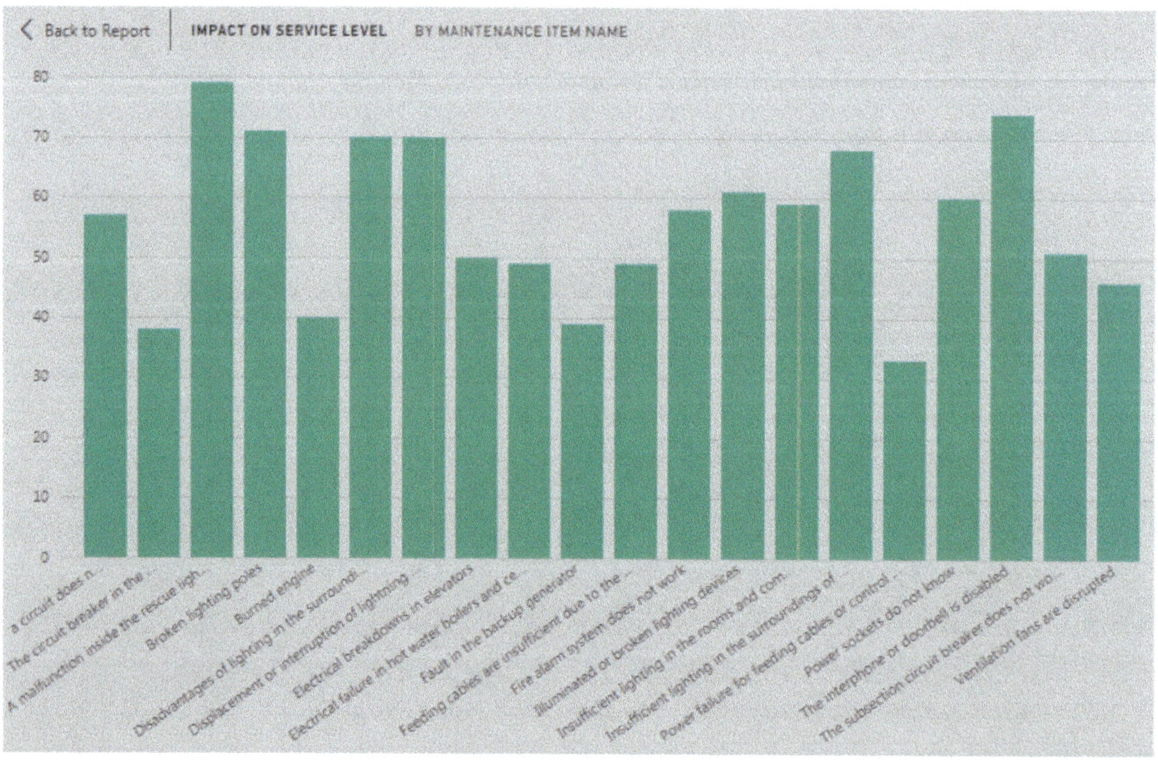

DISCUSSION OF THE RESULTS OF THE STATISTICAL ANALYSIS OF ELECTRICAL COMPONENTS

1- Lighting Fixtures Element: Element of Lighting Fixtures is the most electrical component of the possibility of the occurrence of faults in it as the lighting devices are the most electrical elements used by the occupants of construction and be exposed to breakage or combustion due to weather conditions, but these elements are the least electrical elements of the risk to the building and the cost and time of maintenance are relatively small if compared to the rest of the electrical elements, as well as the impact on the level of service is low. the items of maintenance included in the Lighting Fixtures element:

   a) Disadvantages of lighting in the surroundings of the building: where this item was the most items of electrical maintenance is likely to occur, the presence of lighting outside the building makes them more likely to crash or breakage due to weather conditions.
   b) Broken lighting poles: This item can occur because of accident or result of deliberate sabotage and is treated by replacing broken columns.
   c) Insufficient lighting in the surroundings of the building: This is due to poor electricity supply or poor lighting equipment.
   d) Insufficient lighting in the rooms and common sections.
   e) Illuminated or broken lighting devices: because of an accident or because of the end of service life and shall be replaced.

2- Electrical Equipment Element: This element was the least electrical element likely to be faulted. However, it is the most dangerous electrical element of the building in the event of a failure in it, as well as the most time and cost consuming elements and the most element impact on the level of service in the building due to the importance of electrical equipment that fall within element. the main items of maintenance in the electrical equipment element are:

   a) Burned engine:
   b) Electrical failure in hot water boilers and central heating: This item was the most electrical maintenance items that needed time to fix it.
   c) Displacement or interruption of lightning protection equipment.
   d) Ventilation fans are disrupted.
   e) Electrical breakdowns in elevators: this item was the most electrical maintenance items cost during maintenance and the second most important item of electrical maintenance.
   f) Fire alarm system does not work
   g) Fault in the backup generator: This item is the most important electrical maintenance items because of the importance of the backup generator.
   h) The interphone or doorbell is disabled: This item was the least electrical maintenance items likely to occur as well as the least items in terms of maintenance time and cost where the interferon bell is present in many government buildings, but its use is very little, and in some cases, it is never used.

3- Electric wirings element:

Table 2. Distribution of the evaluation grades for the engineer supervising maintenance

Grade	Description of the Component Condition
1	Failure of the component as the component does not work with the possibility of structural or environmental damage or damage to security and safety at the building.
2	The component works but is frequently disrupted with clear signs of deterioration of the component's status and potential damage to the building
3	The component is medium-state as it works but needs monitoring and needs preventive maintenance
4	The component works well with small and superficial faults that do not affect component work and maintenance can be delayed
5	The component works as if it were new and there are no signs of a malfunction with a new exterior

Main Maintenance Items Included in this Element:

a) Power failure for feeding cables or control equipment: This item was the most dangerous item of electrical maintenance to the building in case it occurs, Where the failure may mean the absence of electrical power for the complete building or important equipment in the building, such as the elevator.
b) The circuit breaker in the main panel does not work.
c) The feeder sub-section does not work.
d) a circuit does not work.
e) Electrical sockets do not work
f) A malfunction inside the rescue lights panel: This item was the least of the electrical maintenance items affecting the service level, but the risk to the establishment is relatively large in the event of a failure where the function of the rescue light is to provide lighting in case of power failure in the event of fire or disaster in construction.
g) Feeding cables are insufficient due to the use of new equipments.

Because of the presence of electric panels and most of the electrical wiring in the basement, therefore the electrical breakdown in the basement is the most serious electrical failures in the building as well as the repair takes time and cost significantly compared to other sections, While the electrical faults in the rooms and their fixtures, which are mainly related to lighting devices less faults in term of the cost of repair, as well as electrical faults that occur in the rooms a are less risk than the faults that occur in the rest of the sections (Alshahrestani et al., 2018; Makdisie et al., 2018; Alhelou et al., 2018; Alhelou et al., 2016; Nadweh et al., 2018; Njenda et al., 2018)

DETERIORATION MODELS

To benefit from the results of the statistical study, deterioration models were developed to predict electrical maintenance work. This predictive model is used for the control and rapid evaluation of buildings [9]. This is achieved by dividing the electrical component into elements, studying each element and writing the deterioration model, and then integrating these elements to obtain a deterioration model of the electrical component for the building completely and use this model in order to predict the state

Table 3. Deterioration model for electrical component elements

Element Name	Deterioration Model
Electrical equipment	$Y1 = 0.52X1 + 0.63X2 + 0.83X3 + 0.62X4 + 0.5X5 + 0.69X6 + 0.46X7 + 0.83X8$
Electric wirings	$Y2 = 0.56X1 + 0.62X2 + 0.67X3 + 0.81X4 + 0.83X5 + 0.83X6 + 0.65X7$
Lighting Fixtures	$Y3 = 0.78X1 + 0.81X2 + 0.81X3 + 0.83X4 + 0.83X5$

of the electrical component of the next year, All this is done through a logical scientific method rather than relying on personal factors, In this way we can maintain the continuity of the construction and the continuation of the electrical network to perform its work perfectly and using a small operating cost. After processing the questionnaire we linked each maintenance item with a number that reflects the importance of this item and its impact on the performance in the building and this number will enter the equation of deterioration model as a constant or coefficient of weighting is smaller or equal to one, The status or condition of the element is the second component of the deterioration model, through the evaluation of the engineer supervising the maintenance work of this component and to obtain this evaluation scientifically, we will rely on a paper approved by (Australasian Association of Higher Education Facilities Officers, 2000).

deterioration models for electrical components:

deterioration model for Lighting Fixtures element:

$$Y3 = m1X1 + m2X2 + m3X3 + m4X4 + \text{m5x5}$$

$$Y3 = 0.78X1 + 0.81X2 + 0.81X3 + 0.83X4 + 0.83X5$$

m1, m2, m3, m4 the weighting coefficients of the maintenance items associated with the Lighting Fixtures element where back to the construction division we note the existence of five maintenance items associated with the Lighting Fixtures element these items are (Disadvantages of lighting in the surroundings of the building, Broken lighting poles, Insufficient lighting in the surroundings of the building, Insufficient lighting in the rooms and common sections, Illuminated or broken lighting devices).

X1, X2, X3, X4: Evaluation of the maintenance engineer to determine the defects associated with the Lighting Fixtures element. Each evaluation is multiplied by the associated weighting factor. After the compensation of the evaluation of the engineer responsible for maintenance in the deterioration model of the Lighting Fixtures element we will obtain the deterioration model value of the Lighting Fixtures element.

The previous equation is valid for a deterioration model for the Lighting Fixtures element in different types of buildings, after compensating the maintenance engineer's assessments in the previous equation, we obtain a value that reflects the condition of the Lighting Fixtures element in the facility in terms of its need for maintenance. In a similar way we get deterioration models of the other elements in the electrical component, Shown in the following table:

Thus, we obtain 3 deterioration models for the electrical components that represent the elements. Using these equations, we can obtain a deterioration model that expresses electrical components where

Table 4. Deterioration model value for dorm buildings in Tishreen University

Element Name	Evaluation of Engineers Supervising Maintenance Maintenance Item- Level								Deterioration Model Value
	X1	X2	X3	X4	X5	X6	X7	X8	Y%
Electrical equipment	1	3	1	1	1	2	1	1	17.58%
Electric wirings	3	3	3	5	3	3	3		47.23%
Lighting Fixtures	3	2	4	3	4				52.04%
Deterioration model for electrical component $Z1 = 0.1758L1 + 0.4723L2 + 0.5204L3$									

the number of variables is equal to the number of elements. In deterioration model for electrical component we consider Y1, Y2, and Y3 values are constants or weight coefficients. L1, L2 AND L3 are the evaluators' assessment of the state of these elements according to maintenance engineer:

$$Z2 = y1L1 + y2L2 + y3L3$$

Knowing the value of the relapse model for the electrical component gives us an idea of an improvement or decline in the electrical component and helps us evaluate the maintenance performance performed in a previous year.

CASE STUDY

To ascertain the validity and reliability of the deterioration models built in the previous paragraphs have been applied to the dorm building in Tishreen University, where we obtained the evaluation of engineers supervising maintenance in the dorm buildings for various electrical components of the building and compensate the value of these assessments in previous deterioration models. In this way, we obtained deterioration models for all electrical components of dorm buildings housing. Where we obtained the deterioration model value of Lighting Fixtures element in the following manner:

$$Y3 = 0.78X1 + 0.81X2 + 0.81X3 + 0.83X4 + 0.83X5$$

$$Y3 = 0.78*3 + 0.81*2 + 0.81*4 + 0.83*3 + 0.83*4 = 13.01$$

This is the degree we obtained from (25) where we have five values for x each evaluated from 5 degrees and therefore the highest result we can get is twenty-five and to make this result percentage must be multiplied by 4 we get the value of the deterioration model for Lighting Fixtures:

Y3= 13.01\25= 52.04%

In a similar way we get deterioration models to other elements in the electrical component Shown in the following table:

DISCUSSION OF DETERIORATION MODELS FOR ELECTRICAL ELEMENTS IN DORM BUILDINGS

As shown in Table (5-6), the values of deterioration models have shown that there is an urgent need for maintenance of the electrical equipment element in the dorm building, which was the value of the model (17.58%), The most urgent equipment for maintenance is the generator in case of power failure due to its great importance and its need for permanent preventive maintenance as well as the elevator which was suffering from permanent breakdowns and water pumping engines with electrical breakdowns in water generation boilers and complete disrupted fans of ventilation and air conditioning in all the dorm buildings And the displacement or interruption of lightning protection equipment, The Electric wirings element was relatively good, with the model value (47.23%), noting the possibility of power failure from the feeder cables, control equipment and the need to provide more feeder cables for the new equipment, The Lighting Fixtures element was relatively good, with a deterioration model (52.04%) with a break in the building's lighting poles and Insufficient lighting in the surroundings of the building.

CONCLUSION

Given that the community will expect high performance levels of university buildings, so the level of maintenance carried out is of great importance to the life of the buildings so maintenance will remain an important part of the work of the construction industry.

In this research we tried to highlight the subject of electrical for university buildings maintenance and the main findings of the research:

1- divide electrical components to elements.
2- Determine the main reasons for electrical maintenance and connect these reasons with the electrical component elements.
3- Get deterioration models by combining two important elements: the condition of the component and the second element is the component importance to the building.
4- Application of deterioration models on dorm buildings.
5- Find the most important elements in the construction that need to be carried out for maintenance according to the budget available for maintenance.

Where deterioration models help giving senior management a comprehensive view of the electrical component in the building and a rapid assessment of the status of this component. In addition to allocating the budget for maintenance according to the priority of the elements.

Thus, we have obtained a scientific tool to assist in evaluating the status of the electrical component in the building and the most important maintenance items and maintenance budget and identify the most important resources necessary for the implementation of maintenance.

REFERENCES

Abdel, A. (2012). *Framework For Monitoring And Maintenance Subways Stations Using Building Information Modeling. Construction Research Congress.*

Akasah. (2007). *Generic Process Model for Maintenance Management of School Buildings.* Malaysia University.

Alhassan, B., & Jrad, F. (2015). *enhancing maintenance management for governmental buildings using building information modelling (BIM)* (Master thesis). Teshreen University, Latakia, Syria.

Alhelou, H., Hamedani-Golshan, M. E., Zamani, R., Heydarian-Forushani, E., & Siano, P. (2018). Challenges and Opportunities of Load Frequency Control in Conventional, Modern and Future Smart Power Systems: A Comprehensive Review. *Energies*, *11*(10), 2497. doi:10.3390/en11102497

Alhelou, H. H. (2018). *Fault Detection and Isolation in Power Systems Using Unknown Input Observer. In Advanced Condition Monitoring and Fault Diagnosis of Electric Machines* (p. 38). Hershey, PA: IGI Global.

Alhelou, H. H., Golshan, M., & Fini, M. (2018). Wind Driven Optimization Algorithm Application to Load Frequency Control in Interconnected Power Systems Considering GRC and GDB Nonlinearities. *Electric Power Components and Syst.*

Alhelou, H. H., & Golshan, M. E. H. (2016, May). Hierarchical plug-in EV control based on primary frequency response in interconnected smart grid. In *Electrical Engineering (ICEE), 2016 24th Iranian Conference on* (pp. 561-566). IEEE. 10.1109/IranianCEE.2016.7585585

Alhelou, H. H., Golshan, M. H., & Askari-Marnani, J. (2018). Robust sensor fault detection and isolation scheme for interconnected smart power systems in presence of RER and EVs using unknown input observer. *International Journal of Electrical Power & Energy Systems*, *99*, 682–694. doi:10.1016/j.ijepes.2018.02.013

Alhelou, H. H., Hamedani-Golshan, M. E., Heydarian-Forushani, E., Al-Sumaiti, A. S., & Siano, P. (2018, September). Decentralized Fractional Order Control Scheme for LFC of Deregulated Nonlinear Power Systems in Presence of EVs and RER. In *2018 International Conference on Smart Energy Systems and Technologies (SEST)* (pp. 1-6). IEEE. 10.1109/SEST.2018.8495858

Alhelou, H. S. H., Golshan, M. E. H., & Fini, M. H. (2015, December). Multi agent electric vehicle control based primary frequency support for future smart micro-grid. In *Smart Grid Conference (SGC)* (pp. 22-27). Academic Press.

Alshahrestani, A., Golshan, M. E. H., & Alhelou, H. H. (2018, November). WAMS Based Online Estimation of Total Inertia Constant and Damping Coefficient for Future Smart Grid Systems. In *Smart Grid Conference (SGC)* (pp. 1-5). Academic Press.

Arditi, D., & Nawakorawit, M. (1999). Designing buildings for maintenance: Designers' perspective. *Journal of Architectural Engineering*, *5*(4), 107–116. doi:10.1061/(ASCE)1076-0431(1999)5:4(107)

Bornand, E. (1991). Le guide de maintenance. Berne.

Fini, M. H., Yousefi, G. R., & Alhelou, H. H. (2016). Comparative study on the performance of many-objective and single-objective optimisation algorithms in tuning load frequency controllers of multi-area power systems. *IET Generation, Transmission & Distribution*, *10*(12), 2915–2923. doi:10.1049/iet-gtd.2015.1334

Frangopol, D. (2003). Reliability Deterioration and Life Time Maintenance Cost Optimization. University of Colorado.

Haddad, R., & Hassan, B. (2001). Develop A Methodology to Analyze the Factors Affecting Maintenance Costs in Government Hospitals. *Tishreen University Magazine*.

Idrus, A., Khamidi, F., & Sodangi, M. (2010). Maintenance management framework for conservation of heritage buildings in Malaysia. *Modern Applied Science*, *4*(11), 66. doi:10.5539/mas.v4n11p66

Makdisie, C., Haidar, B., & Alhelou, H. H. (2018). An Optimal Photovoltaic Conversion System for Future Smart Grids. In *Handbook of Research on Power and Energy System Optimization* (pp. 601–657). IGI Global. doi:10.4018/978-1-5225-3935-3.ch018

Marzouk, M., & Abdel Aty, A. (2012). Maintaining subway infrastructure using BIM. In *Construction Research Congress 2012: Construction Challenges in a Flat World* (pp. 2320-2328). Academic Press. 10.1061/9780784412329.233

Nadweh, S., Hayek, G., Atieh, B., & Haes Alhelou, H. (2018). Using Four – Quadrant Chopper with Variable Speed Drive System Dc-Link to Improve the Quality of Supplied Power for Industrial Facilities. Majlesi Journal of Electrical Engineering.

Njenda, T. C., Golshan, M. E. H., & Alhelou, H. H. (2018, November). WAMS Based Intelligent Under Frequency Load Shedding Considering Online Disturbance Estimation. In *Smart Grid Conference (SGC)* (pp. 1-5). Academic Press.

Njenda, T. C., Golshan, M. E. H., & Alhelou, H. H. (2018, November). WAMS based Under Frequency Load Shedding Considering Minimum Frequency Prediction and Extrapolated Disturbance Magnitude. In *Smart Grid Conference (SGC)* (pp. 1-5). Academic Press.

Olagunju, R.E. (2012a). Predictive Modelling for Sustainable Residential Building Maintenance in Developing Countries: A Nigeria Case. *Interdisciplinary Journal of Contemporary Research in Business*, *4*(6), 1273–1283.

Olagunju, R.E. (2012b). Sustainability of Residential Buildings in Nigeria: An Appraisal of the Factors that Influence Maintenance of Residential Buildings 'Standards. *Civil and Environmental Research*, *2*(4), 20–29.

Zamani, R., Hamedani-Golshan, M. E., Haes Alhelou, H., Siano, P., & Pota, H. (2018). Islanding Detection of Synchronous Distributed Generator Based on the Active and Reactive Power Control Loops. *Energies*, *11*(10), 2819. doi:10.3390/en11102819

KEY TERMS AND DEFINITIONS

Distributed Generation: Electrical generation and storage performed by a variety of small, grid-connected devices referred to as distributed energy resources.

Hydroelectric Power Plant: A power plant used the energy of falling water or fast running water.

Power Grid: An interconnected network designed for transporting electric power from the generation plants to the customer.

Power Plant: An industrial facility for the generation of electric power.

Renewable Energy: Represents the energy produces from an energy source that is naturally replenished after use. Some examples are: sunlight, wind, rain, waves, tides, geothermal, etc.

Selectivity: The property by which only the faulty element of the system is isolated, and the remaining healthy sections are left intact.

Chapter 17
Power Quality and Stability Analysis of Variable-Speed Drive Systems (VSDS)

Safwan Mhrez Nadweh
Tishreen University, Syria

Ola Ahmad Khaddam
Tishreen University, Syria

ABSTRACT

This chapter introduces an analysis of power quality of variable-speed drive systems (VSDS). In this chapter, several issues have been discussed as VSDS in terms of cost and effectiveness, VSDS control loops (open loop, closed loop), and power quality in VSDS. Harmonics standards and harmonics in VSDS were discussed in this chapter in addition to highlighting the effects of harmonics on power factor, crest factor, and other power quality specifications. The solutions used to mitigate the harmonics in VSDS were discussed in detail. Finally, simulation of the conventional VSDS model and VSDS with one harmonic mitigation solution in order to clarify the usefulness of using this solution on power quality specifications were discussed.

INTRODUCTION

Most motors are designed to operate at a constant speed and provide a constant output; however, modern technology applications are very demanding and require motors that can run at different speeds. A variable speed drive (VSD) presents an attractive option that can meet those requirements by regulating the speed and output torque of the mechanical equipment droved by the electrical motor. Some examples of mechanical equipment that incorporate with VSD technology are pumps, fans, compressors and conveyors. In fact, any process that runs variable speed loads is candidate for a VSD (Michael et al., 2010; Raneru, 2013).

DOI: 10.4018/978-1-5225-8030-0.ch017

VSDs can be used to control DC or AC motors. DC drives have been used for many decades and were proper choice for applications that require accurate speed control, controllable torque, reliability and simplicity. One of the problems with DC drives is the high maintenance requirement of the DC motor. Since the 1980s, AC variable speed drives popularity has grown rapidly, mainly due to continuous development in power electronics and digital control technology affecting both the cost and performance of this type of VSD. The main attraction of the AC drives is the reliability and low cost of AC induction motor compared to the DC motor. Induction motors are by far the dominant type of motor used with VSDs.

VSDS IN TERMS OF COST AND EFFECTIVENESS

There are millions of motors employed in a wide range of applications around the world and the majority of them operate only at 100% speed for short periods of time. This often results in significant energy losses during the operation time, especially after considering the electrical power consumed by motor systems that reach more than 60% of electrical power consumed by industry. Significant energy savings can be achieved when the rotational speed of the motor is decreased to match with the load requirement and that can be achievable by installing VSDs systems. VSDs allow the motor speed to change depending on operating conditions, rather than operating continuously at full speed. Because the power drawn is proportional to the cube of its speed, reducing speed can save a lot of energy

The tendency to automate industrial process, and the persistent need for increased efficiencies from plant and machinery, together with the demand for higher performance at a lower cost, contribute to the rapid growth of VSDs market (ABB, 2009; Davari et al., 2015).

Efficiency of the operation can be increased by allowing motors to be operated at the ideal speed for every load condition through the capability of soft start up and over speed, which reflects on the process precision and improves power factor. Due to speed and torque control, motors endure less wear and tear, which leads to lower maintenance costs and downtime and prolong the life of the motor. VSDs also reduce frictional losses related with mechanical or electromechanical adjustable speed technologies and expensive energy-wasting. Furthermore, Applying VSDs can lead to reduction in the amounts of carbon dioxide emissions by reducing the amount of fuel consumed through the operation.

Some examples of energy saving applications that use VSD systems are:

- **Fans:** Varying a fan's speed allows it to match changing load requirements more closely, and because fan power is proportional to the cube of its speed, reducing speed can save a lot of energy.
- **Pumps:** Pumping systems account for nearly 20% of the world's energy consumption by electric motors and 25–50% of the total electrical energy usage in industrial facilities. Many pumping applications offer great potential for energy savings. The savings often go well beyond energy, and may include improved performance, reliability, and reduced life cycle costs. Most existing systems require flow control of bypass lines, throttling valves, or pump speed adjustments. When a pump's speed is reduced, less energy is imparted to the fluid and less energy needs to be throttled or bypassed (Kumar et al., 2014; Kazem, 2013; Latha, 2012)
- **Heating, Ventilating and Air Conditioning (HVAC) Systems:** Building HVAC systems are designed to operate at maximum load conditions; however, they operate at full load only for short periods of time. The load in HVAC systems varies throughout the day and from day to day which

Table 1. Potential VFD Energy Saving

Load Type	Application	Energy saving potentials
Variable torque load - The Power of the load is proportional to the cube of its speed. - Torque is proportional to the square of the speed.	- Centrifugal fans - Centrifugal pumps - Blowers - HVAC systems	Decreased speed results in significant energy savings because power drops with the cube of the speed
Constant torque load - Torque remains constant at all speeds - power of the load proportional to the speed	- Mixers - Conveyors - Compressors - Printing presses	Decreased speed saves energy in direct proportion to the speed reduction
Constant power load - Power constant at all speeds - Torque is inversely proportional to the speed	- Machine tools - Lathes - Milling machines - Punch presses	Decreased speed saves no energy, but some saving can be done by attaining the optimized cutting and machining speeds for the part being produced

results in inefficient operation during long periods of time. The efficiency of such systems can be improved by varying the capacity to match actual load requirements by using VSDs that a run HVAC systems run at partial-load conditions during most of operation time. According to some researches, VSDs can save 15–40% of energy in HVAC application. Table (1) shows some typical loads and their energy savings potential.

Energy consumption of electric motors forms up to 75% of total plant's energy consumption. A small change in motor speed can cause a significant change in energy consumption. Energy savings of motors when installing VSDs can be calculated from the following equation:

$$AES_{VSD} = \left(1 - S_{SR}^{\ 3}\right) \times 100\% \tag{1}$$

The annual cost savings is related to annual energy savings and price of the fuel. The annual bill savings of motor when using VSDs can be calculated using the following equation:

$$\text{Annual Bill Savings}\left(ABS_{VSD}\right) = ABS_{VSD} \times C \tag{2}$$

The payback period is the function of the incremental cost of VSDs divided by the annual bill savings of VSDs in a particular year. Thus Payback period can be expressed mathematically from the following equation:

$$\text{Payback period}\left(\text{years}\right) = \frac{\text{Incremental cost}\left(ICVSD\right)}{\text{Annual bill savings}\left(ABDVSD\right)} \tag{3}$$

It is estimated that replacing conventional motors with adjustable speed motors in appropriate applications would result in saving 41% of the energy used in industrial motors. Power consumption actually drops far more than the drop in motor speed, so the savings can accumulate quickly. For example a 10%

reduction in shaft speed results in a 27% decrease in power consumption (Meyer et al., 2014; Khine, 2016; Alhelou et al., 2018).

Although applying VSDs introduces advantages such as energy saving and efficiency improvement, it has disadvantages. VSDs are nonlinear loads and they draw distorted currents from Ac supply resulting in distorted voltages. This distorted currents and voltages called harmonics. Harmonics pollute the electrical system, which cause problems if harmonic level increases beyond a certain level. Harmonic currents have negative effects such as, overheating of transformers, cables, motors, generators, and capacitors. Large harmonics lead to increase factory downtime and lead to more operating costs. Current waveform distortion is injected back into the power supply system creating several problems such as increase in the electrical losses on the network side of the transformer, decrease efficiency of motor, nuisance tripping of circuit breakers, malfunction or failure of electronic circuits, erroneous reading of metering devices, de-rating of fuses as a result of additional heat generated, and decrease life expectancy of equipment. Another disadvantage of VSDs is the relatively expensive installation costs relatively. The cost ranges between \$3000 for a 5 Hp motor and \$45,000 for a custom-engine 300 hp motor. The costs increase drastically for larger versions. VSD installation time takes 10–70 labor h, depending on system size and complexity; however, payback period ranges from few months to less than 3 years for 25–250 Hp models. Each VSD is capable to drive more than one motor; hence some costs can be consolidated. In addition, savings from reduced maintenance and longer equipment life contribute significantly to achieving a rapid payback and long-term savings.

VSDS CONTROL LOOPS AND POWER QUALITY

VSDS Control Loops

From a control point of view, we can define a VSD as a device that is designed to drive an AC induction motor in a manner that forces the motor's shaft to rotate at a variable speed dictated by an external signal. The VSD uses external signal to control the rotational speed and torque of the induction motor by adjusting the frequency of the power supplied to the motor using PWM inverter. The basic configuration of a typical VSD is illustrated in figure (1).

The main components of a typical VSD system are:

- AC/DC converter, usually a diode rectifier.
- DC Link.
- DC/AC converter, usually a PWM inverter.
- AC motor.
- Droved machine.
- Control circuit.

The main difficulty with the AC variable speed drive has always been the complexity, cost and reliability of the AC frequency inverter circuit.

The VSD control systems modify the output frequency and voltage in order to get the speed required by the motor. The accuracy of the control system and its response is determined by the type of control system used (Hadeed et al., 2016; Hoadley et al., 2008; Alhelou et al., 2015).

Figure 1. The basic configuration of VSD

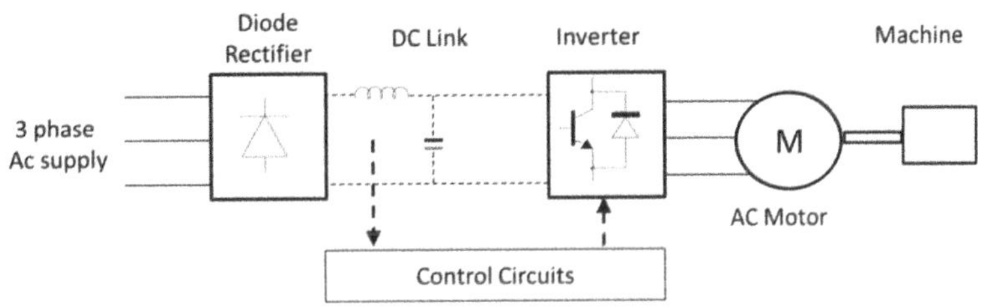

There are two types of control used in VSD systems, they are:

- Open-loop control, no feedback from the process.
- Closed-loop control, feedback of a process variable.

Open-Loop Control

Open-loop control is suitable for steady-state conditions and simple applications, such as pumps, fans or conveyors, in which the time for speed changes from one level to another is not a priority and the consequences of the changes in the process are not severe.

Constant V/f control strategy is an open-loop control that uses the constant ratio of voltage and frequency to control the speed of the AC motor, where the speed reference is taken from external source as shown in figure (2).

Closed-Loop Control

Some applications required a V/f closed-loop control, with speed feedback from either tachogenerator, or a digital shaft encoder. This system provides control of the AC output voltage (via the input rectifier) and frequency (via the switching in the inverter). Figure (3) shows a typical diagram of a V/f closed loop designed to control an induction motor.

The system comprises the following two loops:

1. Inner loop to control the torque indirectly, by regulating the slip speed, and it involves a positive feedback loop.
2. Outer loop to control the speed, and the operation of this loop is straightforward.

Figure 2. Block control diagram of constant V/f open-loop control

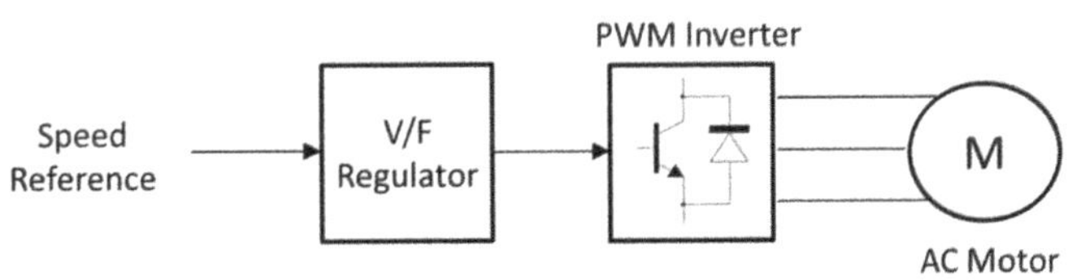

Figure 3. Closed- loop V/f control

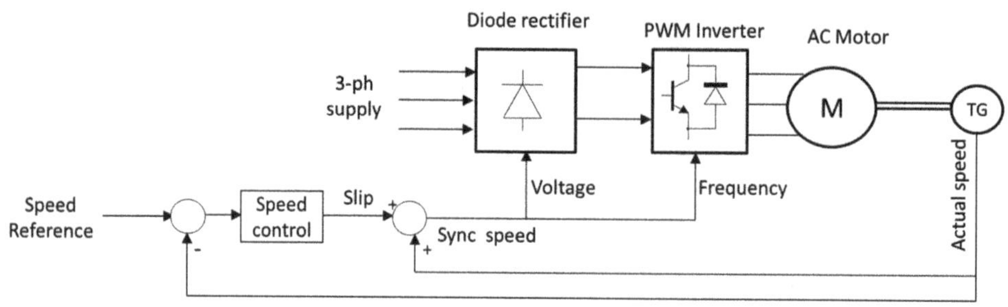

The actual speed (generated by the tachogenerator) is compared with the reference speed and the resulting speed error forms the input to the speed controller. The output of the speed controller serves as a torque reference signal. If we wish to obtain a given torque, we have to add the rotor speed to the slip speed in order to obtain the frequency to be supplied to the stator (Alshahrestani et al., 2018; Makdisie et al., 2018; Alhelou et al., 2018; Alhelou et al., 2016; Nadweh et al., 2018; Njenda et al., 2018).

Torque is directly proportional to the slip speed. In order to control the slip speed, we can directly vary the stator frequency (the synchronous speed), and measure the actual rotor speed. The output from the summing junction directly controls the inverter output frequency (i.e. the synchronous speed), and the amplitude of the inverter output voltage. A constant voltage–frequency ratio is provided through the whole operation in order to guarantee the 'constant flux' condition that is an essential requirement for us to be able to claim that torque is proportional to slip speed (Stephen, 2010; Tim, 2011).

Still the previous closed-loop is not adequate for applications where some demanding tasks require an extremely high dynamic performance, such as high-speed machine tool spindle drives. In such applications a closed-loop vector control is consider to be the most reliable control option (Fini et al., 2016; Alhelou et al., 2018; Zamani et al., 2018; Alhelou et al., 2015; Njenda et al., 2018; Haes Alhelou et al., 2018).

The block diagram for a high performance closed-loop vector control designed to drive AC motor is illustrated in figure (4).

There are two separate control loops involved in this system:

1. Speed loop to control the output frequency, proportional to speed,

Figure 4. Closed-loop vector control block diagram

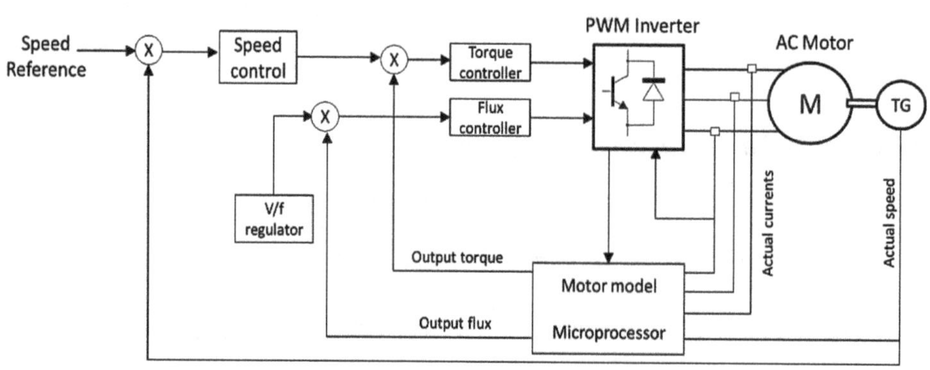

2. Torque loop to control the motor current, proportional to torque (Zare, 2014).

The actual speed generated by tachogenerator is compared with the reference speed and the resulting speed error is passed through a speed controller. The resulting signal is the input or reference signal for the torque controller. This signal is compared to the output torque from the motor circuit and the error signal determines whether the motor is to be accelerated or decelerated. A separate control loop for the flux is applied to provide a constant V/f ratio and consequently a constant flux. Finally, the signal passes to the PWM inverter and controls the switching devices in such a way that the desired voltage and frequency are generated at the output.

POWER QUALITY

Power Quality is defined as "a term that refers to maintaining the near sinusoidal waveform of power distribution bus voltages and currents at rated magnitude and frequency". The ideal situation would be a voltage and current waveforms with a constant amplitude and frequency. Unfortunately, This is not possible due to multiple issues in power quality that can be manifested in the form of deviations in the amplitude and frequency of voltage and current waveform resulting in unexplained equipment trips or shut downs, occasional equipment damage or component failure, random lockups and data, power system component overheating. Consumer concern with power quality has increased in the last years since the use of susceptible equipment has grown rabidly and interruptions caused by disturbances imply a high economical cost.

Power quality problems can be classified into four categories:

- Short duration variations that include sag, swell, and interruption.
- Long duration variations that include under voltage and sustained interruption.
- Transients that include impulsive transient and oscillatory transient.
- Wave distortions that include harmonics, flickering, and notching effect.

However, the main three types that impact the industrial customers the most are voltage sags, harmonics, and transients.

Voltage sags: A short-term reduction in voltage by at least 10% of the nominal voltage that last between half a cycle and a minute. IEEE 1159, "Recommended Practice for Monitoring Electric Power Quality," classifies sags as instantaneous (0.5 cycles to 30 cycles), momentary (30 cycles to 3 seconds), and temporary (>3 seconds to1 minute).figure (5) shows an Instantaneous voltage 60% sag with a duration of 4 cycles.

Transients: An Extremely rapid voltage peak of up to 20000 with duration of 10 microseconds to 100 microseconds. Transients can be either impulsive or oscillatory. Impulsive transients are frequently referred to as surges. Transient are possibly the most damaging type of voltage disturbance. Figure (6) shows an example of impulsive transients.

Harmonics: Harmonics are a form of voltage or current waveform distortion. A harmonic is a waveform with a frequency that is a multiple of the fundamental frequency, general 50 or 60Hz. Harmonics are a relatively steady state condition, unlike momentary conditions such as sags or transients.

Figure 5. Example 60% sag with a duration of 4 cycles

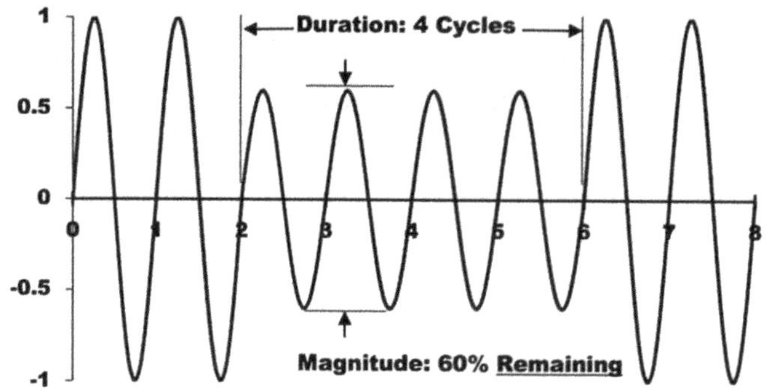

Figure 6. Impulsive transient

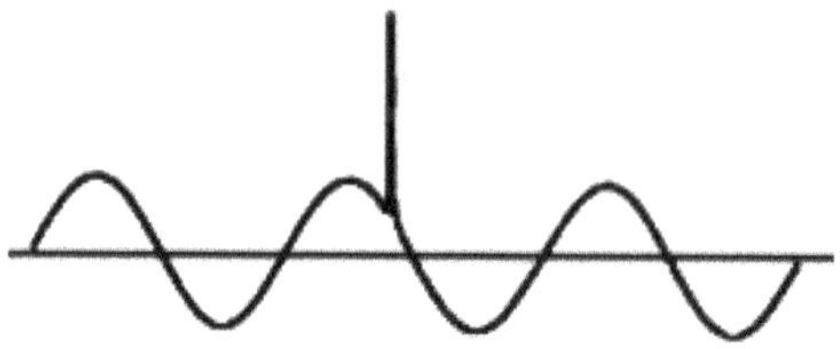

Variable speed drives are susceptible to power quality problems, especially the main three Types mentioned above, and they are source of these problems themselves. For example:

Utility capacitor switching transients disturb the operation of VSDs. During voltage sags, i.e. when the voltage of the line drops to a portion of its nominal value for less than second; the operation of the drive is disturbed. Drives generate harmonics that can resonate with capacitors and inductances in the system disturbing their operation or the operation of nearby devices.

HARMONICS IN VSDS AND HARMONICS STANDARDS

Harmonics in VSDS

Harmonic is a periodic wave with a frequency that is an integer multiple of the fundamental frequency (usually 50Hz). For example, the third harmonic would be 3*50Hz=150Hz.

An AC periodic voltage or current can be presented by a Fourier series of pure sinusoidal wave which contain the basic or the fundamental frequency and its multiples called harmonics. Each harmonic has order denoted as h. The Fourier series is a mathematical way of expressing a complex signal or a waveform as the sum of simpler waveforms (harmonics). Figure (7) illustrates the Decomposition of a distorted waveform into harmonic components using Fourier series.

A common measurement of harmonic distortion of a voltage or current waveform is the Total Harmonic Distortion (THD) which can be defined as the ratio of the root mean square (RMS) of the harmonic content to the root mean square (RMS) of the fundamental component.

THD of a current waveform can be calculated as:

Figure 7. Fourier series for a distorted waveform

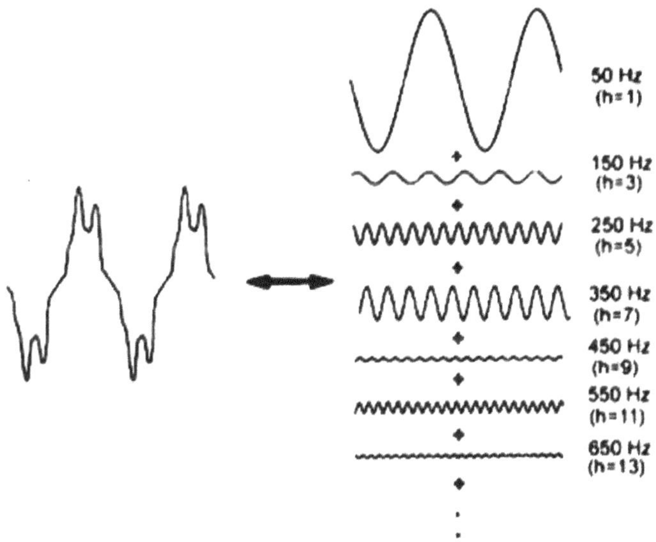

$$THD_I(\%) = \frac{\text{RMS value of harmonics}}{\text{RMS value of its fundamental}} = \frac{\sqrt{I_2^2 + I_3^2 + I_4^2 + \ldots + I_n^2}}{I_1} \times 100 \quad (4)$$

where; I represents the RMS value of nth harmonic.

Similarly, THD of a voltage waveform can be calculated as:

$$THD_v(\%) = \frac{\text{RMS value of harmonics}}{\text{RMS value of its fundamental}} = \frac{\sqrt{v_2^2 + v_3^2 + v_4^2 + \ldots + v_n^2}}{v_1} \times 100 \quad (5)$$

where; V represents the RMS value of nth harmonic.

Nonlinear loads are known to be the source of harmonic distortion in voltage and current waveforms. The term 'nonlinear' refer to the relationship between voltage and current waveforms, i.e., the relationship between voltage and current at every instant of time is not constant. Nonlinear loads draw non-sinusoidal currents, from a perfectly sinusoidal voltage source, called harmonic currents that flow through the system impedance resulting in harmonic voltages. The harmonics order found in three-phase distribution networks are generally odd numbered. The higher order harmonics are normally lower in magnitude. Most industries and utilities monitor harmonic orders 3, 5, 7, 11 and 13 as measurements up-to this order are known to be more than sufficient.

Examples on nonlinear loads are arcing devices such as arc furnaces, saturable devices such as transformers, and power electronic equipment such as variable speed drives (VSDs) and rectifiers or any device that use simeconductors. The Rapid growth of applications that use semiconductor devices lead to a revolution in controlling industrial processes and converting energy. In fact, semiconductor devices are used in all driver circuits, controllers for DC or AC motors, domestic appliances, and photovoltaic systems. The presence of harmonics can cause problems in the power system if the harmonic distortion increases beyond a certain level.

Although using VSDs can make the motor drive systems more efficient, VSD systems are known to be a big source of harmonic distortion due to the wide use of them, especially, when we think of multiple VSD units connected to the point of common coupling (PCC). The switching devices used in the AC/DC converters in VSDs, are the source of harmonics in ASDs. Figure (8) shows a waveform with VSD harmonics, we can see how the fundamental 50Hz waveform changes when harmonics are added.

Harmonics Standards

Because of the negative effects of harmonics, National and international organization have developed standards for the amount of harmonic distortion allowed in the electrical power systems. Applying these standards reduce the potential effects on user and system equipment.

The limits set in these standards should be applied at the point of common coupling (PCC) between the supplier and user. Applying these limits is a mutual responsibility for harmonic control between suppliers and users due to the fact that harmonic currents produced by user flow back through the system and cause voltage harmonics in the voltage supplied to other users (ROBAK, 2016; Sekar, 2012).

the Institute of Electrical and Electronic Engineers (IEEE) has set the standard IEEE 519-1992, "Recommended Practice for Harmonic Control in Electric Power Systems", which provides recommended harmonic current limits for individual customers at the point of common coupling (PCC) with the electric utility. The utility supplying the new facility has specified these harmonic current limits in the contract with the customer.

Standards developed by International Electrical Commission. Such as standard IEC 61000-3-2 which specifies limits of current harmonics produced by low-voltage equipment with rated current lower than 16A, Standard IEC 61000-3-12 specifies limits of current harmonics produced by low-voltage equipment with rated current between 16A and 75A.

Figure 8. Waveform with VSD harmonics

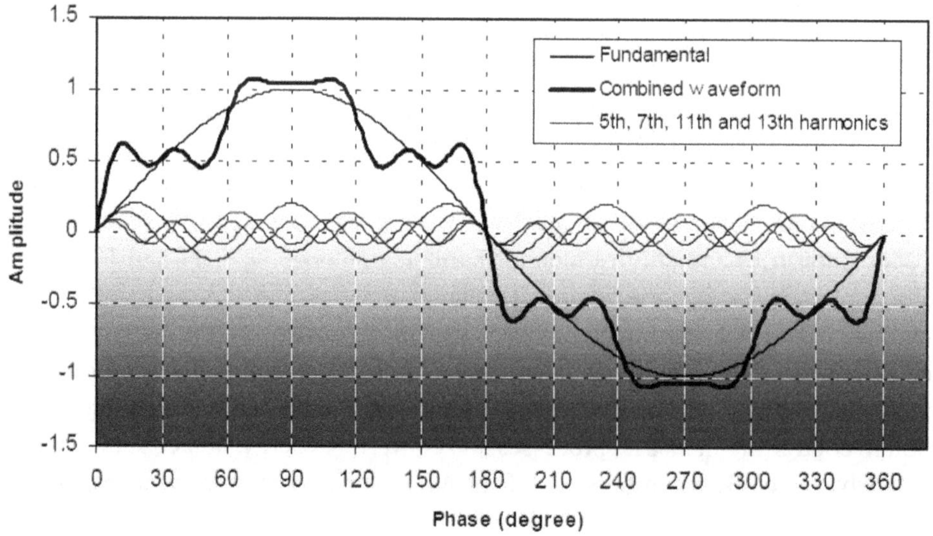

Standard IEC 61000-3-2 defines limits with absolute values, whereas standard IEC 61000-3-12 specifies values which depend on the ratio between the short circuit power of the power system at the point of common coupling (PCC) and the rated power of the equipment. The larger the ratio the higher the harmonics allowed.

Harmonic mitigation can become very expensive if the limits are extremely strict. The harmonics standards should be carefully analysed and applied to minimize the cost of the mitigation solution.

THE EFFECTS OF HARMONICS ON POWER FACTOR, CREST FACTOR, AND OTHER POWER QUALITY SPECIFICATIONS

The effects of harmonics on power factor, crest factor, and other power quality specifications harmonic levels caused by VSDs depends on a large number of variables:

- The magnitude of the current flowing through the converter.
- The configuration of the power electronic circuit (6-pulse, 12-pulse, etc).
- The characteristics and impedances of the connected power supply system.

It does not matter if the rectifier bridge comprises thyristors (controlled rectifier) or diodes (uncontrolled rectifier), they both draw harmonic currents in the same way. Harmonic currents accumulate Through the DC side and cause ripples in DC current. Consequently, the rectifier can be considered to be a source of harmonic currents, which flow back into the power supply resulting in voltage distortion. Power electronic converters do not generate all the harmonics, only certain harmonic currents. The harmonic order and magnitude of the harmonic currents generated by any converter depends on pulse number (during one cycle of the supply voltage), the magnitude of the harmonic currents DC side of the rectifier, and the magnitude of the load voltage DC side of the rectifier.

The performance of other equipment is disturbed by the distortion of supply voltage caused by harmonic currents and this distortion can reach high levels when a harmonic resonance happens between the system reactance and capacitance. Current harmonics also creates additional losses and heating resulting in the aging of the equipment.

Capacitor banks (used for power factor correction) also affected by current harmonics due the fact that a capacitor bank present a low impedance path to high frequency harmonic currents which increases the dielectric losses in the capacitor bank and lead to overloading and eventual failure. Transformers, motors, cables, bus bars and switchgear supplying current to converters should be de-rated to accommodate the additional harmonic currents and the extra losses associated with the high frequency 'skin-effect'. The electronic equipment used for instrumentation, protection, and control is also affected due to the interference coupled into the equipment or communications cables. This affects the reliability and performance of the control system.

When a linear load is connected to AC sinusoidal voltage supply, it draws sinusoidal current from the supply and the magnitude of this current depends on the impedance of the load. Flowing of the current produces active power that is consumed in the load. In the case of resistive loads, the current is in phase with the voltage and the total active power is equal to the apparent power. In this case, the ratio between the active power and the total apparent power is equal to 1.

Figure 9. The radial scheme of power factor

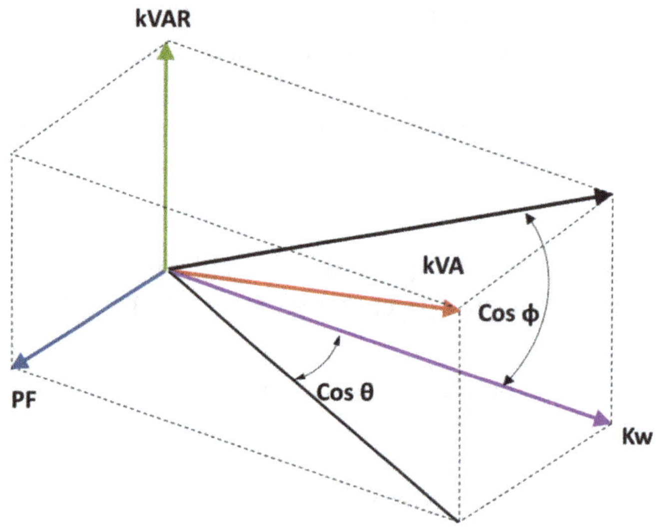

inductive loads, such as electric motors, cause the current to lag behind the supply voltage by an angle between 0° to 90°, which results in a total active power that is lower than apparent power. As a result ratio between the active power and the total apparent power falls to a value less than 1. In purely inductive loads, the ratio between the active power and the total apparent power falls to zero, due to that fact that active power in purely inductive loads is equal to zero.

The ratio between the active power and the total apparent power is known as the power factor and is defined as follows:

$$\text{power factor} = \frac{\text{total active power}}{\text{total apparent power}} \tag{6}$$

Apparent power is equal to the product of the RMS voltage and the RMS current and measured as volt-amperes, and active power is measured as kilo watts.

Figure(9) illustrates the radial scheme of power factor. For purely sinusoidal voltages and currents, the power factor is a function of the phase displacement angle φ between the voltage and the current, and it is equal to the cosine of the phase displacement angle φ.

Cos φ is also referred to as the displacement factor, which has a value between 0 to 1.

Displacement Factor = Cos φ

In those cases where both the voltage and the current are purely sinusoidal:

Power Factor = Displacement Factor = Cos φ

In Non-linear loads, the voltages and currents are distorted and displacement factor is not equal to power factor, because the total active power is no longer related to the displacement angle between the voltage and current. The harmonic components of the current do not do any useful work and are lost as heat in various parts of the power system and the electrical load.

Harmonic components are taken into account by measuring the total RMS volt-amperes, which results in a power factor which is much lower than would be calculated from simply measuring the displacement factor.

The distorted voltages and currents have to be measured by special true RMS reading instruments. In practice, the real power factor with diode converters can be as low as 0.65, even though the measured displacement factor is greater than 0.95.

The real power factor is also affected by circuit components such as the source impedance of the power system and the inductances in the power electronic circuit. In general, the more distorted the current waveform, the lower the real power factor will be.

Although it is quite easy to measure the power factor of an existing circuit, it is quite difficult to calculate the real power factor of a drive system at the design stage. In order to get an accurate analysis, it is necessary to use a computer based circuit analysis program to model the electrical system with respect to various impedances and effects of harmonic frequencies on the inductive components of the electrical system.

The Crest Factor is defined as the peak amplitude of a waveform divided by the RMS value. The purpose of the crest factor calculation is to give an analyst a quick idea of how much impacting is occurring in a waveform. Impacting is often associated with roller bearing wear, cavitation and gear tooth wear.

The Crest Factor is a quick and useful calculation that gives the analyst an idea of how much impacting is occurring in a time waveform. This tells the end user not only the purity of a signal but also the capability for a system to output a particular voltage or current. This is useful information that is lost if one is only viewing a spectrum as the FFT cannot differentiate between impacting and random noise. Impacting in a time waveform may indicate rolling element bearing wear, gear tooth wear or cavitation. Quite often, the Crest Factor is trended over time in order to see if the amount of impacting is increasing or not.

Figure 10. Crest factor definition

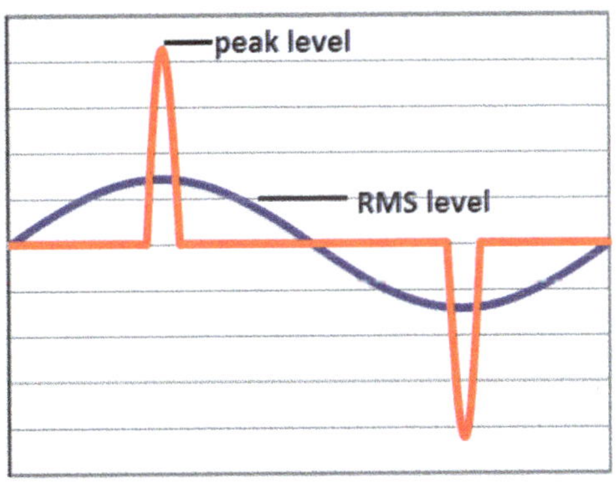

HARMONIC MITIGATION SOLUTIONS

There are many solutions for the reduction of harmonics in industrial plants because of their harmful effects on both the loads and the grid.

Several methods have been used to reduce harmonics and classified according to:

1. Industrial network structure.
2. Other means.

INDUSTRIAL NETWORK STRUCTURE

12-Pulse Rectifier

12-pulse rectifier is one of the solutions using to mitigate the harmonics. This scheme is achieved by connecting two 6-pulses rectifiers in parallel. This structure has a great benefit. it produces more smooth current waves than the 6-pulses rectifier with a 10% smaller total distortion factor and a higher power factor of 9%, but the additional transformer used to achieve this installation made this solution very expensive and increased the overall size of the installation. The components of 18 and 24 pulses were also used and composed of 3 or 4 components with 6 pulses connected in parallel (Alshahrestani et al., 2018; Makdisie et al., 2018; Alhelou et al., 2018; Alhelou et al., 2016; Nadweh et al., 2018; Njenda et al., 2018).

Line Reactors

The reactors added sequentially to the feeding side, called line reactors, and are relatively inexpensive. They reduce harmonics to 50% compared to harmonic levels when no mitigation mechanisms are added, but these reactors have many drawbacks such as reactive power losses, and voltage drop in connecting points.

Figure 11. 12- pulse rectifier with phase shift transformer

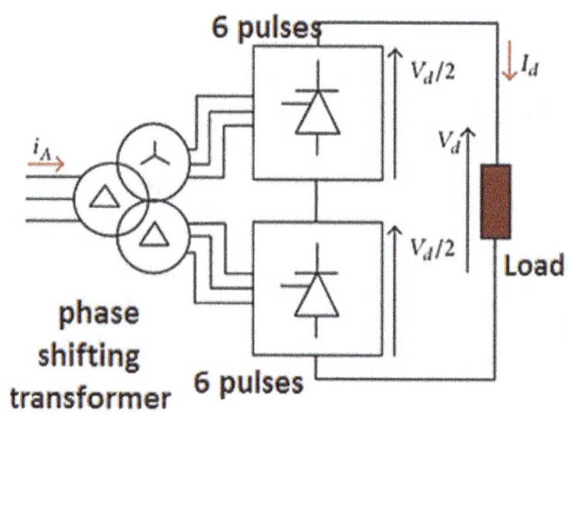

Figure 12. VSDS with ac- line reactor

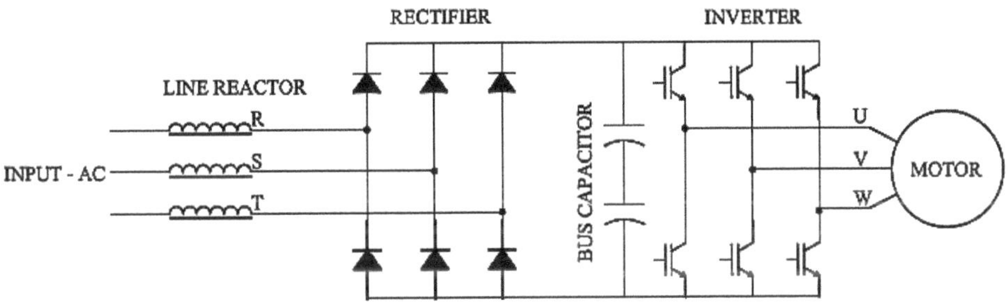

Dc-Chokes

Reactors added to the dc-link instead of the feeding lines called as dc-chokes, are used to smoothing the load current, which smaller in size compared with the alternating reactors. Their disadvantage is their lack of efficiency compared with the line reactors, especially of voltage flickers and increased voltage at the feed source.

PASSIVE FILTERS

These filters consist of inductors and capacitors connected in such way to provide low impedance only for undesirable harmonic elements to reach the ground.

Main features: improve the power factor of the system.

The Most Important Disadvantages

- Not workable with variable loads.
- It has very large size.
- The potential of capacitor exploded which cause out of service.

Figure 13. VSDS with DC choke

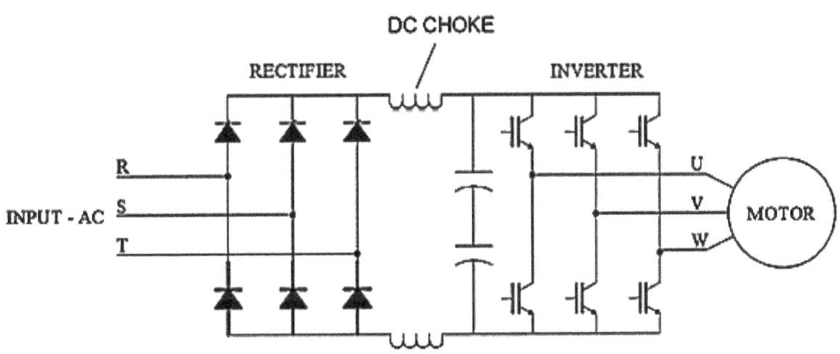

Figure 14. Passive filter

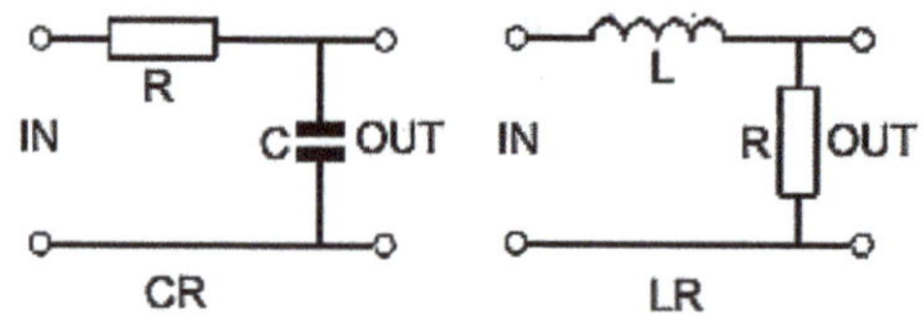

Choppers

To regulate the output voltage of the inverter, either change their gain factor or control their input voltage. Therefore, circuit breakers are used in dc-link.

Boost Converter

Figure 15 shows the use of a boost converter to regulate current in dc-link side. a new technique has been proposed to regulate the main component current and eliminating the higher harmonics. The importance of this method is that the value of the main component has remained constant during different working condition. in addition, the boost chopper raises the voltage, which is appropriate for engine driving applications.

Buck-Boost Chopper

Figure 16 illustrates the use of a buck-boost chopper circuit, consisting of a inductors coil, two transistors, diodes d1 d2, and a capacitor.

Although using dc-link chopper reduces the size of the dc chocks, the load current will pass through more than two electronic elements. Therefore, the conduction losses of the ac drive systems with choppers are slightly higher than the losses in the traditional three-phase ac drive systems.

Different coupling with the files and capacitors has been used and reduced the main current and output voltage impedances. The small size and weight of these cutters is very suitable for integration with engines in variable speed driving systems and easy to add to all existing fittings and circuits. These sections understand different levels of effort and operating patterns and this is what we will do in this research.

Figure 15. Use of a boost converter to regulate current in dc-link side boost chopper

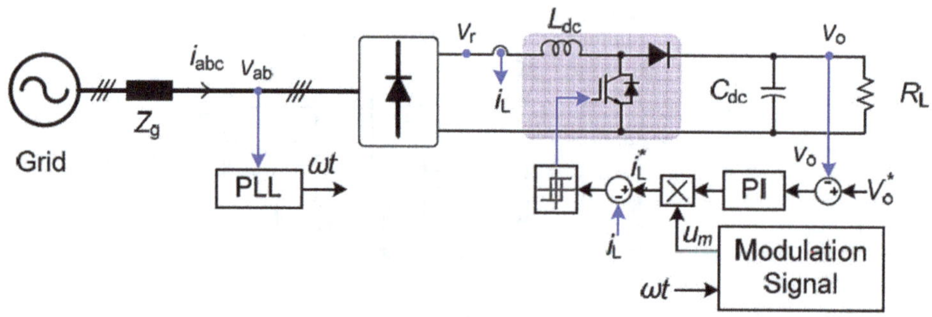

Figure 16. Use buck- boost chopper in DC- LINK to mitigate the harmonics (Buck - Boost chopper)

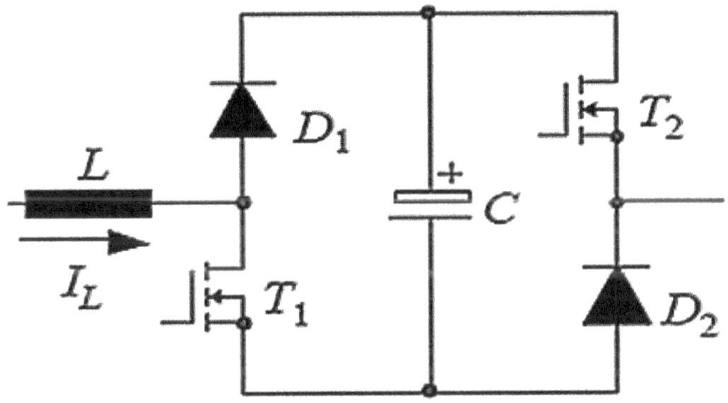

In order to keep on the quality of the performance in the load side, a lot of attempted to obtain a constant value for continuous currents in dc-link have done. a variable voltage source was added between the diode bridge and the output capacitor, but the difficulty of the control in this source stood in the way of its use.

ACTIVE FILTERS

Introduction to Active Filters

Active filters have gained the widest spread in their use as a modern technology to reduce harmonics in industrial plants. The active filter is defined as a device that injects equal in magnitude and opposite in phase noise to the power line. The estimation of active power converters depends on the current amplitude of the distorted current and operates at the specified operating frequency through the desired filtration area.

Uses of Active Filters

As a result of the controller's development, the active filter has been used to eliminate all the harmonic components depending on the nature of the used controller, and there are different techniques for the operation of these controllers such as: SPWM, space vector PWM, third injection PWM (Fini et al., 2016; Alhelou et al., 2018; Zamani et al., 2018; Alhelou et al., 2015; Njenda et al., 2018; Haes Alhelou et al., 2018).

The choice of modification techniques plays a major role in switching losses. So the active filters are used to: compensate the reactive power, reduce harmonics, and eliminate noise in the network.

Active Filters Disadvantages

The active filters have many disadvantages, the most important of which is the low reliability because they have a single working system where is any malfunction in one of its components makes it completely inefficient, in addition to the difficulty of controlling it.

Figure 17. Using active filters for harmonic mitigation

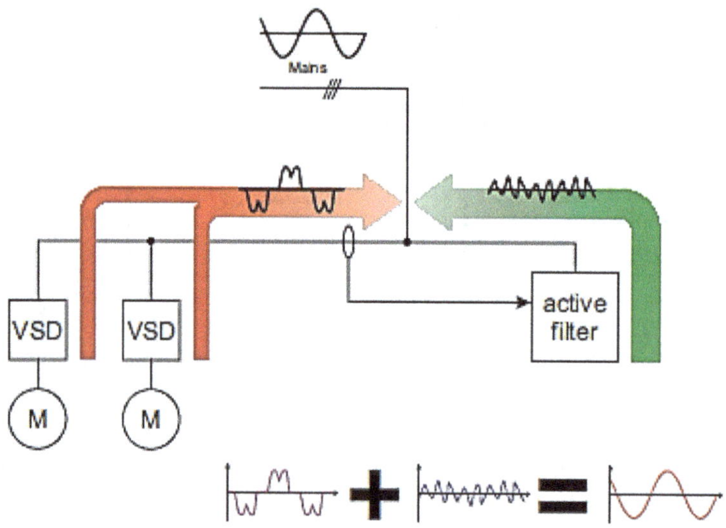

Several plans were made as development on the most effective filters [15]:

- Serial active filters.
- Shunt active filters.
- Multi levels active filters.

The performance of the industrial systems was improved by the reduction of ripples in dc-link by adding a capacitor between the rectifier and the inductor in the variable-speed drive ac power systems, but this capacitor was a subject to many faults that led to its collapse. The collapse of the dc-link is considered a very dangerous issue, especially if the engine is operating at a high speed. It affects the inverter, the motor connected to it, and the electrical grid.

CHOOSING THE BEST HARMONIC MITIGATIONS SOLUTIONS

Several studies have been carried out to compare the most common harmonic mitigations solutions. The analysis criteria are: simplicity, total harmonic distortion factor, efficiency and economic cost. The comparison results show that line reactors are the best solution for applications where high distortions must be filtered, but when harmonic dislocations are not the highest priority. Om the other hand, active filters are the best solution for harmonics mitigation in systems where several variable speed drives work together in parallel and are connected to the network at the PCC.

Figure 18. Conventional variable speed drive system in MATLAB Simulink

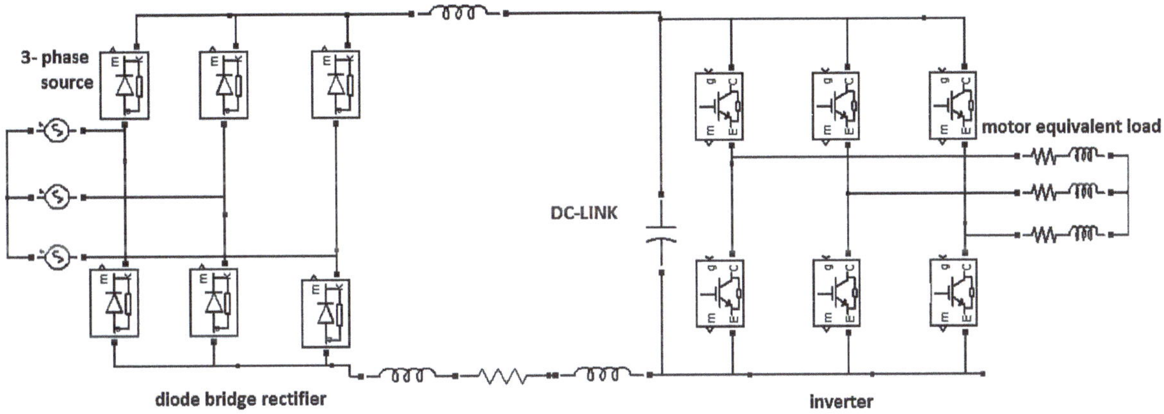

SIMULATION PART

Simulation of the Conventional VSDS Model

Figure 18 shows conventional variable speed drive systems /MATLAB SIMULINK model/ which consist of electrical grid, rectifier, dc-link, inverter, and motor.

The Input currents in conventional VSDs have total harmonic distortion factor Thd% of 36.43% and fundamental of 4.667 at 50 Hz.

VSDS Input current is distorted as shown in figure 20.

Rectifier output voltage has a ripple factor of 75v.

Figure 22 shows that the dc-link current has a large over shoot at the starting time and contains a lot of undesirable oscillation.

Figure 19. Total harmonic distortion factor for conventional VSDS

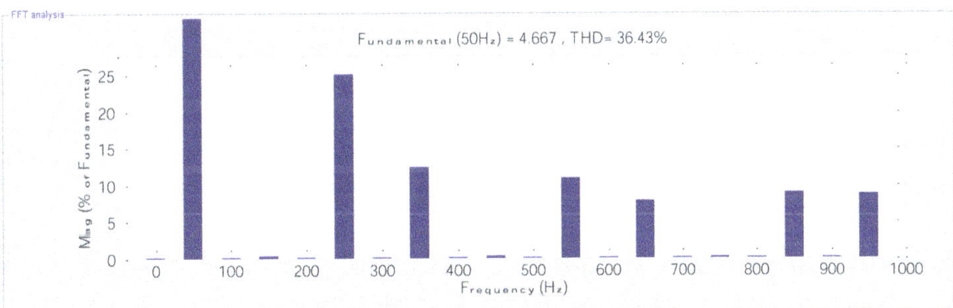

Figure 20. Source current in conventional VSDS

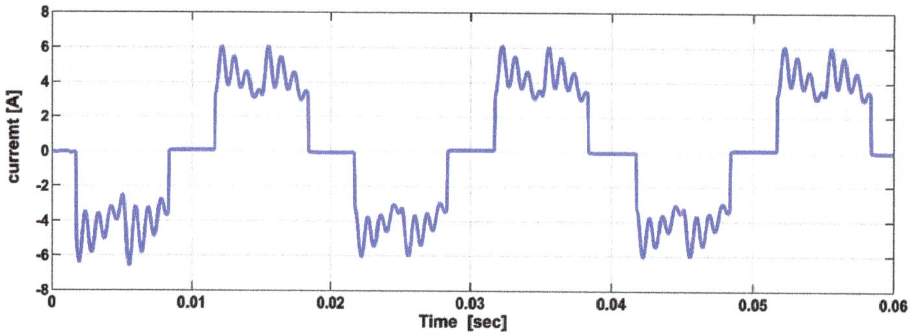

Figure 21. Rectifier output voltage in conventional VSDS

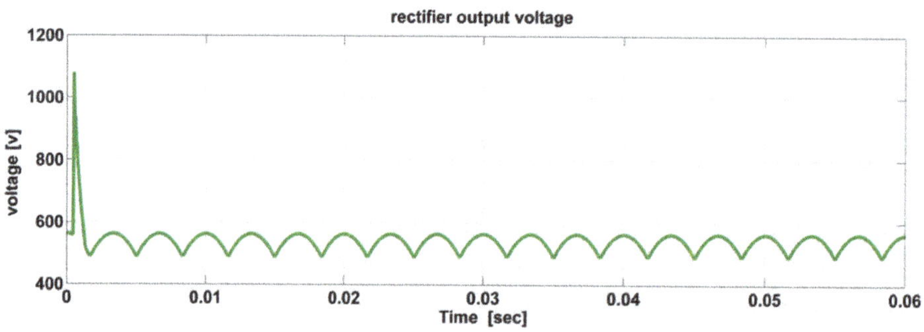

Figure 22. DC-LINK current in conventional VSDS

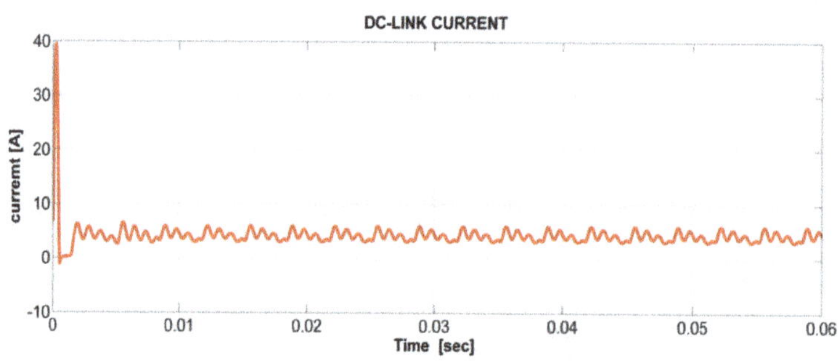

SIMULATION OF VSDS WITH HARMONIC MITIGATION

Figure 23 shows variable speed drive systems with one harmonic mitigation technique, which is adding four- quadrant chopper in dc-link (MATLAB Simulink model).

Figure 24 shows reducing both ripples and the over shoot in dc-link current at the starting time after adding 4- quadrant chopper in dc-link comparing with conventional VSD_S.

Source current is more smooth and less distorted comparing with previous case.

Rectifier output voltage has a ripple factor of 80V as shown in figure 26.

The total harmonic distortion factor Thd% of Input currents is reduced from 36.43% at conventional VSD_S at conventional VSD_S to 28.7% after using 4-quadrant chopper in dc-link as shown in figure 27.

Figure 23. VSDS with 4-quadrant chopper in DC-LINK

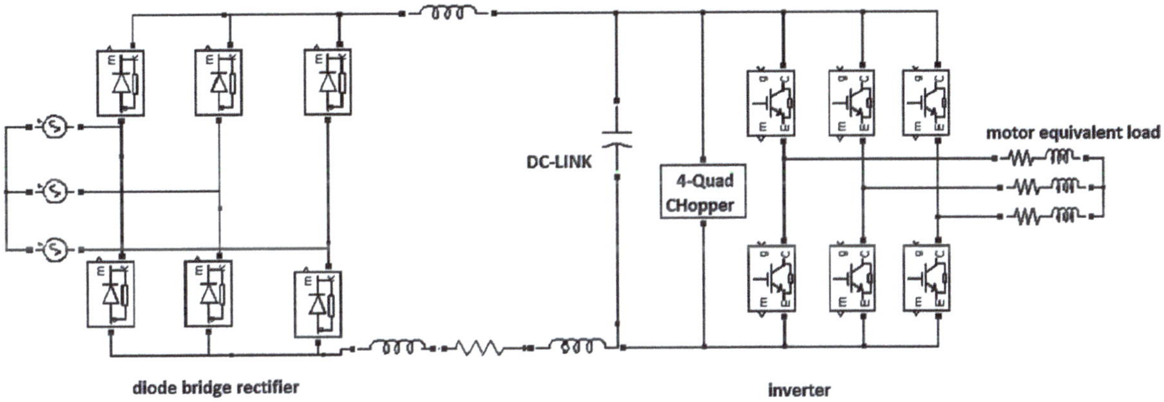

Figure 24. DC-LINK current after adding four quadrant chopper

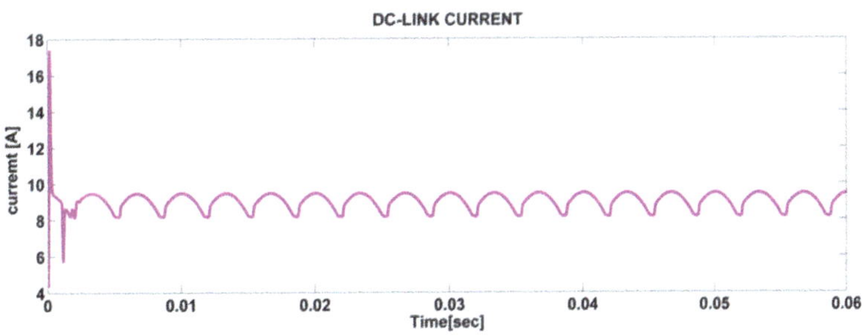

Figure 25. Source current after adding four quadrant chopper

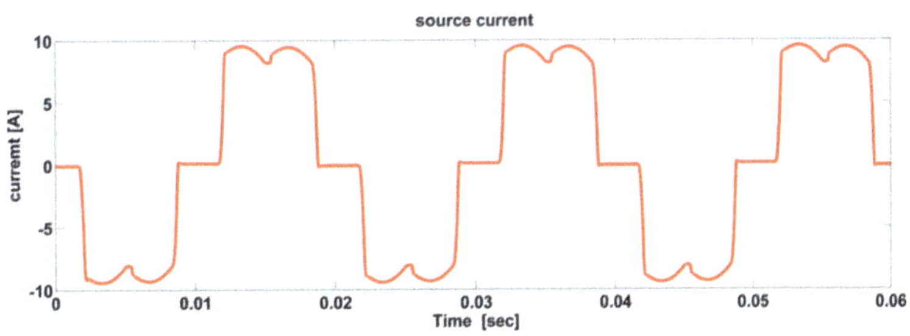

Figure 26. Rectifier output voltage after adding four quadrant chopper

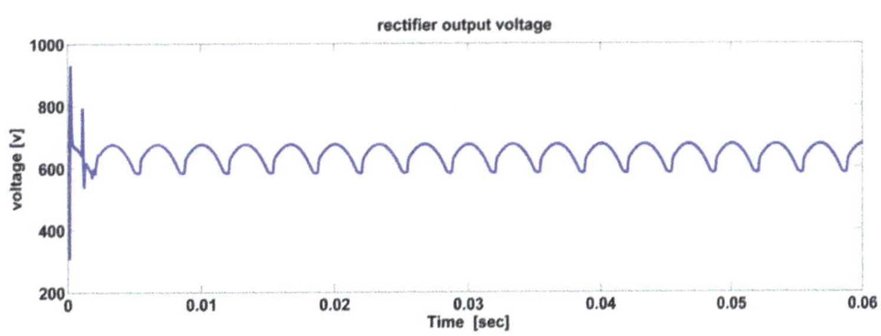

Figure 27. Total harmonic distortion factor after adding four quadrant chopper in dc- link

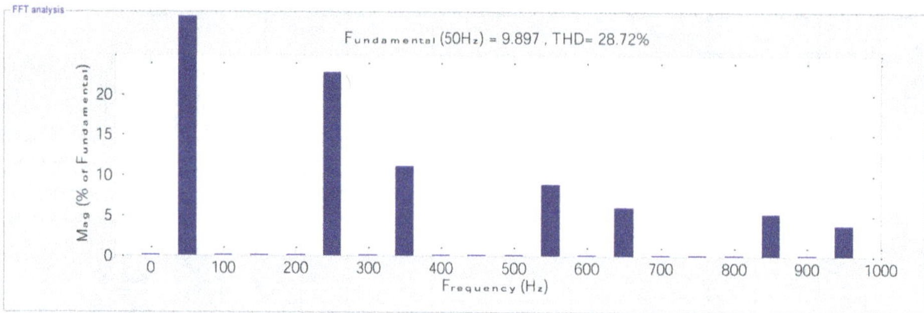

FUTURE RESEARCH DIRECTIONS

- Analysis of the operation of the driving system with new harmonics mitigations.
- Study the various faults that can affect the variable speed drive system.
- Expansion of the study of systems stability.

CONCLUSION

Several issues of power quality for variable speed drive systems VSD_S have been discussed in terms of cost and effectiveness, VSDS control loops (open-loop, close loop) and power quality in VSD_S. The solutions use to mitigate the harmonics in VSD_S were discussed in details. Finally, Simulation of the Conventional VSDS Model and VSD_S with one harmonic mitigation solutions in order clarify the useful of using this solution on power quality specifications.

REFERENCES

ABB. (2009). *Technical note td2 en reducing harmonics caused by variable speed drives*. ABB.

Alhelou, H., Hamedani-Golshan, M. E., Zamani, R., Heydarian-Forushani, E., & Siano, P. (2018). Challenges and Opportunities of Load Frequency Control in Conventional, Modern and Future Smart Power Systems: A Comprehensive Review. *Energies*, *11*(10), 2497. doi:10.3390/en11102497

Alhelou, H. H. (2018). *Fault Detection and Isolation in Power Systems Using Unknown Input Observer. In Advanced Condition Monitoring and Fault Diagnosis of Electric Machines* (p. 38). Hershey, PA: IGI Global.

Alhelou, H. H., Golshan, M., & Fini, M. (2018). Wind Driven Optimization Algorithm Application to Load Frequency Control in Interconnected Power Systems Considering GRC and GDB Nonlinearities. *Electric Power Components and Syst*.

Alhelou, H. H., & Golshan, M. E. H. (2016, May). Hierarchical plug-in EV control based on primary frequency response in interconnected smart grid. In *Electrical Engineering (ICEE), 2016 24th Iranian Conference on* (pp. 561-566). IEEE. 10.1109/IranianCEE.2016.7585585

Alhelou, H. H., Golshan, M. H., & Askari-Marnani, J. (2018). Robust sensor fault detection and isolation scheme for interconnected smart power systems in presence of RER and EVs using unknown input observer. *International Journal of Electrical Power & Energy Systems*, *99*, 682–694. doi:10.1016/j.ijepes.2018.02.013

Alhelou, H. H., Hamedani-Golshan, M. E., Heydarian-Forushani, E., Al-Sumaiti, A. S., & Siano, P. (2018, September). Decentralized Fractional Order Control Scheme for LFC of Deregulated Nonlinear Power Systems in Presence of EVs and RER. In *2018 International Conference on Smart Energy Systems and Technologies (SEST)* (pp. 1-6). IEEE. 10.1109/SEST.2018.8495858

Alhelou, H. S. H., Golshan, M. E. H., & Fini, M. H. (2015, December). Multi agent electric vehicle control based primary frequency support for future smart micro-grid. In *Smart Grid Conference (SGC)* (pp. 22-27). Academic Press.

Alshahrestani, A., Golshan, M. E. H., & Alhelou, H. H. (2018, November). WAMS Based Online Estimation of Total Inertia Constant and Damping Coefficient for Future Smart Grid Systems. In *Smart Grid Conference (SGC)* (pp. 1-5). Academic Press.

Davari, P., Zare, F., & Blaabjerg, F. (2015). A Smart Current Modulation Scheme for Harmonic Reduction in Three- Phase Motor Drive Applications. *Proceedings of the 2015 17th European Conference on Power Electronics and Applications (EPE'15 ECCEEurope)*.

Fini, M. H., Yousefi, G. R., & Alhelou, H. H. (2016). Comparative study on the performance of many-objective and single-objective optimisation algorithms in tuning load frequency controllers of multi-area power systems. *IET Generation, Transmission & Distribution*, *10*(12), 2915–2923. doi:10.1049/iet-gtd.2015.1334

Hadeed, A. S., Khaled, E. A., & Yasin, K. (2016). Effect of short circuited DC link capacitor of an AC–DC–AC inverter on the performance of induction motor. *Journal of King Saud University – Engineering and Science*, *28*, 199–26.

Hoadley, F.L. (2008). Curb the disturbance In Industry Applications. *IEEE Magazine, 14*.

Kazem, H. A. (2013). *Harmonic Mitigation Techniques Applied to Power Distribution Networks Advances in Power Electronics*.

Khine, Z. (2016). Modeling and Harmonic Analysis for Distribution Network In Myanmar. *International Journal for Technological Research in Engineering, 4*(2).

Kumar, D., & Zare, F. (2014). Analysis of harmonic mitigations using hybrid passive filters. *Proc. of PEMC*. 10.1109/EPEPEMC.2014.6980629

Latha, Y. (2012, March). Harmonics Mitigation of Industrial Motor Drives with Active Power Filters in Cement Plant-A Case Study. *International Journal of Power Electronics and Drive System*, *2*(1), 1–8.

Makdisie, C., Haidar, B., & Alhelou, H. H. (2018). An Optimal Photovoltaic Conversion System for Future Smart Grids. In *Handbook of Research on Power and Energy System Optimization* (pp. 601–657). IGI Global.

Meyer, J., Bollen, M., Amarus, H., Blanco, A., De Castro, A., Desmet, J., ... Yang, K. (2014). *Future Work on harmonics – Some Expert Opinions Part II – Supraharmonics, Standards and Measurements. Proc. IEEE.*

Nadweh, S., Hayek, G., Atieh, B., & Haes Alhelou, H. (2018). Using Four – Quadrant Chopper with Variable Speed Drive System Dc-Link to Improve the Quality of Supplied Power for Industrial Facilities. *Majlesi Journal of Electrical Engineering.*

Njenda, T. C., Golshan, M. E. H., & Alhelou, H. H. (2018, November). WAMS Based Intelligent Under Frequency Load Shedding Considering Online Disturbance Estimation. In *Smart Grid Conference (SGC)* (pp. 1-5). Academic Press.

Njenda, T. C., Golshan, M. E. H., & Alhelou, H. H. (2018, November). WAMS based Under Frequency Load Shedding Considering Minimum Frequency Prediction and Extrapolated Disturbance Magnitude. In *Smart Grid Conference (SGC)* (pp. 1-5). Academic Press.

Pacis, Martinez, Jr., & Tecson. (2010). Modelling and Simulation of Active Power Filters for Harmonic Compensation, Voltage Sags and Swells Mitigation and Power Factor Correction. *Proceedings of the World Congress on Engineering and Computer Science Conference, (WCECS), 2.*

Raneru, N. R. (2013). Harmonic Analysis of Small Scale Industrial Loads and Harmonic Mitigation Techniques in Industrial Distribution System. *International Journal of Engineering Research and Applications*, *3*(4), 1511–1540.

Robak, S., Wasilewski, J., Dawidowski, P., & Szewczyk, M. (2016). *Variable Speed Drive (VSD) – towards modern industry and electric power systems*. Academic Press.

Sekar, T. C., & Rabi, B. J. (2012). A Review and Study of Harmonic Mitigation Techniques. *Proc. International Conference on Trends in Electrical Engineering and Energy Management*. 10.1109/ICETEEEM.2012.6494450

Stephen, P. (2010). *Variable frequency drives and energy savings*. Technical report, Siemens.

Tim, W. S. (2011). *Application of line reactors in power electronics*. Technical report, Schaner EMV AG. Retrieved from www.schaffner.com

Zamani, R., Hamedani-Golshan, M. E., Haes Alhelou, H., Siano, P., & Pota, R, H. (. (2018). Islanding Detection of Synchronous Distributed Generator Based on the Active and Reactive Power Control Loops. *Energies*, *11*(10), 2819. doi:10.3390/en11102819

Zare, F. (2014). A novel harmonic elimination method for a three-phase diode rectifier with controlled DC link current. *Proc. of PEMC*, 985-989. 10.1109/EPEPEMC.2014.6980636

KEY TERMS AND DEFINITIONS

Crest Factor: Is a parameter of a waveform, such as alternating current or sound, showing the ratio of peak values to the effective value. In other words, crest factor indicates how extreme the peaks are in a waveform.

Power Factor (PF or cosφ): Is the ratio between the power that can be used in electric circuit (real power, P) and the power from the result of multiplication between the current and voltage circuit (apparent power, S).

Power Quality, Involves Voltage, Frequency, and Waveform: Good power quality can be defined as a steady supply voltage that stays within the prescribed range, steady AC without the proper power an electrical device (or load) may malfunction, fail prematurely or not operate at all.

Chapter 18
Operation and Control of Microgrid

Maheswari M.
Malla Reddy Engineering College (Autonomous), India

Gunasekharan S.
Malla Reddy Engineering College (Autonomous), India

ABSTRACT

The demand for electricity is increasing day by day due to technological advancements. According to the demand, the size of the grid is also increasing rapidly in the past decade. However, the traditional centralized power grid has many drawbacks such as high operating cost, customer satisfaction, less reliability, and security. Distribution generation has less pollution, high energy efficiency, and flexible installation than traditional generation. It also improves the performance of the grid in peak load and reliability of supply. The concept of micro-grid has been raised due to the advent of new technologies and development of the power electronics and modern control theory. Micro-grid is the significant part of the distribution network in the future of smart grid, which has advanced and flexible operation and control pattern, and integrates distributed clean energy.

INTRODUCTION

An electric grid is a network of synchronized power generators and consumers, which are connected by transmission and distribution lines. The demand for electricity is increasing day by day due to technological advancements. According to the demand, the size of the grid is also increasing rapidly in the past decade. However, the traditional centralized power grid has many drawbacks such as high operating cost, customer satisfaction, less reliability and security. The vulnerability of traditional grid exposed during the large block outs occurred in United States in the year 2003 and South snowstorm in the year 2008. To avoid all these problems, a new trend of generating power locally at distribution voltage levels by using renewable sources like solar photovoltaic cells, fuel cells, wind power, biogas, natural gas, combined heat and power (CHP) systems, natural gas and microturbines and also integration of these into

DOI: 10.4018/978-1-5225-8030-0.ch018

the utility distribution network. It is called as distributed generation (DG) and energy sources are called as distributed energy resources (DERs). The major research work carried about the issues in DG by the group of International Council on Large Electric Systems (CIGRE) and the International Conference and Exhibition on Electricity Distribution (CIRED) in late 1990s and discussed by Stevens, et al 2007. According to the researchers, the following are the common attributes accepted universally,

- It is not centrally planned by utility, not distributed centrally
- It is smaller than 50MW
- DG is connected to distribution system, which are typically of voltages 230/415V upto 145kV.

As like conventional sources, the electricity networks are also need transition from passive distribution networks with unidirectional electricity transportation to active distribution networks with bidirectional electricity transportation. The distribution networks without any DG units are passive and it becomes active when DG units are added which leads to bidirectional power flows in the networks. The active networks should have flexible and intelligent control with distributed intelligent systems. To obtain clean energy from DERs, active distribution networks should employ future network technologies leading to smartgrid or microgrid networks (Fini et al., 2016; Alhelou et al., 2018; Zamani et al., 2018; Alhelou et al., 2015; Njenda et al., 2018; Haes Alhelou et al., 2018).

Distribution generation provide less pollution and global warming due to renewable sources. It also provides the scope for co-generation, tri generation or CHP plants by using waste heat in Industries, domestic and commercial applications and thereby increases the overall efficiency of the plant as said by ANG, et al, in 2014. The location of DG sources are very near to its loads. Hence it reduces the transmission and distribution losses. It is possible to connect DERs to the utility distribution network or in the form of Microgrids because DG generating low voltages. DG also provides high energy efficiency and flexible installation than traditional generation. It also improves the performance of the grid in peak load and reliability of supply.

MICRO-GRID

The concept of Micro-grid is proposed by CERTS. A definition was given by CERTS as Micro-grid is a system which composed of load and the Micro-power, while it can provide electricity and heat. The internal power of Micro-grid is mainly responsible for the energy conversion by power electronic devices. Relatively, the external of Micro-grid manifested as a single grid controlled unit, and it meet customer requirements and supply security in power quality.

The concept of Micro-Grid has been raised due to the advent of new technologies and development of the power electronics and modern control theory. It provides technical knowledge about distributed generation and network and it's a booming as a new power system structure in the near future. Micro-grid is the significant part of the distribution network in the future of smart grid, which has advanced and flexible operation and control pattern, and integrates distributed clean energy. Micro - grid has distributed power, load, energy storage devices and control devices to form manageable unit at the customer side.

Microgrids are designed to supply electrical power with low voltage CHP supply networks to the consumers like housing boards, schools, colleges, commercial area, industrial site, municipal areas.

Microgrid is an active distribution network because it is associated with DG systems and different loads. The generators employed in Microgrid are usually renewable sources which are integrated together to generate power at distribution voltage. The microsources are embedded with power electronic interfaces to provide flexible operation and to maintain required energy output and power quality. The flexibility in control of microgrid would allow it as single controlled unit of main utility power system that meets the requirement of the local energy. The major difference between microgrid and conventional power plant are

- Microsources of small capacity compared to large generators in conventional power plants.
- Power generated can be directly fed into the utility distribution network since it is generated at distribution voltage level.
- Microgrids supplies the load with satisfactory voltage and frequency profile with negligible line losses because the microsources are installed nearby the loads.

The features of microgrid shows that it is very much suitable to supply power to remote and hilly areas where supply from national grid is either difficult due to the topology or frequently disrupted due to natural calamities. The advantages of microgrid are

- Ensures reliability and security of the power utility.
- Supplies uninterruptible power with less line losses.
- Reduces environmental pollution and global warming.

The microgrid has the following unique characteristics compared to the conventional grids,

- It consists of miniature power and load and it is very flexible to schedule.
- Micro – Grid constitutes diversity either it involves traditional power or renewable energy
- Based on the system operation, it can choose different mode of operation. So it is more reliable and controllable.
- It transfer generated power to the main electricity network in necessary conditions and at the same time it accepts power from the network also.
- It works independently to ensure local demand for electricity.

A TYPICAL STRUCTURE OF MICRO GRID

Micro-grid is combined with a micro source, the load, the energy storage device, a control device and so on, consisting of a controllable unit to provide heat and power systems. Its basic structure is as follows in the Figure 1.

In the structure, the microsources are associated with plug-and -play features along with PEIs to execute control, metering and protection functions as discussed by Cheng, et al, in 2009. The Microgrid consists of three radial feeders namely A,B and C which supplies electrical and heat loads. It consists of two CHP and two non-CHP microsources and storage devices. Microsources and storage devices are connected to feeder A and C through microsource controllers (MCs). It is assumed that feeder A and C has some priority loads and others are non priority loads. The loads connected to feeder B are all considered as non priority loads. In Figure 1, the microgrid is connected to the main medium voltage utility grid through the point of common coupling (PCC) circuit breaker CB4. CB4 is used to connect or disconnect the microgrid from the main grid as per the selected mode of operation. Circuit breakers CB1, CB2 and CB3 are used to connect or disconnect the feeders A,B and C respectively. In feeders A and C, the micro sources are placed apart from the microgrid bus to ensure less line loss, maintain voltage profile and use of waste heat. The control of power flow and voltage profile along radial feeders is quite complicated when several microsources are connected to a common feeder, but it is necessary to avail the plug-and-play feature of the microsources as conversed by De Brabandere in 2007.

There are two modes of operation for Micro-grid: Grid-connected and Autonomous Modes (Alshahrestani et al., 2018; Makdisie et al., 2018; Alhelou et al., 2018; Alhelou et al., 2016; Nadweh et al., 2018; Njenda et al., 2018). In the normal circumstances, when the Micro-grid runs in network operation, a large power grid can offer voltage and frequency, the inner of the Micro-grid works in voltage or current source state, under the mediation of the energy management system, it can adjust their output power. When the big power grid occur voltage dips, voltage swells, unbalance and harmonics problems or scheduled maintenance, the Micro-grid enters islanding mode, then the internal of micro-grid should adjust the voltage and frequency. Once the fault is cleared, Micro-grid gets into a large grid. This can be achieved by either (i) disconnecting the entire Microgrid by opening CB4 or (ii) disconnecting feeders A and C by opening CB1 and CB3. During the first option the microgrid will operate as an autonomous system with all microsources feeding all loads in feeder A, B and C. In case of option two, feeders A and C will supply only the priority loads while feeder B left to ride through the disturbance.

Figure 1. Typical Structure of Microgrid

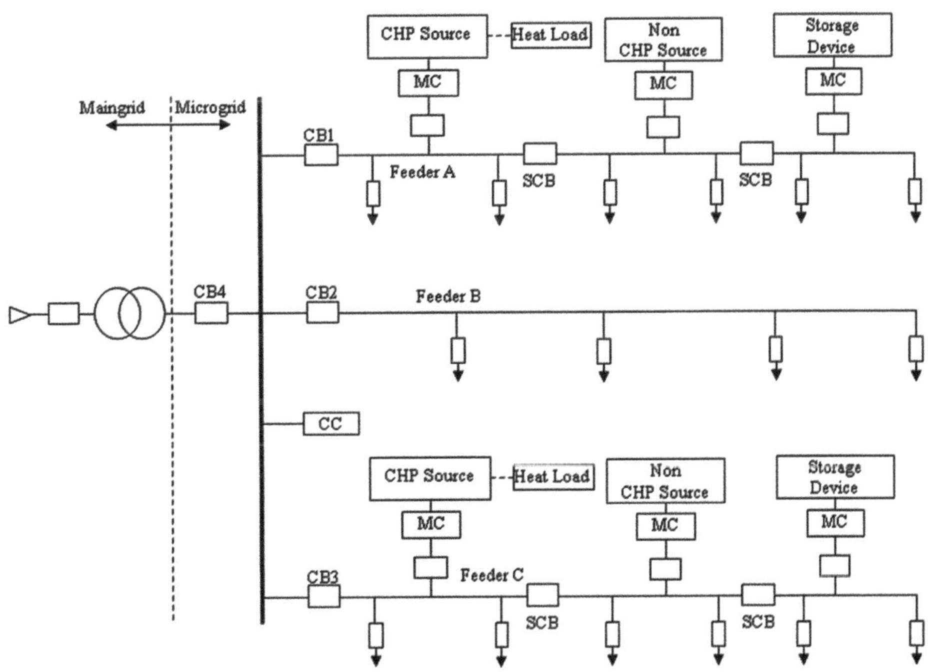

Operation of Micro-Grid

In addition to, Grid-connected, Autonomous Modes, there is the process of converting the transition between the two modes. Mode conversion process is not equaled to power restarting. Inverter continues to work to ensure a smooth transition voltage continuous within the grid by control methods and network topology grid. That is seamlessly switches to ensure that important load of internal power grid unaffected. The transition between the two modes is the conversion of the bridge and link. In transition process, power quality directly affects the operation of the network .It can ensure flexible operation to provide customized services important symbol of power in the Micro-grid system. Micro-grid runs in different operating modes. Micro-power using different control methods and network topology, which resulted Micro-grid in different operating modes and micro-power using different control methods and network topology, which resulted in conversion process diversity and uncertainty of micro power control methods as said by Jin et al, in the year 2012.

From figure 2, during operation, it is possible to switch between several modes with each other and according to the different control methods. Micro-grid will switch to the appropriate operating mode.

The operation of micro grid can be classified as Grid connected and autonomous. In case of grid connected, a large power grid can offer voltage and frequency, the inner of the Micro-grid works in voltage or current source state, under the mediation of the energy management system. It can adjust their output power. Whenever there is any voltage sag/swell or any interruption in the grid, then micro grid enters into autonomous mode or islanding mode. In island operation, the micro power work in the absence of voltage reference and it adjust automatically the voltage magnitude, frequency according to its capacity to supply for the important load as conersed by Katiraei in 2008.

Controllers in Microgrid

The operation and control of microgrid in different modes is controlled by local MCs and the central controllers (CC). The functions of MCs and CCs are listed below as said Lasseter in the year 2006,

Microsource Controller (MC)

The main function of MC is to control independently the power flow and load voltage profile of the microsource with respect to any disturbance and load changes. The term 'independent' denotes that without any communication from CC. MC also concentrates in economic scheduling, load management

Figure 2. Microgrid Operation Modes

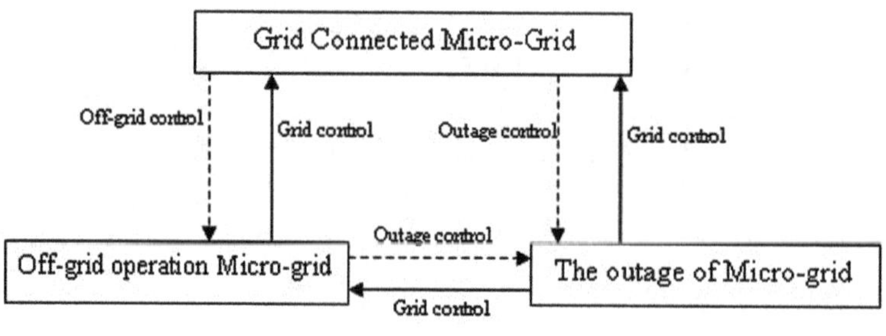

and demand side management through controlling storage devices. It also ensures that each microsource supplies energy to its loads during autonomous mode and returns back to the grid-connected mode automatically with the support of CC. The significant of MC lies in quick respond to locally monitored voltages and currents irrespective of the data from nearby MCs. This control feature of MCs facilitate microsources to act as plug-and-play devices and enables the addition of new microsources at any point of microgrid without disturbing the existing system. MCs will not interact with other MCs in the microgrid and at the same time, it will override the comments from the CC which will cause damages to the microsources as discussed by Lasseter in 2011.

Central Controller

The CC perform the overall control of microgrid operation and protection through the MCs. The main objectives of CCs are

- To maintain specified voltage and frequency at the load side with the help of power-frequency and voltage control
- To make sure energy optimization of microgrid

It afford the power dispatch and voltage set points for all MCs and also carry out protection coordination. It is designed to operate in automatic load as well as manual mode as and when required. There are two main functional modules of CCs such as Energy management Module (EMM) and Protection Co-ordination Module (PCM).

Energy Management Module (EMM)

This provides the set point for maximum and minimum values for active and reactive power, voltage and frequency to each MCs. The maximum and minimum values are set according to the operating condition of the microgrid. It is co-ordinated by the state-of -the-art communication and artificial intelligence techniques. The EMM should ensure that

- Microsources supply heat and electrical loads to customer satisfaction.
- Microgrids operate satisfactorily as per the operational a priori contracts with main grid.
- Microgrids satisfy its obligatory bindings in minimising system losses and emissions of greenhouse gases and particulates.
- Microsources operate at their highest possible efficiencies.

Protection Coordination Module (PCM)

It ensures the protection of microgrid. It gets activated whenever faults occur in main grid or microgrid. It also takes care about the change in fault current levels during changeover from grid-connected to stand-alone mode. It is achieved by the proper communication between the PCM and the MCs and main grid controllers. Whenever fault occurs in main grid, PCM immediately switches over the microgrid to stand-alone mode to supply power to the priority loads. In case of minor faults, the PCM allows the microgrid to operate in grid connected mode for some time and continues if any temporary fault is removed. If the grid fault is severe in nature and it affects the stability of the microgrid then PCM will disconnect the microgrid from main grid and continue to operate in stand-alone mode as discussed by Li, et al, in 2010.

When fault occurs in small portion of the microgrid then that portion gets isolated from the microgrid. It employs under-frequency and under voltage protection schemes to protect the sensitive loads.

The functions of the CC in the grid-connected mode are as follows:

- Monitoring system diagnostics by collecting information from the microsources and loads.
- Performing state estimation and security assessment evaluation, economic generation scheduling and active and reactive power control of the microsources and demand side management functions by using collected information.
- Ensuring synchronised operation with the main grid maintaining the power exchange at priori contract points.

The functions of the CC in the stand-alone mode are as follows:

- Performing active and reactive power control of the microsources in order to maintain stable voltage and frequency at load ends.
- Adopting load interruption/load shedding strategies using demand side management with storage device support for maintaining power balance and bus voltage.
- Initiating a local black start to ensure improved reliability and continuity of service.
- Switching over the Microgrid to grid-connected mode after main grid supply is restored without hampering the stability of either grid.

NEED FOR MICROGRID NETWORK MANAGEMENT

The requirement of network management in microgrid is executed by EMM with support from PCM and individual MCs. EMM performs the following duties

- Microsources generation control
- Domestic process control
- Water heating and chilling optimisation
- Energy storage control
- Maintaining power quality and industrial process
- Ancillary services

Utilities are significantly benefitted by microgrid through peak shave by providing many ancillary services. But initially EMM can execute only basic controls and in future it can be improved by using intelligent electronic devices (IED) and Ethernet communication protocols as discussed by Fan et al in 2013. For successful operation of microgrid, EMM should focus on the following.

Microgrid Structure

Microsource Generation Control

One of the most important goal of microsource generation control is to ensure maximum energy saving. Hence microsources are using various kinds of renewable and low-carbon technologies for electricity generation. These should have minimal fuel cost but should generate maximum capacity. Microsources should be operated most economically when it is running on natural gas, hydrogen, etc., For example, Combined heat and power (CHP) microsources generates heat along with electricity. For such sources, the heat load must also be balanced. So the selection of operating period, operating schedule and power level of microsource is quite difficult and complicated. It also depends on the cost of fuel, cost of electric power and heat. Hence the generation control CHP microsources should have the priority of heat or electric energy because the requirements of these loads might not coincide at all time. At the same time, microsources provides ancillary services like spinning reserve, peak shave, voltage regulation and mitigation of voltage sags. Theses ancillary services also provide sufficient economic benefits and generate revenue as discussed by Lopes, et al, in 2005.

Domestic Process Control

Domestic process control system controls and monitors the air conditioning and heating equipments of a building, like central heating systems, fans, refrigerators, dampers and pumps. The above said equipments are supplied by CHP micrsources of a microgrid. The heat loads of a building includes hot water supply systems, dryer systems and space heaters. Hecne during this situation microsources should optimise their heat generation for these services rather than other ancillary services. The set points in HVAC system to obtain maximum efficiency depends on weather condition, building architecture, occupancy level and the cost of the fuel. Hence EMM has to optimize the generation of heat and electricity production without disturbing the individual controllers of ACs, boilers, heat pumps, blowers etc.,

Morozumi, et al, in 2008 said that the heat generation by conventional boilers and furnace is more economical when natural gas available at low cost. At the same time, running CHP microsources is more economical when electricity cost is high during peak load hours. Hence EMM should collect the real time data from electricity and gas markets for fuel price comparison. Also EMM should plan the operating schedule according to short term weather conditions to obtain optimum efficiency. EMM should monitor the current system to operate it in correct sequence through sending proper comments to all equipments.

EMM should monitor the following control needs of a microgrid with the help of MCs

- Proper scheduling of heat recovery from the sources and their control
- Routing the waste heat to the heat exchanger for proper utilization.
- Monitoring the exhaust gas inlet temperature of heat exchanger to bypass the recovery system when the temperature is too low or too high.
- Monitoring the water temperature in the heat exchanger outlet to prevent overheating and to provide signal to the variable water flow control.

Energy Storage

The successful operation of microgrids depends on the proper operation and control of the storage devices during disturbances and contingencies. Thus EMM should control the operation of energy storage devices like battery, flywheels and ultracapacitors to ensure uninterrupted supply. These storage devices are used to compensate voltage sags in local buses or as backup power source during outages. It should not be used to compensate hourly energy prices or to level out peak loads. It can also support some of the microsources during low voltage transients, motor starts or other short term overloads in case of standalone operation as discussed by Nikos in 2007.

Most of the storage devices stores DC voltage and it should be converted into AC through inverter circuits. But flywheels generate AC voltage directly and able to feed the microgrid directly. Whenever disturbance occurs, these devices should respond rapidly through its local controllers not from EMM commands. storage devices like capacitors stores power at high density and discharges in a short period but flywheels stores power in low density and discharge it for a long period. Hence these devices should be connected with reciprocating engines to provide energy for extended periods.

Regulation and Load Shifting

The load profile of the microgrid has more short term peaks unlike conventional grids due to the nature of its loads. It happens due to domestic loads like water heater, heat pump and oven are running at the same time. This peak will be more in case of commercial customers like hotels and restaurants. Hence to flatten the load profile of the microgrid, the customers with different peak at different times are to be connected together and supplied by a microgrid. When the load aggregate is small in microgrid then its peak will be large and it requires more backup energy storage during autonomous operation in stand-alone mode. In these situations, the charging of energy storage devices can be considered as effective loading to average out the total loads and to make the load profile flatter. The important function of the EMM is smoothening out the electrical load profile through proper scheduling of loads including the charging of the energy storage devices and shifting load peaks. The EMM can also plan to charge the thermal energy storage systems with low cost fuels to flatten the load profile at peak loads.

Regulation means adjusting the generation to meet the load demand in every moment. In microgrid, to obtain short term regulation it is necessary to know about its long term regulation. Short term regulation helps the microgrid to supply the loads during its standalone mode. In such cases, EMM needs to go for load shedding of non-priority loads and supply electrical energy to priority loads otherwise microgrid has to undergo severe voltage dips and it may lead to system collapse. In general, the load profiles of large utilities are smoother than microgrids, hence it should be supported by energy storage devices with very fast power electronic controls. At the same time, EMM should follow the proper load co-ordination by applying load shifting at least for the larger load pockets and should implement functions to predict the larger loads and its influence on the system as discussed by Parikh, R., & Krishnan in 1994.

Control Strategies of Microsource Controller

The MCs are fitted with microsources and storage devices to execute smooth and flexible operation to meet the customer satisfaction and utility requirements. The function of MC highly depends on the power electronic interfaces provided in the microsources and storage devices. It may operate with or without intervention of the CCs. MCs ensures the following in microgrid as discussed by Piagi, P., & Lasseter, in 2006.

- Addition of new microsources without disturbing the existing system configuration
- Connect/disconnect microgrids to/from utilities in seamless fashion
- Control the active power and reactive power independently
- Handle the faults without loss of system stability
- Correct the system voltage sag and unbalances
- Meet the load dynamics of the power utility

The following are the major key points for MC design:

- MCs cannot interact with other microsources without the intervention of CCs. This enables MCs to respond quickly to the system changes without requiring any data from other MCs.
- MCs are designed to communicate and act as per the directions of CC and at the same time it can override the command if it is hazardous to its microsource.

Control Functions for Microsource Controller

The built-in control features of the MCs are as follows:

- Active and reactive power control
- Voltage control
- Storage requirement for fast load tracking
- Load sharing through P-f control.

MCs should also ensure the seamless transition of the microgrid from grid connected to stand alone mode and vice versa with minimum disturbance to both systems as discussed by Shamshiri, et al, in 2012.

Active and Reactive Power Control

The microsources may be DC sources like PV cell, batteries and fuel cells or it may be AC sources like wind turbines and microturbines. When the microsources are DC, then DC power is directly converted into AC and for AC microsources, AC power is converted into DC first then it is converted into AC through converters. Thus voltage source inverter forms the principal component of the power electronic converter.

Figure 3 shows the basic control structure of the MC which consists of microsource and power electronic converter. The voltage source inverter controls both voltage magnitude (V) as well as phase angle (δ_1) at bus 1. The microgrid supplies controlled power to the microgrid bus i.e., bus 2 through inductor X with values of voltage as E and δ_2. The voltage at bus 1 leads the voltage at bus 2 by an angle δ where

Figure 3. Basic control scheme of MC

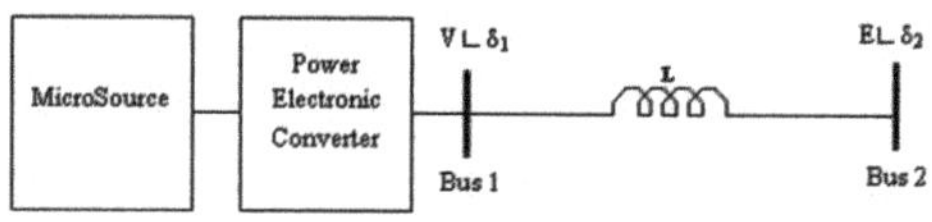

δ is δ_1- δ_2. It is well known that the real power (P) can be controlled by controlling the voltage phase angle (δ) whereas reactive power (Q) is controlled by controlling the voltage magnitude (V) as per the following equations,

$$\mathrm{P} = \frac{3\mathrm{VE}}{2\mathrm{X}} \sin{}' \tag{1}$$

$$Q = \frac{3VE}{2X}\left(V - E\sin{}'\right) \tag{2}$$

Voltage Control

Microgrids requires voltage control for overall stability and reliability operation apart from real and reactive power control. When a microgrid has many number of microsources then it will undergo reactive power oscillations due to improper voltage control. Voltage control functions of MC addresses the large reactive current circulation among microsources such as in large synchronous generators. In utilities, circulating current is restricted by large impedance between generators but in case of microgrids, it becomes prominent problem as it has small impedances between the sources as discussed by Shiki, et al, in 2007. In some situations, circulating currents are more than the rated current of microgrids even with small difference in their voltage set points. The circulating currents can be controlled by using voltage - reactive power (V-Q) droop controllers with droop characteristics as shown in figure 4.

Figure 4. Droop characteristics of V-Q controllers

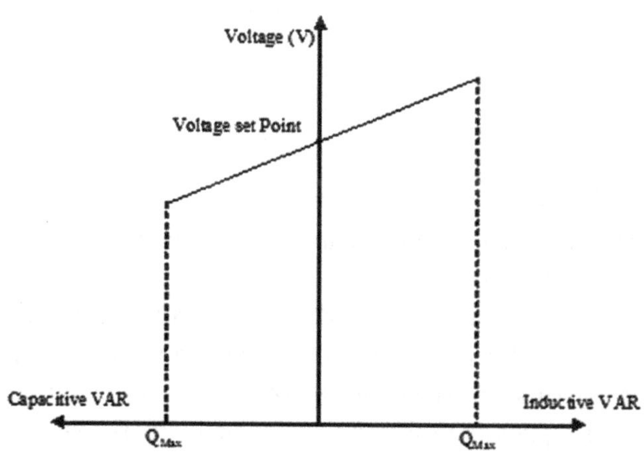

Figure 5. Active power vs. frequency droop characteristics

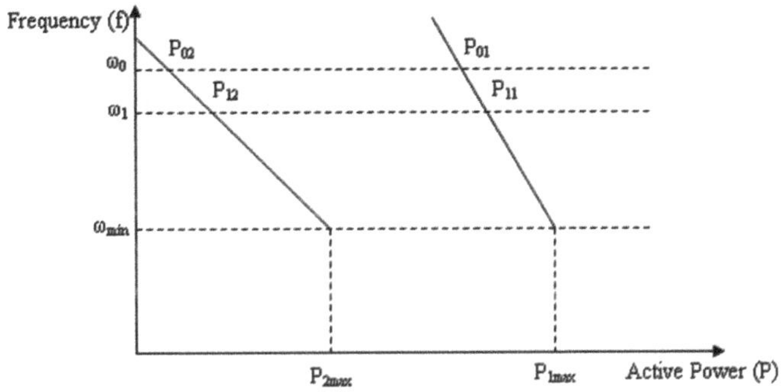

From the droop characteristics, it is clearly understand that the controller has to increase the local set point voltage when the reactive currents become prominently inductive and to decrease the set point when the current become capacitive. The reactive power limit (Q_{max}) is set by the reactive power (S) rating of the inverter and real power (P) output of the microsource as per the following equation,

$$Q_{max} = \sqrt{S^2 - P^2} \tag{3}$$

Load Sharing Through P-F Control

The microgrids operate in two modes such as grid - connected mode and stand - alone mode. The change over from grid - connected mode to stand - alone mode and vice versa is smoothly and automatically done by mocrogrid controllers. During transition to standalone mode, the MC of each microsource make use of local P-f control to change the operating point to achieve local power balance at the new load as discussed by Stevens, et al, in 2007. The MCs has to do this operation autonomously without waiting for any command from the CC or neighbouring MCs through proper load tracking. The drooping P-f characteristics used by the MCs for P-f control is shown in figure 5.

In grid connected mode, microgrid loads are supplied by both utility grid as well as microsources as per the customer needs. Whenever utility supply is interrupted due to any contingency, microgrid is immediately changed to stand alone mode. During change over, the voltage phase angles of the microsource will also change, which in turn drops the power output. Hence, local frequency also changes, in which case each microsource quickly picks up its share of load without any new power dispatch schedule from the CC. For example, two microsources are operating at a common minimum frequency with their maximum capacities P_{1max} and P_{2max}. In grid connected mode they are operating at a base frequency delivering power P_{01} and P_{02} respectively. When the change in load demand occurs, the microsources operate at different frequencies causing change in relative power angle and the frequency of operation drifts to a lower common value according to P-f characteristics as shown in figure 5.

Control Strategies of Central Controller

The CC has two basic control modules such as EMM and PCM. This section describes the control strategies used for EMM and PCM in detail.

Energy Manager Module (EMM)

The EMM has various control functions to obtain optimal operation of the microgrid. This section describes about the basic control operations incorporated in EMM for satisfactory operation of microgrid. To achieve finer and more sophisticated control, number of control functions can be increased in EMM.

Basic Microsource Control Functions

A simplified EMM provides only the active power and voltage set points for the MCs, while the basic microsource control is executed through the MCs only and they are

1. **Voltage Control:** The microgrid loads and power factors are controlled by changing microsource voltage magnitude and phase angle. Microsource voltage and power factor controls are executed by MCs not by EMM to avoid complexity in EMM control scheme. EMM only provides the voltage set points of certain critical buses of the microgrid. When distribution feeders are not fully loaded there may a possibility for voltage rise on the feeders. To avoid this voltage rise, the MCs continuously monitor the local voltage and provide feedback to EMM. Immediately EMM provides necessary voltage set points to the MCs to bring out the required voltage regulation as discussed by Su, et al., in 2010. This control strategy makes the microgrid to appear as an aggregate of microsources and loads as a controlled unit with unity power factor to the main utility.
2. **Power Factor Control:** Microsources do not have any built in power factor control unlike conventional synchronous generators. Power factors are depends upon the loads. But all MCs will have power factor control as a function of load tracking. MCs should incorporate the power factor control so that the control does not require any command from the EMM except the voltage set point. But power electronic interfaces of some microsources may have built-in power factor control to control the phase angle of supplied current and harmonics.
3. **Prime Mover Speed Control:** This control is especially applicable for microsources like wind turbine and microturbines. Whenever load variation occurs in the microgrid within its capacity, the microsource prime mover should change its speed to obtain power balance of new loading. So the constant speed prime mover will change the fuel input. This will affect the efficiency of the prime mover since it depends on fuel usage and speed. Hence prime mover speed control should ensure the optimum power generation for the microsource. For simple design of EMM, this control also executed by MCs.
4. **Frequency Regulation:** The frequency of the generated voltage depends on the speed of the synchronous generators in case of conventional power systems. But microgrid are capable of generating voltages at any frequency with the help of their power electronic interfacing devices. In grid connected mode, MCs will not bother about P-f control because frequency variation will be taken care by the utility. But in standalone mode, MCs should take care of this control to maintain constant system frequency under load variations. EMM will not interfere in the frequency control of the MCs in both mode of operation. But EMM continuously monitor the microgrid frequency. If any changes in frequency occurs and it cannot be restored by the MCs within a preset time, then EMM performs load shedding to achieve power balance in order to ensure microgrid stability.

Control of Heat Loads

In CHP microsources, heat loads are having more priority than electrical loads.

So CHP microsource electrical power output strategy is determined by the heat load requirement of the customers. To manage this priority requirement, EMM should have a priority setting parameter for heat loads in dispatching signals to MCs. However, i some industrial cogeneration systems, electric power has more priority than heat. Hence EMM should set the priority factor as per the relative importance of electrical and heat loads as discussed by Viral, et al, in 2012.

Energy Optimisation With Maximum Efficiency

To supply for a large power pool, the microgrids are interconnected. In such cases, the EMMs of neighbouring microgrids should incorporate a control for achieving energy optimization of the entire interconnected system. In order to obtain maximum efficiency of microsources, the EMMs should ensure running optimum number of microsources at their near rated capacities during light load conditions, instead running all of them at fractional capacities. This control is performed by EMM through prior knowledge of process condition, weather parameters, microsource generation schedule and fuel information like cost, availability and pattern of consumption.

Energy Storage Management

The EMMs controls the non priority loads by shedding them as and when required in interconnected microgrids. So they use these loads as a potential source of long term power reserve within the microgrid. The storage devices supply only the short term power reserve within the microgrid. For long term needs, the power reserve is extracted by shedding the non priority loads without any harmful effect on the microgrid as said by Wang, et al, in 2012.

Optional Control Functions for Intelligent EMM

Intelligent and sophisticated EMMs can be designed in future by incorporating more number of control functions like industrial process control and intelligent control of the microsources. They should have intelligent power electronic interfaces, wide information capacity, communication networks to communicate with neighboring equipment, control algorithms based on artificial intelligence (AI) techniques, remote monitoring and control facilities. In future, features like the Internet and GIS compatibility may also be included in the EMMs.

Intelligent EMMs may be used for (i) providing an overview of the process control systems, (ii) focusing on energy consumption and (iii) analysing energy saving opportunities depending on the time of day, process conditions and weather conditions. They may also be employed to monitor the power consumption for electrical loads in a premise and use this data for assessing thermal equipment efficiencies under full and part loads. From economics point of view, they may even be used to automatically optimise the utilisation of microsources and storage by using real-time market price signal for electricity and fuel as discussed by Zeng, et al, in 2013.

Protection Coordination Module (PCM)

The overall protection of microgrid is taken care by PCM. The protection principle in microgrid is different from conventional distribution networks though both are radial networks because,

- It contains both generators and loads resulting in bidirectional power flow through the protective devices
- Passive distribution network turn into active due to the presence of microsource.
- Short circuit capacity of microgrid changes during the transition from grid connected to stand alone mode and cause severe effect on conventional over current relays.

The main work of the PCM is its ability to distinguish between the protection requirements of the two operating modes and address the contingencies accordingly. Basic requirements for the protection of two modes are discussed in this section by considering basic microgrid configuration as shown in Figure 1. However the additional protection schemes can be incorporated depends upon the customer requirements.

Protection Scheme for Grid-Connected Mode

The PCM identifies and acts for five possible events in grid connected mode. They are (1) normal condition (2) microgrid feeder fault (3) utility fault (4) microgrid bus fault and (5) re-synchronisation. PCM consider the response time of individual microsources as well as that of the Point of common coupling (PCC) circuit breaker CB4.

1. **Normal Condition:** The microgrid remains connected to the utility through the PCC circuit breaker CB4 under normal operating condition. At the same time, the feeders A, B and C are connected to the microgrid bus through circuit breakers CB1, CB2 and CB3 respectively. Under this operation all the breakers will be in closed condition. The loads are supplied by both the microsources and the utility.
2. **Microgrid Feeder Fault:** In case of passive distribution networks, the fault power flows in one direction only i.e., from source to the fault point. Thus feeder faults are cleared by just opening the feeder breaker. Also determination of fault is done by magnitude of fault current not by its direction. But in case of microgrid, feeders have bidirectional power flow. Hence fault power flows from either side of the fault point. If such faults are not cleared quickly, then all microsources are disconnected from that feeder by their own MCs which leads to loss of generation. To avoid this situation, feeders A and C are sectionalized into zones by sectionalized breakers. The breaker have overcurrent relays to detect the faulty zone and clear the fault. If the faulty zone has any microsource, then it will be disconnected from that zone by its own MCs but it continuous to supply its local loads. To perform this operation, the PCM grades the relay settings of all relays such that the fault zone is isolated before all microsources are disconnected by their MCs or entire microgrid is disconnected from the utility. This operation ensures minimum loss of generation and microgrid stability. If fault occurs in feeders A or C, also all microsources are connected at one side of the fault then all microsources are disconnected from the feeder by its own MCs and faulty feeder is disconnected from the microgrid bus by opening its circuit breaker. If fault occurs in feeder B, then

it can be disconnected simply by opening circuit breaker CB2 because feeder B does not have any microsource as discussed by Zongxiang, et al, in 2007.

3. **Utility Fault:** When fault occurs in utility, then the microgrid is disconnected from the utility by opening circuit breaker CB4. It is done by CB4 relay which continuously monitors the current magnitude and direction on each phase, if these values exceeds the preset value then it will send a trip command to CB4. PCM offers this relay setting operation in such a way that the isolation is done without any interruption to priority loads. This scheme also ensures that the microsources are not tripped before the microgrid separates itself from the utility.
4. **Microgrid Bus Fault:** If the fault occurs on the microgrid bus, then it is disconnected from the utility by opening CB4. Also the feeders A and C are disconnected from the bus by opening CB1 and CB3 respectively. If fault occurs within the microgrid, the PCM grades the CB4 relay to co-ordinate with the upstream protection in the utility. CB4 relay is also graded with respect to the protective devices for the microsources to minimize loss of generation, supply interruption and tripping.
5. **Re-synchronisation:** When faults are cleared and utility restores its normal operation, then the responsibility of the PCM is to synchronize and reconnect the microgrid to the utility through synchronism check schemes. This is achieved only after when utility becomes stable and took back its all disconnected loads. But this operation may take time from several seconds to minutes depends on the nature of the feeder and loads. The PCM measures voltage magnitudes, phase angles, frequency and phase sequence on both sides of the breaker CB4 to connect all microsources to utility by synchronization. The PCM have both options for automatic and manual re-synchronization as per the requirement.

Protection Scheme for Stand-Alone Mode

When the microgrid operates in standalone mode, the short circuit level at the microgrid is low compared to grid connected system. Because, microsources with power electronic converters provide only 200% of the load current to the fault. It is very low fault current compared to the grid connected microgrid. Low fault currents will not be picked up by the over current relays used in the conventional protection systems. Hence it will give a huge impact on the fault detection capability of the protective relays of the microgrid. Sometimes standard relays may took more time to pick up or may not respond at all. Thus, alternate fault detection schemes like impedance protection, differential current/voltage relaying, zero sequence current/voltage relaying, directional over current/earth-fault schemes may be adopted for protection of stand-alone Microgrids.

Advantages of Microgrid

The development of microgrid is very much useful to the nation because of the following advantages,

Environmental Issues

The microgrid are having lesser environmental impact compared to the large conventional power stations. Also there are no gaseous and particulate emissions due to the close control of the combustion process. The proximity of microsources to the costomer will improve the awarness about energy saving.

Operation and Investment Issues

The reduction in physical and electrical distance between the microsource and loads will improve the reactive power support of the whole system through enhancement of voltage profile. Also it reduces the T&D losses by 3% and feeder congestion. Hence it reduces the investment cost for the expansion of transmission and generation systems.

Power Quality

Power quality has improved due to decentralized sources, supply meets the demand, reduction in large transmission lines and reduction in downtime operations of microgrid.

Cost Saving

In case of CHP mode of operation, there is no need of substantial infrastructure to heat transmission because heat sources are located nearby the loads. The maximum efficiency of CHP operation is 80% whereas 40% in conventional systems. Due to the plug-and-play mode operation of microsources, there is no need of import/export power to/from the main grid over longer feeders. This will in turn reduces the transmission and distribution cost.

Disadvantages of Microgrid

High Costs of Distributed Energy Sources

The cost for installing microgrids are very high. It can be reduced by providing some form of subsidies from Government bodies for at least transitory period to meet up the environmental goals. It helps us to achieve the global targets of enhance renewable green power generation to 20% by 2020 and reduce carbon emission to 50% by 2050.

Technical Difficulties

These problems are due to the lack of technical experience in controlling the operation of large number of plug-and-play microsources. The research works to be improved in the areas like management, protection, control aspects, sizing and placement of microgrids. at the same time, lack of proper communication infrastructure in rural areas is a major drawback in microgrids.

Absence of Standards

The standards are not available for addressing operation and protection issues since microgrid is a new area. The standard guidelines need to be laid down for different types of sources, standards and protocols for integration of microsources, safety and protection.

Administrative and Legal Barriers

There is no legislation and regulations are available in most of the countries to regulate the operation of microgrids.

Market Monopoly

The microgrids are allowed to supply energy autonomously to priority loads during any main grid contingency. But the problem arises in energy pricing for the above said duration. Microgrids are exploiting monopoly and retail the energy at a very high price. Hence, suitable market infrastructure needs to be designed and implemented for sustainable development of microgrids.

Operational Issues in Microgrid

- Active and reactive power balance has to be maintained within the microgrid to maintain power quality on a short term basis.
- Microgrid should operate as standalone where utility supply is not available.
- The proper planning of generation, supply and storage of energy has to done with respect to load demand and long term energy balance.
- Microgrid MCs and CCs should incorporate SCADA based metering, control and protection functions.
- Generation scheduling, economic dispatch and optimal power flow has to be carried out to ensure economic operation.
- System security to be maintained through contingency analysis and emergency analysis like demand side management, load shedding or islanding.
- The mismatch between generation and load should be improved through load forecasting and demand side management.
- Carrier communication and IEC 61850 communication infrastructure are to be employed for overall energy management, protection and control.

CONCLUSION

The microgrid is an alternative approach of distribution level energy supply and it can be integrated to LV and MV distribution networks. This helps to generate electricity and heat simultaneously to locally supply the electrical and heat loads of the customers. It is an autonomous group of controllable plug-and-play microsources and energy storage devices that can be optimally placed for the benefit of the customers. Microsources are basically renewable/nonconventional resources driven by a set of controllable prime movers. Power electronic based MC controllers are used to maintain energy balance and power quality in the system by executing necessary controls. MCs are capable of working without the intervention of CC. They facilitate connection, disconnection and reconnection of devices without reconfiguring the existing or new equipments. Microgrid helps the main grid by decreasing the congestion, address the immediate needs for generation augmentation, increasing system stability and maintaining the quality of supply to customer satisfaction. However, a large number of technical and regulatory issues need to be addressed carefully, requiring a considerable amount of research and government intervention across the world.

REFERENCES

Alhelou, H., Hamedani-Golshan, M. E., Zamani, R., Heydarian-Forushani, E., & Siano, P. (2018). Challenges and Opportunities of Load Frequency Control in Conventional, Modern and Future Smart Power Systems: A Comprehensive Review. *Energies*, *11*(10), 2497. doi:10.3390/en11102497

Alhelou, H. H. (2018). Fault Detection and Isolation in Power Systems Using Unknown Input Observer. In Advanced Condition Monitoring and Fault Diagnosis of Electric Machines (p. 38). Hershey, PA: IGI Global.

Alhelou, H. H., Golshan, M., & Fini, M. (2018). Wind Driven Optimization Algorithm Application to Load Frequency Control in Interconnected Power Systems Considering GRC and GDB Nonlinearities. *Electric Power Components and Syst.*

Alhelou, H. H., & Golshan, M. E. H. (2016, May). Hierarchical plug-in EV control based on primary frequency response in interconnected smart grid. In *Electrical Engineering (ICEE), 2016 24th Iranian Conference on* (pp. 561-566). IEEE. 10.1109/IranianCEE.2016.7585585

Alhelou, H. H., Golshan, M. H., & Askari-Marnani, J. (2018). Robust sensor fault detection and isolation scheme for interconnected smart power systems in presence of RER and EVs using unknown input observer. *International Journal of Electrical Power & Energy Systems*, *99*, 682–694. doi:10.1016/j.ijepes.2018.02.013

Alhelou, H. H., Hamedani-Golshan, M. E., Heydarian-Forushani, E., Al-Sumaiti, A. S., & Siano, P. (2018, September). Decentralized Fractional Order Control Scheme for LFC of Deregulated Nonlinear Power Systems in Presence of EVs and RER. In *2018 International Conference on Smart Energy Systems and Technologies (SEST)* (pp. 1-6). IEEE. 10.1109/SEST.2018.8495858

Alhelou, H. S. H., Golshan, M. E. H., & Fini, M. H. (2015, December). Multi agent electric vehicle control based primary frequency support for future smart micro-grid. In *Smart Grid Conference (SGC)* (pp. 22-27). Academic Press.

Alshahrestani, A., Golshan, M. E. H., & Alhelou, H. H. (2018, November). WAMS Based Online Estimation of Total Inertia Constant and Damping Coefficient for Future Smart Grid Systems. In *Smart Grid Conference (SGC)* (pp. 1-5). Academic Press.

Ang, X., Su, J., Zhipeng, L., Liu, H., & Li, R. (2014). Overview on Micro-grid Technology. China Electric Power Research Institute.

Cheng, P. T., Chen, C. A., Lee, T. L., & Kuo, S. Y. (2009). A cooperative imbalance compensation method for distributed-generation interface converters. *IEEE Transactions on Industry Applications*, *45*(2), 805–815. doi:10.1109/TIA.2009.2013601

De Brabandere, K., Vanthournout, K., Driesen, J., Deconinck, G., & Belmans, R. (2007, June). Control of microgrids. In *Power Engineering Society General Meeting, 2007. IEEE* (pp. 1-7). IEEE.

Fan, M., Zhang, Z., Su, A., & Su, J. (2013). Enabling technologies for active distribution systems. *Zhongguo Dianji Gongcheng Xuebao, 33*(22), 12–18.

Fini, M. H., Yousefi, G. R., & Alhelou, H. H. (2016). Comparative study on the performance of many-objective and single-objective optimisation algorithms in tuning load frequency controllers of multi-area power systems. *IET Generation, Transmission & Distribution, 10*(12), 2915–2923. doi:10.1049/iet-gtd.2015.1334

Jin, C., Gao, M., Lv, X., & Chen, M. (2012, September). A seamless transfer strategy of islanded and grid-connected mode switching for microgrid based on droop control. In Energy Conversion Congress and Exposition (ECCE), 2012 IEEE (pp. 969-973). IEEE. doi:10.1109/ECCE.2012.6342714

Katiraei, F., Iravani, R., Hatziargyriou, N., & Dimeas, A. (2008). Microgrids management. *IEEE Power & Energy Magazine, 6*(3), 54–65. doi:10.1109/MPE.2008.918702

Lasseter, R. H. (2006). *Control and design of microgrid components.* PSERC Publication.

Lasseter, R. H., Eto, J. H., Schenkman, B., Stevens, J., Vollkommer, H., Klapp, D., & Roy, J. (2011). CERTS microgrid laboratory test bed. *IEEE Transactions on Power Delivery, 26*(1), 325–332. doi:10.1109/TPWRD.2010.2051819

Li, F., Qiao, W., Sun, H., Wan, H., Wang, J., Xia, Y., & Zhang, P. (2010). Smart transmission grid: Vision and framework. *IEEE Transactions on Smart Grid, 1*(2), 168–177. doi:10.1109/TSG.2010.2053726

Lopes, J. P., Moreira, C. L., Madureira, A. G., Resende, F. O., Wu, X. A. W. X., Jayawarna, N. A. J. N., . . . Hatziargyriou, N. A. H. N. (2005, November). Control strategies for microgrids emergency operation. In *Future Power Systems, 2005 International Conference on* (pp. 6-pp). IEEE. 10.1109/FPS.2005.204226

Makdisie, C., Haidar, B., & Alhelou, H. H. (2018). An Optimal Photovoltaic Conversion System for Future Smart Grids. In *Handbook of Research on Power and Energy System Optimization* (pp. 601–657). IGI Global. doi:10.4018/978-1-5225-3935-3.ch018

Morozumi, S., Kikuchi, S., Chiba, Y., Kishida, J., Uesaka, S., & Arashiro, Y. (2008, July). Distribution technology development and demonstration projects in Japan. In Power and Energy Society General Meeting-Conversion and Delivery of Electrical Energy in the 21st Century (pp. 1-7). IEEE.

Nadweh, S., Hayek, G., Atieh, B., & Haes Alhelou, H. (2018). Using Four – Quadrant Chopper with Variable Speed Drive System Dc-Link to Improve the Quality of Supplied Power for Industrial Facilities. Majlesi Journal of Electrical Engineering.

Nikos, H. (2007). Microgrids: an overview of ongoing research, development, and demonstration projects. *IEEE Power & Energy, 5*(4).

Njenda, T. C., Golshan, M. E. H., & Alhelou, H. H. (2018, November). WAMS Based Intelligent Under Frequency Load Shedding Considering Online Disturbance Estimation. In *Smart Grid Conference (SGC)* (pp. 1-5). Academic Press.

Njenda, T. C., Golshan, M. E. H., & Alhelou, H. H. (2018, November). WAMS based Under Frequency Load Shedding Considering Minimum Frequency Prediction and Extrapolated Disturbance Magnitude. In *Smart Grid Conference (SGC)* (pp. 1-5). Academic Press.

Parikh, R., & Krishnan, R. (1994, September). Modeling, simulation and analysis of an uninterruptible power supply. In *Industrial Electronics, Control and Instrumentation, 1994. IECON'94., 20th International Conference on* (Vol. 1, pp. 485-490). IEEE.

Piagi, P., & Lasseter, R. H. (2006, June). Autonomous control of microgrids. In *Power Engineering Society General Meeting* (pp. 8-pp). IEEE.

Shamshiri, M., Gan, C. K., & Tan, C. W. (2012, June). A review of recent development in smart grid and micro-grid laboratories. *In Power Engineering and Optimization Conference (PEDCO) Melaka, Malaysia, 2012 IEEE International* (pp. 367-372). IEEE. 10.1109/PEOCO.2012.6230891

Shiki, A., Yokoyama, A., Baba, J., Takano, T., Gouda, T., & Izui, Y. (2007). Autonomous decentralized control of supply and demand by inverter based distributed generations in isolated microgrid. *IEEJ Transactions on Power and Energy*, *127*(1), 95–103. doi:10.1541/ieejpes.127.95

Stevens, J., Vollkommer, H., & Klapp, D. (2007, June). CERTS microgrid system tests. In 2007 IEEE Power Engineering Society General Meeting (pp. 1-4). IEEE. doi:10.1109/PES.2007.385793

Su, L., Zhang, J. H., Wang, L., Miao, W. S., & Wu, Z. P. (2010). Study on some key problems and technique related to microgrid. *Power System Protection and Control*, *38*(19), 235–239.

Viral, R., & Khatod, D. K. (2012). Optimal planning of distributed generation systems in distribution system*: A review. Renewable & Sustainable Energy Reviews*, *16*(7), 5146–5165. doi:10.1016/j.rser.2012.05.020

Wang, C., Gao, F., Li, P., Huang, B., Ding, C., & Yu, H. (2012, September). Control strategy research on low voltage microgrid. *Zhongguo Dianji Gongcheng Xuebao*, *32*(25), 2–8.

Zamani, R., Hamedani-Golshan, M. E., Haes Alhelou, H., Siano, P., & Pota, H. (2018). Islanding Detection of Synchronous Distributed Generator Based on the Active and Reactive Power Control Loops. *Energies*, *11*(10), 2819. doi:10.3390/en11102819

Zeng, Z., Zhao, R., & Yang, H. (2013). Micro-sources design of an intelligent building integrated with micro-grid. *Energy and Building*, *57*, 261–267. doi:10.1016/j.enbuild.2012.11.018

Zongxiang, L., Caixia, W., Yong, M., Shuangxi, Z., & Jinxiang, L. (2007). Overview on microgrid research. *Dianli Xitong Zidonghua*, *31*(19), 100–107.

KEY TERMS AND DEFINITIONS

Demand Response: In demand response, consumers play a significant role in the operation of the electric grid by reducing or shifting their electricity usage during peak periods.

Islanding: Islanding is a condition in which distribution sector becomes electrically isolated from the main supply power, but it gets continuously energized by DG connected to it.

Renewable Energy: Represents the energy produces from an energy source that is naturally replenished after use. Some examples are: sunlight, wind, rain, waves, tides, geothermal, etc.

Smart Grid: Power grid with smart equipment used for bidirectional communication and control. It refers to the electric power network of the future; the one that will be equipped with intelligent devices and automated control strategies.

Chapter 19
Operation of Microgrid and Control Strategies:
Microgrid Structure and Its Control Schemes

Suma Deepthi Veeraganti
Malla Reddy Engineering College (Autonomous), India

Ramchandra Nittala
St. Martins Engineering College, India

ABSTRACT

Microgrids are the most innovative area in the electric power industry today. A microgrid can operate in grid-connected or islanded mode. In islanded mode, microgrids can provide electricity to the rural areas with lower cost and minimum power losses. Several methods have been proposed in the literature for the successful operation of a microgrid. This chapter presents an overview of the major challenges and their possible solutions for planning, operation, and control of islanded operation of a microgrid. Microgrids are the most innovative area in the electric power industry today. Moreover, microgrids provide local voltage and frequency regulation support and improve reliability and power capacity of the grid. The most popular among the control strategies based on droop characteristics, in addition a central controller is described within a hierarchical control scheme to optimize the operation of the microgrid during interconnected operation. Microgrid control methods, including PQ control, droop control, voltage/frequency control, and current control methods are formulated.

INTRODUCTION

Microgrids are the most innovative area in the electric power industry today. Future microgrids could exist as energy-balanced cells within existing power distribution grids or stand-alone power networks within small communities. The significance of microgrids is growing rapidly. Microgrids have a huge potential in boosting the sustainable growth. A microgrid can operate in grid-connected or islanded mode.

DOI: 10.4018/978-1-5225-8030-0.ch019

In islanded mode, micro grids can provide electricity to the rural areas with lower cost and minimum power losses. In grid-connected mode, microgrids can help in supporting the main grid in many ways with voltage control, frequency control, and can provide more flexibility, control, and reliability. However, successful operation of a micro grid requires proper planning and there are major challenges regarding the operation, control, and protection of micro grids that need to be tackled for successful deployment of microgrids. Depending on the mode of operation either grid connected mode or islanded mode, necessary control strategies and protection schemes are required. Several methods have been proposed in the literature for the successful operation of a microgrid. This chapter presents an overview of the major challenges and their possible solutions for planning, operation, and control of islanded operation of a micro grid. Micro grids are the most innovative area in the electric power industry today. Future microgrids could exist as energy-balanced cells within existing power distribution grids or stand-alone power networks within small communities

During the last decades, the deployment of distributed generation (DG) resources has been growing steadily. In this process, the power distribution utilities have been one of the industry's most concerned stakeholders. The main reason is that DGs are connected primarily within their distribution networks, mainly at Medium Voltage (MV) and High Voltage (HV) level, which have been designed under the paradigm that consumer loads are passive and power flows only from the substations to the consumers and not in the opposite direction. For this reason, many studies on the interconnection of DG within distribution networks have been carried out, ranging from control and protection to voltage stability and power quality among many others. However, different micro-generation technologies, such as micro-turbines (MT), photovoltaic (PV), fuel cells (FC) and wind turbines (WT) with a rated power ranging up to a hundred kWs can be directly connected to the Low Voltage (LV) networks. In this context, micro-generation units, typically located at users' sites, have emerged as a promising option to meet growing customer needs for electric power with an emphasis on reliability and power quality and contribution to different economic, environmental and technical benefits. Furthermore, it has to be recognized that with increased levels of micro-generation penetration, the LV distribution network can no longer be considered as a passive appendage to the transmission network. On the contrary, the impact of micro-generation at LV levels on power balance and grid frequency may become much more significant. Therefore, a control and management architecture is required in order to facilitate full integration of micro-generation and active load management into the system. One promising way to realize the emerging potential of micro-generation is to take a system approach which views generation and associated loads as a subsystem or a Microgrid (Fini et al., 2016; Alhelou et al., 2018; Zamani et al., 2018; Alhelou et al., 2015; Njenda et al., 2018; Haes Alhelou et al., 2018).

WHAT IS A MICROGRID?

Microgrids comprise LV distribution systems with distributed energy resources (microturbines, fuel cells, PV, etc.) together with storage devices (flywheels, energy capacitors and batteries). Such systems can be operated in a non-autonomous way, if interconnected to the grid, or in an autonomous way, if disconnected from the main grid. The operation of micro-sources in the network can provide distinct benefits to the overall system performance, if managed and coordinated efficiently. There are three major messages delivered from this definition, namely as:

1. Microgrid is an integration platform for supply-side (micro-generators) and demand-side resources (storage units and (controllable) loads) located in a local distribution grid. ¾ In the Microgrid concept, there is a focus on local supply of electricity to nearby loads, thus aggregator models that disregard physical locations of generators and loads (such as virtual power plants with cross-regional setups) are not Microgrids. ¾ A Microgrid should contain supply-side resources, and likely demand-side resources are also included in the same time. It is typically located in LV level with total installed micro-generator capacity below MW range, but there can be exceptions.
2. A Microgrid should be capable of handling both normal state (grid-connected) and emergency state (islanded) operations. ¾ The majority of future Microgrids will be operated for most of the time under grid connected condition except for those built on physical islands, thus main benefits of Microgrid concept will arise from grid-connected (i.e. 'normal') operating states. Long-term islanded operation of an entire Microgrid poses high requirements on storage size and capacity ratings of micro-generators. However, reliability benefits of under designed Microgrids (i.e. cannot enter island mode without load shedding measures) can still be quantified from partial islanding of important loads.
3. The difference between a Microgrid and a passive grid penetrated by micro-sources lies mainly in the way of management and coordination of available resources. ¾ A Microgrid operator is more than an aggregator of small generators, or a network service provider, or a load controller, or an emission regulator—it performs all these functionalities and serves multiple economic, technical, and environmental aims. ¾ One major advantage of the Microgrid concept over other 'intelligent' solutions lies in its capability of handling conflicting interests of different stakeholders so as to arrive at a globally optimal operation decision for all players involved.

A microgrid is a localized group of electricity sources and sinks (loads) that typically operates connected to and synchronous with the traditional centralized grid (macrogrid), but can disconnect and maintain operation autonomously as physical and/or economic conditions dictate. A microgrid is a group of interconnected loads and distributed energy resources within clearly defined electrical boundaries that acts as a single controllable entity with respect to the grid. A microgrid can connect and disconnect from the grid to enable it to operate in both grid-connected or island-mode.

Instead, microgrid definitions focus primarily on two features:

- A microgrid is a locally controlled system
- A microgrid can function both connected to the traditional grid (mega grid) or as an electrical island.

There are two major types of microgrids. These include microgrids wholly on one site, akin to a traditional utility customer, which are usually called customer microgrids or true microgrids (μgrids), and microgrids that involve a segment of the legacy regulated grid, often called milligrids (mgrids) (Alshahrestani et al., 2018; Makdisie et al., 2018; Alhelou et al., 2018; Alhelou et al., 2016; Nadweh et al., 2018; Njenda et al., 2018).

Power generation from fossil fuels is associated with several environmental concerns and poor energy efficiency. Use of renewable energy resources as distributed generation (DG) can be a potential solution to these problems. As the penetration of these DGs in the distribution network increases, they create a

Figure 1. Microgrid Overview

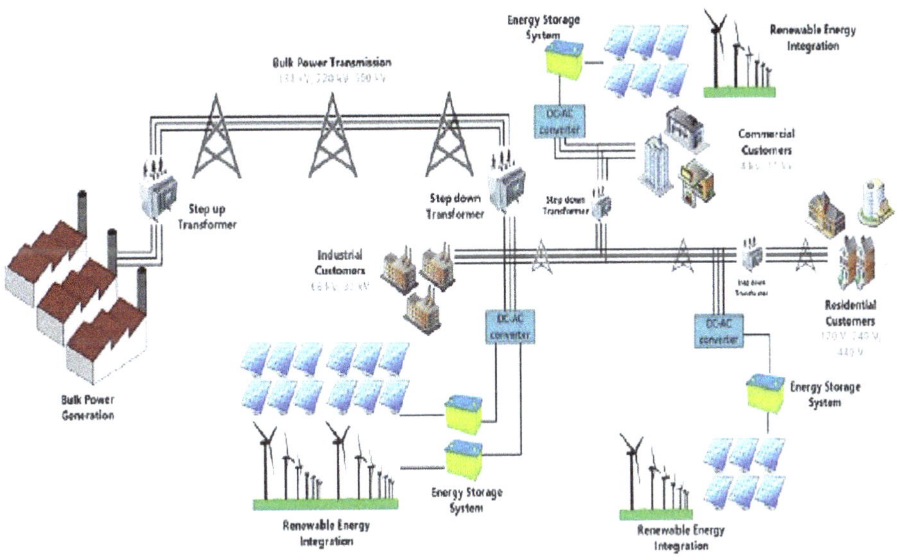

microgrid. A microgrid is a low voltage, the small-scale power grid (on distribution side) with DG, storage devices and controllable loads (Dong et al., 2012). Microgrids can operate independently called the islanded (autonomous) mode of operation or in conjunction with the main grid called the grid connected mode of operation (Dong et al., 2009). Figure 1 shows the typical structure of a microgrid in which there is distributed energy resources (DERs), distribution network and loads (Fini et al., 2016; Alhelou et al., 2018; Zamani et al., 2018; Alhelou et al., 2015; Njenda et al., 2018; Haes Alhelou et al., 2018). A microgrid can be connected or disconnected from the main grid at the point of common coupling (PCC). The limited capacity of microgrids has resulted in the evolution of multi-microgrids. Multi-microgrids are an interconnection of several microgrids which can operate with or without the main grid support.

Microgrids offer several advantages and benefits including increased reliability, improved energy efficiency and resiliency, cost reduction, reduction in transmission losses, CO2 emission reduction, and other environmental benefits. However, they also introduce several major challenges regarding the operation, control, and protection of microgrid. Furthermore, each mode of operation (grid connected or islanded) requires unique control and protection schemes (Alshahrestani et al., 2018; Makdisie et al., 2018; Alhelou et al., 2018; Alhelou et al., 2016; Nadweh et al., 2018; Njenda et al., 2018). In literature, several methods have been proposed for the successful operation of microgrids. According to the IEEE standard 1547, DGs are allowed to be connected to a distribution network but are required to be disconnected in case of a fault. This approach can be applied to very small grid-connected microgrids. However, if the DG penetration is very high, disconnection of all DGs can have adverse effects on the main grid. The major issues and potential solutions in microgrid protection and control include:

- **Bidirectional Power Flows:** The power flow in a conventional distribution system is unidirectional, i.e. from the substation to the loads. Integration of DGs on the distribution side of the grid can cause reverse power flows. As a result, the conventional protection coordination schemes are no longer valid.

- **Short Circuit Capacity:** In the case of inverter based DGs, the fault current is limited (maximum 2 p.u.). Hence, the conventional overcurrent relay cannot sense the fault. Stability issues: Local oscillations may arise as a result of the interaction of the control system of micro generators. Hence, small signal stability analysis and transient stability analysis are required to ensure proper operation in a microgrid.
- **Low Inertia:** In a conventional power system, the bulk power is generated at power plants and hence they have high inertia. Microgrids, on the other hand, have dispersed generation and sizes of the DGs are very small. Consequently, they have a low inertia characteristic, especially for inverter based DGs. Low inertia can result in severe frequency deviations in islanded mode operation. Hence, special control mechanism is required.
- **Intermittent Output:** Microgrids with renewable energy resources (photovoltaic, or wind) as distributed generation are intermittent in their power output. Hence, coordination between DGs and storage devices is essential.

The operation of microgrids offers distinct advantages to customers and utilities, i.e. improved energy efficiency, minimization of overall energy consumption, reduced environmental impact, improvement of reliability of supply, network operational benefits such as loss reduction, congestion relief, voltage control, or security of supply and more cost efficient electricity infrastructure replacement. There is also a philosophical aspect, rooted in the belief that locally controlled systems are more likely to make wise balanced choices, such as between investments in efficiency and supply technologies. Microgrids can coordinate all these assets and present them to the megagrid in a manner and at a scale that is consistent with current grid operations, thereby avoiding major new investments that are needed to integrate emerging decentralized resources. Microgrids have been proposed as a novel distribution network architecture within the SmartGrids concept, capable to exploit the full benefits from the integration of large numbers of small scale distributed energy resources into low-voltage electricity distribution systems.

At the highest level, the Smart Grid has 3 components:

- Improved operation of the legacy high voltage grid, e.g. through use of synchrophasers
- Enhanced grid-customer interaction, e.g. by smart metering or real-time pricing
- New distributed entities that have not existed previously, e.g. microgrids and active distribution networks.

TYPES OF MICROGRIDS

Two key types of microgrids can be distinguished, and two other related types of power systems apply very similar technology.

1. Customer microgrids or true microgrids (μgrids) are self-governed, and usually downstream of a single point of common coupling (PCC). Many of the most well-known demonstrations are of this type. They are particularly easy to imagine because they fit neatly into our current technology and regulatory structure. Just as a traditional customer has considerable leeway in the operation of the power system on its side of the meter, so the restrictions on the nature of a μgrid are relatively

loose. For this reason, one would expect much of the early deployment of microgrid technology to be of this type.

2. Utility or community microgrids or milligrids (mgrids) involve a segment of the regulated grid. There are also existing well known examples. While technically, they not be different from µgrids, they are fundamentally different from a regulatory and business model perspective, primarily because they incorporate traditional utility infrastructure. The corollary of this feature is that utility regulation comes much more significantly into play. In other words, any mgrid must comply with existing utility codes or accommodation must be made in the code.
3. Virtual microgrids (vgrids) cover DER at multiple sites but are coordinated such that they can be presented to the grid as a single controlled entity. Very few demonstrations of vgrids exist, but they have been proposed in the literature. Note that to be consistent with the definition above, the system must be able to operate as a controlled island or coordinated multiple islands.
4. Remote power systems (rgrids) are obviously not able to operate grid-connected, isolated power systems involve similar technology and are closely related. So close that from a research point of view, they are commonly described as microgrids.

How Does a Microgrid Work?

To understand how a microgrid works, you first have to understand how the grid works.

The grid connects homes, businesses and other buildings to central power sources, which allow us to use appliances, heating/cooling systems and electronics. But this interconnectedness means that when part of the grid needs to be repaired, everyone is affected.

This is where a microgrid can help. A microgrid generally operates while connected to the grid, but importantly, it can break off and operate on its own using local energy generation in times of crisis like storms or power outages, or for other reasons.

A microgrid can be powered by distributed generators, batteries, and/or renewable resources like solar panels. Depending on how it's fueled and how its requirements are managed, a microgrid might run indefinitely.

How Does a Microgrid Connect to the Grid?

A microgrid connects to the grid at a point of common coupling that maintains voltage at the same level as the main grid unless there is some sort of problem on the grid or other reason to disconnect. A switch can separate the microgrid from the main grid automatically or manually, and it then functions as an island.

Why Choose to Connect to Microgrids?

A microgrid not only provides backup for the grid in case of emergencies, but can also be used to cut costs, or connect to a local resource that is too small or unreliable for traditional grid use. A microgrid allows communities to be more energy independent and, in some cases, more environmentally friendly.

Operation of Microgrid

A microgrid comes in a variety of designs and sizes. A microgrid can power a single facility like the Santa Rita Jail microgrid in Dublin, California. Or a microgrid can power a larger area. For example, in Fort Collins, Colorado, a microgrid is part of a larger goal to create an entire district that produces the same amount of energy it consumes.

Microgrid is designed to seamlessly connect or disconnect from the distribution system. When connected to the grid, the phase, voltage, frequency, and phase angles of the renewable energy generation sources, batteries and system loads are synchronized to the grid.

1. **Supply to Grid:** The Alectra Utilities Microgrid is connected to the distribution system and is supplying energy to the grid using renewable solar or wind power. Stored electricity from the Sodium Nickel Chloride, Lithium Ion Battery and Lead Acid Battery Systems can also be used to supply energy to the grid. During the Supply to Grid operating mode, the natural gas generator will not be operated.
2. **Supply From Grid:** The Alectra Utilities Microgrid is connected to the distribution system and is taking energy from the grid to power its load. Electricity can also be stored in the Sodium Nickel Chloride, Lithium Ion Battery and Lead Acid Battery Systems for future consumption. During the Supply from Grid operating mode, the solar photovoltaic system and wind turbine system may also be powering the load and charging the batteries, but the natural gas generator will not be operated.
3. **Island (Generator):** The Alectra Utilities Microgrid is designed to operate in isolation from the distribution grid with the Island (Generator) operating mode. During this mode, the natural gas generator will be the primary source of electricity with the renewable solar and wind generators providing supplementary power. Electricity stored in the Lead Acid, Sodium Nickel Chloride and Lithium Ion Battery Systems can also be used at this time.
4. **Island (No Generator):** The Alectra Utilities Microgrid is designed to operate in isolation from the distribution grid with the Island (No Generator) operating mode. During this mode renewable solar and wind generators will be primary source of power. Electricity stored in the Lead Acid, Sodium Nickel Chloride and Lithium Ion Battery Systems can also be used at this time. Since all generation sources are intermittent with this operating mode, low priority Microgrid loads may be disconnected depending on the amount of generation available.
5. **Black Start:** The Alectra Utilities Microgrid is designed to have black start capability that involves using backup systems to help launch the Microgrid's generation sources. During this mode, the Microgrid is not connected to the distribution system and does not have electricity serving its load. The Microgrid will use the backup systems to initiate the renewable generation sources and connect the battery systems to help restore power to system loads.
6. **Unintentional Grid Outage (Generator):** The Alectra Utilities Microgrid is designed to operate in the event of an outage on the distribution system and provide seamless service to its loads. In this scenario, the Microgrid will automatically disconnect from the grid and start drawing electricity from the natural gas generator, renewable energy sources and battery systems.
7. **Unintentional Grid Outage (No Generator):** The Alectra Utilities Microgrid is designed to operate in the event of an outage on the distribution system and provide seamless service to its

loads. In this scenario, the Microgrid will automatically disconnect from the grid and start drawing electricity from renewable energy sources and battery systems. Since only intermittent generation sources are available, low priority Microgrid loads may be disconnected depending on the amount of generation available.

8. **Intentional Grid Outage (Generator):** The Alectra Utilities Microgrid is designed to operate in the event of an outage of the distribution system and provide seamless service to its loads. Utilities from time-to-time have planned outages to allow for maintenance and servicing. In this scenario, the Microgrid will automatically disconnect from the grid and start drawing electricity from the natural gas generator, renewable energy sources and battery systems. Since this operating mode involves a planned outage, the battery systems can be fully charged ahead of time to maximize the amount of power for loads during the outage.
9. **Intentional Grid Outage (No Generator):** The Alectra Utilities Microgrid is designed to operate in the event of an outage on the distribution system and provide seamless service to its loads. Utilities from time-to-time have planned outages to allow for maintenance and servicing. In this scenario, the Microgrid will automatically disconnect from the grid and start drawing electricity from renewable energy sources and battery systems. This operating mode involves a planned outage and hence the battery systems can be fully charged ahead of time to maximize the amount of power for loads during the outage. However, since only intermittent generation sources are available, low priority Microgrid loads may be disconnected depending on the amount of generation available.

Microgrid Control Strategies

A control strategy where the real power, P, and reactive power, Q, are regulated to remain fairly constant is called PQ control. In such control strategy the micro source behaves as a voltage controlled current source. Its terminal quantities (voltage, current, or power) are measured in a

Suitable reference frame, compared with a reference set-point and a control action taken to regulate the output quantity. In this the control strategies are presented

Microgrid Control Methods

Microgrid technologies can be classified into two modes of operation, namely, grid-connected mode or islanded mode. Using microgrid control system has two advantages. One is that the stability of microgrid operation in frequency and voltage is maintained, especially in the load changes or interconnection with other networks (Cvetkovic et al., 2009). The other one is that the active and reactive powers of the converter are regulated. The current survey will consider PQ control, droop control, voltage/frequency control, and current control methods.

PQ Control

In microgrid systems, a public control is used, which is called PQ strategy. PQ controls the voltage output of the inverter by injecting the active and reactive powers. This control is used in cases wherein the microgrid is not required or unable to provide voltage or frequency support as a current-controlled voltage source (Abdel-Rahim, N. M., & Quaicoe, J. E., 1996). With PQ control, the microgrid controller does not change the output settings of its microgrid in response to fluctuations in terminal voltage or frequency (Gu et al., 2006), which requires a reference frequency signal to operate. During grid-connected operation, this reference frequency is supplied by the main grid. Meanwhile, in islanded operation, this reference frequency has to be supplied by another microgrid that operates on droop control.

Droop Control

Droop control method is used in conventional generators of power plants. In islanded operation of converter-based DG systems, this method can be adopted to share total load and control voltage magnitude and frequency in a special range. In this control unit, active power of total load can be shared among sources by drooping the frequency as a function of output active power of converter (Mohan et al., 2007). The applicability of droop control is readily apparent by considering the relationships that dictate the power transfer in a two-inverter system (Figure 2).

Conventional droop control is based on several assumptions such as the output impedance having negligible effect on the power characteristic (output impedance is small compared with line impedance) and no cross coupling occurs in the P–ω or Q–V relationships. These assumptions can significantly influence the appropriate control law. One way to address these issues is by applying a linear rotational transformation matrix and modifying the control law. The linear rotational transformation is applied as follows

$$\begin{pmatrix} P' \\ Q' \end{pmatrix} = \begin{pmatrix} Sin(\theta) & -\cos(\theta) \\ \cos(\theta) & \sin(\theta) \end{pmatrix} \begin{pmatrix} P \\ Q \end{pmatrix} \quad (1)$$

Notably, the power transfer relation is expressed in terms of phase and voltage differences, whereas droop relies on frequency and voltage differences to transfer power. This phenomenon is because the frequency difference will periodically change by a small amount. The wandering droop frequency results in a constant phase difference that is necessary to transfer power.

Voltage/Frequency Control

The microgrid voltage control ensures that the voltage remains within the set point values by adjusting generated or consumed reactive power. A voltage controller is designed with the terminal voltages as input and compared with the reference voltage. The error voltage is filtered using a low pass filter and multiplied by a gain constant to obtain droop control of the voltage source convertor. The output of

Figure 2. A system with two voltage sources

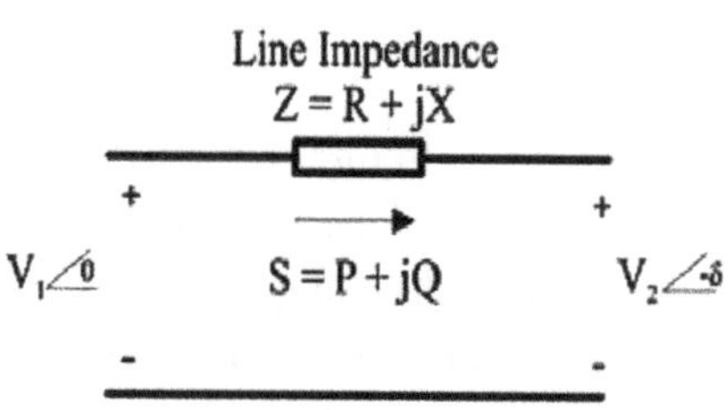

the voltage regulation control block gives the reactive power that needs to be injected to maintain the terminal voltage according to droop set value. Hence, the traditional frequency droop control through active power and voltage droop control through reactive power may not function very well on low-voltage network-based microgrid. Thus, the voltage control should be implemented through active power and frequency control through reactive power generated in an inverter. The inverter of the PV array consists of two major control loops, namely, the reactive power control and the DC voltage control (control of active power). A control scheme is proposed to maintain frequency and voltage during islanded mode in microgrid with wind turbine, photovoltaic system, and battery inverter. A power system level MVAR controller is designed, which issues reference VAR commands over the communication channel to each microgrid controllable asset controller. Then, the microgrid controller regulates the VARs locally in each component to obtain the desired voltage and VAR output at the grid connecting feeder. Proper voltage controllers are required to avoid the collapse of voltage. Conventional power grid has high (X/R) ratio, that is, the grid is predominantly inductive. In such system, reactive power cannot be transported over long electrical distances because of the drop in terminal voltage. However, given that the electrical distances are relatively small in a microgrid, the reactive power can be transported from generators to the loads without a significant drop in the voltage. The voltage at a bus in any system is an important operating parameter, which is required to be within certain limits. Various disturbances, such as line outages, load switching, and line faults, lead to a drop in voltage at buses. In the present power system, when a large reactive load switches into the system, the consumer is required to notify the utility to install reactive power compensator(s) at the consumer site. However, the electrical distances between the sources and loads are small in a microgrid. Thus, the reactive power can be transported from the source to the load directly. This stabilizer can be used to coordinate the consumption of reactive power in a microgrid system. The MGVS is based on the inputs from load buses, runs a subroutine, and gives a signal to the DGs to kick in the reactive power to support the new reactive load that switches into the microgrid system. When the microgrid is in islanded mode, the voltages at the load buses drop. In this case, the MGVS can function as a fast acting voltage controller by sensing the voltage drop immediately and pushing the distributed generators to generate more active and reactive powers. During short circuit faults, large currents flow from generator buses to the faulted load buses. Therefore, the MGVS can help the microgrid to use the reserved reactive power efficiently and improve the dynamic voltage stability. Meanwhile, the microgrid frequency control keeps the frequency of the microgrid around the set point value. In a large power system, the generators have storage in their inertia. Therefore, when a new load comes into the system, the frequency decreases slightly and the generator supports the load during the dynamic state. Meanwhile, distributed generators have slow response and are almost inertia-less. Therefore, designing a microgrid along with the frequency controllers to provide some storage is important. Power versus frequency drop characteristics vary depending on the nature of the DG and one of the critical microgrid performance curves. At the instant of islanding, the voltage phase angles at DG buses change, resulting in an apparent reduction in local frequency. This drop in frequencies of the DGs forces them to raise their power output and share the load proportionally. A dynamic analysis of generation control scheme consisting of active power-frequency and reactive power-voltage controllers for the inverter-based DGs is presented. In addition, guidelines are provided for the design of active power-frequency and reactive power-voltage controllers. Fast and accurate voltage and frequency control are essential requirements for successful islanding operation of weak low-voltage network-based microgrid. To manage with the fast changing frequency and voltage in normal range, the microgrid should include one relatively large capacity inverter-based energy storage unit, which would address this rapid control. A microgrid with synchronous machine and an electronically interfaced DG unit can maintain the angle stability even after most severe islanding transients, primarily through its fast real power control. In addition, such system can enhance voltage quality at specific buses, mainly through its fast reactive power control. In the absence of synchronous machines, inverters are responsible for maintaining and controlling the frequency of the microgrid. A small-signal state-space model of autonomous operation of inverter-based microgrid is presented in. In this model, each distributed generation inverter has an outer power loop based on the droop control to share the real and reactive powers with other distributed generations. Voltage and current controllers are also used in inverter controls to reject high-frequency disturbances and damp the output filter to prevent any resonance with external network.

Current Control

Current mode control (CMC) (Figure 3) is typically used to track the set point derived from droop or other control approaches. Notably, in this control approach, both loops have finite gain, the inner (current) loop acts as a disturbance that introduces a voltage and phase offset proportional to load current. Therefore, changes in the load impedance will change the system transfer function, thereby changing the amplitude and phase at a given frequency. This phenomenon is apparent from the system transfer function. Current mode control has been shown to have better dynamic response and tracking compared with the basic droop approach. Current mode control also has inherent line feed-forward and current limit. This method has been successfully used as the control strategy in parallel inverters. Two fundamental issues with nested control loops are instability because of simultaneous perturbations and uncertainty. Both of these issues are addressed by designing a μ-controller, which can be used to guarantee robust stability over all specified values of uncertainty. Given that uncertainty bounds may be specified for μ-controllers, the load maybe specified as an uncertainty. The resulting controller would allow the use of the inverter as long as the load is within the specified range. The current sharing ability of inverters with CMC may be considered. The difference in currents is also referred to as circulating current. If these equations and the impedance "seen" by each inverter are considered, then the difference in line impedances playing an important role in current sharing becomes apparent. At this point, the output impedance of each inverter maybe considered. The output impedance of an inverter is typically ignored and the line impedance is considered. However, the output impedance is considered to be in series with the impedances and represented as z1 and z2.

Figure 3. Two loop (current mode) control with inductor feedback

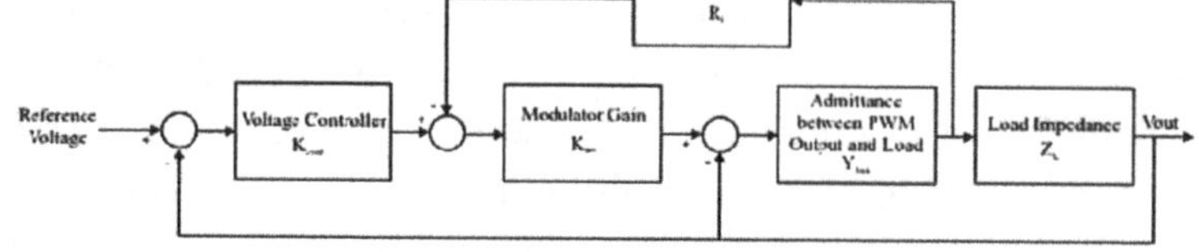

MICROGRID CONTROL ISSUES

Islanded Mode

During islanded operation, storage devices and load-shedding technique have major functions to avoid large frequency deviations. A control strategy for voltage, frequency, and protection during microgrid islanding is defined in. Voltage, frequency, and protection control strategies were initially defined separately. Then, the control strategies were coordinated by using an islanding detection algorithm to improve the response of the overall system. In addition, a black start procedure within islanded mode including DGs has been developed to increase the reliability of the microgrid. Moreover, the PQ inverter controls the active and reactive powers set point. Furthermore, the VSI control maintains the voltage and frequency set point. The input to PQ controller is the loading conditions of the system. This controller is also responsible for load shedding, if generation and demand are mismatched. The VSI controller acts as a synchronous machine in a power system to the microgrid by giving the required voltage and frequency reference. The following issues usually occur within the islanded mode. Firstly, when the DGs supply all the load demand within the microgrid, a proper sharing is necessary. However, considering the presence of DGs of unequal capacity, uniform load sharing is impossible. Therefore, the total load is shared in proportion to the capacity of each source. Apart from the normal load sharing, the DGs need to share harmonic current and the compensation efforts for load unbalance and non-linearity. Meanwhile, given the difference in the line impedance, the output voltage of the inverters is different from each other. This phenomenon makes sharing the reactive power equally without a steady state error difficult. Furthermore, the inverters supply excess reactive current to compensate the error in reactive power sharing, which suffers from reduced capacity to supply active power. Secondly, the of a DG in isolated microgrid allows exploring the scope of using an advanced load shedding, smart loads, and battery unit that can keep the critical loads running even under severe power outages. A few control challenges that are commonly faced using either load shedding or smart load methods are as follows:

- If the fault is unclear within the critical clearance time, the wind turbine in the microgrid can trip, resulting in more power shortage in the microgrid. This phenomenon requires the load shedding or the smart load method to act fast.
- Load shedding has to be conducted in proper stages by optimal disconnection of less critical loads, with switching points lying in a hysteresis band to avoid oscillations.
- Microgrids consisting of a synchronous generator can suffer from voltage instability because of heavy load shedding. This phenomenon is caused by the excess generation of reactive power by the synchronous generator to stabilize voltage in response to the sudden load shedding.

Figure 4. Current sharing for inverter

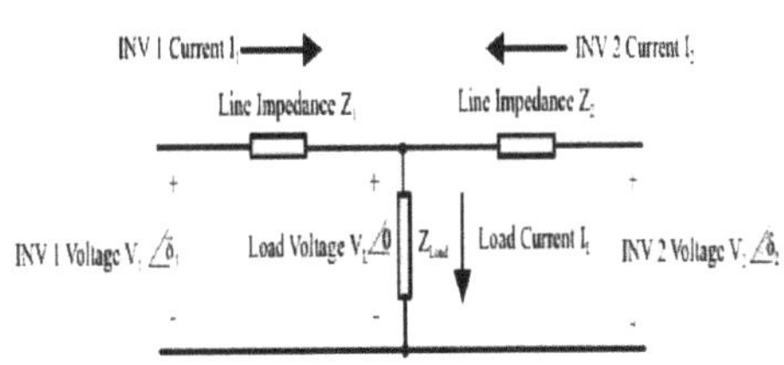

- With a large power shortage or presence of a huge load, the smart load control fails to stabilize the frequency of the isolated microgrid and results in frequency decay. Third, the presence of non-linear load or unbalanced makes ensuring stability in an islanded microgrid more challenging.

A few control challenges are listed as follows:

- The presence of unbalanced loads result in second order harmonic component in the load current, which distorts the DG/MS output voltages. The non-linear load results in higher order harmonics. Hence, current and voltage are more distorted. In these scenarios, normal voltage and frequency regulation schemes are ineffective. Therefore, a control scheme that can maintain the stability of the system without removing the unbalanced and non-linear loads from the network is required.
- Low-frequency variations of the DG output also vary the distorted load currents, thereby making tracking and stabilizing more difficult.
- The conventional method of using negative sequence current for voltage unbalance compensation only addresses the voltage unbalance at the DG terminal. Meanwhile, unbalance compensation at the load bus is more important because all the sensitive loads are connected to it.
- With droop control, DG units supply sinusoidal voltage with fundamental frequency, which can form a short circuit path for harmonic current. This phenomenon results in an improper sharing of harmonic current among the DGs. Therefore, the DG in close proximity to the unbalanced load supplies the highest harmonic current, thereby reducing its capacity to supply other loads.

Stability

Microgrid stability issues may arise from various causes, such as islanding and reconnection to the grid, different types of faults, changes in network parameters, mismatch in the power

Generation demand, and sudden connection or disconnection of DGs or loads in the system. With these issues, over or under voltages and frequencies may occur. Therefore, considerable research works were conducted to develop voltage controllers, frequency controllers, and schemes of islanding operation or reconnection. In addition, power electronic-based DGs provide flexibility to the microgrid control. Connecting the distributed generation sources to the electrical network has various major challenges. These barriers include technical issues, business practices, and regulatory issues.

Unbalanced Voltages

In a three-phase power system, the phase voltages are imperfectly sinusoidal in nature and have some degree of unbalance. Therefore, a microgrid needs to be designed to operate under unbalanced conditions. The voltage unbalance occurs because of various reasons such as the spacing of the overhead transmission lines, three-phase loads with unbalanced impedances, and a fault in the power system. Voltage unbalance can be defined as unequal voltage magnitudes at fundamental system frequency (under voltages and over voltages), fundamental phase angle deviation, and unequal levels of harmonic distortion between the phases. Most literature works consider microgrids as a balanced system and very few papers have addressed the problems regarding unbalanced conditions. As mentioned previously, unbalanced load is the main contributor to unbalanced voltages. Proper voltage balancing techniques, such as feeder switching operations, can distribute the load among the three phases. Distribution transformers with balanced impedances can also remove voltage unbalances. Voltage unbalance compensation can also be achieved by using passive power filters or power electronic devices. Thus, uncontrolled excess generation results in distortion of the voltage profile of the islanded microgrid. The following control strategies are recommended to overcome these challenges:

- The characteristic difference between the DGs (PV and wind turbine) in the microgrid is necessary to develop different control strategies to regulate their maximum output power injected into the network
- In scenarios wherein the AC loads, DC loads, and storage units are unable to consume the active power produced by the DG units, a power oscillation can be developed in the islanded microgrid.

FUTURE RESEARCH DIRECTIONS

The presence of non-linear and the unbalanced loads in the distribution system causes power quality issues in the Microgrid system. This chapter explores and reviews different control strategies developed in the literature for the future power quality enhancement in microgrids.

CONCLUSION

This chapter presented a comprehensive overview of microgrid control strategies. The current key issues of technical challenges especially in the microgrid control methods and its control issues are introduced. These challenges can be addressed by collecting the information for an optimal performance from the microgrid controller and the load conditions. However, a good control strategy is required to manage the power from each DG while optimizing the use of renewable energy sources and meeting the energy demand of all loads. Several research challenges need to be solved in the future to support the integration of the microgrid with utility network, such as obtaining information about energy demand of the load without any communication with the microgrid centre controller. Meanwhile, the multi resources within a microgrid can be controlled by using multi object controllers, which can be divided into number of controllers, such as voltage, frequency, and current controls.

REFERENCES

Abdel-Rahim, N. M., & Quaicoe, J. E. (1996). Analysis and design of a multiple feedback loop control strategy for single-phase voltage-source UPS inverters. *IEEE Transactions on Power Electronics*, *11*(4), 532–541. doi:10.1109/63.506118

Alhelou, H., Hamedani-Golshan, M. E., Zamani, R., Heydarian-Forushani, E., & Siano, P. (2018). Challenges and Opportunities of Load Frequency Control in Conventional, Modern and Future Smart Power Systems: A Comprehensive Review. *Energies*, *11*(10), 2497. doi:10.3390/en11102497

Alhelou, H. H. (2018). Fault Detection and Isolation in Power Systems Using Unknown Input Observer. In Advanced Condition Monitoring and Fault Diagnosis of Electric Machines (p. 38). Hershey, PA: IGI Global.

Alhelou, H. H., Golshan, M., & Fini, M. (2018). Wind Driven Optimization Algorithm Application to Load Frequency Control in Interconnected Power Systems Considering GRC and GDB Nonlinearities. *Electric Power Components and Syst.*

Alhelou, H. H., & Golshan, M. E. H. (2016, May). Hierarchical plug-in EV control based on primary frequency response in interconnected smart grid. In *Electrical Engineering (ICEE), 2016 24th Iranian Conference on* (pp. 561-566). IEEE. 10.1109/IranianCEE.2016.7585585

Alhelou, H. H., Golshan, M. H., & Askari-Marnani, J. (2018). Robust sensor fault detection and isolation scheme for interconnected smart power systems in presence of RER and EVs using unknown input observer. *International Journal of Electrical Power & Energy Systems, 99*, 682–694. doi:10.1016/j.ijepes.2018.02.013

Alhelou, H. H., Hamedani-Golshan, M. E., Heydarian-Forushani, E., Al-Sumaiti, A. S., & Siano, P. (2018, September). Decentralized Fractional Order Control Scheme for LFC of Deregulated Nonlinear Power Systems in Presence of EVs and RER. In *2018 International Conference on Smart Energy Systems and Technologies (SEST)* (pp. 1-6). IEEE. 10.1109/SEST.2018.8495858

Alhelou, H. S. H., Golshan, M. E. H., & Fini, M. H. (2015, December). Multi agent electric vehicle control based primary frequency support for future smart micro-grid. In *Smart Grid Conference (SGC)* (pp. 22-27). Academic Press.

Alshahrestani, A., Golshan, M. E. H., & Alhelou, H. H. (2018, November). WAMS Based Online Estimation of Total Inertia Constant and Damping Coefficient for Future Smart Grid Systems. In *Smart Grid Conference (SGC)* (pp. 1-5). Academic Press.

Cvetkovic, I., Thacker, T., Dong, D., Francis, G., Podosinov, V., Boroyevich, D., . . . Lesko, J. (2009, September). Future home uninterruptible renewable energy system with vehicle-to-grid technology. In Energy Conversion Congress and Exposition, 2009. ECCE 2009. IEEE (pp. 2675-2681). IEEE. doi:10.1109/ECCE.2009.5316064

Dong, D., Thacker, T., Burgos, R., Boroyevich, D., Wang, F., & Giewont, B. (2009, September). Control design and experimental verification of a multi-function single-phase bidirectional PWM converter for renewable energy systems. In *Power Electronics and Applications, 2009. EPE'09. 13th European Conference on* (pp. 1-10). IEEE.

Dong, D., Thacker, T., Cvetkovic, I., Burgos, R., Boroyevich, D., Wang, F., & Skutt, G. (2012). Modes of operation and system-level control of single-phase bidirectional PWM converter for microgrid systems. *IEEE Transactions on Smart Grid, 3*(1), 93–104. doi:10.1109/TSG.2011.2167352

Fini, M. H., Yousefi, G. R., & Alhelou, H. H. (2016). Comparative study on the performance of many-objective and single-objective optimisation algorithms in tuning load frequency controllers of multi-area power systems. *IET Generation, Transmission & Distribution, 10*(12), 2915–2923. doi:10.1049/iet-gtd.2015.1334

Gu, H., Yang, Z., Wang, D., & Wu, W. (2006, August). Research on control method of double-mode inverter with grid-connection and stand-alone. In *Power Electronics and Motion Control Conference, 2006. IPEMC 2006. CES/IEEE 5th International* (Vol. 1, pp. 1-5). IEEE.

Makdisie, C., Haidar, B., & Alhelou, H. H. (2018). An Optimal Photovoltaic Conversion System for Future Smart Grids. In *Handbook of Research on Power and Energy System Optimization* (pp. 601–657). IGI Global. doi:10.4018/978-1-5225-3935-3.ch018

Mohan, N., & Undeland, T. M. (2007). *Power electronics: converters, applications, and design*. John Wiley & Sons.

Nadweh, S., Hayek, G., Atieh, B., & Haes Alhelou, H. (2018). Using Four – Quadrant Chopper with Variable Speed Drive System Dc-Link to Improve the Quality of Supplied Power for Industrial Facilities. Majlesi Journal of Electrical Engineering.

Njenda, T. C., Golshan, M. E. H., & Alhelou, H. H. (2018, November). WAMS Based Intelligent Under Frequency Load Shedding Considering Online Disturbance Estimation. In *Smart Grid Conference (SGC)* (pp. 1-5). Academic Press.

Njenda, T. C., Golshan, M. E. H., & Alhelou, H. H. (2018, November). WAMS based Under Frequency Load Shedding Considering Minimum Frequency Prediction and Extrapolated Disturbance Magnitude. In *Smart Grid Conference (SGC)* (pp. 1-5). Academic Press.

Zamani, R., Hamedani-Golshan, M. E., Haes Alhelou, H., Siano, P., & Pota, H. (2018). Islanding Detection of Synchronous Distributed Generator Based on the Active and Reactive Power Control Loops. *Energies*, *11*(10), 2819. doi:10.3390/en11102819

KEY TERMS AND DEFINITIONS

Droop Speed Control: Droop speed control is a speed control mode of a prime mover driving a synchronous generator connected to an electrical grid. This mode allows synchronous generators to run in parallel, so that loads are shared among generators in proportion to their power rating.

Islanding Mode: Islanding is the condition in which a distributed generator (DG) continues to power a location. In this case, the circuit detached by the outage becomes an "island". This is a function of their construction; some THD is a natural side-effect of the switched-mode power supply circuits most inverters are based on.

Microgrid: A microgrid is a small-scale power grid that can operate independently or collaboratively with other small power grids. The practice of using microgrids is known as distributed, dispersed, decentralized, district, or embedded energy production.

PQ Control: PQ control is the use of grid capable frequency and voltage drops. From then it is given to grid or the load that is connected to the circuit. The reference voltage of the boost inverter is provided from the PQ controller that is used to control both the active and reactive power.

Compilation of References

Aamir, M., & Mekhilef, S. (2017). An online transformerless uninterruptible power supply system with a smaller battery bank for low power applications. *IEEE Transactions on Power Electronics, 32*(1), 233–247.

ABB. (2009). *Technical note td2 en reducing harmonics caused by variable speed drives.* ABB.

Abdel, A. (2012). *Framework For Monitoring And Maintenance Subways Stations Using Building Information Modeling. Construction Research Congress.*

Abdel-Rahim, N. M., & Quaicoe, J. E. (1996). Analysis and design of a multiple feedback loop control strategy for single-phase voltage-source UPS inverters. *IEEE Transactions on Power Electronics, 11*(4), 532–541. doi:10.1109/63.506118

Abraham, R. J., Das, D., & Patra, A. (2007). Effect of TCPS on oscillations in tie-power and area frequencies in an interconnected hydrothermal power system. *IET Generation, Transmission & Distribution, 1*(4), 632–639. doi:10.1049/iet-gtd:20060361

Abu-Mouti, F. S., & El-Hawary, M. E. (2009). Modified artificial bee colony algorithm for optimal distributed generation sizing and allocation in distribution systems. Proceedings of IEEE Conference on Electrical Power & Energy, 1-9. doi:10.1109/EPEC.2009.5420915

Abu-Mouti, F. S., & El-Hawary, M. E. (2011). Optimal Distributed Generation Allocation and Sizing in Distribution Systems via Artificial Bee Colony Algorithm. *IEEE Transactions on Power Delivery, 26*(4), 2090–2101. doi:10.1109/TPWRD.2011.2158246

Abusara, M. A., Guerrero, J. M., & Sharkh, S. M. (2014). Line-interactive UPS for micro-grids. *IEEE Transactions on Industrial Electronics, 61*(3), 1292–1300.

Acharya, N., Mahat, P., & Mithulananthan, N. (2006). An analytical approach for DG allocation in primary distribution network. *Electrical Power and Energy Systems, 28*(10), 669–678. doi:10.1016/j.ijepes.2006.02.013

Adomavicius, V., Kharchenko, V., Valickas, J., & Gusarov, V. (2013). RES-based microgrids for environmentally friendly energy supply in agriculture. *Proceeding of 5th International Conference Trends in Agriculture Engineering*, 51-55.

AER. (2007). *Final the events of 16 January 2007 investigation report.* Retrieved from https://www.aer.gov.au

Afshar, H., Moravej, Z., & Niasati, M. (2013). Modeling and optimization of micro-grid considering emissions. *Proc. Smart Grid Conf. (SGC)*, 225-229. 10.1109/SGC.2013.6733812

Agarwal, T., & Cui, S. (2011). *Noncooperative games for autonomous con-sumer load balancing over smart grid.* Available at: http://arxiv.org/abs/1104.3802

Agnetis, A., de Pascale, G., Detti, P., & Vicino, A. (2013). Load Scheduling for Household Energy Consumption Optimization. *IEEE Transactions on Smart Grid, 4*(4), 2364–2373. doi:10.1109/TSG.2013.2254506

Agnetis, A., Dellino, G., Detti, P., Innocenti, G., de Pascale, G., & Vicino, A. (2011), Appliance operation scheduling for electricity consumption optimization. *Proceedings of the 50th IEEE Conference on Decision and Control and European Control Conference (CDC-ECC)*, 5899–5904. 10.1109/CDC.2011.6160450

Ahmad, A., Saqib, A. S., & Kashif, A. R. (2016). Impact of wide-spread use of uninterruptible power supplies on Pakistan's power system. *Energy Policy*, *98*, 629–636.

Ahmed, O., & Bleijs, J. (2015). An overview of DC–DC converter topologies for fuel cellultracapacitor hybrid distribution system. *Renewable & Sustainable Energy Reviews*, *42*, 609–626.

Akasah. (2007). *Generic Process Model for Maintenance Management of School Buildings*. Malaysia University.

Akhter, S. R., & Biswas, M. M. (2016). Roadmap of smart grid for Bangladesh based on functions and technological challenges. *Electric Power Components and Systems*, *44*(8), 864–872. doi:10.1080/15325008.2016.1138159

Albadi, M. H. E. F. E.-S. (2007). *Demand Response in Electricity Markets: An Overview*. IEEE.

Alharbi, H., & Bhattacharya, K. (2014). Optimal sizing of battery energy storage systems for micro-grids. *Proc. IEEE Electr. Power Energy Conf.*, 275-280. 10.1109/EPEC.2014.44

Alhassan, B., & Jrad, F. (2015). *enhancing maintenance management for governmental buildings using building information modelling (BIM)* (Master thesis). Teshreen University, Latakia, Syria.

Alhelou, H. H. (2018). Fault Detection and Isolation in Power Systems Using Unknown Input Observer. In Advanced Condition Monitoring and Fault Diagnosis of Electric Machines (p. 38). Hershey, PA: IGI Global.

Alhelou, H. H. (2018c). Fault Detection and Isolation in Power Systems Using Unknown Input Observer. In Advanced Condition Monitoring and Fault Diagnosis of Electric Machines. IGI Global.

Alhelou, H. H., & Golshan, M. E. H. (2016, May). Hierarchical plug-in EV control based on primary frequency response in interconnected smart grid. In *Electrical Engineering (ICEE), 2016 24th Iranian Conference on* (pp. 561-566). IEEE.

Alhelou, H. H., Golshan, M., & Fini, M. (2018). Wind Driven Optimization Algorithm Application to Load Frequency Control in Interconnected Power Systems Considering GRC and GDB Nonlinearities. *Electric Power Components and Syst.*

Alhelou, H. H., Golshan, M., & Fini, M. (2018). Wind Driven Optimization Algorithm Application to Load Frequency Control in Interconnected Power Systems Considering GRC and GDB Nonlinearities. *Electric Power conponents and Syst.*

Alhelou, H. S. H., Golshan, M. E. H., & Fini, M. H. (2015, December). Multi agent electric vehicle control based primary frequency support for future smart micro-grid. In *Smart Grid Conference (SGC)* (pp. 22-27). Academic Press.

Alhelou, H. H. (2018). *Fault Detection and Isolation in Power Systems Using Unknown Input Observer. In Advanced Condition Monitoring and Fault Diagnosis of Electric Machines* (p. 38). Hershey, PA: IGI Global.

Alhelou, H. H. (2018). Fault detection and isolation in power systems using unknown input observer. In *Advanced Condition Monitoring and Fault Diagnosis of Electric Machines*. Hershey, PA: IGI Global publisher.

Alhelou, H. H., & Golshan, M. E. H. (2016). Hierarchical plug-in EV control based on primary frequency response in interconnected smart grid. *24th Iranian Conference on Electrical Engineering (ICEE)*, 561-566. 10.1109/IranianCEE.2016.7585585

Alhelou, H. H., Golshan, M. H., & Askari-Marnani, J. (2018). Robust sensor fault detection and isolation scheme for interconnected smart power systems in presence of RER and EVs using unknown input observer. *International Journal of Electrical Power & Energy Systems*, *99*, 682–694. doi:10.1016/j.ijepes.2018.02.013

Alhelou, H. H., Hamedani-Golshan, M. E., Heydarian-Forushani, E., Al-Sumaiti, A. S., & Siano, P. (2018, September). Decentralized Fractional Order Control Scheme for LFC of Deregulated Nonlinear Power Systems in Presence of EVs and RER. In *2018 International Conference on Smart Energy Systems and Technologies (SEST)* (pp. 1-6). IEEE. 10.1109/SEST.2018.8495858

Alhelou, H. S. H., Golshan, M. E. H., & Fini, M. H. (2015). Multi agent electric vehicle control based primary frequency support for future smart micro-grid. *Smart Grid Conference (SGC)*, 22-27.

Alhelou, H. S. H., Golshan, M. E. H., & Fini, M. H. (2015, December). Multi agent electric vehicle control based primary frequency support for future smart micro-grid. In *Proc. of Smart Grid Conference (SGC)*. Iran University of Science and Technology.

Alhelou, H., Hamedani-Golshan, M. E., Zamani, R., Heydarian-Forushani, E., & Siano, P. (2018). Challenges and opportunities of load frequency control in conventional, modern and future smart power systems: A comprehensive review. *Energies*, *11*(10), 1–35. doi:10.3390/en11102497

Alhelou, H., Hamedani-Golshan, M. E., Zamani, R., Heydarian-Forushani, E., & Siano, P. (2018). Challenges and Opportunities of Load Frequency Control in Conventional, Modern and Future Smart Power Systems: A Comprehensive Review. *Energies*, *11*(10), 2497.

Alshahrestani, A., Golshan, M. E. H., & Alhelou, H. H. (2018, November). WAMS Based Online Estimation of Total Inertia Constant and Damping Coefficient for Future Smart Grid Systems. In *Smart Grid Conference (SGC)* (pp. 1-5). Academic Press.

An Introduction to Smart Grids. (n.d.). *ABB Back Ground information: Smart Grids.* Retrieved from: https://mycourses.aalto.fi/pluginfile.php/365574/mod_folder/content/0/Smart_grids_introduction.pdf?forcedownload=1

Ang, X., Su, J., Zhipeng, L., Liu, H., & Li, R. (2014). Overview on Micro-grid Technology. China Electric Power Research Institute.

Antony, S. (1997). *Use of genetic algorithms in electromagnetics* (Bachelor thesis). University of Queensland.

Arboleya, P., Gonzalez-Moran, C., & Coto, M. (2012, September). A hybrid central-distributed control applied to microgrids with droop characteristic based generators. In *Power Electronics and Motion Control Conference (EPE/PEMC), 2012 15th International* (pp. LS7a-5). IEEE. 10.1109/EPEPEMC.2012.6397503

Arditi, D., & Nawakorawit, M. (1999). Designing buildings for maintenance: Designers' perspective. *Journal of Architectural Engineering*, *5*(4), 107–116. doi:10.1061/(ASCE)1076-0431(1999)5:4(107)

Armstrong, K, (2007). *EMC for Printed Circuit Boards – Basic and Advanced.* Academic Press.

Armstrong, K. (2006). Design Techniques for EMC, in six parts. *The EMC Journal, 62*. Retrieved from www.theemcjournal.com

Arsad, N., & Ali, U. (2017). An analysis of the effects of residential uninterruptable power supply systems on Pakistan's power sector. *Energy for Sustainable Development*, *36*, 16–21.

Arya, L. D., Choube, S. C., & Shrivastava, M. (2008). Technique for voltage stability assessment using newly developed line voltage stability index. *Energy Conversion and Management*, *49*(2), 267–275. doi:10.1016/j.enconman.2007.06.018

Atputharajah, A., & Tapan Kumar, S. (2009). Power system blackouts - Literature review. *Fourth International Conference on Industrial and Information Systems (ICIIS).*

Attia, H., El-Sobki, M., & Wahadan, S. (2006). *Priority ranking of Industrial Loads and Application of Demand Side Management (DSM) Technique*. The Eleventh International Middle East Power System Conference, ElMinia, Egypt.

Australian Energy Regulator. (2007). *Final the events of 16 January 2007 investigation report*. Retrieved from https://www.aer.gov.au

Aziz, A., Oo, A. T., & Stojcevski, A. (2018). Analysis of frequency sensitive wind plant penetration effect on load frequency control of hybrid power system. Int. *J Elect Power Energy Syst.*, *99*, 603–617. doi:10.1016/j.ijepes.2018.01.045

Badri Ram, & Vishwakarma, B. N. (2011). *Power system protection and switchgear.* McGraw Higher Ed.

Bahgaat, N. K., Ibrahim, M., Hassan, M. A. M., & Bendary, F. M. A. (2018). Load frequency Control Based on Modern Techniques in Two Areas Power Systems. In Advances in System Dynamics and Control. IGI Global. doi:10.4018/978-1-5225-4077-9.ch006

Balijepalli, M., & Pradhan, K. (2011). *Review of Demand Response under Smart Grid Paradigm*. IEEE PES Innovative Smart Grid Technologies. doi:10.1109/ISET-India.2011.6145388

Balls, C., Battaglini, A., Haas, A., & Lilliestam, J. (2008). *The SuperSmart Grid*. Retrieved from http://www.supersmartgrid.net/wp-content/uploads/2008/06/battaglini-lilliestam-2008- supersmart-grid-tallberg1.pdf

Barbato, A., & Capone, A. (2014). Optimization Models and Methods for Demand-Side Management of Residential Users: A Survey. *Energies*, *7*(9), 5787–5824. doi:10.3390/en7095787

Barisal, A.K., Panigrahi, T.K., & Mishra, S. (2017). A hybrid PSO-LEVY flight algorithm based fuzzy PID controller for automatic generation control of multi-area power systems. *Int. J of Energy Optimization and Engg.*, *6*(2), 42-63.

Barker & Unruh. (2010). *A Standard Glossary of Power Quality Terminology*. IEEE.

Basu, Bhattacharya, Chowdhury, & Chowdhury. (2012). Planned scheduling for economic power sharing in a CHP-based micro-grid. *IEEE Trans. Power Syst.*, *27*(1), 30-38.

Bekiarov, S. B., & Emadi, A. (2002).Uninterruptible power supplies: classification, operation, dynamics, and control. *APEC Seventeenth Annual IEEE Applied Power Electronics Conference and Exposition*, 597-604.

Belenov, A. T., Daus, Yu. V., Rakitov, S. A., Yudaev, I. V., & Kharchenko, V. V. (2016). The Experience of Operation of the Solar Power Plant on the Roof of the Administrative Building in the Town of Kamyshin, Volgograd Oblast. *Applied Solar Energy*, *52*(2), 151–156. doi:10.3103/S0003701X16020092

Berker, T., & Throndsen, W. (2017). Planning story lines in smart grid road maps (2010–2014): Three types of maps for coordinated time travel. *Journal of Environmental Policy and Planning*, *19*(2), 214–228. doi:10.1080/1523908X.2016.1207159

Bessa, R. J., & Matos, M. A. (2012). Economic and technical management of an aggregation agent for electric vehicles: A literature survey. *European Transactions on Electrical Power*, *22*(3), 334–350. doi:10.1002/etep.565

Bhatti, T. S., Al-Ademi, A. A. F., & Bansal, N. K. (1997). Load frequency control of isolated wind-diesel-microhydro power systems (WDMHPS). *Energy*, *22*(5), 461–470. doi:10.1016/S0360-5442(96)00148-X

Bhatt, J., Shah, V., & Jani, O. (2014). An instrumentation engineer's review on smart grid: Critical applications and parameters. *Renewable & Sustainable Energy Reviews*, *40*, 1217–1239. doi:10.1016/j.rser.2014.07.187

Bichler, M. (2013). *Smart Grids and the Energy Transformation: Mapping Smart Grid Activities in Germany*. Heinrich Böll Foundation.

Bojoi, R. I., Limongi, L. R., Roiu, D., & Tenconi, A. (2011). Enhanced power quality control strategy for single-phase inverters in distributed generation systems. *IEEE Transactions on Power Electronics*, *26*(3), 789–806.

Bolandi, T. G., Haghifam, M. R., & Khederzadeh, M. (2017). Real-time monitoring of zone 3 vulnerable distance relays to prevent maloperation under load encroachment condition. *IET Generation, Transmission & Distribution*, *11*(8), 1878–1888. doi:10.1049/iet-gtd.2016.0486

Bornand, E. (1991). Le guide de maintenance. Berne.

Bortolini, M., Gamberi, M., & Graziani, A. (2014). Technical and economic design of photovoltaic and battery energy storage system. *Energy Conversion and Management*, *86*, 81–92.

Branco, C. G. C., Torrico-Bascope, R. P., Cruz, C. M. T., & Lima, F. K. de A. (2013). Proposal of three-phase high-frequency transformer isolation UPS topologies for distributed generation applications. *IEEE Transactions on Industrial Electronics*, *60*(4), 1520–1531.

Brown, D. R., & Chvala, W. D. (2009). Flywheel Energy Storage: An Alternative to Batteries for UPS Systems. *Energy Engineering*, *102*(5), 7–26.

Brownlie, d. (1988). Battery requirements for uninterruptible power-supply applications. *Journal of Power Sources*, *23*, 211–220.

Bühler, R. (2010). *Integration of renewable energy sources using microgrids, virtual power plants and the energy hub approach* (MSc Master thesis). Swiss Federal Institute of Technology (ETH), Zurich, Switzerland.

Burr MT, Zimmer MJ, Meloy B, Bertrand J, Levesque W, Warner G. et al. (2013) Minnesota microgrids; 2013.

Buxton, J. (2011). *Creating a smart grid roadmap. Black & Veatch building a world of difference.* Retrieved from https: // www.bv.com/docs/articles/creating–a–smart–grid-roadmap.pdf

Camacho, E. F., & Bordons, C. (2007). *Model Predictive Control*. New York: Springer-Verlag.

Camblong, H., Sarrb, J., Niangc, A. T., Curea, O., Alzola, J. A., Sylla, E. H., & Santos, M. (2009). Micro-grids project, Part 1: Analysis of rural electrification with high content of renewable energy sources in Senegal. *Renewable Energy*, *34*(10), 2141–2150. doi:10.1016/j.renene.2009.01.015

Cao, H. (2016). Economic dispatch of micro-grid based on improved particle-swarm optimization algorithm. *Proc. North Amer. Power Symp. (NAPS)*, 1-6. 10.1109/NAPS.2016.7747875

Capehart, B. L., Turner, W. C., & Kennedy, W. J. (2006). *Guide to Energy Management* (5th ed.). The Fairmont Press.

Centeno, V., Phadke, A. G., Edris, A., Benton, J., Gaudi, M., & Michel, G. (1997). An adaptive out-of-step relay for power system protection. *IEEE Transactions on Power Delivery*, *12*(1), 61–71. doi:10.1109/61.568226

Center for Energy, Marine Transportation and Public Policy (CEMP) at Columbia University. (2010). Microgrids: an assessment of the value, opportunities and barriers to deployment in New York State. New York State Energy Research and Development Authority.

Chakraborty, Breaz, Rathore, & Gao. (2017). Parasitics-assisted soft-switching and secondary modulated snubberless clamping currentfed bidirectional voltage doubler for fuel cell vehicles. *IEEE Trans. Veh. Technol.*, *66*(2), 1053-1062.

Chalise, S. (2016). *Power management of remote microgrids considering battery lifetime*. Academic Press.

Chalise, S., Sternhagen, J., Hansen, T. M., & Tonkoski, R. (2016). Energy management of remote microgrids considering battery lifetime. *The Electricity Journal, 29*(6), 1–10. doi:10.1016/j.tej.2016.07.003

Changsong, C., Shanxu, D., Tao, C., & Bangyin, L. (2009). Energy trading model for optimal micro-grid scheduling based on genetic algorithm. *Proc. IEEE 6th Int. Power Electron. Motion Control Conf.*, 2136-2139.

Chattopadhyay, S., Mitra, M., & Sengupta, S. (2011). *Electric Power Quality*. Department of Applied Physics. University of Calcutta. doi:10.1007/978-94-007-0635-4

Chauhan, A., & Saini, R. (2014). A review on integrated renewable energy system based power generation for stand-alone applications: Configurations, storage options, sizing methodologies and control. *Renewable & Sustainable Energy Reviews, 38*, 99–120.

Chedid, R. B., Karaki, S. H., & El-Chamali, C. (2000). Adaptive Fuzzy Control for Wind-Diesel Weak Power Systems. *IEEE Transactions on Energy Conversion, 15*(1), 71–78. doi:10.1109/60.849119

Chedid, R., Mrad, F., & Basma, M. (1999). Intelligent Control of a Class of Wind Energy Conversion Systems. *IEEE Transactions on Energy Conversion, 14*(4), 1597–1604. doi:10.1109/60.815111

Chellali, F., Khellaf, A., Belouchrani, A., & Recioui, A. (2011). A Contribution to the actualization of the wind map of Algeria. *Renewable and Sustainable Energy Reviews, Elsevier., 15*(2), 993–1002. doi:10.1016/j.rser.2010.11.025

Chellappan, M. V., Todorovic, M. H., & Enjeti, P. N. (2008). Fuel Cell Based Battery-less UPS System. *IEEE Industry Applications society Annual meeting*, 1-8.

Chen, J., Wang, J., Chen, Q., & Wu, D. (2015). Optimal dispatch of medium voltage micro-grid using an adaptive PSO algorithm. *Proc. 7th Int. Conf. Intell. Human-Mach. Syst. Cybern.*, 324-329. 10.1109/IHMSC.2015.48

Chen, D., Zhang, J., & Qian, Z. (2013). An improved repetitive control scheme for grid-connected inverter with frequency-adaptive capability. *IEEE Transactions on Industrial Electronics, 60*(2), 814–823.

Cheng, P. T., Chen, C. A., Lee, T. L., & Kuo, S. Y. (2009). A cooperative imbalance compensation method for distributed-generation interface converters. *IEEE Transactions on Industry Applications, 45*(2), 805–815. doi:10.1109/TIA.2009.2013601

Chen, X., Fu, Q., & Wang, D. (2009). Performance analysis of PV grid-connected power conditioning system with UPS. *4th IEEE Conference on Industrial Electronics and Applications (ICIEA)*, 2172-2176.

Chiang, H., Ma, T., Cheng, Y., Chang, J., & Chang, W. (2010). Design and implementation of a hybrid regenerative power system combining grid–tie and uninterruptible power supply functions. *IET Renewable Power Generation, 4*, 85–99.

Chiu, W.-Y., Sun, H., & Poor, H. V. (2012). Demand-side energy storage system management in smart grid. *2012 IEEE Third International Conference on Smart Grid Communications (Smart Grid Comm.),* 73,. 10.1109/SmartGridComm.2012.6485962

Chiu, W.-Y., Sun, H., & Poor, H. V. (2013). Energy Imbalance Management Using a Robust Pricing Scheme. *IEEE Transactions on Smart Grid, 4*(2), 896–904. doi:10.1109/TSG.2012.2216554

Choi, J. H., Kwon, J. M., Jung, J. H., & Kwon, B. H. (2005). High performance online UPS using three-leg-type converter. *IEEE Transactions on Industrial Electronics, 52*(3), 889–897.

Choi, W., Howze, J. W., & Enjeti, P. (2006). Fuel-cell powered uninterruptible power supply systems: Design considerations. *Journal of Power Sources, 157*, 311–317.

Colak, I., Kabalci, E., Fulli, G., & Lazarou, S. (2015). A survey on the contributions of power electronics to smart grid systems. *Renewable & Sustainable Energy Reviews*, *47*, 562–579.

Coley David, A. (1999). *An introduction to genetic algorithms for scientists and engineers*. World Scientific. doi:10.1142/3904

Colson, C. M., Nehrir, M. H., & Gunderson, R. W. (2011, August). Multi-agent microgrid power management. In *18th IFAC world congress* (pp. 3678-3683). Academic Press. 10.3182/20110828-6-IT-1002.01188

Cortes, P., Kazmierkowski, M. P., Kennel, R. M., Quevedo, D. E., & Rodriguez, J. (2008). Predictive Control in Power Electronics and Drives. *IEEE Transactions on Industrial Electronics*, *55*(12), 4312–4324.

Cvetkovic, I., Thacker, T., Dong, D., Francis, G., Podosinov, V., Boroyevich, D., . . . Lesko, J. (2009, September). Future home uninterruptible renewable energy system with vehicle-to-grid technology. In Energy Conversion Congress and Exposition, 2009. ECCE 2009. IEEE (pp. 2675-2681). IEEE. doi:10.1109/ECCE.2009.5316064

Das, D., Aditya, S. K., & Kothari, D. P. (1999). Dynamics of diesel and wind turbine generators on an isolated power system. Int. *J Elect Power Energy Syst.*, *21*(3), 183–189. doi:10.1016/S0142-0615(98)00033-7

Dash, P., Saikia, L. C., & Sinha, N. (2015). Automatic generation control of multi area thermal system using Bat algorithm optimized PD–PID cascade controller. Int. *J Elect Power Energy Syst.*, *68*, 364–372. doi:10.1016/j.ijepes.2014.12.063

Daus, Yu. V., Kharchenko, V. V., & Yudaev, I. V. (2018). Solar Radiation Intensity Data as Basis for Predicting Functioning Modes of Solar Power Plants. In V. Kharchenko & P. Vasant (Ed.), Handbook of Research on Renewable Energy and Electric Resources for Sustainable Rural Development (pp. 283-310). Academic Press. doi:10.4018/978-1-5225-3867-7.ch012

Daus, Yu. V., Kharchenko, V. V., & Yudaev, I. V. (2016). Evaluation of Solar Radiation Intensity for the Territory of the Southern Federal District of Russia when Designing Microgrids Based on Renewable Energy Sources. *Applied Solar Energy*, *52*(2), 124–129. doi:10.3103/S0003701X16020109

Davari, P., Zare, F., & Blaabjerg, F. (2015). A Smart Current Modulation Scheme for Harmonic Reduction in Three- Phase Motor Drive Applications. *Proceedings of the 2015 17th European Conference on Power Electronics and Applications (EPE'15 ECCEEurope).*

De Almeida, A., Moreira, L., & Delgado, J. (2011). *Power Quality Problems and New Solutions*. University of Coimbra.

De Brabandere, K., Vanthournout, K., Driesen, J., Deconinck, G., & Belmans, R. (2007, June). Control of microgrids. In *Power Engineering Society General Meeting, 2007. IEEE* (pp. 1-7). IEEE.

De Brabandere, K., Bolsens, B., Van den Keybus, J., Woyte, A., Driesen, J., & Belmans, R. (2007). A voltage and frequency droop control method for parallel inverters. *IEEE Transactions on Power Electronics*, *22*(4), 1107–1115. doi:10.1109/TPEL.2007.900456

Delassi, A., Arif, S., & Mokrani, L. (2018). Load frequency control problem in interconnected power systems using robust fractional PI^{λ} D controller. *Ain Shams Engg J.*, *9*(1), 77–88. doi:10.1016/j.asej.2015.10.004

Delfino, B., Massucco, S., Morini, A., Scalera, P., & Silvestro, F. (2001, July). Implementation and comparison of different under frequency load-shedding schemes. In *2001 Power Engineering Society Summer Meeting. Conference Proceedings (Cat. No. 01CH37262)* (Vol. 1, pp. 307-312). IEEE. 10.1109/PESS.2001.970031

Deng, Q., Gao, X., & Zhou, H. (2011). System modeling and optimization of micro-grid using genetic algorithm. *Proc. 2nd Int. Conf. Intell. Control Inf. Process.*, 540-544.

Dhal, P. K., & Rajan, C. A. (2015). Power Quality Improvement Based on Uninterruptible Power Supply (UPS) in Distribution System. *IEEE Sponsored 2nd International Conference on Electronics and Communication System,* 286-290.

Dipak, P., Mallikarjuna, B., Jayakumar Reddy, R., Jaya Bharata Reddy, M., & Mohanta, D. K. (2017). Synchrophasor assisted adaptive relaying methodology to prevent zone-3 mal-operation during load encroachment. *IEEE Sensors Journal, 17*(23), 7713–7722.

Dirkman, J. (2014). *Developing a Roadmap to a Smarter Utility.* Retrieved from https://www.schneider-electric.com.vn/en/download/document/998-2095-10-28-14AR0_EN/

Dogger, J. D., Roossien, B., & Nieuwenhout, F. D. J. (2011). Characterization of Li-ion batteries for intelligent management of distributed grid-connected storage. *IEEE Transactions on Energy Conversion, 26*(1), 256–263. doi:10.1109/TEC.2009.2032579

Dong, D., Thacker, T., Burgos, R., Boroyevich, D., Wang, F., & Giewont, B. (2009, September). Control design and experimental verification of a multi-function single-phase bidirectional PWM converter for renewable energy systems. In *Power Electronics and Applications, 2009. EPE'09. 13th European Conference on* (pp. 1-10). IEEE.

Dong, D., Thacker, T., Cvetkovic, I., Burgos, R., Boroyevich, D., Wang, F., & Skutt, G. (2012). Modes of operation and system-level control of single-phase bidirectional PWM converter for microgrid systems. *IEEE Transactions on Smart Grid, 3*(1), 93–104. doi:10.1109/TSG.2011.2167352

Dostert, K. M. (1990). Frequency-hopping spread-spectrum modulation for digital communications over electrical power lines. *IEEE Journal on Selected Areas in Communications, 8*(4), 700–710. doi:10.1109/49.54466

Du, Y., Turner, M., Liao, Y., Sunkara, M., DeRouen, J., Greenwell, A., . . . Gardner, J. (2012, January). Kentucky smart grid roadmap initiative. In Innovative Smart Grid Technologies (ISGT), 2012 IEEE PES (pp. 1-6). IEEE. doi:10.1109/ISGT.2012.6175628

Dugan. (2004). Electrical Power Systems Quality (2nd ed.). McGraw-Hill.

Edward, J. N., Ramadan, A., & Shatshat, E. (2010). Multi-Microgrid Control Systems (MMCS). *IEEE Power and Energy Society General Meeting, 25,* 1-6.

Elamine, D. O., Nfaoui, E. H., & Jaouad, B. (2014). Multi-agent architecture for smart micro-grid optimal control using a hybrid BP-PSO algorithm for wind power prediction. *Proc. 2nd World Conf. Complex Syst. (WCCS),* 554-560. 10.1109/ICoCS.2014.7060951

Elci, H., Longman, R. W., Phan, M. Q., Juang, J., & Ugoletti, R. (2002). Simple learning control made practical by zero-phase filtering: Applications to robotics. *IEEE Transactions on Circuits and Systems. I, Fundamental Theory and Applications, 49*(6), 753–767.

Eldessouky, A. S., & Gabbar, H. A. (2015). Micro grid renewables dynamic and static performance optimization using genetic algorithm. *Proc. IEEE Int. Conf. Smart Energy Grid Eng. (SEGE),* 1-6. 10.1109/SEGE.2015.7324596

Electric Power Research Institute (EPRI). (1993). *Principles and Practice of DSM.* TR-102556, Project # 2342-16. EPRI.

El-Sadek, M. Z. (1998). Preventive measures for voltage collapses and voltage failures in the Egyptian power system. *Electric Power Systems Research, 44*(3), 203–211. doi:10.1016/S0378-7796(97)01200-5

Elsobki, M. S., & Wahadan, S. (1999). *A Demand Side Management (DSM) Priority Selection Technique – its Design and Implementation.* CIRED, International Conference & Exhibition on Electricity Distribution, Nice, France.

El-Sobki, M. S., Jr. (1997). Tariff as a Demand Side Management Tool-its Design and Impact in View of Demand Reduction and Energy Savings technologies. Academic Press.

Elsobki, M. S. (1996). DSM Strategy Options – An Optimal Based formulation. *Proceedings AUPTDECIRED International Symposium*, 58-65.

EN/IEC 61800-5-2, *Adjustable speed electrical power drive systems* – Part 5-2: Safety requirements – Functional.

Ettihir, Boulon, & Agbossou. (2016). Energy management strategy for a fuel cell hybrid vehicle based on maximum efficiency and maximum power identification. *IET Electr. Syst. Transp., 6*(4), 261-268.

European Commission. (2011). *Standardization Mandate to European Standardisation Organisations (ESOs) to support European Smart Grid deployment.* Author.

Fan, Z. (2012). A distributed demand response algorithm and its application to PHEV charging in smart grids. *IEEE Trans. Smart Grid, 3*(3), 1280-1290.

Fan, M., Zhang, Z., Su, A., & Su, J. (2013). Enabling technologies for active distribution systems. *Zhongguo Dianji Gongcheng Xuebao, 33*(22), 12–18.

Fanuel, M., Khan, B., Singh, N., & Singh, P. (2018). Energy Production in Smart Cities by Utilization of Kinetic Energy of Vehicles over Speed Breaker. *International Journal of Civic Engagement and Social Change, 5*(2), 1–35. doi:10.4018/IJCESC.2018040101

Farhangi, H. (2010). The path of the smart grid. *Power Energy and Magazine. IEEE, 8*, 18–28.

Favuzza, Graditi, & Sanseverino. (2006). Adaptive and dynamic ant colony search algorithm for optimal distribution systems reinforcement strategy. *Appl. Intell., 24*(1), 31-42.

Ferracci, P. (2001). *Cahiers Techniques Schneider Electric no 199, Power Quality. Andreas Eberhard.*

Fini, M. H., Yousefi, G. R., & Alhelou, H. H. (2016). Comparative study on the performance of many-objective and single-objective optimisation algorithms in tuning load frequency controllers of multi-area power systems. *Generation, Transmission & Distribution, 10*(12), 2915-2923.

Fini, M. H., Yousefi, G. R., & Alhelou, H. H. (2016). Comparative study on the performance of many-objective and single-objective optimisation algorithms in tuning load frequency controllers of multi-area power systems. *IET Generation, Transmission & Distribution, 10*(12), 2915-2923.

Fini, M. H., Yousefi, G. R., & Alhelou, H. H. (2016). Comparative study on the performance of many-objective and single-objective optimisation algorithms in tuning load frequency controllers of multi-area power systems. *IET Generation, Transmission & Distribution, 10*(12), 2915–2923. doi:10.1049/iet-gtd.2015.1334

Finnigan, J. (2014). *Smart Planning for a Successful Smart Grid Roll-Out. Environmental Defense Fund-EFD.* Retrieved from http://blogs.edf.org/energyexchange/2014/01/30/smart-planning-for-a-successful-smart-grid-roll-out/

Frangopol, D. (2003). Reliability Deterioration and Life Time Maintenance Cost Optimization. University of Colorado.

Fu, Q., Montoya, L. F., Solanki, A., Nasiri, A., Bhavaraju, V., Abdallah, T., & ... . (2012). Microgrid generation capacity design with renewables and energy storage addressing power qualityand surety. *IEEE Transactions on Smart Grid, 3*, 2019–2027.

Gamarra & Guerrero. (2015). Computational optimization techniques applied to micro-grids planning: A review. *Renew. Sustain. Energy Rev., 48*, 413-424.

Gampa, S. R., & Das, D. (2015). Real power and frequency control of a small isolated power system. Int. *J Elect Power Energy Syst.*, *64*, 221–232. doi:10.1016/j.ijepes.2014.07.037

Ganguly, S., Mahto, T., & Mukherjee, V. (2017). Integrated frequency and power control of an isolated hybrid power system considering scaling factor based fuzzy classical controller. *Swarm and Evolutionary Computation*, *32*, 184–201. doi:10.1016/j.swevo.2016.08.001

Geidl, M., Koeppel, G., Favre-Perrod, P., Klockl, B., Andersson, G., & Frohlich, K. (2007). Energy hubs for the future. *IEEE Power & Energy Magazine*, *5*(1), 24–30. doi:10.1109/MPAE.2007.264850

Gellings, C. W. (1985). The concept of Demand Side Management for Electric Utilities. *IEEE, 73*(10).

Gellings, C. W., & Chamberlin, J. H. (1993). *DSM Concepts and Methods. The Fairmont Press.*

Global Smart Grid Impact Report. (n.d.). Retrieved fromhttps://www.smartgrid.gov/files/global_smart_grid_impact_report_2013.pdf

Golovanov, N., Postolache, P., Iordănescu, I., Toader, C., Popescu, S., Porumb, R., & Lipan, L. (2008). Instalații Electroenergetice ȘI Elemente De Audit Industrial. Bucharest, Romania: Academic Press.

Goncalves, M. S., Lopez, R. H., & Miguel, L. F. F. (2015). Search group algorithm: A new metaheuristic method for the optimization of truss structures. *Computers & Structures*, *153*, 165–184. doi:10.1016/j.compstruc.2015.03.003

Goodwin, G. C., Seron, M. M., & Dona, J. D. (2004). *Constrained Control and Estimation*. New York: Springer-Verlag.

Goudarzi, H., Hatami, S., & Pedram, M. (2011), Demand-side load scheduling incentivized by dynamic energy prices. *Proceedings of the IEEE International Conference on Smart Grid Communications (SmartGridComm)*, 351–356. 10.1109/SmartGridComm.2011.6102346

Graditi, Ippolito, Telaretti, & Zizzo. (2016). Technical and economical assessment of distributed electrochemical storages for load shifting applications: An Italian case study. *Renew. Sustain. Energy Rev.*, *57*, 515-523.

Gribkov, S. V. (2017). Vetroehnergeticheskoe oborudovanie kompleksy garantirovannogo ehlektrosnabzheniya maloj moshchnosti nauchno-inzhenernogo centra «Vindek» [Wind power equipment complexes of guaranteed power supply of low power of the scientific and engineering center "Vindek"]. *Ehnergetika i ehlektrooborudovanie, 6*(44), 24-27.

Griffiths Anthony, J. (1993). *An introduction to genetic analysis*. New York: W. H. Freeman and Company.

Guerrero, J. M., Hang, L., & Uceda, J. (2008). Control of distribution uninterruptible power supply systems. *IEEE Transactions on Industrial Electronics*, *55*(8), 2845–2859.

Gu, H., Yang, Z., Wang, D., & Wu, W. (2006, August). Research on control method of double-mode inverter with grid-connection and stand-alone. In *Power Electronics and Motion Control Conference, 2006. IPEMC 2006. CES/IEEE 5th International* (Vol. 1, pp. 1-5). IEEE.

Guha, D., Roy, P. K., & Banerjee, S. (2018). Robust Optimization Algorithms for Solving Automatic Generation Control of Multi-Constrained Power System. In Handbook of Research on Power and Energy System Optimization. IGI Global. doi:10.4018/978-1-5225-3935-3.ch003

Guha, D., Roy, P. K., & Banerjee, S. (2016a). Load frequency control of interconnected power system using grey wolf optimization. *Swarm and Evolutionary Computation*, *27*, 97–115. doi:10.1016/j.swevo.2015.10.004

Guha, D., Roy, P. K., & Banerjee, S. (2016b). Oppositional biogeography-based optimisation applied to SMES and TCSC-based load frequency control with generation rate constraints and time delay. *Int J of Power and Energy Conversion, 7*(4), 391–423. doi:10.1504/IJPEC.2016.079887

Guha, D., Roy, P. K., & Banerjee, S. (2017). Multi Verse Optimizatio: A novel method for solution of load frequency control problem in power system. *IET Generation, Transmission & Distribution, 11*(14), 3601–3611. doi:10.1049/iet-gtd.2017.0296

Guha, D., Roy, P. K., & Banerjee, S. (2018). Optimal tuning of 3 degree-of-freedom proportional-integral-derivative controller for hybrid distributed power system using dragonfly algorithm. *Computers & Electrical Engineering, 72*, 137–153. doi:10.1016/j.compeleceng.2018.09.003

Guo, Y., Pan, M., & Fang, Y. (2012). Optimal power management of residential customers in the smart grid. *IEEE Transactions on Parallel and Distributed Systems, 23*(9), 1593–1606. doi:10.1109/TPDS.2012.25

Gusarov, V. A., & Gusarova, E. V. (2017). Mikrogazoturbinnaya ustanovka dlya avtonomnogo ehnergosnabzheniya [Micro-gas turbine installation for autonomous power supply]. *Gazoturbinnye tekhnologii, 6*(149), 8-12.

Gusarov, V. A., & Kharchenko, V. V. (2018). Mikrogazoturbinnaya ustanovka GTEH-10S [Micro-Gas Turbine Installation GTE-10S]. *Vestnik VIEHSKH, 1*(30), 49–55.

Gusarov, V. A., Kharchenko, V. V., & Lapshin, S. A. (2013). Al'ternativnaya ehnergetika i ehkologiya [Alternative energy and ecology]. *Nauchno-tekhnicheskij centr. TATA, 7*, 15–18.

Haddad, R., & Hassan, B. (2001). Develop A Methodology to Analyze the Factors Affecting Maintenance Costs in Government Hospitals. *Tishreen University Magazine.*

Hadeed, A. S., Khaled, E. A., & Yasin, K. (2016). Effect of short circuited DC link capacitor of an AC–DC–AC inverter on the performance of induction motor. *Journal of King Saud University – Engineering and Science, 28*, 199–26.

Hadjipaschalis, I., Poullikkas, A., & Efthimiou, V. (2009). Overview of current and future energy storage technologies for electric power applications. *Renewable & Sustainable Energy Reviews, 13*, 1513–1522.

Haes, H. S. A., Golshan, M. E. H., & Fini, M. H. (2015). Multi agent electric vehicle control based primary frequency support for future smart micro-grid. *2015 Smart Grid Conference (SGC)*, 22-27. 10.1109/SGC.2015.7857385

Hajar, K., Hably, A., Elrafhi, A., Obeid, Z., & Bacha, S. (2015). Optimization of a micro-grid with renewable energy and distributed generation: A case study. *Proc. 19th Int. Conf. Syst. Theory, Control Comput. (ICSTCC)*, 662-665. 10.1109/ICSTCC.2015.7321369

Hamilton, B. A., Miller, J., & Renz, B. (2010). *Understanding the benefits of the smart grid-smart grid implementation strategy*. United States Department of Energy's National Energy Technology Laboratory.

Haneyoshi, T., Kawamura, A., & Hoft, R. G. (1988). Waveform compensation of PWM inverter with cyclic fluctuating loads. *IEEE Transactions on Industry Applications, 24*(4), 582–589.

Hao, G., Cong, R., & Zhou, H. (2014). PSO applied to optimal operation of a micro-grid with wind power. *Proc. 6th Int. Symp. Parallel Architectures, Algorithms Programm.*, 46-51. 10.1109/PAAP.2014.61

Hatziargyriou, N., Asano, H., Iravani, R., & Marnay, C. (2007). Microgrids. *IEEE Power & Energy Magazine, 5*(4), 78–94. doi:10.1109/MPAE.2007.376583

Haupt, R. L., & Ellen, H. S. (2004). *Practical genetic algorithms* (2nd ed.). John Wiley & Sons, Inc.

Heidari, A. A., & Pahlavani, P. (2017). An efficient modified grey wolf optimizer with Lévy flight for optimization tasks. *Applied Soft Computing*, *60*, 115–134. doi:10.1016/j.asoc.2017.06.044

Heuberger, D. (2015, November). Roadmap for Smart Grids: Four Steps to an intelligent electrical distribution grid. In *International ETG Congress 2015; Die Energiewende-Blueprints for the new energy age; Proceedings of* (pp. 1-6). VDE.

Heydt, G. T. (2000). *Contemporary Topics in Electric Power Quality*. Arizona State University.

Hoadley, F.L. (2008). Curb the disturbance In Industry Applications. *IEEE Magazine, 14*.

Holland, J. H. (1975). *Adaptation in Natural and Arti_cial Systems*. Ann Arbor, MI: Univ. Michigan Press.

Hosseini, & Karami, & KarimiMadahi. (2011). Optimal capacity, location and number of distributed generation at 20 kv substations. *Australian Journal of Basic and Applied Sciences*, *5* (10), 1051-1061.

Hosseini, S. A., Karami, M. S., & Madahi, S. K. (2011). Optimal capacity, location and number of distributed generation at 20 kv substations. *Australian Journal of Basic and Applied Sciences*.

Howlader, A. M., Izumi, Y., Uehara, A., Urasaki, N., Senjyu, T., Yona, A., & Saber, A. Y. (2012). A minimal order observer based frequency control strategy for an integrated wind-battery-diesel power system. *Energy*, *46*(1), 168–178. doi:10.1016/j.energy.2012.08.039

Hsu, C. T., Chuang, H. J., & Chen, C. S. (2011). Adaptive load shedding for an industrial petroleum cogeneration system. *Expert Systems with Applications*, *38*(11), 13967–13974.

Hu, Jiang, Egardt, & Cao. (2015). Advanced power-source integration in hybrid electric vehicles: Multicriteria optimization approach. *IEEE Trans. Ind. Electron.*, *62*(12), 7847-7858.

Huang, J. Y., Jiang, C. W., & Xu, R. (2008). A review on distributed energy resources and Microgrid. *Renewable & Sustainable Energy Reviews*, 2472–2483.

Huh. (2017). *Smart Grid Test Bed Using OPNET and Power Line Communication: Emerging Research and Opportunities*. IGI-Global.

Hussein, A. A. (2010). Optimal Based Demand Side Management DSM Formulation. In *Proceedings of the 14th International Middle East Power Systems Conference (MEPCON'10)*. Cairo University.

Hu, X., Karimi, H. R., Wu, L., & Guo, Y. (2014). Model predictive control-based non-linear fault tolerant control for air-breathing hypersonic vehicles. *IET Control Theory & Applications*, *8*(13), 1147–1153.

Ibrahim, H., Ilinca, A., & Perron, J. (2008). Energy storage systems—characteristics and comparisons. *Renewable & Sustainable Energy Reviews*, *12*, 1221–1250.

Idrus, A., Khamidi, F., & Sodangi, M. (2010). Maintenance management framework for conservation of heritage buildings in Malaysia. *Modern Applied Science*, *4*(11), 66. doi:10.5539/mas.v4n11p66

IEA. (2014a). Energy Technology Roadmaps: A Guide to Development and Implementation. OECD/IEA.

IEC 61000-6-4, *Electromagnetic compatibility (EMC) – Part 6-4: Generic standards – Emission standard for industrial environments.*

IEC TS 61000-1-2,(2008), *basic safety publication, draft second edition, Electromagnetic Compatibility (EMC) – Part 1-2: Design and Layout Techniques*, Nutwood/ Armstrong.

IEC/TR 61000-2-5, Electromagnetic compatibility *(EMC) – Part 2-5: Environment – Description and classification* of electromagnetic environments

IEEE PES-PSRC-SPSWG C-6. (2002). *Wide Area Protection and Emergency Control.* Retrieved from http://www.pes-psrc.org

IEEE PSRC WG D6. (2005). *Power swing and out-of-step considerations on transmission lines.* New York: IEEE.

IEEE Standard 1250-1995, *IEEE Guide for Service to Equipment Sensitive to Momentary Voltage Disturbances.*

IEEE Standard 142-1991, *IEEE Recommended Practice for Grounding of Industrial and Commercial Power Systems.*

IEEE Standard 446-1995, *IEEE Recommended Practice for Emergency and Standby Power Systems for Industrial and Commercial Applications.*

IEEE Standard 519-1992, *IEEE Recommended Practices and Requirements for Harmonic Control in Electrical Power Systems*

IEEE Standards Board, IEEE Std. 1159-1995, *"IEEE Recommended Practice for Monitoring Electric Power Quality"*. New York: IEEE, Inc. June 1995.

International Electrotechnical Commission. (2010). *IEC smart grid standardization roadmap.* Author.

International Energy Agency. (2015). *How 2 Guide for Smart Grids in Distribution Networks Roadmap Development and Implementation.* Available at https://www.iea.org/publications/freepublications/publication/TechnologyRoadmapHow2GuideforSmartGridsinDistributionNetworks.pdf

International Energy Agency's (IEA) Global EV Outlook. (GEVO). (2018). Retrieved from https://www.iea.org/gevo2018/

Jariso, M., Khan, B., Tesfaye, D., & Singh, J. (2017). Modelling and Designing of Stand-Alone Photovoltaic System (Case Study: Addis Boder Health Center south west Ethiopia). *IEEE International conference on Electronics, communication, and aerospace Technology, ICECA 2017*, 168-173.

Jayadev, V., & Swarup, K. S. (2013). Optimization of micro-grid with demand side management using Genetic Algorithm. *Proc. IET Conf. Power Unity, Whole Syst. Approach*, 1-6. 10.1049/ic.2013.0124

Jiang, S., Cao, D., Li, Y., Liu, J., & Peng, F. Z. (2012). Low-THD, fast-transient, and cost-effective synchronous-frame repetitive controller for three-phase UPS inverters. *IEEE Transactions on Power Electronics, 27*(6), 2994–3005.

Jin, C., Gao, M., Lv, X., & Chen, M. (2012, September). A seamless transfer strategy of islanded and grid-connected mode switching for microgrid based on droop control. In Energy Conversion Congress and Exposition (ECCE), 2012 IEEE (pp. 969-973). IEEE. doi:10.1109/ECCE.2012.6342714

Jin, M., & Sidhu, T. S. (2008). Adaptive load encroachment prevention scheme for distance protection. *Electric Power Systems Research, 78*(10), 1693–1700. doi:10.1016/j.epsr.2008.02.016

Jithin, K. K., Hareesh, S. V., Raja, P., & Selvan, M. P. (2017). Design and implementation of an algorithm for diagnosis of load encroachment in EHV Lines. *Energy Procedia, 117*, 519–526. doi:10.1016/j.egypro.2017.05.178

Joffe, E. (2010). Grounds for Grounding. A Circuit to System Handbook. Wiley-IEEE.

Johnson, J. M., & Rahmat-Samii, Y. (1997). Genetic algorithms in engineering electromagnetics. *IEEE Antennas & Propagation Magazine, 39*(4), 7–21. doi:10.1109/74.632992

Jonsson, M., & Daalder, J. E. (2003). An adaptive scheme to prevent undesirable distance protection operation during voltage instability. *IEEE Transactions on Power Delivery*, *18*(4), 1174–1180. doi:10.1109/TPWRD.2003.817501

Jose, T., Biswal, M., Venkatanagaraju, K., & Malik, O. P. (2018). Integrated approach based third zone protection during stressed system conditions. *Electric Power Systems Research*, *161*, 199–211. doi:10.1016/j.epsr.2018.04.011

Kaliannan, J., Baskaran, A., & Dey, N. (2017). Automatic Generation Control of Thermal-Thermal-Hydro Power Systems with PID controller using Ant Colony Optimization. In Renewable and Alternative Energy: Concepts, Methodologies, Tools, and Applications. IGI-Global. doi:10.4018/978-1-5225-1671-2.ch023

Kamwa, I. (1990). Dynamic modeling and robust regulation of a no-storage wind-Diesel hybrid power system. *Electric Power Systems Research*, *18*(3), 219–233. doi:10.1016/0378-7796(90)90054-7

Karshenas, H. R., & Niroomand, M. (2005). Design and implementation of a single phase inverter with sine wave tracking method for emergency power supply with high performance. *International conference on electrical machines and systems*, 1232-1237.

Karve, S. (2002). Three of a kind. *IEE Review*, *46*(2), 27–31.

Katiraei, F., Iravani, R., Hatziargyriou, N., & Dimeas, A. (2008). Microgrids management. *IEEE Power & Energy Magazine*, *6*(3), 54–65. doi:10.1109/MPE.2008.918702

Katz, M. (2010). Demand Side Management: Reflections of an Irreverent Regulator.. Oregon Public Utility Commission.

Kazem, H. A. (2013). *Harmonic Mitigation Techniques Applied to Power Distribution Networks Advances in Power Electronics*.

Kema, D.N.V. (2014). *Microgrids – benefits, models, barriers and suggested policy initiatives for the commonwealth of Massachusetts*. Academic Press.

Kempf, C. J., & Kobayashi, S. (1999). Disturbance observer and feed forward design for a high-speed direct-drive positioning table. *IEEE Transactions on Control Systems Technology*, *7*(5), 513–526.

Kennedy, J., & Eberhart, R. C. (1995). Particle swarm optimization. *Proc. IEEE Int. Conf. Neural Netw.*, 1942-1948. 10.1109/ICNN.1995.488968

Khan, B., & Singh, P. (2017). Optimal Power Flow Techniques under Characterization of Conventional and Renewable Energy Sources: A Comprehensive Analysis. *Journal of Engineering*.

Khan, B., & Singh, P. (2017). Selecting a Meta-Heuristic Technique for Smart Micro-Grid Optimization Problem: A Comprehensive Analysis. IEEE Access, 5, 13951-13977.

Khan, B., & Singh, P. (2017a). Optimal Power Flow Techniques under Characterization of Conventional and Renewable Energy Sources: A Comprehensive Analysis. *Journal of Engineering*.

Khan, B., & Singh, P. (2017b). Selecting a Meta-Heuristic Technique for Smart Micro-Grid Optimization Problem: A Comprehensive Analysis. IEEE Access, 5, 13951-13977.

Khan, B., Agnihotri, G., & Gupta G. (2013). A multipurpose matrices methodology for transmission usage, loss and reliability margin allocation in restructured environment. *Electrical & Computer Engineering: An International Journal*, *2*(3).

Khan, B., Agnihotri, G., & Gupta, G. (2013). A multipurpose matrices methodology for transmission usage, loss and reliability margin allocation in restructured environment. *Electrical & Computer Engineering: An International Journal*, *2*(3).

Khan, B., & Agnihotri, G. (2012). A Novel Transmission Loss Allocation Method based on Transmission Usage. *IEEE Fifth Power India Conference*. 10.1109/PowerI.2012.6479479

Khan, B., Agnihotri, G., Gupta, G., & Rathore, P. (2014). A Power Flow Tracing based Method for Transmission Usage, Loss & Reliability Margin Allocation. *AASRI Procedia*, *7*, 94–100. doi:10.1016/j.aasri.2014.05.035

Khan, B., Agnihotri, G., & Mishra, A. S. (2016). An Approach for Transmission Loss and Cost Allocation by Loss Allocation Index and Cooperative Game Theory. *J. Inst. Eng. India Ser. B*, *97*(1), 41–46. doi:10.100740031-014-0165-1

Khan, B., Agnihotri, G., Rathore, P., Mishra, A., & Naidu, G. (2014). A Cooperative Game Theory Approach for Usage and Reliability Margin Cost Allocation under Contingent Restructured Market. *International Review of Electrical Engineering*, *9*(4), 854–862.

Khan, B., & Singh, P. (2017c). The Current and Future States of Ethiopia's Energy Sector and Potential for Green Energy: A Comprehensive Study. *International Journal of Engineering Research in Africa*, *33*, 115–139. doi:10.4028/www.scientific.net/JERA.33.115

Kharchenko, V. V., Nikitin, B. A., & Tikhonov, P. V. (2010). Estimation and forecasting of PV cells and modules parameters on the basis of the analysis of interaction of a sunlight with a solar cell material. *Proceedings of 4th International Conference TAE 2010, Trends in Agricultural Engineering,* 307-310.

Khattam, W. E., & Salama, M. M. A. (2004). Distributed generation technologies, definitions and benefits. *Electric Power Systems Research*, *71*(2), 119–128. doi:10.1016/j.epsr.2004.01.006

Khine, Z. (2016). Modeling and Harmonic Analysis for Distribution Network In Myanmar. *International Journal for Technological Research in Engineering, 4*(2).

Khosrow-Pour. (2018). *Incorporating Nature-Inspired Paradigms in Computational Applications*. IGI Global.

Kifle, Y., Khan, B., & Singh, J. (2018). Designing and Modelling Grid Connected Photovoltaic System: (Case Study: EEU Building at Hawassa city). *Int. J. of Convergence Computing*, *3*(1), 20–34. doi:10.1504/IJCONVC.2018.091113

Kifle, Y., Khan, B., & Singh, P. (2018). Assessment and Enhancement of Distribution System Reliability by Renewable Energy Sources and Energy Storage. *Journal of Green Engineering*, *8*(3), 219–262. doi:10.13052/jge1904-4720.832

Kim, C. H., Heo, J. Y., & Aggarwal, R. K. (2005). An enhanced zone 3 algorithm of a distance relay using transient components and state diagram. *IEEE Transactions on Power Delivery*, *20*(1), 39–46. doi:10.1109/TPWRD.2004.837827

Kim, E.-H., Kwon, J.-M., & Kwon, B.-H. (2009). Transformerless three-phase on-line UPS with high performance. *IET Power Electronics*, *2*(2), 103–112.

Kim, S. K., Park, C. R., Yoon, T. W., & Lee, Y. I. (2015). Disturbance observer based model predictive control for output voltage regulation of three-phase inverter for uninterruptible power supply applications. *European Journal of Control*, *23*, 71–83.

Kirubakaran, A., Jain, S., & Nema, R. (2009). A review on fuel cell technologies and power electronic interface. *Renewable & Sustainable Energy Reviews*, *13*, 2430–2440.

Kissaoui, M., Al Tahir, A. A. R., Abouloifa, A., Chaoui, F. Z., Abouelmahjoub, Y., & Giri, F. (2016). Output-Feedback Nonlinear Adaptive Control Strategy of Three-phase AC/DC Boost Power Converter for On-line UPS Systems. *IFAC-Papers On Line*, *49*(13), 324–329.

Kissaoui, M., Chaoui, F. Z., Abouloifa, A., Giri, F., & Abouelmahjoub, Y. (2014). Adaptive control of uninterruptible power supply based on AC/AC Power Converter. *2014 International Conference on Multimedia Computing and Systems (ICMCS)*, 1557-1562.

Kollimalla, S. K., Mishra, M. K., & Narasamma, N. L. (2014). Design and Analysis of Novel Control Strategy for Battery and Super-capacitor Storage System. *IEEE Transaction on Sustainable Energy*, *5*, 1137–1144.

Koohi-Kamali, S., Tyagi, V., Rahim, N., Panwar, N., & Mokhlis, H. (2013). Emergence of energy storage technologies as the solution for reliable operation of smart power systems: A review. *Renewable & Sustainable Energy Reviews*, *25*, 135–165.

Krein, P. T. (2007). Battery management for maximum performance in plug-in electric and hybrid vehicles. *Proceedings of the IEEE VPPC 2007, Vehicle Power and Propulsion Conference*, 2–5. 10.1109/VPPC.2007.4544086

Kumar, A., & Shankar, G. (2017). Quasi-oppositional harmony search algorithm based optimal dynamic load frequency control of a hybrid tidal–diesel power generation system. *IET Generation, Transmission & Distribution*, *12*(5), 1099–1108. doi:10.1049/iet-gtd.2017.1115

Kumar, D., & Zare, F. (2014). Analysis of harmonic mitigations using hybrid passive filters. *Proc. of PEMC*. 10.1109/EPEPEMC.2014.6980629

Kumari, S., & Shankar, G. (2018). Novel application of integral-tilt-derivative controller for performance evaluation of load frequency control of interconnected power system. *IET Generation, Transmission & Distribution*, *12*(14), 3550–3560. doi:10.1049/iet-gtd.2018.0345

Kundu, P., & Pradhan, A. K. (2014). Synchrophasor-assisted zone-3 operation. *IEEE Transactions on Power Delivery*, *29*(2), 660–667. doi:10.1109/TPWRD.2013.2276071

Kundur, P. (1994). *Power System Stability and Control*. New York: Mc-Graw-Hill.

Kwon, B. H., Choi, J. H., & Kim, T. W. (2001). Improved single-phase line-interactive UPS. *IEEE Transactions on Industrial Electronics*, *48*(4), 804–811.

Laaksonen, H. (2011). Technical solutions for low-voltage microgrid concept. *Acta Wasaensia*, *241*, 251.

Lachuga, Y. U. F., & Strebkov, D. S. (2009). *EHnergeticheskaya strategiya sel'skogo hozyajstva Rossii na period do 2020 g* [Energy strategy of agriculture in Russia for the period until 2020]. Moscow, Russia: VIEHSKH.

Lahyani, A., Venet, P., Guermazi, A., & Troudi, A. (2013). Battery/super-capacitors combination in uninterruptible power supply (UPS). *IEEE Transactions on Power Electronics*, *28*, 1509–1522.

Lajoie, B., Debarre, R., Fayet, Y., & Dreyfus, G. (2015). *Beyond smart meters Introduction to Smart Grids*. A.T. Kearney Energy Transition Institute. Retrieved from: http://www.energy-transition-institute.com/_/media/Files/ETI/Summary%20Introduction%20to%20Smart%20Grids.pdf

Lamarre, L. (1997). At home with telecommunications. *EPRI Journal*, *22*(1), 6–16.

Lapshin, S. A. (2013). Sistema upravleniya generaciej i raspredeleniem ehnergii v lokal'nyh setyah nizkogo napryazheniya s ispol'zovaniem vozobnovlyaemyh i ne vozobnovlyaemyh vidov ehnergii [Control system for generation and distribution of energy in local low-voltage networks using renewable and non-renewable types of energy]. Materialy X Mezhdunarodnoj ezhegodnoj konferencii "Vozobnovlyaemaya i malaya ehnergetika 2013".

Laria, A. (2009). Survey on microgrids: analysis of technical limitations to carry out new solutions. *Proceeding of the 13th European Conference on Power Electronics and Applications 2009 (EPE'09)*, 1-8.

Lasseter, R. H. (2002). Microgrids. In *Power Engineering Society Winter Meeting, 2002. IEEE* (Vol. 1, pp. 305-308). IEEE.

Lasseter, R. H. (2006). *Control and design of microgrid components.* PSERC Publication.

Lasseter, R. H. (2011). Smart distribution: Coupled micro-grids. *Proceedings of the IEEE, 99*(6), 1074–1082.

Lasseter, R. H., Eto, J. H., Schenkman, B., Stevens, J., Vollkommer, H., Klapp, D., & Roy, J. (2011). CERTS microgrid laboratory test bed. *IEEE Transactions on Power Delivery, 26*(1), 325–332. doi:10.1109/TPWRD.2010.2051819

Latha, Y. (2012, March). Harmonics Mitigation of Industrial Motor Drives with Active Power Filters in Cement Plant-A Case Study. *International Journal of Power Electronics and Drive System, 2*(1), 1–8.

Laureri, F., Puliga, L., & Robba, M. (2016). An optimization model for the integration of electric vehicles and smart grids: Problem definition and experimental validation. *Proc. IEEE Int. Smart Cities Conf. (ISC)*, 1-6.

Lee, D. J., & Wang, L. (2008). Small-Signal Stability Analysis of an Autonomous Hybrid Renewable Energy Power Generation/Energy Storage System Part I: Time-Domain Simulations. *IEEE Transactions on Energy Conversion, 23*(1), 311–320. doi:10.1109/TEC.2007.914309

Letendre, S., & Watts, R. A. (2009). *Effects of Plug-in Hybrid Electric Vehicles on the Vermont Electric Transmission System.* Transportation Research Board Annual Meeting, Washington, DC.

Li, D., Jayaweera, S., & Naseri, A. (2011). Auctioning game based demand-response scheduling in smart grid. *Online Conf. Green Commun. (GreenCom).* 10.1109/GreenCom.2011.6082508

Li, Wu, Hu, Li, Hossain, & Ghoneim. (2017). HELOS: Heterogeneous load scheduling for electric vehicle-integrated micro-grids. *IEEE Trans. Veh. Technol., 66*(7), 5785-5796.

Liang, G., Liyuan, P., Ruihuan, L., Fen, Z., Jinhui, L., & Xin, W. (2014). Study on economic operation for micro-grid based on scenario and PSO. *Proc. Int. Conf. Power Syst. Technol.*, 3152-3156. 10.1109/POWERCON.2014.6993642

Liang, H., & Zhuang, W. (2014). Stochastic modeling and optimization in a microgrid: A survey. *Energies, 7*(4), 2027–2050. doi:10.3390/en7042027

Liao, G. C. (2010). Using chaotic quantum genetic algorithm solving environmental economic dispatch of smart micro-grid containing distributed generation system problems. *Proc. Int. Conf. Power Syst. Technol.*, 1-7. 10.1109/POWERCON.2010.5666468

Li, F., Qiao, W., Sun, H., Wan, H., Wang, J., Xia, Y., & Zhang, P. (2010). Smart transmission grid: Vision and framework. *IEEE Transactions on Smart Grid, 1*(2), 168–177. doi:10.1109/TSG.2010.2053726

Lin & Su. (2012). Using an Efficient Artificial Bee Colony Algorithm for Protein Structure Prediction on Lattice Models. *International Journal of Innovative Computing, Information and Control, 8*(3B), 2049-2064.

Litos Strategic Communication. (2008). *The Smart Grid: An introduction.* U.S. Department of Energy.

Liu, X., & Su, B. (2008). Microgrids – An Integration of RE Technologies. *Proceeding of China International Conference on Electricity Distribution,* 1-7.

Liu, J., Laghrouche, S., Ahmed, F. S., & Wack, M. (2014). PEM fuel cell air-feed system observer design for automotive applications: An adaptive numerical differentiation approach. *International Journal of Hydrogen Energy, 39*(30), 17210–17221.

Liu, J., Laghrouche, S., & Wack, M. (2014). Observer-based higher order sliding mode control of power factor in three-phase ac/dc converter for hybrid electric vehicle applications. *International Journal of Control, 87*(6), 1117–1130.

Liu, S., Zhuang, S., Xu, Q., & Xiao, J. (2016). Improved voltage shift islanding detection method for multi-inverter grid-connected photovoltaic systems. *IET Generation, Transmission & Distribution*, *10*(13), 3163–3169.

Lopes, J. P., Moreira, C. L., Madureira, A. G., Resende, F. O., Wu, X. A. W. X., Jayawarna, N. A. J. N., . . . Hatziargyriou, N. A. H. N. (2005, November). Control strategies for microgrids emergency operation. In *Future Power Systems, 2005 International Conference on* (pp. 6-pp). IEEE. 10.1109/FPS.2005.204226

Lopes, J. P., Moreira, C. L., & Madureira, A. G. (2006). Defining control strategies for microgrids islanded operation. *IEEE Transactions on Power Systems*, *21*(2), 916–924. doi:10.1109/TPWRS.2006.873018

Luo, X., Wang, J., Dooner, M., & Clarke, J. (2015). Overview of Current Development in Electric Energy Storage Technologies and the Application Potential in Power Systems Operation. *Applied Energy*, *137*, 511–536. doi:10.1016/j.apenergy.2014.09.081

Lurie, B. J. (1994). *Three parameters tunable tilt-integral derivative (TID) controller*. US Patent, US5371670.

Machowski, J., Janus, W. B., & Bumby, J. R. (1988). *Power System Dynamics*. Academic Press.

Maciejowski, J. M. (2001). *Predictive Control with Constraints*. Englewood Cliffs, NJ: Prentice-Hall.

MacKay, D. J. C. (2009). *Sustainable Energy: Without the Hot Air*. Cambridge, UK: UIT.

Madrigal, M., Uluski, R., & Gaba, M. K. (2017). Practical Guidance for Defining a Smart Grid Modernization Strategy: The Case of Distribution (Revised Edition). The World Bank.

Mahari, A., & Seyedi, H. (2015). High impedance fault protection in transmission lines using a WPT-based algorithm. *Electric Power and Energy Systems*, *67*, 537–545. doi:10.1016/j.ijepes.2014.12.022

Mahat, P., Chen, Z., & Bak-Jensen, B. (2010). Underfrequency load shedding for an islanded distribution system with distributed generators. *IEEE Transactions on Power Delivery*, *25*(2), 911–918. doi:10.1109/TPWRD.2009.2032327

Mahto, T., & Mukherjee, V. (2017). A novel scaling factor based fuzzy logic controller for frequency control of an isolated hybrid power system. *Energy*, *130*, 339–350. doi:10.1016/j.energy.2017.04.155

Makdisie, C., Haidar, B., & Alhelou, H. H. (2018). An Optimal Photovoltaic Conversion System for Future Smart Grids. In *Handbook of Research on Power and Energy System Optimization* (pp. 601–657). IGI Global.

Makdisie, C., Haidar, B., & Alhelou, H. H. (2018). An optimal photovoltaic conversion system for future smart grids. In *Handbook of Research on Power and Energy System Optimization* (pp. 601–657). IGI Global. doi:10.4018/978-1-5225-3935-3.ch018

Mallikarjuna, B., Vardhan Varma, P. V., Samir, S. P., Jaya bharatareddy, J., & Mohanta, D. K. (2017). An adaptive supervised wide-area backup protection scheme for transmission lines protection. *Protection and Control of Modern Power Systems, 2*(22), 1-16.

Marei, M. I., Abdallah, I., & Ashour, H. (2011). Transformerless Uninterruptible Power Supply with Reduced Power Device Count. *Electric Power Components and Systems*, *39*(11), 1097–1116.

Mariam, L., Basu, M., & Conlon, M. F. (2013). A Review of Existing Microgrid Architectures. *Journal of Engineering*. doi:10.1155/2013/937614

Martin, K. E., Benmouyal, G., Adamiak, M. G., Begovic, M., Burnett, R. O., Carr, K. R., ... Michel, G. L. (1998). IEEE standard for synchrophasors for power systems. *IEEE Transactions on Power Delivery*, *13*(1), 73–77. doi:10.1109/61.660853

Marzouk, M., & Abdel Aty, A. (2012). Maintaining subway infrastructure using BIM. In *Construction Research Congress 2012: Construction Challenges in a Flat World* (pp. 2320-2328). Academic Press. 10.1061/9780784412329.233

Mastromauro, R. A., Liserre, M., Kerekes, T., & Dell'Aquila, A. (2009). A single-phase voltage-controlled grid-connected photovoltaic system with power quality conditioner functionality. *IEEE Transactions on Industrial Electronics*, *56*(11), 4436–4444.

Mattavelli, P. (2005). An improved deadbeat control for UPS using disturbance observers. *IEEE Transactions on Industrial Electronics*, *52*, 206–212.

Mckenzie, M. (n.d.). *Developing a standards road map for smart grid Deployment in Australia*. Retrieved from http://www.standards.org.au/OurOrganisation/Events/Documents/Smart%20Grid%20Forum%20-%20Presentation%204%20-%20Mark% 20 McKenzie.pdf

Mekhilef, S., Saidur, R., & Safari, A. (2012). Comparative study of different fuel cell technologies. *Renewable & Sustainable Energy Reviews*, *16*, 981–989.

Melhem, F. Y., Moubayed, N., & Grunder, O. (2016). Residential energy management in smart grid considering renewable energy sources and vehicleto- grid integration. *Proc. IEEE Electr. Power Energy Conf. (EPEC)*, 1-6.

Melo, V. V. D., & Banzhaf, W. (2017). *Drone Squadron Optimization: a novel self-adaptive algorithm for global numerical optimization. Neural Comput and Applic.* doi:10.100700521-017-2881-3

Meyer, J., Bollen, M., Amarus, H., Blanco, A., De Castro, A., Desmet, J., ... Yang, K. (2014). *Future Work on harmonics – Some Expert Opinions Part II – Supraharmonics, Standards and Measurements. Proc. IEEE.*

Michael, B., Münch, V., Aichinger, M., Kuhn, M., Weymann, M., & Schmid, G. (2011). Power blackout risks. *CRO Forum.*

Millner, A. (2010). Modeling Lithium Ion battery degradationin electric vehicles. *Proceedings of the IEEE CITRES 2010 Conference on Innovative Technologies for an efficient and reliable electricity supply*, 349 – 356. 10.1109/CITRES.2010.5619782

Mirjalili, S., & Lewis, A. (2016). The whale optimization algorithm. *Advances in Engineering Software*, *95*, 51–67. doi:10.1016/j.advengsoft.2016.01.008

Mohamed, F. A., & Koivo, H. N. (2007). System modelling and online optimal management of MicroGrid using multiobjective optimization. *Proc. Int. Conf. Clean Electr. Power*, 148-153. 10.1109/ICCEP.2007.384202

Mohan, N., & Undeland, T. M. (2007). *Power electronics: converters, applications, and design*. John Wiley & Sons.

Mohsenian-Rad, A., Wong, V., Jatskevich, J., Schober, R., & Leon-Garcia, A. (2010). Autonomous demand-side management based on game-theo-retic energy consumption scheduling for the future smart grid. *IEEE Transactions on Smart Grid*, *1*(3), 320–331. doi:10.1109/TSG.2010.2089069

Molla, T., Khan, B., & Singh, P. (2018). A Comprehensive Analysis of Smart Home Energy Management System Optimization Techniques. *Journal of Autonomous Intelligence*, *1*(1).

Mondal, D., Chakraborty, A., & Sengupta, A. (2014). *Power System small signals stability and Control.* Academic Press.

Monfared, M., Golestan, S., & Guerrero, J. M. (2014). Analysis, design, and experimental verification of a synchronous reference frame voltage control for single-phase inverters. *IEEE Transactions on Industrial Electronics*, *61*, 258–269.

Moore, T. (1999). Utility communications go into orbit. *EPRI Journal*, *24*(3), 18–27.

Morales-Morales, Cervantes, & Cano-Castillo. (2015). On the design of robust energy management strategies for FCHEV. *IEEE Trans. Veh. Technol., 64*(5), 1716-1728.

Morison, K., Wang, L., & Kundur, P. (2004). Power system security assessment. *IEEE Power & Energy Magazine*, *2*(5), 30–39. doi:10.1109/MPAE.2004.1338120

Moriyama, A., Ando, I., & Takahashi, I. (1998). Sinusoidal voltage control of a single phase uninterruptible power supply by a high gain PI circuit. In *Proceedings of 24th Annual Conference of Industrial Electronics Society*. IEEE.

Morozumi, S., Kikuchi, S., Chiba, Y., Kishida, J., Uesaka, S., & Arashiro, Y. (2008, July). Distribution technology development and demonstration projects in Japan. In Power and Energy Society General Meeting-Conversion and Delivery of Electrical Energy in the 21st Century (pp. 1-7). IEEE.

Morsali, J., Zare, K., & Hagh, M. T. (2017). Applying fractional order PID to design TCSC-based damping controller in coordination with automatic generation control of interconnected multi-source power system. *Engineering Science and Technology, an Int. J, 20*(1), 1-17.

Muller, S., Ammann, U., & Rees, S. (2005). New time-discrete modulation scheme for matrix converters. *IEEE Transactions on Industrial Electronics*, *52*(6), 1607–1615.

Mulualem, T., & Khan, B. (2017). Design of an off-grid hybrid PV/wind power system for remote mobile base station: A case study. *AIMS Energy*, *5*(1), 96–112. doi:10.3934/energy.2017.1.96

Naduvathuparambil, B., Valenti, M. C., & Feliachi, A. (2002). Communication delays in wide area measurement systems. In *System Theory, 2002. Proceedings of the Thirty-Fourth Southeastern Symposium on* (pp. 118-122). IEEE. 10.1109/SSST.2002.1027017

Nadweh, S., Hayek, G., Atieh, B., & Haes Alhelou, H. (2018). Using Four – Quadrant Chopper with Variable Speed Drive System Dc-Link to Improve the Quality of Supplied Power for Industrial Facilities. Majlesi Journal of Electrical Engineering

Nadweh, S., Hayek, G., Atieh, B., & Haes Alhelou, H. (2018). Using Four – Quadrant Chopper with Variable Speed Drive System Dc-Link to Improve the Quality of Supplied Power for Industrial Facilities. *Majlesi Journal of Electrical Engineering*.

Nadweh, S., Hayek, G., Atieh, B., & Haes Alhelou, H. (2019). Using Four – Quadrant Chopper with Variable Speed Drive System Dc-Link to Improve the Quality of Supplied Power for Industrial Facilities. Majlesi Journal of Electrical Engineering.

Nale, R., & Biswal, M. (2017).Comparative assessment of passive islanding detection techniques for microgrid. *International Conference on Innovations in Information, Embedded and Communication System*, 1-5.

Nasiraghdam & Jadid. (2012). Optimal hybrid PV/WT/FC sizing and distribution system reconfiguration using multi-objective artificial bee colony (MOABC) algorithm. *International Journal of Solar Energy, 86*(10), 3057–3071.

Nasiri, A. (2007). Digital control of three-phase series-parallel uninterruptible power supply systems. *IEEE Transactions on Power Electronics*, *22*, 1116–1127.

Nasiri, A., Nie, Z., Bekiarov, S. B., & Emadi, A. (2008). An on-line UPS system with power factor correction and electric isolation using BIFRED converter. *IEEE Transactions on Industrial Electronics*, *55*(2), 722–730.

Nasser, A., & Reji, P. (2016). Optimal planning approach for a cost effective and reliable micro-grid. *Proc. Int. Conf. Cogeneration, Small Power Plants District Energy (ICUE)*, 1-6. 10.1109/COGEN.2016.7728966

National Energy Technology Laboratory. (2009). *The Transmission Smart Grid Imperative.* Retrieved from https://www.netl.doe.gov/File%20Library/research/energy%20efficiency/smart%20grid/whitepapers/The-Transmission-Smart-Grid-Imperative_2009_09_29.pdf

National Institute of Standards and Technology. (2010). *NIST Framework and Roadmap for Smart Grid Interoperability Standards, Release 1.0.* NIST Special Publication 1108.

Navigant. (2015). *Ontario Smart Grid Assessment and Roadmap.* Report of Navigant. Retrieved from https://www.ontarioenergyreport.ca/pdfs/Navigant-Smart-Grid-Assessment-and-Roadmap-Final-Report-.pdf

Nayak, P. K., Pradhan, A. K., & Bajpai, P. (2015). Secured zone 3 protection during stressed condition. *IEEE Transactions on Power Delivery, 30*(1), 89–96. doi:10.1109/TPWRD.2014.2348992

Nayar, C. V., Ashari, M., & Keerthipala, W. (2000). A grid-interactive photovoltaic uninterruptible power supply system using battery storage and a back -up diesel generator. *IEEE Transactions on Energy Conversion, 15*, 348–353.

NERC. (2003). *NERC recommendations to blackout-prevent and mitigate the impacts of future cascading blackouts.* Retrieved from https://www.nerc.com

NERC. (2005a). *Protection system review program - Beyond zone 3.* Retrieved from https://www. nerc.com

NERC. (2005b). *Rationale for the use of local and remote (zone 3) protective relaying backup systems - A report on the implications and uses of zone-3 relays.* Retrieved from https://www.nerc.com

NERC-SPCS. (2011). *Reliability guideline: Transmission system phase backup protection.* Retrieved from https://www.nerc.com

New York State Smart Grid Consortium. (2010). *Smart Grid Roadmap for the State of New York.* Report to New York State Smart Grid Consortium. Retrieved from https://www.energy.gov/sites/prod/files/oeprod/DocumentsandMedia/NYSSGC_Attachment.pdf

Ngamroo, I. (2012). Application of electrolyzer to alleviate power fluctuation in a standalone microgrid based on an optimal fuzzy PID control. *Int. J Elect Power Energy Syst.*, 969-976.

Nikitin, O. (2012). Mikroturbiny v bor'be za potrebitelya [Microturbines in the struggle for the consumer]. *Ehkologicheskie sistemy, 11.*

Nikos, H. (2007). Microgrids: an overview of ongoing research, development, and demonstration projects. *IEEE Power & Energy, 5*(4).

Niroomand, M., & Karshenas, H. (2010). Review and comparison of control methods for uninterruptible power supplies. IEEE Power Electronic & Drive Systems & Technologies Conference (PEDSTC), 18-23.

NIST-Office of the National Coordinator for Smart Grid Interoperability & Office of the National Coordinator for Smart Grid Interoperability. (2018). *NIST framework and roadmap for smart grid interoperability standards, release 1.0.* US Department of Commerce, National Institute of Standards and Technology.

Njenda, T. C., Golshan, M. E. H., & Alhelou, H. H. (2018, November). WAMS Based Intelligent Under Frequency Load Shedding Considering Online Disturbance Estimation. In *Smart Grid Conference (SGC)* (pp. 1-5). Academic Press.

Njenda, T. C., Golshan, M. E. H., & Alhelou, H. H. (2018, November). WAMS based Under Frequency Load Shedding Considering Minimum Frequency Prediction and Extrapolated Disturbance Magnitude. In *Smart Grid Conference (SGC)* (pp. 1-5). Academic Press.

NRPC. (2010). *Report of the inquiry committee on grid disturbance in Northern region on 2nd January 2010.* Retrieved from http://www.nrpc.gov.in

Odeim, Roes, & Heinzel. (2016). Power management optimization of a fuel cell/battery/supercapacitor hybrid system for transit bus applications. *IEEE Trans. Veh. Technol., 65*(7), 5783-5788.

Olagunju, R.E. (2012a). Predictive Modelling for Sustainable Residential Building Maintenance in Developing Countries: A Nigeria Case. *Interdisciplinary Journal of Contemporary Research in Business, 4*(6), 1273–1283.

Olagunju, R.E. (2012b). Sustainability of Residential Buildings in Nigeria: An Appraisal of the Factors that Influence Maintenance of Residential Buildings 'Standards. *Civil and Environmental Research, 2*(4), 20–29.

Olivares, D. E., Cañizares, C. A., & Kazerani, M. (2011, July). *A centralized optimal energy management system for microgrids.* IEEE Power and Energy Society General Meeting.

Ortega, M., del Río, P., & Montero, E. A. (2013). Assessing the benefits and costs of renewable electricity. The Spanish case. *Renewable & Sustainable Energy Reviews, 27*, 294–304. doi:10.1016/j.rser.2013.06.012

Oudalova, A. (2011). *Advanced architectures and control concepts for more micro-grids.* Retrieved from www.microgrids.eu/documents/654.pdf

Pacis, Martinez, Jr., & Tecson. (2010). Modelling and Simulation of Active Power Filters for Harmonic Compensation, Voltage Sags and Swells Mitigation and Power Factor Correction. *Proceedings of the World Congress on Engineering and Computer Science Conference, (WCECS), 2.*

Pan, I., & Das, S. (2015). Fractional-order load-frequency control of interconnected power systems using chaotic multi-objective optimization. *Applied Soft Computing, 29*, 328–344. doi:10.1016/j.asoc.2014.12.032

Parhizi, S., Lotfi, H., Khodaei, A., & Bahramirad, S. (2015). State of the Art in Research on Microgrids: A Review. *IEEE Access: Practical Innovations, Open Solutions, 3*(1), 890–925. doi:10.1109/ACCESS.2015.2443119

Parikh, R., & Krishnan, R. (1994, September). Modeling, simulation and analysis of an uninterruptible power supply. In *Industrial Electronics, Control and Instrumentation, 1994. IECON'94., 20th International Conference on* (Vol. 1, pp. 485-490). IEEE.

Park, J. K., Kwon, J. M., Kim, E. H., & Kwon, B. H. (2008). High-performance transformerless online UPS. *IEEE Transactions on Industrial Electronics, 55*(8), 2943–2953.

Partha, K., & Chanda, C. K. (2013). A simple and fast approach for allocation and size evaluation of distributed generation. *International Journal of Energy and Environmental Engineering, 4*(1), 1–9.

Passey, R., Spooner, T., MacGill, I., Watt, M., & Syngellakis, K. (2011). The potential impacts of grid-connected distributed generation and how to address them: A review of technical and non-technical factors. *Energy Policy, 39*(10), 6280–6290. doi:10.1016/j.enpol.2011.07.027

Paterakis, Erdinç, Pappi, Bakirtzis, & Catalão. (2016). Coordinated operation of a neighborhood of smart households comprising electric vehicles, energy storage and distributed generation. *IEEE Trans. Smart Grid, 7*(6), 2736-2747.

Patterson, Macia, & Kannan. (2015). Hybrid micro-grid model based on solar photovoltaic battery fuel cell system for intermittent load applications. *IEEE Trans. Energy Convers., 30*(1), 359-366.

Paul, C. R. (2006). *Introduction to Electromagnetic Compatibility* (2nd ed.). Wiley Series in Microwave and Optical Engineering.

Pawan & Baseem. (2017). Smart Microgrid Energy Management Using a Novel Artificial Shark Optimization. Complexity.

Peças Lopes, J. A., Soares, F. J., & Rocha Almeida, P. M. (2009). Identifying management procedures to deal with connection of electric vehicles in the grid. *Proceedings of the IEEE PowerTech*. 10.1109/PTC.2009.5282155

Pecas, L. J. A., Soares, F. J., & Almeida, P. M. R. (2011). Integration of electric vehicles in the electric power system. *Proceedings of the IEEE*, *99*(1), 168–183. doi:10.1109/JPROC.2010.2066250

Peterson, S. B., Apt, J., & Whitacre, J. F. (2010). Lithium-ion battery cell degradation resulting from realistic vehicle and vehicle-to-grid utilization. *Journal of Power Sources*, *195*(8), 2385–2392. doi:10.1016/j.jpowsour.2009.10.010

Pfeiffer, B., & Mulder, P. (2013). Explaining the diffusion of renewable energy technology in developing countries. *Energy Econ.*, *40*, 285–296. doi:10.1016/j.eneco.2013.07.005

Piagi, P., & Lasseter, R. H. (2006, June). Autonomous control of microgrids. In *Power Engineering Society General Meeting* (pp. 8-pp). IEEE.

Pillai, J.R., & Bak-Jensen, B. (2011). Integration of vehicle to grid in the western Danish power system. *IEEE Trans Sustain Energy, 2*(1), 12–9.

Power Systems Relaying Committee. (2007). IEEE Guide for the Application of Protective Relays Used for Abnormal Frequency Load Shedding and Restoration. *IEEE Std C, 37*, c1–c43.

Powergrid. (2012). *Report of the enquiry committee on grid disturbance in northern region on 30th July 2012 and in Northern, Eastern and North-Eastern Region on 31st July 2012*. Tech Rep. Grid_ENQ_REP_16_8_12, 2012, New Delhi, India.

Prabaakaran, K., Chitra, N., & Kumar, A. S. (2013). Power quality enhancement in microgrid—A survey. *Proceedings of the IEEE International Conference on Circuits, Power and Computing Technologies (ICCPCT)*, 126–131.

Purcell, E. M. (2012). *Electricity and magnetism* (3rd ed.). Cambridge, UK: Cambridge Univ. Press.

Racine, M. S., Parham, J. D., & Rashid, M. (2005). An overview of uninterruptible power supplies. *Proceedings of the 37th Annual North American Power Symposium*, 159-64.

Radford, D. (1996). Spread spectrum data leap through AC power wiring. *IEEE Spectrum*, *33*(11), 48–53. doi:10.1109/6.542275

Rahman, A., Saikia, L. C., & Sinha, N. (2015). Load frequency control of a hydro-thermal system under deregulated environment using biogeography-based optimised three degree-of-freedom integral-derivative controller. *IET Generation, Transmission & Distribution*, *9*(15), 2284–2293. doi:10.1049/iet-gtd.2015.0317

Rajbongshi, R., & Saikia, L. C. (2018). Combined voltage and frequency control of a multi-area multisource system incorporating dish-Stirling solar thermal and HVDC link. *IET Renewable Power Generation*, *12*(3), 323–334. doi:10.1049/iet-rpg.2017.0121

Raju, M., Saikia, L. C., & Sinha, N. (2018). Maiden application of two degree of freedom cascade controller for multi-area automatic generation control. *Int Trans Electr Energ Syst., e2586*, 1-20.

Ramírez-Murillo, Restrepo, Calvente, Romero, & Giral. (2015). Energy management of a fuel-cell serial_parallel hybrid system. *IEEE Trans. Ind. Electron., 62*(8), 5227-5235.

Ramonas, C., Adomavicius, V., & Kepalas, V. (2009). Research of the power conversion processes in the system of power supply from a number of wind turbines over the one grid-tied inverter. *Proceedings of International Conference Electrical and Control Technologies*, 368-373.

Raneru, N. R. (2013). Harmonic Analysis of Small Scale Industrial Loads and Harmonic Mitigation Techniques in Industrial Distribution System. *International Journal of Engineering Research and Applications*, *3*(4), 1511–1540.

Rao & Obulesh. (2015). ABC and ANN based minimization of power loss for distribution system stability. *Proceedings of 10th IEEE conference on Industrial Electronics and Applications (ICIEA)*, 772-777.

Rao, G. S., & Obulesh, Y. P. (2013). Optimal Location of DG for Maintaining Distribution System Stability: A Hybrid Technique. *International Journal of Power and Energy Conversion*, *4*(4), 387–403. doi:10.1504/IJPEC.2013.057036

Rathmann, S., & Warner, H. A. (1996). New generation UPS technology, the delta conversion principle. *IAS '96. Conference Record of the 1996 IEEE Industry Applications Conference Thirty-First IAS Annual Meeting.*

Ravichandran, A., Malysz, P., Sirouspour, S., & Emadi, A. (2013). The critical role of microgrids in transition to a smarter grid: A technical review. *Proceedings of the IEEE Transportation Electrification Conference and Expo (ITEC)*, 1–7. 10.1109/ITEC.2013.6573507

Recioui, A., Djehaiche, M., & Boumezrag, A. (2016). *Load Side Management in Smart Grids using a Global Optimizer.* Presented at the 8th International Exergy, Energy and Environment Symposium, Antalya, Turkey.

Recioui, A. (2012). *Mutiple antenna systems over MIMO channels: Multiple antenna communication systems.* LAP Lambert Academic Publishing.

Redfern, M. A., Usta, O., & Fielding, G. (1993). Protection against loss of utility grid supply for a dispersed storage and generation unit. *IEEE Transactions on Power Delivery*, *8*(3), 948–954.

Ren, G., Ma, G., & Cong, N. (2015). Review of electrical energy storage system for vehicular applications. *Renewable & Sustainable Energy Reviews*, *41*, 225–236.

REO Guide on Good EMC Engineering Practices in the Design and Construction of Industrial Cabinets. (n.d.). Available from www.reo.co.uk/

Richards, G., Noble, B., & Belcher, K. (2012). Barriers to renewable energy development: A case study oflarge-scale wind energy in Saskatchewan, Canada. *Energy Policy*, *42*, 691–698. doi:10.1016/j.enpol.2011.12.049

Robak, S., Wasilewski, J., Dawidowski, P., & Szewczyk, M. (2016). *Variable Speed Drive (VSD) – towards modern industry and electric power systems*. Academic Press.

Rodríguez, J., Pontt, J., Silva, C., Salgado, M., Rees, S., Ammann, U., ... Cortes, P. (2004). Predictive control of a three-phase inverter. *Electronics Letters*, *40*(9), 561–562.

Rudez, U., & Mihalic, R. (2016). WAMS-based underfrequency load shedding with short-term frequency prediction. *IEEE Transactions on Power Delivery*, *31*(4), 1912–1920. doi:10.1109/TPWRD.2015.2503734

Saadat, H. (1999). *Power System Analysis McGraw-Hill Series in Electrical Computer Engineering*. Academic Press.

Saad, W., Han, Z., Poor, H. V., & Basar, T. (2012). Game-theoretic methods for the smart grid: An overview of microgrid systems, demand-side management, and smart grid communications. *IEEE Signal Processing Magazine*, *29*(5), 86–105. doi:10.1109/MSP.2012.2186410

Saber, A. Y., & Venayagamoorthy, G. K. (2013). Smart micro-grid optimization with controllable loads using particle swarm optimization. *Proc. IEEE Power Energy Soc. General Meet.*, 1-5. 10.1109/PESMG.2013.6672780

Saleh, M. S., Althaibani, A., Esa, Y., Mhandi, Y., & Mohamed, A. A. (2015). Impact of clustering micro grids on their stability and resilience during blackouts. *2015 International Conference on Smart Grid and Clean Energy Technologies (ICSGCE)*, 195–200. 10.1109/ICSGCE.2015.7454295

Samadi, P., Schober, R., & Wong, V. (2011). Optimal energy consumption scheduling using mechanism design for the future smart grid. *Proc. IEEE Int. Conf. Smart Grid Commun. (Smart Grid Comm.)*. 10.1109/SmartGridComm.2011.6102349

Sankaran, C. (2002). *Power quality*. Academic Press.

Santis, E. D., Rizzi, A., Sadeghiany, A., & Mascioli, F. M. F. (2013). Genetic optimization of a fuzzy control system for energy flow management in micro-grids. *Proc. Joint IFSA World Congr. NAFIPS Annu. Meet. (IFSA/NAFIPS)*, 418-423. 10.1109/IFSA-NAFIPS.2013.6608437

Sato, T., Kammen, D. M., Duan, B., Macuha, M., Zhou, Z., Wu, J., & Asfaw, S. A. (2015). *Smart grid standards: specifications, requirements, and technologies*. John Wiley & Sons. doi:10.1002/9781118653722

Seethalekshmi, K., Singh, S. N., & Srivastava, S. C. (2012). A classification approach using support vector machines to prevent distance relay maloperation under power swing and voltage instability. *IEEE Transactions on Power Delivery*, *27*(3), 1124–1133. doi:10.1109/TPWRD.2011.2174808

Sekar, T. C., & Rabi, B. J. (2012). A Review and Study of Harmonic Mitigation Techniques. *Proc. International Conference on Trends in Electrical Engineering and Energy Management*. 10.1109/ICETEEEM.2012.6494450

Semi, S., Nasri, S., Zafar, B., & Cherif, A. (2017). Multi-Input Single-Output State Space for Hybrid Power System Approach Using PEMFC: Fuel Cell and Applications. *International Journal of Energy Optimization and Engineering*, *6*(4), 35–50. doi:10.4018/IJEOE.2017100103

Senjyu, T., Nakaji, T., Uezato, T., & Funabashi, T. (2005). A Hybrid Power System Using Alternative Energy Facilities in Isolated Island. *IEEE Transactions on Energy Conversion*, *20*(2), 406–414. doi:10.1109/TEC.2004.837275

Setiawan, A. A., Zhao, Y., Lee, R. S., & Nayar, C. V. (2009). Design, Economic Analysis and Environmental Considerations of Mini-Grid Hybrid Power System with Reverse Osmosis desalination Plant for Remote Areas. *Renewable Energy-Elsevier.*, *34*(2), 374–383. doi:10.1016/j.renene.2008.05.014

Seyedfarshi, S., & Pourmostadam, K. (2013). *The challenges in developing a smart grid roadmap for the distribution network of Iran*. Academic Press.

Shabanzadeh, M., & Moghaddam, M. P. (2013, November). What is the smart grid? definitions, perspectives, and ultimate goals. *28th International Power System Conference (PSC)*.

Shamshiri, M., Gan, C. K., & Tan, C. W. (2012, June). A review of recent development in smart grid and micro-grid laboratories. *In Power Engineering and Optimization Conference (PEDCO) Melaka, Malaysia, 2012 IEEE International* (pp. 367-372). IEEE. 10.1109/PEOCO.2012.6230891

Shandilya, S. (2016). *Handbook of Research on Emerging Technologies for Electrical Power Planning, Analysis and Optimization*. IGI Global. doi:10.4018/978-1-4666-9911-3

Shankar, G., & Mukjerjee, V. (2016). Load frequency control of an autonomous hybrid power system by quasi-oppositional harmony search algorithm. *Int. J Elect Power Energy Syst.*, *78*, 715–734. doi:10.1016/j.ijepes.2015.11.091

Sharifzadeh, M., Lesaniand, H., & Sanaye-Pasand, M. (2014). A new algorithm to stabilize distance relay operation during voltage-degraded conditions. *IEEE Transactions on Power Delivery*, *29*(4), 1639–1647. doi:10.1109/TPWRD.2013.2285502

Sharma, Bhattacharjee, & Bhattacharya. (2016). Grey wolf optimisation for optimal sizing of battery energy storage device to minimise operation cost of micro-grid. *IET Generat., Transmiss. Distrib., 10*(3), 625-637.

Shehu, A. A. (2009). *Voltage stability and distance protection zone (Master's thesis).* Chalmers University of Technology. Retrieved from http://webfiles.portal.chalmers.se/et/MSc/ShehuAbbaAliyuMSc.pdf

Shen, J., Jiang, C., Liu, Y., & Wang, X. (2016). A Microgrid Energy Management System and Risk Management Under an Electricity Market Environment. IEEE Access, 4, 2349-2356. doi:10.1109/ACCESS.2016.2555926

Shi, K., Li, H., Hu, C., & Xu, D. (2015). Topology of super uninterruptible power supply with multiple energy sources. *2015 9th International Conference on Power Electronics and ECCE Asia (ICPE-ECCE Asia),* 1742-1749.

Shi, Z., Peng, Y., & Wei, W. (2014). Optimal sizing of DGs and storage for micro-grid with interruptible load using improved NSGA-II. *Proc. IEEE Congr. Evol. Comput. (CEC),* 2108-2115. 10.1109/CEC.2014.6900414

Shiki, A., Yokoyama, A., Baba, J., Takano, T., Gouda, T., & Izui, Y. (2007). Autonomous decentralized control of supply and demand by inverter based distributed generations in isolated microgrid. *IEEJ Transactions on Power and Energy, 127*(1), 95–103. doi:10.1541/ieejpes.127.95

Short, W., & Denholm, P. (2006). *A Preliminary Assessment of Plug-in Hybrid Electric Vehicles on Wind Energy Markets.* National Renewable Energy Laboratory, Golden, CO. Tech. Rep. NREL/TP-620-39729.

Shuai, D., Nian, L., & Yingmiao, C. (2013). Optimal operation of the island micro-grid with renewable energy and desalination. *Proc. Int. Conf. Mech. Sci., Electric Eng. Comput. (MEC),* 3718-3722.

Siemens and RWE Energy. (2008). *First virtual power plant operated by Siemens and RWE Energy.* Retrieved from http://w1.siemens.com/press/en/pressrelease

Silva, E. I., McGrath, B. P., Quevedo, D. E., & Goodwin, G. C. (2007). Predictive control of a flying capacitor converter. *Proceedings of American Control Conference*, 3763-3768.

Silvestre, Di, Graditi, & Sanseverino. (2014). A generalized framework for optimal sizing of distributed energy resources in micro-grids using an indicator-based swarm approach. *IEEE Trans Ind. Informat., 10*(1), 152-162.

Singh, O. (2017). Automatic generation control of multi-area interconnected power systems using hybrid evolutionary algorithm. In Handbook of Research on Soft Comput and Nature-Inspired Algorithms. IGI-Global. doi:10.4018/978-1-5225-2128-0.ch010

Singh, P., & Khan, B. (2017). Smart Microgrid Energy Management Using a Novel Artificial Shark Optimization. Complexity. doi:10.1155/2017/2158926

Singh, V. P., Mohanty, S. R., Kishor, N., & Ray, P. K. (2013). Robust H-infinity load frequency control in hybrid distributed generation system. *Int. J Elect Power Energy Syst., 46*, 294–305. doi:10.1016/j.ijepes.2012.10.015

Siqi, C. (2014). Optimal coordinated operation for micro-grid with hybrid energy storage and diesel generator. *Proc. Int. Conf. Power Syst. Technol.*, 3207-3212.

SMB Smart Grid Strategic Group (SG3). (2010). *IEC Smart Grid Standardization Roadmap.* International Electrotechnical Commission.

Sodhi, R., Shrivastav, S. C., & Singh, S. N. (2012). A simple scheme for wide area detection of impending voltage instability. *IEEE Transactions on Smart Grid*, *3*(2), 818–827. doi:10.1109/TSG.2011.2180936

Solter, W. (2002). A new international UPS classification by IEC 62040-3. *IEEE INTELEC 24th Annual International Telecommunications Energy Conference*, 541-555.

Song, H., Lee, B., & Ajarappu, V. (2009). Control strategies against voltage collapse considering undesired relay operations. *IET Generation, Transmission & Distribution*, *3*(2), 164–172. doi:10.1049/iet-gtd:20080055

Sortomme, E., & El-Sharkawi, M. A. (2011). Optimal Charging Strategies for Unidirectional Vehicle-to-Grid. *IEEE Transactions on Smart Grid*, *2*(1), 131–138. doi:10.1109/TSG.2010.2090910

Specht, M., Rohjans, S., Trefke, J., Uslar, M., & Vázquez, J. M. G. (2013). International Smart Grid Roadmaps and their Assessment. *EAI Endorsed Trans. Energy Web*, *1*(1), e2.

Standard EN 50160 Voltage characteristics of electricity supplied by public distribution systems. issued by CENELEC, November 1994.

Steenstra, H. T., & van Deursen, A. P. J. (2008). Reduction of conducted interference by steel armor in buried cables: Measurements and modelling. *IEEE Transactions on Electromagnetic Compatibility*, *50*(3), 678–686. doi:10.1109/TEMC.2008.926902

Stephen, P. (2010). *Variable frequency drives and energy savings.* Technical report, Siemens.

Stevens, J., Vollkommer, H., & Klapp, D. (2007, June). CERTS microgrid system tests. In 2007 IEEE Power Engineering Society General Meeting (pp. 1-4). IEEE. doi:10.1109/PES.2007.385793

Storn, R., & Price, K. (1997). Differential evolution: A simple and efficient heuristic for global optimization over continuous spaces. *J. Global Optim.*, *11*(4), 341-359.

Strbac, G. (2008). Demand side management: Benefits and challenges. *Energy Policy*, *36*(12), 4419–4426. doi:10.1016/j.enpol.2008.09.030

Strebkov, D. S., & Irodionov, A. E. (2004). Global Solar Power System. In *Eurosun 2004. 14 Intern. Sonnen Forum, Frebing, Germany. Vol. 3, PV systems and PV Cells* (pp. 336 – 343). Academic Press.

Strebkov, D. S., & Irodionov, A. E. (2005). PV research and technological development in Russia. In *Proceedings of the 2005 Solar World Congress.* American Solar Energy Society, International Solar Energy Society CD, ISES.

Strebkov, D., Nekrasov, A., Trubnikov, V., & Nekrasov, An. (2018). Single-Wire Resonant Electric Power Systems for Renewable-Based Electric Grid. In V. Kharchenko & P. Vasant (Eds.), Handbook of Research on Renewable Energy and Electric Resources for Sustainable Rural Development (pp. 449–474). Academic Press. doi:10.4018/978-1-5225-3867-7.ch019

Strebkov, D.S., & Kharchenko, V.V. (2011). Rol' i mesto VIEH v razvitii global'noj ehnergetiki [The role and place of renewable energy in the development of global energy]. *ZH. Malaya ehnergetika*, *3*(3), 3-12.

Strebkov, D. S., Tver'yanovich, EH. V., Tyuhov, I. I., & Irodionov, A. E., & Yarcev, N.V. (2002). Solnechnye koncentratornye tekhnologii dlya ehnergoobespecheniya zdanij [Solar concentrator technologies for power supply of buildings]. *Geliotekhnika*, *3*, 64–68.

Su, L., Zhang, J. H., Wang, L., Miao, W. S., & Wu, Z. P. (2010). Study on some key problems and technique related to microgrid. *Power System Protection and Control*, *38*(19), 235–239.

Sun, Liu, Cai, Zhang, Zhu, & Wang. (2016). Frequency based power management for photovoltaic/battery/fuel cell-electrolyser stand-alone micro-grid. *IET Power Electron., 9*(13), 2602-2610.

Surachet, T. (1994). *Analysis of power system disturbances due to relay hidden failures* (PhD thesis). Virginia Polytechnic Institute.

Tah, A., & Das, D. (2016). Operation of small hybrid autonomous power generation system in isolated, interconnected and grid connected modes. *Sustainable Energy Technologies and Assessments, 17*, 11–25. doi:10.1016/j.seta.2016.07.001

Tang, J., Liu, J., Ponci, F., & Monti, A. (2013). Adaptive load shedding based on combined frequency and voltage stability assessment using synchrophasor measurements. *IEEE Transactions on Power Systems, 28*(2), 2035–2047. doi:10.1109/TPWRS.2013.2241794

Tang, Q., Liu, N., & Zhang, J. (2014). Optimal operation method for micro-grid with wind/PV/diesel generator/battery and desalination. *Journal of Applied Mathematics, 2014*(Jun), 857541.

Terzija, V. V., Valverde, G., Cai, D., Regulski, P., Madani, V., Fitch, J., ... Phadke, A. G. (2011). Wide-area monitoring, protection, and control of future electric power networks. *Proceedings of the IEEE, 99*(1), 80–93. doi:10.1109/JPROC.2010.2060450

Thale, Wandhare, & Agarwal. (2015). A novel reconfigurable micro-grid architecture with renewable energy sources and storage. *IEEE Trans. Ind. Appl., 51*(2), 1805-1816.

Thompson, Oikonomou, Etemadi, & Sorger. (2016). Optimization of data center battery storage investments for micro-grid cost savings, emissions reduction, and reliability enhancement. *IEEE Trans. Ind. Appl., 52*(3), 2053-2060.

Tim, W. S. (2011). *Application of line reactors in power electronics*. Technical report, Schaner EMV AG. Retrieved from www.schaffner.com

Tiwari, N., & Srivastava, L. (2016). Generation scheduling and micro-grid energy management using differential evolution algorithm. *Proc. Int.Conf. Circuit, Power Comput. Technol. (ICCPCT)*, 1-7. 10.1109/ICCPCT.2016.7530218

Tofis, Y., Yiasemi, Y., & Kyriakides, E. (2013, September). A plug and play, approximation-based, selective load shedding mechanism for the future electrical grid. In *International Workshop on Critical Information Infrastructures Security* (pp. 74-83). Springer. 10.1007/978-3-319-03964-0_7

Torriti, Hassan, M. G., & Leach, M. (2010). Demand response experience in Europe: Policies, programmes and implementation. *Energy, 35*(4), 1575–1583. doi:10.1016/j.energy.2009.05.021

Tosun. (2014). *Artificial Bee Colony Algorithm. In Encyclopedia of Business Analytics and Optimization* (pp. 1–14). IGI Global.

Turunen, S. I. (2016). *Protection of Microgrids and Distributed Energy Resources based on IEC 61850*. Academic Press.

Tzou, Y. Y., Ou, R. S., Jung, S. L., & Chang, M. Y. (1997). High-performance programmable ac power source with low-harmonic distortion using DSP-based repetitive control technique. *IEEE Transactions on Power Electronics, 2*(4), 715–725.

U.S. Energy Information Administration. (2014). *Independent Statistics and Analysis*. U.S. Energy Information Administration.

U.S.Energy Information Administration (EIA). (2013). International Energy Outlook, Report No.: DOE/EIA-0484(2013). Washington, DC: Office of Energy Analysis, U.S. Department of Energy.

UCTE. (2004). *Final report on the August 14, 2003 blackout in the United States and Canada*. Retrieved from https://www.energy.gov

UCTE. (2007). *Final report system disturbance on 4 November 2006*. Retrieved from https://www.ucte.org

UCTE-RAE. (2004). *Final report of the investigation committee on the 28 September 2003 blackout in Italy*. Retrieved from http://www.rae.gr

Uhlen, K., Foss, B. A., & Gjosaeter, O. B. (1994). Robust Control and Analysis of a Wind-Diesel Hybrid Power Plant. *IEEE Transactions on Energy Conversion*, *9*(4), 710–708. doi:10.1109/60.368338

US-CPSOTF. (2004). *Final report on the August 14, 2003 Blackout in the United States and Canada*. Author.

van der Werff, E., Perlaviciute, G., & Steg, L. (2016). Transition to smart grids: A psychological perspective. In *Smart Grids from a Global Perspective* (pp. 43–62). Cham: Springer. doi:10.1007/978-3-319-28077-6_4

Varesi, K. (2011). Optimal allocation of dg units for power loss reduction and voltage profile improvement of distribution networks using PSO algorithm. *World Academy of Science, Engineering and Technology*, *60*, 1938–1942.

Vargas, R., Rodriguez, J., Ammann, U., & Wheeler, P. W. (2008). Predictive current control of an induction machine fed by a matrix converter with reactive power control. *IEEE Transactions on Industrial Electronics*, *55*(12), 4362–4371.

Veloza, O. P., & Santamaria, F. (2016). Analysis of major blackouts from 2003 to 2015: Classification of incidents and review of main causes. *The Electricity Journal*, *29*(7), 42–49. doi:10.1016/j.tej.2016.08.006

Verbong, G. P. J., Verkade, N., Verhees, B., Huijben, J. C. C. M., & Höffken, J. I. (2016). Smart Business for Smart Users: A Social Agenda for Developing Smart Grids. In *Smart grids from a global perspective* (pp. 27–42). Cham: Springer. doi:10.1007/978-3-319-28077-6_3

Vergara, P. P., Torquato, R., & da Silva, L. C. P. (2015). Towards a real-time energy management system for a micro-grid using a multi-objective genetic algorithm. *Proc. IEEE Power Energy Soc. General Meet.*, 1-5. 10.1109/PESGM.2015.7285956

Vidyanandan, K. V., & Senroy, N. (2016). Frequency regulation in a wind diesel powered micro-grid using flywheels and fuel cells. *IET Generat., Transmiss. Distrib.*, *10*(3), 780-788.

Vinogradov, A., Vasiliev, A., Bolshev, V., Semenov, A., & Borodin, M. (2018). Time Factor for Determination of Power Supply System Efficiency of Rural Consumers. In V. Kharchenko & P. Vasant (Eds.), Handbook of Research on Renewable Energy and Electric Resources for Sustainable Rural Development (pp. 394–420).Academic Press. doi:10.4018/978-1-5225-3867-7.ch017

Vinogradov, A., Bolshev, V., Vinogradova, A., Kudinova, T., Borodin, M., Selesneva, A., & Sorokin, N. (2019). A System for Monitoring the Number and Duration of Power Outages and Power Quality in 0.38 kV Electrical Networks. In P. Vasant, I. Zelinka, & G. W. Weber (Eds.), *Intelligent Computing & Optimization. ICO 2018. Advances in Intelligent Systems and Computing* (Vol. 866, pp. 1–10). Cham: Springer; doi:10.1007/978-3-030-00979-3_1

Vinot, Reinbold, & Trigui. (2016). Global optimized design of an electric variable transmission for HEVs. *IEEE Trans. Veh. Technol.*, *65*(8), 6794-6798.

Viral, R., & Khatod, D. K. (2012). Optimal planning of distributed generation systems in distribution system*: A review. Renewable & Sustainable Energy Reviews*, *16*(7), 5146–5165. doi:10.1016/j.rser.2012.05.020

Walter, K., Sibylle, K., Johanna, K., Ina, R., & Silas, S. (2010). *DIN EN 16001: Energy Management Systems in Practice: A Guide for Companies and Organisations. Federal Ministry for the Environment, Nature Conservation and Nuclear Safety*. BMU.

Wang, G., Xin, H., Gan, D., Li, N., & Wang, Z. (2012, July). An investigation into WAMS-based Under-frequency load shedding. In *Power and Energy Society General Meeting, 2012 IEEE* (pp. 1-7). IEEE.

Wang, C., Gao, F., Li, P., Huang, B., Ding, C., & Yu, H. (2012, September). Control strategy research on low voltage microgrid. *Zhongguo Dianji Gongcheng Xuebao, 32*(25), 2–8.

Wang, W., Kliber, J., Zhang, G., Xu, W., Howell, B., & Palladino, T. (2007). A power line signaling based scheme for anti-islanding protection of distributed generators-part II: Field test results. *IEEE Transactions on Power Delivery, 22*(3), 1767–1772.

Wang, X., Li, X., Wang, J., Fang, X., & Zhu, X. (2016). Data-driven model-free adaptive sliding mode control for the multi degree-of-freedom robotic exoskeleton. *Information Sciences, 327*, 246–257.

Weile, D. S., & Eric, M. (1997). Genetic algorithm optimization applied to electromagnetics; A review. IEEE Transactions on Antenna and Propagation, 45(3). doi:10.1109/8.558650

Williams, K., & Armstrong, M. (2012). Newnes EMC for Systems and Installations. Academic Press.

WSCC. (1996). *Disturbance report for the power system outage that occurred on the western interconnection August 10, 1996. 1548 PAST*. Salt Lake City, UT: The Council.

Xu, Bishop, Oikarinen, & Hao. (2016). Application and modeling of battery energy storage in power systems. *CSEE Journal of Power and Energy Systems, 2*(3), 82-90.

Xu, Y. (2015). Optimal distributed charging rate control of plug-in electric vehicles for demand management. *IEEE Trans. Power Syst., 30*(3), 1536-1545.

Xu, W., Zhang, G. C., Wang, W., Wang, G., & Kliber, J. (2007). A power line signaling based technique for anti-islanding protection of distributed generators-part 1: Scheme and analysis. *IEEE Transactions on Power Delivery, 22*(3), 1758–1766.

Yafaoui, A., Wu, B., & Kouro, S. (2012). Improved active frequency drift anti-islanding detection method for grid connected photovoltaic systems. *IEEE Transactions on Power Electronics, 27*(5), 2367–2375.

Yang, H.-T., Yang, C.-T., Tsai, C.-C., Chen, G.-J., & Chen, S.-Y. (2015). Improved PSO based home energy management systems integrated with demand response in a smart grid. *Proc. IEEE Congr. Evol. Comput. (CEC)*, 275-282. 10.1109/CEC.2015.7256902

Yang, X.-S. (2009). Firefly algorithms for multimodal optimization. In Stochastic Algorithms: Foundations and Applications (SAGA). Berlin, Germany: Springer.

Yao, Lim, & Tsai. (2017). A real-time charging scheme for demand response in electric vehicle parking station. *IEEE Trans. Smart Grid, 8*(1), 52-62.

Yee Wei, L., & Marimuthu, P. (2013). WAKE: Key management scheme for wide-area measurement systems in smart grid. *IEEE Communications Magazine, 51*(1), 34–41. doi:10.1109/MCOM.2013.6400436

Yook, J. K., Tilbury, D. M., & Soparkar, N. R. (2000). A design methodology for distributed control systems to optimize performance in the presence of time delays. In *American Control Conference, 2000. Proceedings of the 2000* (Vol. 3, pp. 1959-1964). IEEE. 10.1109/ACC.2000.879544

You, S., Træholt, C., & Poulsen, B. (2010). *Developing virtual power plant for optimized distributed energy resources operation and integration.* Academic Press.

Yu, Zhong, Xie, Yuen, Gjessing, & Zhang. (2016). Balancing power demand through EV mobility in vehicle-to-grid mobile energy networks. *IEEE Trans. Ind. Informat., 12*(1), 79-90.

Zamani, R., Hamedani-Golshan, M. E., Haes Alhelou, H., Siano, P., & Pota, R, H. (. (2018). Islanding Detection of Synchronous Distributed Generator Based on the Active and Reactive Power Control Loops. *Energies, 11*(10), 2819. doi:10.3390/en11102819

Zaminga, A. (2011). House energy demand optimization in single and multi-user scenarios. *Proceedings of the IEEE International Conference on Smart Grid Communications (SmartGridComm)*, 345–350.

Žaneta, E., & Anton, B. (2008). Blackout in the power system. *AT&P Journal PLUS2.*

Zare, F. (2014). A novel harmonic elimination method for a three-phase diode rectifier with controlled DC link current. *Proc. of PEMC*, 985-989. 10.1109/EPEPEMC.2014.6980636

Zare, K., Hagh, M. T., & Morsali, J. (2015). Effective oscillation damping of an interconnected multi-source power system with automatic generation control and TCSC. Int. *J Elect Power Energy Syst., 65*, 220–230. doi:10.1016/j.ijepes.2014.10.009

Zeng, B., Shaojie, O., Jianhua, Z., Hui, S., Geng, W., & Ming, Z. (2015). An analysis of previous blackouts in the world: Lessons for china's Power industry. *Renewable & Sustainable Energy Reviews, 42*, 1151–1163. doi:10.1016/j.rser.2014.10.069

Zeng, Z., Zhao, R., & Yang, H. (2013). Micro-sources design of an intelligent building integrated with micro-grid. *Energy and Building, 57*, 261–267. doi:10.1016/j.enbuild.2012.11.018

Zhang, Mou, & Yang. (2016). An energy management study on hybrid power of electric vehicle based on aluminum air fuel cell. *IEEE Trans. Appl. Supercond., 26*(7), 1-6.

Zhang, K., Kang, Y., Xiong, J., & Chen, J. (2003). Direct repetitive control of SPWM inverter for UPS purpose. *IEEE Transactions on Power Electronics, 18*(3), 784–792.

Zhang, W., Xu, D., Li, X., Xie, R., & Li, H. (2013). Seamless transfer control strategy for fuel cell uninterruptible power supply system. *IEEE Transactions on Power Electronics, 28*, 717–729.

Zhan, Y., Guo, Y., Zhu, J., & Li, Li. (2015). Performance comparison of input current ripple reduction methods in UPS applications with hybrid PEM fuel cell/super-capacitor power sources. *International Journal of Electrical Power & Energy Systems, 64*, 96–103.

Zhan, Y., Guo, Y., Zhu, J., & Li, Li. (2015). Power and energy management of grid/PEMFC/ battery/super-capacitor hybrid power sources for UPS applications. *International Journal of Electrical Power & Energy Systems, 67*, 598–612.

Zhan, Y., Guo, Y., Zhu, J., & Wang, H. (2008). Intelligent uninterruptible power supply system with back-up fuel cell/battery hybrid power source. *Journal of Power Sources, 179*, 745–753.

Zhao, B., Song, Q., Liu, W., & Xiao, Y. (2013). Next-generation multi-functional modular intelligent UPS system for smart grid. *IEEE Transactions on Industrial Electronics, 60*, 3602–3618.

Zhao, Z., Lee, W. C., Shin, Y., & Song, K. B. (2013). An optimal power scheduling method for demand response in home energy management system. *IEEE Transactions on Smart Grid, 4*(3), 1391–1400. doi:10.1109/TSG.2013.2251018

Zhou, Z. J., Zhang, X., Xu, P., & Shen, W. X. (2008). Single-phase uninterruptible power supply based on Z-source inverter. *IEEE Transactions on Industrial Electronics*, *55*(8), 2997–3004.

Zin, A. M., Hafiz, H. M., & Wong, W. K. (2004, November). Static and dynamic under-frequency load shedding: a comparison. In *Power System Technology, 2004. PowerCon 2004. 2004 International Conference on* (Vol. 1, pp. 941-945). IEEE. 10.1109/ICPST.2004.1460129

Zin, A. M., Hafiz, H. M., & Aziz, M. S. (2004, November). A review of under-frequency load shedding scheme on TNB system. In *Power and Energy Conference, 2004. PECon 2004. Proceedings. National* (pp. 170-174). IEEE. 10.1109/PECON.2004.1461637

Zolfaghari, M., Habil, H. B., Abyaneh, H. A., & Abedi, M. (2016). Load sharing improvement between parallel-connected inverter based DGs using a GA based optimization control strategy in micro-grids. *Proc. IEEE PES Asia-Pacific Power Energy Eng. Conf. (APPEEC)*, 320-323. 10.1109/APPEEC.2016.7779520

Zongxiang, L., Caixia, W., Yong, M., Shuangxi, Z., & Jinxiang, L. (2007). Overview on microgrid research. *Dianli Xitong Zidonghua*, *31*(19), 100–107.

About the Contributors

Hassan Haes Alhelou received a BSc degree in Electrical Power engineering (with honor-ranked first) from the Tishreen University, Lattakia, Syria, in 2011. the M.Sc. (Hons.) degree in Elec. Eng. from Isfahan University of Technology, Isfahan, Iran, in 2014, and the Ph.D. candidate degree in Electrical Engineering - Power from Isfahan University of Technology, Isfahan, Iran, in 2017, respectively. He was the recipient of the Outstanding Reviewer Award for Energy Conversion and Management Journal in 2016, ISA Transactions Journal in 2018 and many other Awards form prestigious journals and publishers. Also, He has been an editor of several special issues and books in the field of smart grid and micro-grid systems. H. H. Alhelou was the recipient of Sentinels of Science and Research Award 2018, Peer-review Award 2018, Named as one of the top 1% reviewers in the world (Engineering), Publons, 12 Sep. 2018. His main research areas are Smart grid, Mico-grid, Electric vehicle integration to power systems, Power system operation and control, Power system dynamic, control and stability, intelligent control, fuzzy control, automation and process control, preventive maintenance, condition monitoring, FACTS, Fault diagnosis.

Ghassan Hayek was born in Lattakia in 1958, received a BSc degree in Electrical Power engineering from the Aleppo University, Aleppo, Syria, in 1985 and the Ph.D. degree in Electrical Engineering - Automation and Control from TU Dresden, Dresden, Germany, in 1989, respectively. He was the recipient of the Outstanding Reviewer Award ISA Transactions Journal in 2018 and many other Awards form prestigious journals and publishers. Also, He has been an editor of several special issues and books in the field of power systems. His main research areas are Electric drive, Mico-grid, Power system operation and control, Power system dynamic, control and stability, intelligent control, fuzzy control, automation and process control, preventive maintenance, condition monitoring, Fault diagnosis.

* * *

Subrata Banerjee is presently working as a Professor in the Electrical Engineering Department at NIT-Durgapur. He has about 30 years of professional experience. He has published a numbers of research papers in national/international journals and conference records. His area of interest includes controller design, intelligent control, optimization techniques, control of switch-mode converters, nonlinear dynamics of converters, renewable energy, etc.

Zeina Barakat was born in Syria in 1989. She received the M.Sc. degree in electrical Power engineering from Tishreen University Lattakia, Syria, in 2016. Currently, she is Engineer in Maintenance Dept. Latakia Electrical Substation – Latakia, Syria. Her research interests include power system protection, AI, renewable energy.

Monalisa Biswal is working as an Assistant Professor in the Department of Electrical Engineering, NIT Raipur.

Vadim Bolshev is a research at Federal Scientific Agroengineering Centre VIM and specializes in research in the field of power supply for rural consumers. Vadim studied at the Orel Technical College as an electrical technician from 2004 to 2008, at the Orel State Agrarian University as an electrical engineer from 2008 to 2012. In 2016 he graduated from the postgraduate course at the Orel State Agrarian University. Vadim has a great professional experience in the field of working with electrical equipment. So he worked in the water and sewer utility "Orelvodokanal" for 3.5 years as an electrician (2010 - 2013), in the scientific and clinical multidisciplinary center named after Z.I. Krugloi for 3 years as an engineer (2013-2016) and 1.5 years as chief engineer (2016-2018). To date, he works in the Federal Scientific Agro-Engineering Center of VIM as a research fellow. The field of scientific activity is to develop methods and tools aimed at improving power supply efficiency. This includes, first of all, the development of methods and devices for monitoring power quality and the technical state of power supply system elements as well as the automation of these systems.

Ruchi Chandrakar is pursuing her M.Tech degree in the Department of Electrical Engineering in NIT Raipur.

Cosmin Darab received his BSc degree in Systems Engineering in 2008, two years later he graduated the M.Sc. degree at Technical University of Cluj-Napoca. In 2013 he received his Ph.D. degree and he is now a lecturer at Technical University of Cluj-Napoca. The research topics of interests are renewable energy, smart grid and electric vehicle grid integration.

Dipayan Guha was born in Barasat, North 24 Parganas, West Bengal, India, in 1986. He has completed his B.Tech in Electrical Engineering, M.Tech in Electrical Engineering (Specialization: Control System and Automation), and PhD in the year 2009, 2013, and 2017 respectively. He has published more than 20 research papers in National/International conference records/Journals. He is currently working as an Assistant Professor in Department of Electrical Engineering at Motilal Neheru National Institute of Technology, Allahabad, India. His research interest includes power system operation and control, soft computing, FACTS controllers and nonlinear control system.

Valentin Gusarov, Ph.D., is a graduate of the Faculty of Civil Engineering (Industrial and Civil Engineering) of the Moscow Engineering and Construction Institute. He has worked with wind power since 1998. He has worked with VIESH as the head of the department (in Istra) since 2000. Currently He is the leading researcher, head of the laboratory "Power equipment on renewable energy sources". The direction of his scientific work is renewable autonomous power engineering of agriculture objects.

His scientific research is aimed at reducing energy intensity, increasing processes productivity and increasing the energy supply reliability of the agro-industrial complex. He is the author of more than 120 scientific papers.

Baseem Khan received his Bachelor of Engineering degree in Electrical Engineering from the Rajiv Gandhi Technological University, Bhopal, India in 2008. He received his Master of Technology and Doctor of Philosophy degrees in Electrical Engineering from the Maulana Azad National Institute of Technology, Bhopal, India, in 2010 and 2014, respectively. Currently, he is working as an Assistant Professor in the School of Electrical and Computer Engineering, Hawassa University Institute of Technology, Hawassa University, Hawassa, Ethiopia. His research interest includes power system restructuring, power system planning, smart grid technologies, meta-heuristic optimisation techniques, reliability analysis of renewable energy system, power quality analysis and renewable energy integration.

Maheswari M. is currently working as Professor and Head in Malla Reddy Engineering College (Autonomous), Hyderabad. She completed her B.E.(EEE) from K.S.Rangaswamy College of Technology, Thiruchengode, Tamilnadu and M.E.(Power System Engineering) from Government College of Technology, Coimbatore, Tamilnadu in the year 2000 and 2006 respectively. She completed her doctoral degree from Anna University, Chennai in the year 2015 in the area of Power Quality Improvement. She has 18 years of teaching experience under various position. She has published 15 papers in international journals and 13 papers in International Conferences.

Ram Nittala has completed Ph.D. in BITS Pilani Hyderabad Campus in the Department of Electrical Engineering. He has completed M.Tech in Specialization of Power Systems from the Department of Electrical and Electronics Engineering in KL University, Vaddeswaram, Andhra Pradesh. He has done B. Tech. in Electrical and Electronics Engineering from V.R. Siddhartha Engineering College (an Autonomous Engineering College) under the jurisdiction of Acharya Nagarjuna University, Nagarajuna Nagar, Andhra Pradesh. His Research interests are power systems, Facts devices and Robotics.

Y. P. Obulesh received his B.E. degree in Electrical and Electronics Engineering in 1995 from Andhra University, India, M.Tech degree from Indian Institute of Technology, Khargapur, India in 1998 and PhD from Jawaharlal Nehru Technological University, Hyderabad, India in 2006..His fields of interests are Power electronics and drives, renewable energy systems and active filters.

B. Rao was born in Ramabhadrapuram, India in 1978. He received his Bachelor degree in Electrical and Electronics Engineering from College of Engineering, Gandhi Institute of Technology And Management(GITAM) Visakapatnam, Andhra Pradesh, India in 2000, the Master degree in Electrical Power Engineering from the College of Engineering, JNTU, Hyderabad in 2007 and He received his Doctoral degree from Jawaharlal Nehru Technological University, Hyderabad in 2015. He is presently working as Associate Professor in the Department of Electrical and Electronics Engineering, V R Siddhartha Engineering College, Vijayawada. His research interests are Power system stability analysis, FACTS devices, Power system control and power system optimization. He has published several research papers in national and international conferences and journals. He is a member of ISTE and IE.

Gummadi Rao received his Bachelor degree in Electrical and Electronics Engineering in 2000 from University of Madras, India, the Master degree in Electrical Power Systems from the College of Engineering, JNTA, Anantapur in 2005 and received his Doctoral degree from Jawaharlal Nehru Technological University, Hyderabad in 2015. He is presently working as Associate Professor in the Department of Electrical and Electronics Engineering, V R Siddhartha Engineering College, Vijayawada. His fields of interests are Distributed Generating Units, Deregulated Power Systems, Optimization and Power Quality.

Abdelmadjid Recioui is a Professor at the Institute of Electrical Engineering and Electronics University of Boumerdes, Algeria. He obtained a PhD degree in electrical and electronic engineering option telecommunications from the Institute of Electrical Engineering and Electronics, University of Boumerdes in 2011. He holds also Master (Magister) Electronic System Engineering degree which has been achieved at the Institute of Electrical Engineering and Electronics, University of Boumerdes in 2006. In June 2002, he finished his engineering studies at the institute of Electrical Engineering and Electronics, University of Boumerdes. He is a research Assistant at the laboratory signals and systems from January 2008 to Present in Laboratory: signals and systems, Inst. of Electrical Engineering and electronics, University of Boumerdes. His research interests include: Antennas, Wireless Communication Systems, antenna array synthesis and design, capacity enhancement, system optimization, smart antennas, power system protection, power system optimization, power system communications.

Provas Kumar Roy is currently working as a Professor in the Electrical Engineering Department at Kalyani Government Engineering College, West Bengal, India. He has published a numbers of research papers in national/international journals and conference records. His research interest includes economic load dispatch, optimal power flow, FACTS, automatic generation control, radial distribution network, evolutionary computing techniques, etc.

Gunasekharan S. is currently working as Professor and Head in Malla Reddy Engineering College. He has an experience of 18 years in teaching as well as industry.

Pawan Singh has received the B.E. (Computer Science and Engineering) from CCS University, Meerut, India, M.Tech. (Information Technology) from GGSIPU, New Delhi, India and Ph.D. (Computer Science) from Magadh University, Bodh Gaya, India in 2013. Currently, he is serving in Department of Computer Science & Engineering, Amity School of Engineering and Technology, Amity University, Lucknow Campus, India. His research interests include software metrics, software testing, software cost estimation, web structure mining, energy-aware scheduling, energy management, nature-inspired meta-heuristic optimization techniques and its applications. He has authored and co-authored a number of research papers in the journals of international reputation.

Sudeep Tanwar is working as Associate Professor in Computer Engineering Department at Institute of Technology, Nirma University, Ahmedabad, Gujarat, India. He received his Bachelor's of Technology in Computer Engineering from Kurukshetra University, Kurukshetra, and Master's of Technology with Honours in Information Technology from Guru Gobind Singh Indraprastha University, Delhi (USIT Campus). He received his Doctoral Degree (Ph.D.) in Computer Science & Engineering with specialization in Wireless Sensor Network. He has authored more than 50 technical research papers published in leading International Conferences and Peer-reviewed International journals from the IEEE, Elsevier, Springer, John Wiley, etc. Some of his research findings are published in top-cited journals such as the Journal of Network and Computer Application, Pervasive and Mobile Computing of Elsevier, International Journal of Communication System of Wiley, Telecommunication System of Springer and IEEE Systems Journal. He has guided many students leading to M.E./M.Tech and guiding students leading to Ph.D. He is Associate Editor of "Security and Privacy Journal, (Wiley). He has also authored three books on Role of Heterogeneity in WSN, Big Data Analytics and Mobile Computing with International/National Publishers for students of UG/PG. He has visited London and Singapore for academic collaboration under CSIR Grant in 2015 and 2017, respectively. His current research interest includes Routing Issues in WSN, IoT, Integration of sensor with the cloud, Computational aspects of Smart Grid and Assessment of Fog Computing in BASN.

Suma Veeraganti works as an Assistant Professor in the Department of Electrical and Electronics in Malla Reddy Engineering College. She has completed her M.Tech from KL University, Guntur. She pursued Bachelor of Technology from Acharya Nagarjuna University. Her Research interests are Power Quality, Facts Devices and Power System Deregulation.

Kasimala Venkatanagaraju is pursuing his Ph.D. degree in the Department of Electrical Engineering, NIT Raipur.

Index

www.ingramcontent.com/pod-product-compliance
Lightning Source LLC
LaVergne TN
LVHW060255011026
849161LV00011B/2716